THE CHEMICAL FORMULARY

The Chemical Formulary

A Collection of Valuable, Timely, Practical Commercial Formulae and Recipes for Making Thousands of Products in Many Fields of Industry

VOLUME VI

Editor-in-Chief

H. BENNETT

1943

CHEMICAL PUBLISHING CO., INC.

212 FIFTH AVENUE NEW YORK 10, N. Y.

PREFACE

Chemistry as taught in our schools and colleges is confined to synthesis, analysis and engineering—and properly so. It is part of the proper foundation for the education of the chemist.

Many a chemist on entering an industry soon finds that the bulk of the products manufactured by his concern are not synthetic or definite chemical compounds but are mixtures, blends or highly complex compounds of which he knows little or nothing. The literature in this field, if any, may be meagre, scattered or antiquated.

Even chemists, with years of experience in one or more industries, spend considerable time and effort in acquainting themselves on entering a new field. Consulting chemists, similarly, have problems brought to them from industries foreign to them. A definite need has existed for an up-to-date compilation of formulae for chemical compounding and treatment. Since the fields to be covered are many and varied, an editorial board was formed, composed of chemists and engineers in many industries.

Many publications, laboratories, manufacturing companies and individuals have been drawn upon to obtain the latest and best information. It is felt that the formulae given in this volume will save chemists and allied workers much time and effort.

Manufacturers and sellers of chemicals will find in these formulae new uses for their products. Non-chemical executives, professional men and others, who may be interested, will gain from this volume a "speaking acquaintance" with products which they may be using, trying, or with which they are in contact.

It often happens that two individuals using the same ingredients in the same formula get different results. This may be the result of slight deviations or unfamiliarity with the intricacies of a new technique. Accordingly, repeated experiments may be necessary to get the best results. Although many of the formulae given are being used commercially many have been taken from patent specifications and the literature. Since these sources are often subject to various errors and omissions, due regard must be given to this factor. Wherever possible it is advisable to consult with other chemists or technical workers regarding commercial production. This will save time and money and avoid "headaches."

It is seldom that any formula will give exactly the results which one requires. Formulae are useful as starting points from which to work out one's own ideas. Formulae very often give us ideas which may help us in our specific problems. In a compilation of this kind errors of omission, commission and printing may occur. We shall be glad to receive any constructive criticism in this, our first attempt.

To the layman, it is suggested that he arrange for the services of a chemist or technical worker familiar with the specific field in which he is interested. Although this involves an expense it will insure quicker and better formulation without wastage of time and materials.

H. BENNETT

PREFACE TO VOLUME VI

Additional new formulae have been gathered to compile a sixth volume of the *Chemical Formulary*—an addition which will broaden and bring up-to-date the contents of volumes I, II, III, IV and V. Because the board of editors feels that information of this nature, to be most helpful, should be released as soon as possible and since we have had hundreds of inquiries as to when Volume VI would be ready, an early publication date was decided upon.

Special elementary formulae of direct and indirect military interest have been included. A chapter on substitutes for scarce materials is an innovation which may be of interest and use to many.

Schools and colleges in increasing numbers seem to find it advisable to use the *Chemical Formulary* as an aid in promoting a practical interest in chemistry. By its use, students learn to make cosmetics, inks, polishes, insecticides, paints and countless other products. The result is that chemistry becomes an extremely interesting practical and useful subject. This interest often continues even when the students reach the theoretical or more difficult phases of this subject.

Since some mature users of this book have not had the good fortune to have had previous training or experience in the art of chemical compounding, the simple introductory chapter of directions and advice has been repeated. This chapter should be studied carefully by all beginners (and some more experienced workers) and some of the preparations given therein should be made before attempting to duplicate the more complex formulae in the succeeding chapters.

An enlarged directory of sources of chemicals and supplies has been added. This should prove useful in locating new as well as old materials and products.

It is a sincere pleasure to acknowledge the valuable assistance of the members of the board of editors and others who have given of their time and knowledge in contributing the special formulae which have made this volume possible.

H. BENNETT

NOTE

All the formulae in volumes I, II, III, IV, V and VI (except in the introduction) are different. Thus, if you do not find what you are looking for in this volume, you may find it in one of the others.

TABLE OF CONTENTS

ABBREVIATIONS

amp. ....ampere
amp./dm^2 ....amperes per square decimeter
amp./sq. ft. ....amperes per square foot
anhydr. ....anhydrous
avoir. ....avoirdupois
Bé. ....Baumé
b.p. ....boiling point
C. ....Centigrade
°C. ....degrees Centigrade
cc. ....cubic centimeter
c.d. ....current density
cm. ....centimeter
cm^3 ....cubic centimeter
conc. ....concentrated
c.p. ....chemically pure
cps. ....centipoises
cu. ft. ....cubic foot
cu. in. ....cubic inch
cwt. ....hundredweight
d. ....density
dil. ....dilute
dm. ....decimeter
dm^2 ....square decimeter
dr. ....dram
E. ....Engler
F. ....Fahrenheit
°F. ....degrees Fahrenheit
f.f.c. ....free from chlorine
f.f.p.a. ....free from prussic acid
fl. dr. ....fluid dram
fl. oz. ....fluid ounce
f.p. ....freezing point
ft. ....foot
ft.^{2} ....square foot
g. ....gram
gal. ....gallon
gr. ....grain
hl. ....hectoliter
hr. ....hour
in. ....inch
kg. ....kilogram
l. ....liter
lb. ....pound
liq. ....liquid
m. ....meter
min. ....minim, minute
ml. ....milliliter—cubic centimer
mm. ....millimeter
m.p. ....melting point
N. ....normal
N.F. ....National Formulary
oz. ....ounce
pH ....hydrogen-ion concentration
p.p.m. ....parts per million

pt. ....pint
pwt. ....pennyweight
q.s. ....a quantity sufficient to make
qt. ....quart
r.p.m. ....revolutions per minute
S.A.E. ....Society of Automotive Engineers
sec. ....second
sp. ....spirits
sp. gr. ....specific gravity
sq. dm. ....square decimeter
tech. ....technical
tinc. ....tincture
tr. ....tincture
Tw. ....Twaddell
U.S.P. ....United States Pharmacopeia
v. ....volt
visc. ....viscosity
vol. ....volume
wt. ....weight
x. ....extra

CHAPTER I

INTRODUCTION

At the suggestion of a number of teachers of chemistry and home economics the following introductory matter has been included.

The contents of this section are written in a simple way so that anyone, regardless of technical education or experience, can start making simple products without any complicated or expensive machinery. For commercial productions, however, suitable equipment is necessary.

Chemical specialties en masse are composed of pigments, gums, resins, solvents, oils, greases, fats, waxes, emulsifying agents, water, chemicals of great diversity, dyestuffs, and perfumes. To compound certain of these with some of the others requires certain definite and well-studied procedure, any departure from which will inevitably result in failure. The successful steps are given with the formulas. Follow them explicitly. If the directions require that A should be added to B, carry this out literally, and not in reverse fashion. In making an emulsion, the job is often quite as tricky as the making of mayonnaise. In making mayonnaise, you add the oil to the egg, *slowly*, with constant and even and regular stirring. If you do it correctly, you get mayonnaise. If you depart from any of these details: if you add the egg to the oil, or pour the oil in too quickly, or fail to stir regularly, the result is a complete disappointment. The same disappointment might be expected if the prescribed procedure of any other formula is violated.

The next point in importance is the scrupulous use of the proper ingredients. Substitutions are sure to result in inferior quality, if not in complete failure. Use what the formula calls for. If a cheaper product is desired, do not obtain it by substituting a cheaper material for the one prescribed: resort to a different formula. Not infrequently a formula will call for some ingredient which is difficult to obtain: in such cases, either reject the formula or substitute a similar material only after preliminary experiment demonstrates its usability. There is a limit to which this rule may reasonably be extended. In some instances the substitution of an equivalent ingredient may legitimately be made. For example: when the formula calls for *white wax* (beeswax), yellow wax can be used, if the color of the finished product is a matter of secondary importance. Yellow beeswax can often replace white beeswax, making due allowance for color: but paraffin will *not* replace beeswax, even though its light color recommends it above yellow beeswax.

And this leads to the third point: the use of good quality ingredients, and ingredients of the correct quality. Ordinary lanolin is not the same thing as *anhydrous* lanolin: the replacement of one for the other, weight for weight, will give discouragingly different results. Use exactly what the formula calls for: if you are unacquainted with the material and a doubt arises as to just what is meant, discard the formula and use one that you understand. Buy your materials from reliable sources. Many ingredients are obtainable in a number of different grades: if the formula does not designate the grade, it is understood that the best grade is to be used. Remember that a formula and the directions can tell you only a part of the story. Some skill is often required to attain success. Practice with a small batch in such cases until you are sure of your technique. Many instances can be cited. If the formula calls for steeping quince seed for 30 minutes in cold water, your duplication of this procedure may produce a mucilage of too thin a consistency. The originator of the formula may have used a fresher grade of seed, or his conception of what "cold" water means may be different from yours. You should have a feeling for the right degree of mucilaginousness, and if steeping the seed for 30 minutes fails to produce it, steep them longer until you get the right kind of mucilage.

If you do not know what the right kind is, you will have to experiment until you find out. Hence the recommendation to make small experimental batches until successful results are arrived at. Another case is the use of dyestuffs for coloring lotions, and the like. Dyes vary in strength: they are all very powerful in tinting value: it is not always easy to state in quantitative terms how much to use. You must establish the quantity by carefully adding minute quantities until you have the desired tint. Gum tragacanth is one of those products which can give much trouble. It varies widely in solubility and bodying power: the quantity prescribed in the formula may be entirely unsuitable for *your* grade of tragacanth. Hence a correction is necessary, which can only be made after experiments to determine *how much* to correct.

In short, if you are completely inexperienced, you can profit greatly by gaining some experience through recourse to experiment. Such products as mouth washes, hair tonics, astringent lotions, need little or no experience, because they are as a rule merely mixtures of simple liquid and solid ingredients, the latter dissolving without difficulty and the whole being a clear solution that is ready for use when mixed. On the other hand, face creams, tooth pastes, lubricating greases, wax polishes, etc., which require relatively elaborate procedure and which depend for their usability on a definite final viscosity, must be made with the exercise of some skill, and not infrequently some experience.

Figuring

Some prefer proportions expressed by weight, volume or in terms of percentages. In different industries and foreign countries various systems of weights and measures are used. For this reason no one set of units could be satisfactory for everyone. Thus divers formulae appear with different units in accordance with their sources of origin. In some cases, parts instead of percentages or weight or volume is designated. On the pages preceding the index, tables of weights and measures are given. These are of use in changing from one system to another. The following examples illustrate typical units:

Ink for Marking Glass

Glycerin	40	Ammonium Sulphate	10
Barium Sulphate	15	Oxalic Acid	8
Ammonium Bifluoride	15	Water	12

Here no units are mentioned. When such is the case it is standard practice to use parts by weight, using the same system throughout. Thus here we may use ounces or grams as desired. But if ounces are used for one item then ounces must be the unit for all the other items in the particular formula.

Flexible Glue

Glue, Powdered	30.9 %	Glycerin	5.15%
Sorbitol (85%)	15.45%	Water	48.5 %

Where no units of weight or volume but percentages are given then forget the percentages and use the same instructions as given under Example No. 1.

Example No. 3

Antiseptic Ointment

Petrolatum	16 parts	Benzoic Acid	1 part
Coconut Oil	12 parts	Chlorthymol	1 part
Salicylic Acid	1 part		

The same instructions as given under Example No. 1 apply to Example No. 3.

It is not wise in many cases to make up too large a quantity of material until one has first made a number of small batches to first master the necessary technique and also to see whether it is suitable for the particular outlet for which it is intended. Since, in many cases, a formula may be given in proportions as made up on a commercial factory scale, it is advisable to reduce the proportions accordingly. Thus, taking the following formula:

Example No. 4

Neutral Cleansing Cream

Mineral Oil	80 lb.	Water	90 lb.
Spermaceti	30 lb.	Glycerin	10 lb.
Glyceryl Monostearate	24 lb.	Perfume	to suit

Here, instead of pounds, grams may be used. Thus this formula would then read:

Mineral Oil	80 g.	Water	90 g.
Spermaceti	30 g.	Glycerin	10 g.
Glyceryl Monostearate	24 g.	Perfume	to suit

Reduction in bulk may also be obtained by taking the same fractional part or portion of each ingredient in a formula. Thus in the following formula:

Example No. 5

Vinegar Face Lotion

Acetic Acid (80%)	20	Alcohol	440
Glycerin	20	Water	500
Perfume	20		

We can divide each amount by ten and the finished bulk is only 1/10th of the original formula. Thus it becomes:

Acetic Acid (80%)	2	Alcohol	44
Glycerin	2	Water	50
Perfume	2		

Apparatus

For most preparations pots, pans, china and glassware, such as is used in every household, will be satisfactory. For making fine mixtures and emulsions a "malted-milk" mixer or egg-beater is necessary. For weighing, a small, low priced scale should be purchased from a laboratory supply house. For measuring of fluids, glass graduates or measuring glasses may be purchased from your local druggist. Where a thermometer is necessary a chemical thermometer should be obtained from a druggist or chemical supply house.

Methods

To better understand the products which you intend making, it is advisable that you read the complete section covering such products. Very often an important idea is thus gotten. You may learn different methods that may be used and also avoid errors which many beginners are prone to make.

Containers for Compounding

Where discoloration or contamination is to be avoided (as in light colored, or food and drug products) it is best to use enameled or earthenware vessels. Aluminum, as well, is highly desirable in such cases but it should not be used with alkalies as the latter dissolve and corrode this metal.

Heating

To avoid overheating, it is advisable to use a double boiler when temperatures below 212° F. (temperature of boiling water) will suffice. If a double boiler is not at hand, any pot may be filled with water and the vessel containing the ingredients to be heated is placed therein. The pot may then be heated by any flame without fear of overheating. The water in the pot, however, should be replenished from time to time as necessary—it must not be allowed to "go dry." To get uniform higher temperatures, oil, grease or wax is used in the outer container in place of water. Here of course care must be taken to stop heating when thick fumes are given off as these are inflammable. When higher uniform temperatures are necessary, molten lead may be used as a heating medium. Of course, where materials melt uniformly and stirring is possible, direct heating over an open flame is permissible.

Where instructions indicate working at a certain temperature, it is important that the proper temperature be attained—not by guesswork, but by the use of a thermometer. Deviations from indicated temperatures will usually result in spoiled preparations.

Temperature Measurements

In Great Britain and the United States, the Fahrenheit scale of temperature measurement is used. The temperature of boiling water is 212° Fahrenheit (212° F.); the temperature of melting ice is 32° Fahrenheit (32° F.).

In scientific work and in most foreign countries the Centigrade scale is used. On this scale of temperature measurement, the temperature of boiling water is 100 degrees Centigrade (100° C.) and the temperature of melting ice is 0 degrees Centigrade (0° C.).

The temperature of liquids is measured by a glass thermometer. The latter is inserted as deeply as possible in the liquid and is moved about until the temperature remains steady. It takes a little time for the glass of the thermometer to come to the temperatures of the liquid. The thermometer should not be placed against the bottom or side of the container, but near the center of the liquid in the vessel. Since the glass of the bulb of the thermometer is very thin, it can be broken easily by striking it against any hard surface. A cold thermometer should be warmed gradually (by holding over the surface of a hot liquid) before immersion. Similiarly the hot thermometer when taken out should not be put into cold water suddenly. A sharp change in temperature will often crack the glass.

Mixing and Dissolving

Ordinary solution (e.g. sugar in water) is hastened by stirring and warming. Where the ingredients are not corrosive, a clean stick, bone or composition fork or spoon is used as a mixing device. These may also be used for mixing thick creams or pastes. In cases where most efficient stirring is necessary (as in making mayonnaise, milky polishes, etc.) an egg-beater or a malted-milk mixer is necessary.

Filtering and Clarification

When dirt or undissolved particles are present in a liquid, they are removed by settling or filtering. In the former the solution is allowed to stand and if the particles are heavier than the liquid they will gradually sink to the bottom. The upper liquid may be poured or siphoned off carefully and in some cases is then of sufficient clarity to be used. If, however, the particles do not settle out then they must be filtered off. If the particles are coarse they may be filtered or strained through muslin or other cloth. If they are very small particles then filter paper is used. Filter papers may be obtained in various degrees of fineness. Coarse filter paper filters rapidly but will not, of course, take out extremely fine particles. For the latter, it is necessary to use a very fine grade of filter paper. In extreme cases even this paper may not be fine enough. Here it will be necessary to add to the liquid 1-3% of infusorial earth or magnesium carbonate. The latter clog up the pores of the filter paper and thus reduce their size and hold back undissolved material of extreme fineness. In all such filtering, it is advisable to take the first portions of the filtered liquid and pour them through the filter again as they may develop cloudiness in standing.

Decolorizing

The most commonly used decolorizer is decolorizing carbon. The latter is added to the liquid to the extent of 1-5% and heated with stirring for ½ hour to as high a temperature as is feasible. It is then allowed to stand for a while and filtered. In some cases bleaching must be resorted to. Examples of this are given in this book.

Pulverizing and Grinding

Large masses or lumps are first broken up by wrapping in a clean cloth and placing between two boards and pounding with a hammer. The smaller pieces are then pounded again to reduce their size. Finer grinding is done in a mortar with a pestle.

Spoilage and Loss

All containers should be closed when not in use to prevent evaporation or contamination by dust; also because, in some cases, air affects the material adversely. Many materials attack or corrode the metal containers in which they are received. This is particularly true of liquids. The latter, therefore, should be transferred to glass bottles which should be as full as possible. Corks should be covered with aluminum foil (or dipped in melted paraffin wax when alkalies are present).

Materials such as glue, gums, oilve oil or other vegetable or animal products may ferment or become rancid. This produces discoloration or unpleasant odors. To avoid this, suitable antiseptics or preservatives must be used. Too great stress cannot be placed on cleanliness. All containers must be cleaned thoroughly before use to avoid various complications.

Weighing and Measuring

Since, in most cases, small quantities are to be weighed, it is necessary to get a light scale. Heavy scales should not be used for weighing small amounts as they are not accurate for this type of weighing.

For measuring volume (liquids) measuring glasses or cylinders (graduates) should be used. Since this glassware cracks when heated or cooled suddenly it should not be subjected to sudden changes of temperature.

Caution

Some chemicals are corrosive and poisonous. In many cases they are labeled as such. As a precautionary measure, it is advised not to smell bottles directly, but only to sniff a few inches from the cork or stopper. Always work in a well ventilated room when handling poisonous or unknown chemicals. If anything is spilled, it should be wiped off and washed away at once.

Where to Buy Chemicals and Apparatus

Many chemicals and most glassware can be purchased from your druggist. A list of suppliers of all products will be found at the end of this book.

ADVICE

This book is the result of co-operation of many chemists and engineers who have given freely of their time and knowledge. It is their business to act as consultants and, for a fee, to give advice on technical matters. As publishers, we do not maintain a laboratory or consulting service to compete with them.

Please, therefore, do not ask us for advice or opinions, but confer with a chemist in your vicinity.

Extra Reading

Keep up with new developments of new materials and methods by reading technical magazines. Many technical publications are listed under references in the back section of this book.

Calculating Costs

Purchases of raw materials, in small quantities, are naturally higher in price than when bought in large quantities. Commercial prices, as given in the trade papers and catalogs of manufacturers, are for quantities such as barrels, drums or sacks. For example, a pound of epsom salts, bought at retail, may cost 10 or 15 cents. In barrel lots its price today is about 2 to 3 cents per pound.

Typical Costing Calculation
Formula for Beer- or Milk Pipe Cleaner

Soda Ash	25 lb. @	.02½ per lb. =	$0.63
Sodium Perborate	75 lb. @	.16 per lb. =	12.00
Total	100 lb.	Total	$12.63

If 100 lb. cost $12.63, 1 lb. will cost $12.63 divided by 100 or about $0.126 per lb. for raw materials, assuming no loss.

Always weigh the amount of finished product and use *this* weight in calculating costs. Most compounding results in some loss of material because of spillage, sticking to apparatus, evaporation, etc. Costs of making experimental lots are always high and should not be used for figuring costs. To meet competition, it is necessary to buy in larger units and costs should be based on the latter.

Elementary Preparations

The recipes that follow have been gotten up in a very simple way. Only one of each type is given so as to avoid confusion. These have been selected because of their importance and because they can be made readily.

The succeeding chapters go into greater detail and give many different types and modifications of these and other recipes for home and commercial use.

Cleansing Creams

Cleansing creams as the name implies serve as skin cleaners. Their basic ingredients are oils and waxes which are rubbed into the skin. When wiped off they carry off dirt and dead skin. The liquefying type of cleansing cream contains no water and melts or liquefies when rubbed on the skin. To suit different climates and likes and dislikes harder or softer products can be made.

Cleansing Cream (Liquefying)

Liquid Petrolatum (White Mineral Oil)	5½ oz.
Paraffin Wax	2½ oz.
Petrolatum (Vaseline)	2 oz.

Melt together with stirring in an aluminum or enamelled dish and allow to cool. Then stir in a dash of perfume oil. Allow to stand until a haziness appears and then pour into jars, which should be allowed to stand *undisturbed* over night.

Cold Creams

The most important facial cream is cold cream. This type of cream consists of a mineral oil and wax which are emulsified in water with a little borax or glycosterin. The function of a cold cream is to furnish a greasy film which takes up dirt and waste tissue which are removed when the skin is wiped thoroughly. Many modifications of this basic cream are enountered in stores. They vary in color, odor, and in claims but, essentially, they are no more useful than this simple cream. The latest type of cold cream is the non-greasy cold cream which is of particular interest because it is non-alkaline and therefore non-irritating to sensitive skins.

Cold Cream

Liquid Petrolatum (White Mineral Oil)	52 g.
White Beeswax	14 g.

Heat the above in an aluminum or enamelled double boiler (the water in the outer pot should be brought to a boil). In a separate aluminum or enamelled pot dissolve.

Borax	1 g.
Water	33 c.c.

and bring this to a boil. Add this in a thin stream, to the melted wax, while stirring vigorously in one direction only, to the melted wax mixture. Use a fork for stirring. When the mixture turns to a smooth thin cream, immerse the bottom of the thermometer in it from time to time, stirring continuously. When the temperature drops to 140° F. add ½ c.c. of perfume oil and continue stirring until the temperature drops to 120° F. At this point pour into jars where the cream will "set" after a while. If a harder cream is desired, reduce the amount of liquid petrolatum. If a softer cream is wanted increase it.

Cold Cream (Non-Greasy)

White Paraffin Wax	1¼ oz.
Petrolatum (Vaseline)	1½ oz.
Glycosterin or Glyceryl Monostearate	2¼ oz.
Liquid Petrolatum (White Mineral Oil)	3 oz.

Heat the above in an aluminum or enamelled double boiler (the water in the outer pot should be boiling). Stir until clear. To this slowly add, while stirring vigorously with a fork,

Water (boiling)	10 oz.

Continue stirring until smooth and then add with stirring, a little perfume oil. Pour into jars at 110-130° F. and cover the jars as soon as possible.

Vanishing Creams

Vanishing creams are non-greasy creams, soapy in nature. Some are white and others have a very beautiful pearly appearance. This type of cream depends on its soapiness for its cleansing character and is useful as a powder base.

Vanishing Cream

Stearic Acid	18 oz.

Melt the above in an aluminum or enamelled double boiler (the water in the outer pot must be boiling). To the above add, in a thin stream, while stirring vigorously with a fork, the following boiling solution made in an aluminum or enamelled pot:

Potassium Carbonate	¼ oz.
Glycerin	6½ oz.
Water	5 lb.

Continue stirring until the temperature falls to 135° F., then stir in a little perfume oil and stir from time to time until cold. Allow to stand over night and stir again the next day. Pack into jars which should be closed tightly.

Hand Lotions

Hand lotions are usually clear or milky liquids or salves which are useful in protecting the skin from roughness and redness because of exposure to cold, hot water, soap and other materials. "Chapped" hands are a common occurrence. The use of a good hand lotion keeps the skin smooth, soft, and in a normally healthy condition. The lotion is best applied at night, rather freely, and cotton gloves may be worn to prevent soiling. During the day it should be put on sparingly and the excess wiped off.

Hand Lotion (Salve)

Boric Acid	1 oz.
Glycerin	6 oz.

Warm the above in an aluminum or enamelled dish and stir with a clean wooden stick until dissolved (clear). Then allow to cool and work into the following mixture with a potato masher, or rounded stick, adding only a little of the above liquid at a time to the mixture below and not adding a further portion until it is fully absorbed.

Lanolin	6 oz.
Petrolatum or "Vaseline"	8 oz.

If it is desired to impart a pleasant odor to this lotion a little perfume may be added and worked in.

Hand Lotion (Milky Liquid)

Lanolin	¼ teaspoonful
Glycosterin or Glyceryl Monostearate	1 oz.
Tincture of	

Benzoin	2 oz.
Witch Hazel	25 oz.

Melt the first two items together in an aluminum or enamelled double boiler. If no double boiler is at hand improvise one by standing the dish in a small pot containing boiling water. When the mixture becomes clear remove from the double boiler and add slowly, while stirring vigorously with a fork or stick, the tincture of benzoin and then the witch hazel. Continue stirring until cool and then put into one or two large bottles and shake vigorously. The finished lotion is a beautiful milky liquid comparable to the best hand lotions on the market sold at high prices.

Brushless Shaving Creams

Brushless or latherless shaving creams are soapy in nature and do not require lathering or water. The formula given below is of the latest type being free from alkali and non-irritating. It should be borne in mind, however, that certain beards are not softened by this type of cream and require the old-fashioned lathering shaving cream.

Brushless Shaving Cream

White Mineral Oil	10 oz.
Glycosterin or Glyceryl Monostearate	10 oz.
Water	50 oz.

Heat the first two ingredients together in a pyrex or enamelled dish to 150° F. and into this run slowly, while stirring with a fork, the water which has been heated to boiling. Allow to cool to 105° F. and while stirring add a few drops of perfume oil. Continue stirring until cold.

Mouth Washes

Mouth washes and oral antiseptics are of practically negligible value. Many, however, insist on their use because of their refreshing taste and deodorizing value.

Mouth Wash

Benzoic Acid	⅝ oz.
Tincture of Rhatany	3 oz.
Alcohol	20 oz.
Peppermint Oil	⅛ oz.

Just shake together in a dry bottle until it is dissolved and it is ready. A teaspoonful is used to a small wine-glassful of water.

Tooth Powders

Tooth powders depend for their cleansing action on soap and mild abrasives such as precipitated chalk and magnesium carbonate. The antiseptic present is practically of no value. The flavoring ingredients mask the taste of the soap and give the user's mouth a pleasant after-taste.

Tooth Powder

Magnesium Carbonate	420 g.
Precipitated Chalk	565 g.
Sodium Perborate	55 g.
Sodium Bicarbonate	45 g.
Soap, Powdered White	50 g.
Sugar, Powdered	90 g.
Wintergreen Oil	8 cc.
Cinnamon Oil	2 cc.
Menthol	1 g.

Dissolve the last three ingredients together and then rub well into the sugar. Add the soap and perborate mixing in well. Add the chalk with good mixing and then the sodium bicarbonate and magnesium carbonate. Mix thoroughly and sift through a fine wire screen. Keep dry.

Foot Powders

Foot powders consist of a filler such as talc or starch with or without an antiseptic or deodorizer. In the following formula the perborates liberate oxygen when in contact with perspiration which tends to destroy unpleasant odors. The talc acts as a lubricant and prevents friction and chafing.

Foot Powder

Sodium Perborate	3 oz.
Zinc Peroxide	2 oz.
Talc	15 oz.

Shake together thoroughly in a dry container until uniformly mixed. This powder must be kept dry or it will spoil.

Liniments

Liniments usually consist of an oil and an irritant such as methyl salicylate or turpentine. The oil acts as a solvent and tempering agent for the irritant. The irritant produces a rush of blood and warmth which is often slightly helpful.

Liniment, Sore Muscle

Olive Oil	6 fl. oz.
Methyl Salicylate	3 fl. oz.

Shake together and keep in a well stoppered bottle. Apply externally but do not apply to chafed or cut skin.

Chest-Rubs

In spite of the fact that chest-rubs are practically useless countless sufferers use them. Their action is similar to that of liniments and they differ only in that they are in the form of a salve.

"Chest-Rub" Salve

Yellow Petrolatum or Yellow Vaseline	1 lb.
Paraffin Wax	1 oz.
Eucalyptus Oil	2 fl. oz.
Menthol	½ oz.
Cassia Oil	⅛ fl. oz.
Turpentine	½ fl. oz.

Melt the vaseline and paraffin wax together in a double boiler and then add the menthol. Remove from the heat, stir, and cool a little; then stir in the oils, turpentine, and acid. When it begins to thicken pour into tins and cover.

Insect Repellents

Preparations of this type may irritate sensitive skins. Moreover, they will not always work. Psychologically they often are helpful, even though they may not keep insects away, because they give one confidence of protection.

Mosquito Repelling Oil

Cedar Oil	2 fl. oz.
Citronella Oil	4 fl. oz.
Spirits of Camphor	8 fl. oz.

Just shake together in a dry bottle and it is ready for use. This preparation may be smeared on the skin as often as is necessary to repel mosquitoes and other insects.

Fly Sprays

Fly sprays usually consist of deodorized kerosene, perfuming material, and an active insecticide. In some cases they merely stun the flies who may later recover and begin buzzing again.

Fly Spray

Deodorized Kerosene	89 fl. oz.
Methyl Salicylate	1 fl. oz.
Pyrethrum Powder	10 oz.

Mix thoroughly by stirring from time to time; allow to stand covered over night and then filter through muslin.

Caution: This spray is inflammable and should not be used near open flames.

Deodorant Spray

(For public buildings, sick-rooms, lavatories, etc.)

Pine Needle Oil	2 oz.
Formaldehyde	2 oz.
*Acetone	6 oz.
*Isopropyl Alcohol	20 oz.

One ounce of the above is mixed with a pint of water for spraying.

Cresol Disinfectant

†Caustic Soda	25½ g.
Water	140 cc.

Dissolve the above in a pyrex or enamelled dish and warm it. To this add slowly the following warmed mixture:

†Cresylic Acid	500 cc.
Rosin	170 g.

Stir until dissolved and add water to make 1000 cc.

Ant Poison

Sugar	1 lb.
Water	1 qt.
‡Arsenate of Soda	125 g.

Boil and stir until uniform; strain through muslin; add a spoonful of honey.

Bedbug Exterminator

*Kerosene	90 fl. oz.
Clove Oil	5 fl. oz.
§Cresol	1 fl. oz.
Pine Oil	4 fl. oz.

Simply shake and bottle.

Mothproofing Fluid (Non-Staining)

Sodium Aluminum Silicofluoride	½ oz.
Water	98 oz.
Glycerin	½ oz.
Sulfatate (Wetting Agent)	¼ oz.

Stir until dissolved.

Fly Paper

Rosin	32 oz.
Rosin Oil	20 oz.
Castor Oil	8 oz.

Heat the above in an aluminum or enamelled pot on a gas stove with stirring until all the rosin has melted and dissolved. While hot pour on firm paper sheets of suitable size which have been brushed with soap water just before coating. Smooth out the coating with a long knife or piece of thin flat wood and allow to cool. If a

* Inflammable.
† Do not get this on skin as it is corrosive.
‡ Poison.
§ Corrosive to skin.

heavier coating is desirable increase the amount of rosin used. Similarly a thinner coating is gotten by reducing the amount of rosin. The finished paper should be laid flat and not exposed to undue heat.

Household Baking Powder

Bicarbonate of Soda	28 oz.
Mono Calcium Phosphate	35 oz.
Corn Starch	27 oz.

Mix the above powders thoroughly in a dry can by shaking and rolling for a half hour. Pack into dry air-tight tins as moisture will cause lumping.

Malted Milk Powder

Malt Extract, Powdered	5 oz.
Skim Milk, Powdered	2 oz.
Sugar, Powdered	3 oz.

Mix thoroughly by shaking and rolling in a dry can. Pack in an air-tight container.

Cocoa Malt Powder

Corn Sugar	55 oz.
Malt, Powdered, Mild	19 oz.
Skim Milk, Powdered	12½ oz.
Cocoa	13 oz.
Vanillin	⅛ oz.
Salt, Powdered	⅜ oz.

Mix thoroughly and then run through a fine wire sieve.

Sweet Cocoa Powder

Cocoa	17½ oz.
Sugar, Powdered	32½ oz.
Vanillin	¾ g.

Mix thoroughly and sift.

Pure Lemon Extract

Lemon Oil U.S.P.	6½ fl. oz.
Alcohol	121½ fl. oz.

Shake together in a gallon jug till dissolved.

Artificial Vanilla Flavor

Vanillin	¾ oz.
Coumarin	¼ oz.
Alcohol	2 pt.

Stir the above in a glass or china pitcher until dissolved. Then stir in the following solution which has been made by stirring in another pitcher.

Sugar	12 oz.
Water	5¼ pt.
Glycerin	1 pt.

Color brown by adding sufficient "burnt" sugar coloring.

Canary Bird Food

Yolk of Eggs, Dried and Chopped	2 oz.
Poppy Heads (Coarse Powder)	1 oz.
Cuttlefish Bone (Coarse Powder)	1 oz.
Granulated Sugar	2 oz.
Soda Crackers, Powdered	8 oz.

Mix well together.

Writing Ink (Blue-Black)

Naphthol Blue Black	1 oz.
Gum Arabic, Powdered	½ oz.
Carbolic Acid	¼ oz.
Water	1 gal.

Stir together in a glass or enamelled vessel until dissolved.

Laundry Marking Ink (Indelible)

A. Soda Ash	1 oz.
Gum Arabic, Powdered	1 oz.
Water	10 fl. oz.

Stir the above until dissolved.

B. Silver Nitrate	4 oz.
Gum, Arabic Powdered	4 oz.
Lampblack	2 oz.
Water	40 fl. oz.

Stir this in a glass or porcelain dish until dissolved. Do not expose this to strong light or it will spoil. Finally pour into a brown glass bottle. In using these solutions wet the cloth with solution A and allow to dry. Then write on it with solution B using a quill pen.

Marking Crayon (Green)

Ceresin	8 oz.
Carnauba Wax	7 oz.
Paraffin Wax	4 oz.
Beeswax	1 oz.
Talc	10 oz.
Chrome Green	3 oz.

Melt the first four ingredients in any container and then add the last two slowly while stirring. Remove from the heat and continue stirring until thickening begins. Then pour into molds. If other color crayons are desired, other pigments may be used. For example for black, use carbon or bone-black; for blue, Prussian blue; for red, orange chrome yellow.

Antique Coloring for Copper

Copper Nitrate	4 oz.
Acetic Acid	1 oz.
Water	2 oz.

Dissolve by stirring together in a

glass or porcelain vessel. Pack in glass bottles.

To Use: Wet the copper to be colored and apply the above solution hot.

Blue-Black Finish on Steel

a. Place object in molten sodium nitrate (700–800° F.) for 2–3 minutes. Remove and allow to cool somewhat; wash in hot water; dry and oil with mineral or linseed oil.

b. Place in following solution for 15 minutes:

Copper Sulphate	½ oz.
Iron Chloride	1 lb.
Hydrochloric Acid	4 oz.
Nitric Acid	½ oz.
Water	1 gal.

Then allow to dry for several hours; place in above solution again for 15 min.; remove and dry for 10 hours. Place in boiling water for ½ hour; dry and scratch brush very lightly. Oil with mineral or linseed oil and wipe dry.

Rust Prevention Compound

Lanolin	1 oz.
*Naphtha	2 oz.

Mix until dissolved.

The metal to be protected is cleaned with a dry cloth and then coated with the above composition.

Metal Polish

Naphtha	62 oz.
Oleic Acid	⅓ oz.
Abrasive	7 oz.
Triethanolamine Oleate	⅓ oz.
Ammonia (26°)	1 oz.
Water	1 gal.

In one container mix together the naphtha and oleic acid to a clear solution. Dissolve the triethanolamine oleate in water separately, stir in the abrasive, if it is of a clay type, and then add the naphtha solution. Stir the resulting mixture at a high speed until a uniform creamy emulsion results. Then add the ammonia and mix well, but do not agitate as vigorously as before.

Glass Etching Fluid

Hot Water	12 fl. oz.
†Ammonium Bifluoride	15 oz.
Oxalic Acid	8 oz.
Ammonium Sulfate	10 oz.
Glycerin	40 oz.
Barium Sulfate	15 oz.

* Inflammable—keep away from flames.
† Corrosive.

Warm the washed glass slightly before writing on it with this fluid. Allow the fluid to act on the glass for about two minutes.

Leather Preservative

Neatsfoot Oil (Cold Pressed)	10 oz.
Castor Oil	10 oz.

Just shake together.

This is an excellent preservative for leather book bindings, luggage and other leather goods.

White Shoe Dressing

Lithopone	19 oz.
Titanium Dioxide	1 oz.
Shellac (Bleached)	3 oz.
Ammonium Hydroxide	¼ fl. oz.
Water	25 fl. oz.
Alcohol	25 fl. oz.
Glycerin	1 oz.

Dissolve the last four ingredients by mixing in a porcelain vessel. When dissolved stir in the first two pigments. Keep in stoppered bottles and shake before using.

Waterproofing for Shoes

Wool Grease	8 oz.
Dark Petrolatum	4 oz.
Paraffin Wax	4 oz.

Melt together in any container. Apply this grease warm but never hotter than the hand can bear.

Polishes

Polishes are usually used to restore the original lustre and finish of a smooth surface. As a secondary purpose they are expected to clean the surface and also to prevent corrosion or deterioration. There is no one polish which will give good results on all surfaces.

Most polishes depend on oil or wax for their lustering or polishing properties. Oil polishes are applied easily but the surfaces on which they are used attract dust and show finger marks. Wax polishes are more difficult to apply but are more lasting.

Oil or wax polishes are of two types: waterless and with water. The former are clear or translucent and the latter are milky in appearance.

For use on metals abrasives of various kinds such as tripoli, silica dust or infusorial earth are incorporated to grind away oxide films or corrosion products present.

Shoe Polish (Black)

Carnauba Wax	5½ oz.
Crude Montan Wax	5½ oz.

Melt together in a double boiler (the water in outer container should be at a boil) then stir in the following melted and dissolved mixture:

Stearic Acid	2 oz.
Nigrosine Base	1 oz.

Then stir in

Ceresin	15 oz.

Remove all flames and run in slowly, while stirring

Turpentine	90 fl. oz.

Allow mixture to cool to 105° F. and pour into air-tight tins which should be allowed to stand undisturbed over night.

Auto Polish (Clear Oil Type)

Paraffin (Mineral) Oil	5 pt.
Raw Linseed Oil	2 pt.
China Wood Oil	½ pt.
*Benzol	¼ pt.
Kerosene	¼ pt.
Amyl Acetate	1 tbsp.

Shake together in a glass jug and keep stoppered.

Auto and Floor Wax (Paste Type)

Yellow Beeswax	1 oz.
Ceresin	2½ oz.
Carnauba Wax	4½ oz.
Montan Wax	1¼ oz.
*Naphtha or Mineral Spirits	1 pt.
*Turpentine	2 oz.
Pine Oil	½ oz.

Melt the waxes together in a double boiler. Turn off the heat and run in the last three ingredients in a thin stream and stir with a fork. Pour into cans; cover and allow to stand undisturbed overnight.

Furniture Polish (Oil and Wax Type)

Thin Paraffin (Mineral Oil	1 pt.
Carnauba Wax, Powdered	¼ oz.
Ceresin Wax	⅛ oz.

Heat together until all of the wax is melted. Allow to cool and pour into bottles before mixture turns cloudy.

Polishing Wax (Liquid)

Beeswax, Yellow	1 oz.
Ceresin Wax	4 oz.

Melt together and then cool to 130° F.; turn off all flames and stir in slowly.

*Turpentine	17 fl. oz.
Pine Oil	½ fl. oz.

Pour into cans or bottles which are closed tightly to prevent evaporation.

* Inflammable—Keep away from flames.

Floor Oil

Mineral Oil	46 fl. oz.
Beeswax	½ oz.
Carnauba Wax	1 oz.

Heat together in double boiler until dissolved (clear). Turn off flame and stir in

*Turpentine	3 fl. oz.

Lubricants

Lubricants in the form of oils or greases are used to prevent friction and wearing of parts which rub together. Lubricants must be chosen to fit specific uses. They consist of oils and fats often compounded with soaps and other unctuous materials. For heavy duty heavy oils or greases are used and light oils for light duty.

Gun Lubricant

White Petrolatum	15 oz.
Bone Oil (Acid Free)	5 oz.

Warm gently and mix together.

Graphite Grease

Ceresin	7 oz.
Tallow	7 oz.

Warm together and gradually work in, with a stick

Graphite	3 oz.

Stir until uniform and pack in tins when thickening begins.

Penetrating Oil

(For freeing rusted bolts, screws, etc.)

Kerosene	2 oz.
Thin Mineral Oil	7 oz.
Secondary Butyl Alcohol	1 oz.

Shake together and keep in a stoppered bottle.

Molding Material

White Glue	13 lb.
Rosin	13 lb.
Raw Linseed Oil	⅓ qt.
Glycerin	1 qt.
Whiting	19 lb.

This mixture is prepared by cooking the white glue until it is dissolved. Then cook separately the rosin and raw linseed oil until they are dissolved. Add the rosin, oil, and glycerin to the cooked glue, stirring in the whiting until the mass makes up to

* Inflammable.

the consistency of putty. Keep the mixture hot.

Place this putty mass in the die, pressing it firmly into the same and allowing it to cool slightly before removing. The finished product is ready to use within a few hours after removal. Suitable colors can be added to secure brown, red, black or other color.

In applying ornaments made of this composition to a wood surface, they are first steamed to make them flexible; in this condition they can be glued to the wood surface easily and securely. They can be bent to any shape, and no nails are required for applying them.

Grafting Wax

Wool Grease	11 oz.
Rosin	22 oz.
Paraffin Wax	6 oz.
Beeswax	4 oz.
Japan Wax	1 oz.
Rosin Oil	9 oz.
Pine Oil	1 oz.

Melt together until clear and pour into tins. This composition can be made thinner by increasing the amount of rosin oil and thicker by decreasing it.

Candles

Paraffin Wax	30 oz.
Stearic Acid	17½ oz.
Beeswax	2½ oz.

Melt together and stir until clear. If colored candles are desired a pinch of any oil soluble dye is dissolved at this stage. Pour into vertical molds in which wicks are hung.

Adhesives

Adhesives are sticky substances used to unite two surfaces. Adhesives are specifically called glues, pastes, cements, mucilages, lutes, etc. For different uses different types are required.

Wall Patching Plaster

Plaster of Paris	32 oz.
Dextrin	4 oz.
Pumice Powder	4 oz.

Mix thoroughly by shaking and rolling in a dry container. Keep away from moisture.

Cement Floor Hardener

Magnesium Fluosilicate	1 lb.
Water	15 pt.

Mix until dissolved.

In using this, the cement should first be washed with clean water and then drenched with the above solution.

Paperhanger's Paste

Use a cheap grade of rye or wheat flour, mix thoroughly with cold water to about the consistency of dough or a little thinner, being careful to remove all lumps. Stir in a tablespoonful of powdered alum to a quart of flour, then pour in boiling water, stirring rapidly until the flour is thoroughly cooked. Let this cool before using and thin with cold water.

a.	White or Fish Glue	4 oz.
	Cold Water	8 oz.
b.	Venice Turpentine	2 fl. oz.
c.	Rye Flour	1 lb.
	Cold Water	16 fl. oz.
d.	Boiling Water	64 fl. oz.

Soak the 4 oz. of glue in the cold water for 4 hours. Dissolve on a water bath (glue-pot) and while hot stir in the Venice turpentine. Make up *c* into a batter free from lumps and pour into *d.* Stir briskly, and finally add the glue solution. This makes a very strong paste, and it will adhere to a painted surface, owing to the Venice turpentine in its composition.

Aquarium Cement

Litharge	10 oz.
Plaster of Paris	10 oz.
Powdered Rosin	1 oz.
Dry White Sand	10 oz.
Boiled Linseed Oil	Sufficent

Mix all together in the dry state, and make into a stiff putty with the oil when wanted for use.

Do not fill the aquarium for three days after cementing. This cement hardens under water, and will stick to wood, stone, metal, or glass, and, as it resists the action of sea-water, it is useful for marine aquaria. The linseed oil may have an addition of drier to the putty made up four or five hours before use, but after standing fifteen hours, however, it loses its strength when in the mass.

Wood Dough Plastic

*Collodion	86 g.
Ester Gum, Powdered	9 g.
Wood Flour	30 g.

Allow first two ingredients to stand until dissolved, stirring from time to time. Then while stirring add the wood flour a little at a time until

* Inflammable.

uniform. This product can be made softer by adding more collodion.

Putty

Whiting	80 oz.
Raw Linseed Oil	16 oz.

Rub together until smooth. Keep in closed container.

Wood Floor Bleach

Sodium Metasilicate	90 oz.
Sodium Perborate	10 oz.

Mix thoroughly and keep dry in a closed can. Use 1 pound to a gallon of boiling water. Mop or brush on the floor, allow to stand ½ hour, then rub off and rinse well with water.

* Paint Remover

Benzol	5 pt.
Ethyl Acetate	3 pt.
Butyl Acetate	2 pt.
Paraffin Wax	½ lb.

Stir together until dissolved.

Soaps and Cleaners

Soaps are made from a fat or fatty acid and an alkali. They lather and produce a foam which entraps dirt and grease which is washed away with water. There are numerous kinds of soaps depending on the uses to which they are to be put.

Cleaners consist of solvent such as naphtha with or without a soap. Abrasive cleaners are soap pastes containing powdered pumice, stone, silica, etc.

Liquid Soap (Concentrated)

Water	11 oz.
†Caustic Potash (Solid)	1 oz.
Glycerin	4 oz.
Red Oil (Oleic Acid)	4 oz.

Dissolve the caustic in water, add the glycerin and bring to a boil in an enamelled pot. Remove from heat, add the red oil slowly while stirring. If a more neutral soap is wanted, use a little more red oil.

Saddle Soap

Beeswax	5 oz.
†Caustic Potash	0.8 oz.
Water	8 oz.

Boil for 5 minutes while stirring. In another vessel heat

Castile Soap	1.6 oz.
Water	8 oz.

* Inflammable.
† Do not get on skin as it is corrosive.

Mix the two with good stirring; remove from heat and add

Turpentine	12 oz.

while stirring.

Mechanics Hand Soap Paste

Water	1.8 qt.
White Soap Chips	1.5 lb.
Glycerin	2.4 oz.
Borax	6 oz.
Dry Sodium Carbonate	3 oz.
Coarse Pumice Powder	2.2 lb.
Safrol	enough to scent

Dissolve the soap in ⅔ of the water by heat. Dissolve the last three in the rest of the water. Pour the two solutions together and stir well. When it begins to thicken, sift in the pumice, stirring constantly till thick, then pour into cans. Vary amount of water, for heavier or softer paste (water cannot be added to the finished soap).

Dry Cleaning Fluid

Glycol Oleate	2 fl. oz.
Carbon Tetrachloride	60 fl. oz.
Varnoline (Naphtha)	20 fl. oz.
Benzine	18 fl. oz.

An excellent cleaner that will not injure the finest fabrics.

Wall Paper Cleaner

Whiting	10 lb.
Magnesia Calcined	2 lb.
Fuller's Earth	2 lb.
Pumice Powder	12 oz.
Lemenone or Citronella Oil	4 oz.

Mix well together.

Household Cleaner

Soap Powder	2 oz.
Soda Ash	3 oz.
Trisodium Phosphate	40 oz.
Finely Ground Silica	55 oz.

Mix well and put up in the usual containers.

Window Cleanser

Castile Soap	2 oz.
Water	5 oz.
Chalk	4 oz.
French Chalk	3 oz.
Tripoli Powder	2 oz.
Petroleum Spirits	5 oz.

Mix well and pack in tight containers.

Straw Hat Cleaner

Sponge the hat with a solution of

Sodium Hyposulphite	10 oz.
Glycerin	5 oz.

Alcohol	10 oz.
Water	75 oz.

Lay aside in a damp place for 24 hours and then apply

Citric Acid	2 oz.
Alcohol	10 oz.
Water	90 oz.

Press with a moderately hot iron after stiffening with gum water if necessary.

Grease, Oil, Paint & Lacquer Spot Remover

Alcohol	1 oz.
Ethyl Acetate	2 oz.
Butyl Acetate	2 oz.
Toluol	2 oz.
Carbon Tetrachloride	3 oz.

Place garment with spot over a piece of clean paper or cloth and wet with the above fluid; rub with clean cloth toward center of spot. Use a clean section of cloth for rubbing and clean paper or cloth for each application of the fluid. The above product is inflammable and should be kept away from flames. Use of cleaners of this type should be out-of-doors or in well-ventilated rooms as the fumes are toxic.

Paint Brush Cleaner

Mix (1)

Kerosene	2 pt.
Oleic Acid	1 pt.

Mix (2)

Strong Liquid Ammonia, 28%	¼ pt.
Denatured Alcohol	¼ pt.

Slowly stir 2 into 1 until a smooth mixture results. To clean brushes, pour into a can and stand the brushes in it overnight. In the morning, wash out with warm water.

Rust & Ink Remover

Immerse portion of fabric with rust or ink spot alternately in Solution A and B, rinsing with water after each immersion.

Solution A

Ammonium Sulphide Solution	1 oz.
Water	19 oz.

Solution B

*Oxalic Acid	1 oz.
Water	19 oz.

Javelle Water (Laundry Bleach)

Bleaching Powder	2 oz.
Soda Ash	2 oz.
Water	5 gal.

* Poisonous.

Mix well until reaction is completed. Allow to settle overnight and siphon off the clear liquid.

Laundry Blue (Liquid)

Prussian Blue	1 oz.
Distilled Water	32 oz.
*Oxalic Acid	¼ oz.

Dissolve by mixing in a crock or wooden tub.

"Glassine" Paper

Paper is coated with or dipped in the following solution and then hung up to dry.

Gum Copal	10 oz.
Alcohol	30 fl. oz.
Castor Oil	1 fl. oz.

Dissolve by letting stand overnight in a covered jar and stirring the next day.

Waterproofing Paper and Fibreboard

The following composition and method of application will render uncalendered paper, fibreboard, and similar porous material waterproof and proof against the passage or penetration of water.

Paraffin (Melting Point about 130° F.)	22.5 oz.
Trihydroxyethylamine Stearate	3.0 oz.
Water	74.5 oz.

The paraffin wax is melted and the stearate added to same. The water is then heated to nearly boiling and then vigorously agitated with a suitable mechanical stirring device while the above mixture of melted wax and emulsifier is slowly added. This mixture is cooled while it is stirred.

The paper or fibreboard is coated on the side which is to be in contact with water. This is then quickly heated to the melting point of the wax, which then coalesces into a continuous film that does not soak into the paper which is preferentially wetted by the water. This method works most effectively on paper pulp moulded containers and possesses the advantages of being much cheaper than dipping in melted paraffin as only about a tenth as much paraffin is needed. In addition, the outside of the container is not greasy, and can be printed upon after treatment which is not the case when treated with melted wax.

Waterproofing Liquid

Paraffin Wax	⅝ oz.

* Poison.

Gum Dammar	1⅕ oz.
Pure Rubber	⅛ oz.
Benzol	13 oz.
Carbon Tetrachloride to make	1 gal.

Dissolve rubber in benzol; add other ingredients and allow to dissolve. (Inflammable.)

The above is suitable for wearing apparel and wood. It is applied by brushing on two or more coats, allowing each to dry before applying another coating. Apply outdoors as vapors are inflammable and toxic.

Waterproofing Heavy Canvas

Raw Linseed Oil	1 gal.
Beeswax, Crude	13 oz.
White Lead	1 lb.
Rosin	12 oz.

Heat the above, while stirring, until all lumps are gone and apply warm to upper side of canvas; wetting the canvas with a sponge on the underside before applying.

Cement Waterproofing

Chinawood Oil Fatty Acids	10 oz.
Paraffin Wax	10 oz.
Kerosene	2½ gal.

Stir until dissolved. This is painted or sprayed on cement walls, which must be dry.

Oil and Greaseproofing Paper and Fibreboard

This solution applied by brush, spray, or dipping will leave a thin film which is impervious to oils and grease. Applied to paper or fibre containers, it will enable them to retain oils and greases. All the following ingredients are by weight:

Starch	6.6 oz.
Caustic Soda	0.1 oz.
Glycerin	2.0 oz.
Sugar	0.6 oz.
Water	90.5 oz.
Sodium Salicylate	0.2 oz.

The caustic soda is dissolved in the water and then the starch is made into a thick paste by adding a portion of this solution. This paste is then added to the water. This mixture is placed in a water jacket and heated to about 85° C. until all the starch granules have broken and the temperature maintained for about half an hour longer. The other substances are then added and thoroughly mixed and the composition is completed and ready for application. A smaller water content may be used if applied hot and a thicker coating will result. Two coats will result in a very considerable resistance to oil penetration.

Fireproof Paper

Ammonium Sulphate	8 oz.
Boric Acid	3 oz.
Borax	1¾ oz.
Water	100 fl. oz.

Mix together in a gallon jug, by shaking, until dissolved.

The paper to be treated is dipped into this solution in a pan, until uniformly saturated. It is then taken out and hung up to dry. Wrinkles can be prevented by drying between cloths in a press.

Fireproofing Canvas

Ammonium Phosphate	1 lb.
Ammonium Chloride	2 lb.
Water	½ gal.

Impregnate with above; squeeze out excess and dry. Washing or exposure to rain will remove fireproofing salts.

Fireproofing Light Fabrics

Borax	10 oz.
Boric Acid	8 oz.
Water	1 gal.

Impregnate; squeeze and dry. Fabrics so impregnated must be treated again after washing or exposure to rain as the fireproofing salts wash out easily.

Dry Fire Extinguisher

Ammonium Sulphate	15 oz.
Sodium Bicarbonate	9 oz.
Ammonium Phosphate	1 oz.
Red Ochre	2 oz.
Silex	23 oz.

Use powdered materials only; mix well and pass through a fine sieve. Pack in tight containers to prevent "lumping."

Fire Extinguishing Liquid

Carbon Tetrachloride	95 oz.
Solvent Naphtha	5 oz.

The inclusion of the naphtha minimizes production of toxic fumes when extinguishing fires.

Fire Kindler

Rosin or Pitch	10 oz.
Sawdust	10 or more oz.

Melt, mix, and cast in forms.

Solidified Gasoline

*Gasoline	½ gal.

* Inflammable.

White Soap (Fine Shaved) 12 oz.
Water 1 pt.
Household Ammonia 5 oz.

Heat the water, add soap, mix and when cool add the ammonia. Then slowly work in the gasoline to form semi-solid mass.

Boiler Compound

Soda Ash	87 oz.
Trisodium Phosphate	10 oz.
Starch	1 oz.
Tannic Acid	2 oz.

Use powdered materials, mixing well and then pass through a fine sieve.

Anti-Freezes

The materials listed below are the basic ingredients used in all good anti-freeze liquids. Of these, alcohol is the only one that evaporates. Radiators containing alcohol should be tested from time to time to be sure of protection. A hydrometer for testing alcohol solution strength can be bought from sellers of denatured alcohol.

Anti-Freeze Liquids

Pints of anti-freeze per gal. of water for protection at:

	+10° F.	0° F.	—10° F.	—20° F.
Denatured Alcohol 180° proof	3.4	4.9	6.5	8.3
Denatured Alcohol 188° proof	3.3	4.7	6.0	7.7
Glycerin 95%	3.3	5.3	7.1	9.0
Radiator Glycerin 60%	10.0	18.7	39.0	106.5
Ethylene Glycol 95%	2.7	4.0	5.1	6.5

Specific gravity for protection at:

	+10° F.	0° F.	—10° F.	—20° F.	—30° F.
Denatured Alcohol	0.968	0.959	0.950	0.942	0.921
Glycerin	1.090	1.112	1.131	1.147	1.158
Ethylene Glycol	1.038	1.048	1.056	1.064	1.069

Soldering Flux (Non-corrosive)

Rosin, Powdered	1 oz.
Denatured Alcohol	4 oz.

Soak overnight and mix well.

Photographic Solutions

Developing Solution

Stock Solution A

Dissolve the following, separately, in glass or enamel dishes.

Pyro	4 oz.
Sodium Bisulphite, Pure	280 gr.
Potassium Bromide	32 gr.
Distilled Water	64 oz.

Stock Solution B

Sodium Sulphite, Pure	7 oz.
Sodium Carbonate, Pure	5 oz.
Distilled Water	64 oz.

To use take the following proportions:

Stock Solution A	2 oz.
Stock Solution B	2 oz.
Distilled Water	16 oz.

At a temperature of 65° F. this developer requires about 8 minutes.

Acid Hardening Fixing Bath

A. Sodium Hyposulphite 32 oz.
Distilled Water 8 oz.

Stir until dissolved and then add the following chemicals in the order given below, stirring each until dissolved:

B. Distilled Water (Warm) 2½ oz.
Sodium Sulphite, Pure ½ oz.
Acetic Acid (28%), Pure 1½ oz.
Potassium Alum Powder ½ oz.

Add Solution B to A and store in dark bottles away from light.

Chapter Two

ADHESIVES

AQUEOUS ADHESIVES

Flexible Glues

Formula No. 1

For general bindery use, utilizing waste roller composition:

Glue	20.80
Glycerin	16.60
Waste Roller Composition	9.80
Water	52.50
Beta Naphthol	.15
Terpineol	.15

No. 2

For use on gathering, stitching, and covering machines:

Glue	36.30
Glycerin	16.60
Water	46.80
Beta Naphthol	.15
Terpineol	.15

No. 3

For use on Perfect Binding Machine:

Glue	39.90
Glycerin	33.30
Water	26.50
Beta Naphthol	.15
Terpineol	.15

No. 4

Tablet composition:

Glue	26.50
Glycerin	26.50
Water	46.70
Beta Naphthol	.15
Terpineol	.15

No. 5

Tablet composition utilizing waste roller composition:

Glue	17.00
Waste Roller Composition	42.50
Water	39.60
Zinc Oxide	.90

No. 6

For general bindery use:

Glue	22.60
Diethylene Glycol	22.10
Water	55.00
Beta Naphthol	.15
Terpineol	.15

No. 7

For use on gathering, stitching, and covering machines:

Glue	36.30
Glycerin	8.30
Diethylene Glycol	8.30
Water	46.80
Beta Naphthol	.15
Terpineol	.15

No. 8

Tablet composition:

Glue	26.50
Glycerin	13.25
Diethylene Glycol	13.25
Water	46.70
Beta Naphthol	.15
Terpineol	.15

No. 9

For general bindery use:

Glue	22.60
Sorbitol Syrup	20.50
Water	56.60
Beta Naphthol	.15
Terpineol	.15

No. 10

For use on gathering, stitching, and covering machines:

Glue	36.40
Sorbitol Syrup	16.60
Water	46.70
Beta Naphthol	.15
Terpineol	.15

No. 11

Tablet composition:

Glue	26.50
Sorbitol Syrup	26.50
Water	46.70
Beta Naphthol	.15
Terpineol	.15

No. 12

For gluing-off large, thick books previous to rounding and backing:

Glue	22.50
Sorbitol Syrup	25.90
Water	51.30
Beta Naphthol	.15
Terpineol	.15

No. 13

For bookbinders, telephone books:

Gelatin	35.0
Glycerin	14.0
Glucose	5.0
Phenol	0.4
Water	45.6

Add glycerin and glucose to water. Mix gelatin into solution. Let stand in the cold for several hours. Heat to 140° F. and maintain at that temperature till all the gelatin is dissolved. When clear, add phenol. Pour into suitable forms and allow to cool.

This glue is melted when needed. Water may be added to suit requirements.

Insolubilizing Glue and Casein (Patented)

Glue

Glue	20
Water	30

When the water has been absorbed by the glue, glycerine formal (10) and 10 per cent sulfuric acid (1) are added. After curing this product at 50°–70° C. it is insoluble in boiling water following a preliminary soaking of three hours in cold water.

Casein

Water	30
Glycerin Formal	2.4
Sulphuric Acid (10%)	1
Casein	10

This mixture after thorough mixing may be pressed into a mold and cured at 50°–100° C. to form a water resistant product.

Other acids than sulfuric acid may be used and the results will depend upon the pH of the mixture. In general a low pH results in faster curing and lower curing temperatures.

Shoe Cement

Glue, Viscous Carpenter's	1.00
Glycerin	0.30
Acetic Acid	0.04
Water	0.50–0.60

Apply warm.

Transparent Colorless Adhesive
British Patent 525,620

Gelatin	1 lb.
Water	1½ gal.
Potassium Chlorate (26° Bé. Solution)	0.15 gal.
Boric Acid	¼ lb.

"Cellophane" Adhesive

Gum Arabic	16½
Aquaresin	30
Water	50
Formaldehyde	4½

Envelope Adhesive

The gum used by the United

States Government on postage stamps is probably one of the best that could be used not only for envelopes but for labels as well. It will stick to almost any surface, and is prepared with:

Gum Arabic	1
Starch	1
Sugar	4

Water, sufficient to give the desired consistency.

The gum arabic is first dissolved in some water, the sugar added, then the starch, after which the mixture is boiled for a few minutes in order to dissolve the starch, after which it is thinned down to the desired consistency. To cut the expense, dextrine may be substituted for the gum arabic, glucose for the sugar, and boric acid added to preserve and help stiffen the paste.

Mucilage for Secret Message Envelopes and Stamps

Mucilage for sealing letters and documents, or for use in stamps and revenue tags which shows the effects of tampering and steaming is prepared by adding beta-methylumbelliferone to ordinary paper mucilage or glue base. When the stamp or seal has been tampered with, or an attempt to steam open has been made, tell-tale traces of the attempt will be seen by a bright blue coloration on the paper or about the stamp which can only be seen in ultraviolet light and not in white light.

Plywood Adhesive
Australian Patent 110,458

Formula	No. 1	No. 2
Casein	100	100
Urea	15	15½
Water	460	480
Formaldehyde	7½	8
Trisodium Phosphate	—	4

Oil and Vapor Proof Seal
Australian Patent 113,458

Glue	3
Water	10
Turkey-Red Oil	11
Caustic Potash	1
Graphite	15
Chalk	40
Asbestos	10
Barytes	10

Neutral Tin Paste
U. S. Patent 2,133,098

Casein (18% solution)	100.0
Ammonium Fluoride	90.0
Sodium Borophosphate	5.0
Glycerin	0.6
Water	4.4

Dilute with water as desired.

Dry Paste
Formula No. 1
U. S. Patent 2,204,384

Sugar	55 –45 %
Starch	43.5–53.5%
Citric Acid	1.5%

No. 2
U. S. 2,231,050

Starch	100
Lactic Acid	25
Sodium Chlorate	1

Heat to 110° C. until soluble in hot water.

Metal Container Sealing Dope
British Patent 517,037

Barytes	10.1
Asbestine	5.0
Gum Karaya	0.3
Colloidal Clay	1.9
Glue	4.5
Glycerin	2.5

Casein	1.8
Ammonia (28%)	0.7
Rubber Latex (Solids)	2.0
Vulcanized Rubber Latex (Solids)	5.0
Water	66.2

Viscous Water Soluble Sticky Liquid

Sodium Alginate (1% Solution)	50
Abopon (Sodium Borophosphate)	50

Mix together to produce a material having great body, stickiness and "length."

Gasket Paste

Formula No. 1

Potash Soap	18.25%
Glycerin	4.6 %
Water	2 %
Castor Oil, Bodied to Make	100 %

Heat gradually with mixing to 250° F. to get a smooth emulsion. Continue stirring while cooling to 100° F. and then pack.

No. 2

Soap	20%
Water	2%
Glycerin	5%
Blown Castor Oil, to Make	100%

No. 3

Potash Soap	20%
Diglycol Laurate	5%
Glycerin	5%
Castor Oil, Blown	59%

Heat and mix at 250° F. until uniform. Stop and heat and mix until it cools to 100° F. and package.

Adhesive for Window Glass

Gum Arabic	18
Water	52
Glycerin	30

Dissolve the gum arabic in water and add the glycerin.

Paper to Glass Adhesive

Water	150
Zinc Chloride	2
Glycerin	10
Cassava Flour or Tapioca	12½

Mix thoroughly and heat to boiling, stirring vigorously, all the time, during the heating. If tapioca is in lumps, the mixture should soak overnight before heating. Other starches and flour can be used in place of the cassava and tapioca, but the products are much poorer in quality.

Hot Enamel Lithographer's Adhesive

Formula No. 1

Soak 400 g. gelatin in 500 ml. water and heat until dissolved. Add 66 ml. water, 30 ml. concentrated hydrochloric acid and 4 g. pepsin. Heat 10–15 hours at 45° then to 90° to inactivate the pepsin, and add alcoholic solution of thymol. Final mixture shows pH 4.1 and can be kept indefinitely. For enameling, 375 g. of suspension is mixed with 13.5 g. ammonium dichromate and 1200 ml. water. The enamel is transparent and has good adhesive properties.

No. 2

From 20 to 25 ml. of a 10% solution of aluminum sulfate is added to a 1:6 solution of bone glue in water. The mixture is neutralized with a lime solution.

Carton Sealing Glue

White Corn Dextrin (Medium Viscosity)	35.0
Borax	5.0

Sodium Hydroxide (25%)	3.0
Formaldehyde	0.3
Water	56.7

Mix the dextrin in 45 parts of cold water till free of lumps. Heat to 180° F. In balance of water dissolve the borax. Add borax solution to dextrin solution; finish with the caustic and formaldehyde.

For most work this glue will stand considerable dilution.

Tube Glue

For reels, tubes, cones and similar material:

Canary Corn Dextrin	50.0
Borax	5.0
Caustic Soda (25%)	2.0
Formaldehyde	0.3
Water	42.7

Warm water, as dextrin is being added. By the time all the dextrin is added the solution should be at 180° F. Keep stirring till free of lumps. Then sift in the borax and maintain the heat and the stirring till all the borax is dissolved. Finish with the caustic and formaldehyde.

Stripping Glue

For paper box making:

White British Gum	28.0
Borax	5.0
Phenol	0.3
Caustic Soda (30%)	1.8
Water	64.9

Dissolve the gum in the water. Heat to 180° F. Add the borax, phenol, and caustic.

Envelope Gum—Heavy Paper

Canary Tapioca Dextrin	65.0
Glucose	5.0
Glycerin or Glycol	1.0
Formaldehyde	0.2
Water	28.8

Procedure as for envelope paste.

Envelope Gum—Kraft Paper

Canary Tapioca Dextrin	60.0
Glucose	5.0
Phosphoric Acid (80%)	1.0
Formaldehyde	0.2
Water	33.8

Heat water, add dextrin while stirring. At the end of dextrin addition the temperature should be 190° F. Maintain temperature till free of lumps. Add glucose and cool to 110° F. Add formaldehyde and phosphoric acid.

Bottle Labelling Glue

Formula No. 1

Canary Corn or Tapioca Dextrin	52.0
Glucose	22.5
Formaldehyde	0.3
Water	25.2

Add glucose to water and heat. Add dextrin while heating. When dextrin is all added the temperature should be 180° F. Continue heating while stirring till the solution is clear. Cool, finish with formaldehyde.

This glue is suitable for automatic labeling machines. For hand work dilute with water.

No. 2

Iceproof type—suitable for beer and soda bottles kept in ice water:

Sago Starch	32.0
Caustic Soda (25%)	18.0
Nitric Acid	11.0
Phenol	0.3
Water	38.7

Heat water to 130° F. Add starch and stir till smooth. Add small amounts of caustic till mass is smooth and opaque. Quickly add balance of caustic. Stir till uniform. Discontinue stirring for five hours. Add acid till just neutral. Finish with phenol.

Bench Paste

Soft tacky paste for bench work:

White Corn Dextrin	24.0
Chlorinated Starch	4.0
Glucose	20.0
Borax	3.5
Phenol	0.2
Water	48.3

Mix glucose, corn, dextrin, and chlorinated starch in cold water. When smooth heat to 180° F. then dissolve in it the borax and phenol. Draw off in containers and age two weeks before using.

Library Paste

White Potato Dextrin	25.0
Glucose	22.0
Borax	5.0
Phenol	0.4
Water	47.6

Mix glucose and dextrin in the water warmed to 110° F. When free of lumps raise temperature to 180° F. When the solution is clear, add the borax and phenol.

Non-Warp Glue

For high grade boxes, loose leaf pad covers, etc.:

Low Grade Gelatin	30.0
Glycerin	10.0
Glucose	25.0
Phenol	0.4
Water	34.6

Prepare as flexible glue.

Adhesive

For fixing mirrors to microscopes and other scientific instruments:

Gelatin (Medium Grade)	30.0
Glycerin	14.0
Phenol	0.4
Water	55.6

Prepare as flexible glue. Use warm.

Tightwrap Glue

Formula No. 1

Canary Tapioca Dextrin	40.0
Glucose	10.0
Borax	5.0
Caustic Soda (25%)	1.0
Formaldehyde	0.3
Water	43.7

Dissolve the dextrin and glucose in the water. Heat to 180° F. Dissolve the borax in the solution. Finish with the caustic soda and formaldehyde.

This glue is slow drying and very tacky.

No. 2

Canary Corn Dextrin (Completely converted high viscosity)	36.0
Glucose	10.0
Glycerin or Glycol	4.0
Caustic Soda (25%)	1.0
Formaldehyde	0.3
Water	48.7

Method of manufacture same as above.

Envelope Paste

Canary Tapioca Dextrin (Completely converted)	60.0
Formaldehyde	0.3
Water	39.7

Heat water, add dextrin while stirring. At the end of the dextrin addition, the temperature should be 190° F. Maintain temperature till free of lumps. Cool, add formaldehyde. More concentrated pastes can be made if desired.

WATER INSOLUBLE ADHESIVES

Acid Soluble Oil Well Cement

Portland Cement	1
Calcium Carbonate, Precipitated	1

Acid Resisting Cement

Sulphur	95
Polybutene Vistanex	5

Acid Proof Cement—Silicate Type

A cement capable of setting in the cold to a hard infusible mortar, resistant to acids at all concentrations and temperature; not resistant to boiling water or alkali.

Mix thoroughly 2 parts powder to 1 part binder, with allowable variations to obtain a desirable troweling consistency. Setting time 20 minutes to 1 hour.

A. Powder
 1. Filler: 93 to 97
  Silica or pulverized quartz, 50 to 325 mesh, preferably.
 2. Setting Agent: 3 to 7
  a) Aryl or alkal-aryl sulfo chlorides such as benzene sulfochloride, p-toluene sulfochloride, etc.
  or b) Silico fluorides, such as sodium silico fluoride, etc.
  or c) Mixtures of a) and b)

B. Binder
 Sodium or potassium silicate water glass, 30° to 40° Baume

Acid Proof Cement—Synthetic Resin Type

A cement capable of setting in the cold to a hard, infusible mortar, resistant to acids except the highly oxidizing ones, oils, solvents, greases, water and mild alkalies of temperatures up to 300 to 330 degrees Fahrenheit.

Mix thoroughly 2 to 2.5 parts powder to 1 part binder to obtain desirable troweling consistency. Setting time at 70° C. is under 1 hour. Final cure within a day.

A. Powder
 1. Filler: 93 to 96
  a) Silica preferably 50 to 325 mesh
  or b) Carbon preferably 50 to 325 mesh
 2. Setting Agent: 4 to 7
 Aromatic sulfonic acids, such as benzene sulfonic acid, p-toluene sulfonic acid, etc.

B. Binder
 1. A phenol-formaldehyde, or homologues of these two products, reacted by means of an acid or alkaline catalyst, and arrested at a stage wherein the partially polymerized resin has a viscosity of 25 to 70 seconds at 25° C. in a Gardner-Holt tube: 75
 2. Glycerin: 10
 3. Water: 15

Water and Alkali Resistant Adhesive

This adhesive is specially useful for bonding a dry surface to a wet sub-surface, and in maintaining adhesion where exposed to water and alkali.

Formula No. 1

Para-Coumarone Resin Hard (M.P. 100–115° C.) 90% Soft (M.P. 20–30° C.) 10%	27.8
Portland Cement, Finely ground and quick setting	57.2
Asbestos or Asbestine	4.6
Solvent (Water miscible) Acetone 20% Methyl Ethyl Ketone 80%	10.4

No. 2

Phenol-Aldehyde Resin (In liquid stage)	28.5

Portland Cement, Finely ground	57.5
Asbestos or Asbestine	4.5
Solvent (Water miscible)	10.5
Cellosolve 15%	
Ethyl Alcohol 85%	

Water and Acid Resistant Glue for Plywood

This is valuable in that it utilizes waste from phenol-formaldehyde molding wastes. It develops a bond stronger than the wood itself.

A. Glue

Cold Setting Water Soluble Urea Formaldehyde Resin	58
Water	29
Finely Ground Bakelite Waste	13

(C stage phenol-formaldehyde resin)

This may be spread on one surface. On the other surface the hardening agent is spread and pressure is applied to the contacting surfaces 'til cured.

B. Hardening Agent

This may be organo phosphate esters or ammonium salts such as the chloride, thio cyanate, nitrate.

Waterproof Adhesive

Gutta Percha	100
Pine Tar Pitch	100

Boat Plug Compound

Portland Cement	75
Lanolin or Sublan	25

This is plugged into holes in boats, with or without oakum or rags to keep out water.

Vacuum Holding Wax Adhesive

Beeswax	10
Rosin	10

"Nylon" Adhesives

(for obtaining good anchorage between Nylon fabric and rubber or synthetic rubber coatings)

Saturate fabric with solutions of
Hexamethylene Diisocyanate
Monomethylene Diisocyanate

Cement for Vinyl Coatings

Formula No. 1

Perbunan	40
BRT No. 7	5
Zinc Oxide	2
Sulphur	1
Resin 1112 (Reichold)	10
Altax	1
Diphenylguanidine	1
Catalpo	30
Colors	To suit

Dissolve in:

Methyl Ethyl Ketone	80
2 Nitropropane	20

No. 2

Olive Drab Color

Hycar	24 lb.
Neoprene	6 lb.
Whiting	25 lb.
Kadox	1 lb. 8 oz.
Ultramarine Blue	3¼ oz.
Lampblack	1 oz.
Red Oxide LL	2 oz.
Amberol Resin 801	3 lb.
Sulphur	15 oz.
Santicizer	3 lb.

Churn in:

2 Nitropropane	4 pt.
Methyl Ethyl Ketone	4 pt.
Toluol	1 pt.
Butyl Acetate	1 pt.

No. 3

For Polyvinyl Chloride Surfaces

Solution A

Hycar OR	100
Zinc Oxide	2½
Butyl 8	C

Dissolve one pound of stock in

ethylene dichloride and dilute to form one gallon.

Solution B

Hycar OR	100
Zinc Oxide	2½
Sulphur	3

Dissolve one pound of stock in ethylene dichloride and dilute to form one gallon.

Mix equal parts of A and B when ready for use. Add 10% benzyl alcohol. This cement gives films which are tacky for long periods of time. Cure: 5 hours at 158° F.

No. 4

Hycar OR	15.20
Dibutyl Phthalate	4.90
Cumar P 25	4.90
Whiting	15.15
Iron Oxide, Black	3.60
Iron Oxide, Yellow	3.60
Zinc Oxide	3.60
Sulphur	0.35
Nitroethane	18.90
Methyl Ethyl Ketone	14.90
Ethyl Acetate	14.90

Add before use:

Captax	0.27

Mill Hycar 20–30 minutes; add sulphur, pigments and filler. Mill together for 3–5 minutes. Soak stock in nitroethane for at least 1 hour followed by mixing and adding of dibutyl phthalate and cumar.

Thin this heavy paste with rest of solvents while mixing thoroughly. Add the Captax as a 10% solution in methyl ethyl ketone before using.

Nitrocellulose Cement

Nitrocellulose (15 sec.)	8.7
Camphor	4.0
Acetone	37.8
Butyl Acetate	44.0
Methanol	5.0
Nigrosin (Dye)	0.5

Instrument Cross-Hair Cement

Pyroxylin (5–6 sec.)	30 g.
Acetone	100 cc.
Amyl Acetate	42 cc.
Butyl Acetate	18 cc.
Ethyl Acetate	15 cc.
Theop (Plasticizer)	½ cc.
Ethyl Abietate	1 cc.

Dental Cement

This is used in dentistry by mixing, just before use, 1 g. of the powder and 0.2 cc. of the liquid:

Zinc Acetate	0.1
Zinc Stearate	1.0
Zinc Oxide	70.0
Rosin (Water White)	28.5

Powder the rosin, incorporate an equal weight of zinc oxide until thoroughly mixed; sift through a 100 mesh sieve. Regrind the unsieved material with more zinc oxide and sift again, repeating the process until all passes through the sieve. Mix the other zinc salts with zinc oxide and pass through a 100 mesh sieve.

Eugenol	85 cc.
Cottonseed Oil	15 cc.

Temporary Dental Cement

Zinc oxide U.S.P. is mixed with just enough eugenol U.S.P. to make it plastic and workable.

RUBBER CEMENT FOR PAPER

Masticated Rubber	8 g.
Solvent	100 cc.
Standard Viscous Oil No. 32	3 g. or more

The solvent may be of variable composition according to conflicting requirements of safety and expense. For a fire-proof solvent use 80 cc. of carbon tetrachloride and

20 cc. of light petroleum thinner. For a fire-resistant product of low hazard, use 50–50 proportions of tetrachloride and thinner. For a cheap solvent regardless of fire hazard, use straight petroleum solvents, which will be satisfactory mechanically. Commercial light "rubber solvents" are available; these contain relatively large fractions of the aromatic solvents benzene and possibly toluene.

The viscous oil gives the product a tacky character, permitting easier stripping. Rosin is often used instead. Varying proportions should be tried to give the desired degree of tackiness.

If only common crepe rubber, not sticky to the touch, is available instead of the relative tacky masticated product, it will be necessary to use a mechanical stirrer for some hours to break down the jelly-like structure obtained by such a formula.

Rubber Sole Cement

This cement hot vulcanizes rubber soles to leather without bonding with gutta percha.

Synthetic Rubber	100
Lampblack	80
Rosin	20
Stearic Acid	2
Zinc Oxide	1
Sulphur	2.6
Tetramethylthiuram Disulphide	0.3
1–Mercaptobenzothiazole	1.2

Fiber Container Adhesive (Thermoplastic Adhesive)

U. S. Patent 2,259,490

Polymerized Vinyl Acetate	70
Dammar	30

Melt together and apply above 300° F. This retains its adhesiveness at high temperatures.

Hycar (Butadine Rubber) Cement

Hycar OR Compound	1.6 lb.
Methyl Ethyl Ketone	0.7 gal.
Chlorobenzene	0.3 gal.

General Purpose Air Curing "Hycar" Cement High Tensile Strength

	A	B
Hycar OR	100	100
Channel Black	35	35
Catalpo "X"	15	15
Zinc Oxide	5	5
Agerite Resin	5	5
Nevoll	30	25
Cumar P25	25	26
Sulphur	4	—
Butyl 8	—	8

Dissolve each part separately in:

Chlorobenzene	70
2 Nitropropane	20
Methyl Ethyl Ketone	10

Use same weight per gallon.

To use, add equal volumes of A and B.

THERMOSETTING CEMENTS

Formula No. 1

The use of Glyceryl Phthalate is suggested to replace shellac in alcoholic phenol-formaldehyde thermosetting cements for lamp and radio tube bases. Less glyceryl phthalate is required than shellac and it gives better thermoplastic effects before heat setting.

No. 2

Canadian Patent 395,145

Ethyl Cellulose	62.0
Alcohol	11.6
Ethyl Acetate	10.0
Dibutyl Phthalate	12.4
Carnauba Wax	4.0

Pressure Sensitive Adhesive

Nitrocellulose	35
Flexalyn (Diglycol Abietate)	45
Dibutyl Phthalate	25

Adhesive, Bitumen to Rubber
British Patent 521,401

Rubber, Powdered	50
Petroleum Bitumen	900

Heat and digest until dissolved. Then add a solution of

Zinc Oleate	10
Light Hydrocarbon Oil	100

Mix well until uniform.

Wood Joining Adhesive

Celluloid	1
Butyl Acetate	5

Wood Laminating Adhesive
U. S. Patent 2,290,833
Formula No. 1

Microcrystalline Wax	10
Rosin	13
Oxidized Asphalt	77

No. 2

Oxidized Asphalt	80
Nuba Rosin No. 2	20

Laminating Adhesive
U. S. Patent 2,229,028

Zein	100
Alcohol (95%)	275
Benzol	10
Castor Oil	10
Formaldehyde	5

Dip sheets in above solution; dry and heat under pressure to laminate.

Laminating and Stiffening Adhesive
U. S. Patent 2,120,054

An adhesive which may be used in laminating and coating fabrics to produce a permanent stiffness not affected by washing or ironing contains polymerized metacrylic acid 100, ortho-dibutyl phthalate 10, and chlorinated rubber 10–40 pints.

Cellulose Acetate Compositing Cement

Acetone	3
Ethyl Lactate	1

Celluloid to Wood or Cloth, Cement

Transparent pyroxylin scrap	30 lb.
Acetone	16 gal.
Butyl Acetate	4 gal.

Luminous Laminating Adhesive
British Patent 530,065

Acetone	100 cc.
Cellulose Acetate	10 g.
Calcium Sulphide, Luminescent	20 g.

Use between two sheets of transparent cellulose acetate.

CHEAP BELTING PASTE

Black Treacle (Molasses)	2 lb.
Rosin, Powdered	½ lb.
Whiting	¾ lb.
Black Soot	2 oz.

Mix thoroughly and rub a little inside the belt.

ASPHALT ROOFING CEMENT
(A Troweling Asphalt Composition)

Asphalt, Cutback	66
Slate Flour	16
"Short Fiber" Asbestos	21
"Long Fiber" Asbestos	7

The materials are added in the order given, using a hoe and mortar box or mechanical mixing. The first three ingredients should be well mixed before adding the "long fiber" asbestos. All materials should be at room temperature. The ratio of reagents may be varied in accordance with the grade of asphalt used.

CEMENT FOR PIPE LEAKS

A modified glycerin-litharge cement that is water-proof, resistant to fairly high temperatures and that sets under water may be used to repair leaks in pipes. To prepare this compound for use, equal parts of cement and litharge are thoroughly mixed together and then a volume of glycerin equal to half the volume of the mixed powders is added and the whole thoroughly mixed with a spatula or similar flat-blade tool.

When repairing a leaky pipe, the hole is filled with the cement and bound in place with cheese cloth, then a quantity of the cement is daubed on the cloth wrapping and the whole is tightly bound with iron wire.

Although the powders may be mixed ready for use, the glycerin should be added only when the cement is needed for immediate use.

Metal Thread Connection Compound

Graphite	20
Rheolan (Special Pitch)	20
White Lead	120
Mineral Oil (SAE 60)	20
Petrolatum	20–30

Stone Table Repair Cement

Mix litharge and manganese dioxide to match color of table top. Add glycerin until a soft putty is formed and mix thoroughly and apply at once. This sets within an hour.

This cement is useful on pipe joints, caulking seams in stones or metal and sealing in sight glasses, etc.

Adhesive for Polythenes to Metal or Ceramics

British Patent 544,359

Polythene	1–2
Cyclorubber	1

Hot roll together and apply hot.

Chloroprene to Metal Adhesive

U. S. Patent 2,227,991

Chloroprene (30% Chlorine)	100
Sulphur	50
Dibutyl Phthalate	75
Accelerator	3
Magnesium Oxide	10
Lead Oxide	10

Joint Seal for Liquid Cooled Engines

N-Butyl Methacrylate Polymer	40–46
Toluene	54–60

FLEXIBLE ADHESIVE, NON-STICKY METAL FINISH

Formula No. 1

Asphalt	47.5
F. Rosin	47.5
Paraffin Wax (M.P. 175° F.)	5.0
Calcium Carbonate	100.0
Gasoline	200–400.0

No. 2

Rosin	50
Gilsonite	25
Petropol	25
Calcium Carbonate	100
Gasoline	200–400

CEMENT FOR GLASS OR METAL

The following composition has a very low viscosity when melted and can be pressed into a thin film which will have excellent adherence when cold:

Paraffin Wax	95
Piccolyte Resin	5

WATERPROOF CEMENT FOR GLASS TO GLASS

Grind surface for better adhesion. Coat surfaces with 50° Bé Sodium silicate and apply 10 lb. pressure for 24 hours. Dry in oven 4 hours at 50° C. followed by four additional hours at 105° C. Immerse overnight in 5% Acetic acid.

Sealing Composition
British Patent 517,037
Formula No. 1

This composition is most suitable for food containers comprising a body of one or more layers of a fibrous material and a metallic end.

Barytes	10.1
Asbestine	5.0
Gum Karaya	0.3
Bentonite	1.9
Glue	4.5
Glycerin	2.5
Casein	1.8
Ammonia (28%)	0.7
Rubber Latex (Solid Basis)	2.0
Vulcanized Rubber Latex	5.0
Water	66.2

No. 2
French Patent 853,364

"Celluloid"	2.5	g.
Acetone	7	l.
Pigment	20	kg.
Vinyl Resin	80	g.

ADHESIVE FOR TISSUE PAPER TO ALUMINUM

For geological survey and other types of map work, base maps are very often printed on thin sheets of aluminum. Topographic maps of this type require the names of rivers, towns, etc. to be printed on thin tissue paper which is then fastened to the base map at the proper spot by an adhesive. It is very important in work of this kind that the tissue paper does not tear off, become wrinkled or discolored, so that photoreproduction is not made difficult or impossible.

Flexo Wax C Light	40	g.
Turpentine	57	cc.

The wax is melted in a porcelain dish, the turpentine is added and the solution mixed well and bottled. It solidifies on cooling, forming a tacky wax which can be remelted for use.

"Bakelite" Putty

"Bakelite" Lacquer	100
Barytes, Kaolin or Graphite	200–280

Apply and heat at 80–100° C. for 12 hours.

Putty, Acid Resistant
German Patent 702,739

Water glass (sodium oxide 20, silicon dioxide 60, water 20%) 2 parts, sodium fluosilicate 8.5 parts, clay 1.5 parts, and quartz flour 70 parts. The size of the quartz particle depends on the purpose of the putty.

Putty for Joining Magnesium
French Patent 850,700

Linseed Oil	28
Manganese Borate	1
Strontium Chromate	28
Strontium Fluoride	28
Magnesium Oxide	15

ASPHALT CAULKING MATERIAL

Asphalt	50
Powdered Clay	16
Slate Flour	34

The asphalt is melted, then the solids are added with stirring. The product may then be used by pouring into the desired crevice or may be allowed to solidify and may be melted and used as needed.

Fastening Cork to Metal

In fastening cork to iron and brass, even when these are lacquered, a good sealing wax containing shellac will be found to serve the purpose nicely. Wax prepared with rosin is not suitable. The cork surface is painted with the melted sealing wax. The surface of the metal is heated with a spirit flame entirely free from soot, until the sealing wax melts when pressed upon the metal surface. The wax is held in the flame until it burns, and it is then applied to the hot surface of the metal. The cork surface painted with sealing wax is now held in the flame, and as soon as the wax begins to melt, the cork is pressed firmly on the metallic surface bearing the wax.

Gas Tight Valve Seal
British Patent 540,352

Vermiculite	40
Grease	50
Mica or Talc	10

Vacuum Tube Vitreous Seal
U. S. Patent 2,223,031

Silica	28
Sodium Silicate	53
Ferric Oxide	19

Attaching Canvas to Walls

Before attaching ordinary canvas, muslin or burlap to plaster walls, the painter must first see to it that the plaster is smooth, clean and dry. Cracks should be dug out in the approved wedge-shaped manner and filled with patching plaster. All dents and surface imperfections must be brought up level after a preliminary size or primer has been applied over the entire wall and allowed to dry. Spackle may be used over the primer to provide a smooth, even surface. After sanding down the walls and dusting them thoroughly, a coat of white lead and lead mixing oil paint should be put on. This paint is made by mixing white lead with an equal volume of lead mixing oil.

When the paint is dry and hard, the adhesive should be applied. This adhesive is sometimes made of white lead (heavy paste) and good spar varnish but many painters and decorators prefer a white lead and lead mixing oil combination—about two gallons of the oil to the hundred pounds of lead. This heavy-bodied paint is spread evenly over the wall with an ordinary wide wall brush.

Before the adhesive sets up, the fabric—canvas, muslin, burlap, etc.—is hung, much in the manner of wall paper. In large public rooms or halls, when it is necessary to work from scaffolds, the canvas is first rolled on rug or linoleum poles, then hoisted to the scaffold. The top edge of the cloth is fastened at the top of the wall, using strips of wood nailed so as to hold the canvas in place. Then the roll of cloth is dropped so the lower edge reaches the bottom of the wall. The canvas is then pressed firmly to the white lead cement, using a stiff smoothing brush or some other suitable implement. All air bubbles and wrinkles should be very carefully worked out to the edges of the fabric.

With one vertical strip of fabric in place, the next strip is cemented to the wall in the same manner, the edge of this second strip overlapping the edge of the first strip by two to four inches, preparatory to producing a tight butted joint. After al-owing sufficient time for the cement to partly set, a straightedge is held against the center of the overlapped joint and, using a very sharp blade, the double-lapped joint is cut through to the plaster.

Cement (Household) or Celluloid Adhesive

A waterproof cement for making repairs on many different materials (it is especially good for leather) is easily and cheaply made by dissolving celluloid in "Cellosolve," amyl acetate, or ethyl acetate, or a similar moderately high-boiling solvent. Use about one part of celluloid to two parts of solvent. A few days are generally necessary for the solution to form and the material, during making and storing, must be kept in tightly closed vessels. Many different easily available objects will furnish celluloid, e.g., tooth brush handles, parasol handles, some of the "horn" spectacle frames, some novelties, as balls, cards, etc. Some produce clear and some opaque cements, and many different colors are available. This cement is similar to various "household" cements marketed in metal tubes.

Flypaper Ribbon

Rubber	10
Rapeseed Oil	8
Wool Fat	2
Honey	1
Benzol	79

Dissolve rubber in benzol. Mix honey with the wool fat and this mixture with the rapeseed oil. Add the honey, wool fat, and oil mixture to the rubber solution. Distill off the benzol and spread the residue on properly sized paper.

Flypaper Sheets

Rosin	50
Blown Castor Oil	50

Melt rosin in oil by heating in a steam jacketed kettle; spread on properly sized paper.

Flypaper Adhesive

Formula No. 1

Rosin	52
Pine Oil	15
Rosin Oil	15
Thin Mineral Lubricating Oil	15
Glyceryl Bori-Borate ("Aguaresin")	2
Glycerin	1

No. 2

1. Rosin "H"	54
2. Mineral Oil (d. 0.885–0.900)	25
3. Rapeseed or Castor Oil	11
4. Gum Mastic	5
5. Paraffin Wax	3
6. Gutta Percha	2

2, 3, and 4 are warmed together and mixed until dissolved. Do the same to 1 and 6 and then add to the former.

No. 3

Rubber	4
Rosin	60
Oleic Acid	34

Dissolve the rubber in the oleic acid with heat. Add the rosin and heat until it is melted. Mix well and spread hot on paper.

Chapter Three

BEVERAGES

Grape Wine

Measure out 1 quart clean Concord grapes in a cup. Place these in a 1 gallon jug and add a sugar solution composed of 3 to 3½ pounds of sugar, depending on what sweetness is wanted. For a medium wine, use 1 quart of water and 3¼ pounds of sugar dissolved therein. Now pour this over the grapes in the jug, add a small piece of a yeast cake dissolved in 1 cup of water, and fill the jug to within 4–5 inches of the top. Next shake the jug to mix the grapes, water, sugar water, and yeast and then tie a piece of cloth over the top of the jug to keep out gnats, and let it ferment. *Do not stopper the jug during fermentation.* After about 3–4 weeks the fermentation has about ceased, if the jug and contents have been kept in a warm place. Next, strain off the grapes, press out the remaining juice and pomace and let work or ferment for another 2 weeks. At the end of this time filter or strain through a double folded cloth, bottle, and set away to age. At the end of 5–6 months nearly a full gallon of ruby red grape wine will be ready.

Golden Dynamite or Cornmeal Wine

Place 1 lb. of yellow cornmeal in a crock and add 1 gallon of water. Squeeze out the juice of 3 lemons and 3 oranges and add to the crock. Now add 1 lb. of ground raisins, 4 lb. of sugar, and 1 yeast cake. Set aside in a warm place to ferment and stir once a day for at least 30 days. Strain at the end of this time, and bottle. This makes a very potent, powerful wine.

Mint Wine

Mint Leaves 1 qt.
Juice and Pulp of 3 Oranges
Juice and Pulp of 3 Lemons

Mash all the above together, add 1 quart of warm water and let set over night—or 24 hours.

Now add 4 lb. of sugar, 3 more quarts of water, ½ of a yeast cake and set away to ferment. After about 4 to 5 weeks when the fermentation is complete, and no more bubbles are seen rising in the mixture, strain and bottle. This makes a light amber colored, mint flavored wine.

Stabilizing Alcoholic Beverages
U. S. Patent 2,075,653

Non-distilled alcoholic beverages, particularly wines, are stabilized against hazes due to iron, aluminum, copper and calcium by addition of 0.1–1 gal. of soluble sodium phosphite. The liquor is agitated, and the precipitate which has appeared in about 12–18 hours later is removed by filtration.

General Formula for Flavors

Terpeneless Oil

(essential)	3/8	fl. oz.
Alcohol	10	fl. oz.
Distilled Water	6	fl. oz.

For usual purposes, 2 fl. oz. can be used per gal. of syrup.

Non-Alcoholic Flavors
U. S. Patent 2,180,932

Lemon Flavor

Lemon Oil	5
Monoethylglycerin	95

Orange Flavor

Sweet Orange Oil	5
Diethyl Glycerin	95

Imitation Vanilla Flavor

Vanillin	1½	oz.
Coumarin	¼	oz.
Monoethylglycerin	10	oz.
Caramel	1	oz.
Water, To make	1	gal.

Vanilla Bean Flavor

Mix 30 parts of monoethylglycerin with 70 parts of water and use this diluted solvent as a menstruum for the percolation of vanilla beans.

Imitation Raspberry Flavor

Monoethylglycerin	96
Orris Concrete	½
Benzyl Acetate	2
Vanillin	⅛
Rhodinol	⅛
Ionone	¼
Amyl Butyrate	2
Ethyl Pelargonate	1½
Ethyl Benzoate	½
Amyl Acetate	4
Raspberry Juice	40

Imitation Banana Flavor

Ethyl Pelargonate	2½
Ethyl Butyrate	3¼
Amyl Butyrate	11¼
Ethyl Acetate	2¾
Amyl Acetate	12
Sweet Orange Oil	3
Lavender Flowers Oil	1
Amly Valerianate	¼
Diethyl Glycerin	91

Apple Juice

Select sound firm-ripe autumn or winter apples.

Wash carefully.

Crush apples in grinder.

Press in barrel or hydraulic press.

Strain juice through muslin or cheesecloth.

Heat juice promptly and rapidly to 170°–175° F. in upper part of double boiler.

Without cooling, fill immediately, into fruit enamel-lined cans, or into hot sterile glass jars, hot crown-closure bottles, or hot screw-top glass jugs, taking care to remove foam and to fill each container completely.

Close each container immediately.

Invert cans or place bottles or jars on their sides for 3 to 5 minutes.

Place in hot water (160°–165°F.) and cool with running water.

Grape Juice

Select, preferably, fully ripe Concord, or Concord type grapes. Catawba and Ives also give excellent products.

Wash carefully.

Remove grapes from stems.

Crush grapes in grape crusher.

Heat crushed grapes to 140°–145° F. in upper part of double boiler.

Press hot grapes in barrel press or hydraulic press.

Strain juice through muslin or cheesecloth.

Heat juice rapidly to 170°–175° F. in upper part of double boiler.

Without cooling, fill immediately into large hot fruit jars or hot screw-top glass jugs taking care to remove foam and to fill each container completely.

Close each container immediately.

Invert container or lay on side for three to five minutes. Place in water at 160°–165° F., and cool bottle in running water.

Allow to stand in cool cellar 3 or more months.

Carefully siphon supernatant juice from sediment, and pass juice from bottom (containing sediment) through heavy muslin cloth.

Heat juice rapidly to 170° F. in upper part of double boiler.

Without cooling, fill immediately into hot, sterile fruit jars, hot crown-closure bottles, or hot screw-top glass jugs, taking care to remove the foam and to fill each container completely.

Close immediately and cool as directed above.

Cherry Juice—Cold Pressed

Use sound, fully ripe, Montmorency or English Morello cherries or a mixture of the two varieties.

Wash carefully.

Crush cherries in grape crusher.

Press in barrel-press or hydraulic press.

Strain juice through muslin or cheesecloth.

Heat juice promptly and rapidly to 170° F. in upper part of double boiler.

Without cooling, fill immediately into fruit-enamel-lined cans, or into hot, sterile, glass jars, hot crown-closure bottles, or hot screw-top glass jugs, taking care to remove foam and to fill each container completely.

Close each container immediately, invert or place on side for three to five minutes. Place in water at 160°–165° F., and

Cool with running water.

Cherry Juice—Hot Pressed

Use sound, fully ripe, Montmorency or English Morello cherries or a mixture of the two varieties.

Wash carefully.

Crush cherries in grape crusher.

Heat crushed cherries to 175°–180° F. in upper part of double boiler.

Press hot cherries in barrel press or hydraulic press.

Strain juice through muslin or cheesecloth.

Heat juice promptly and rapidly to 170° F. in upper part of double boiler.

Without cooling, fill immediately into fruit-enamel-lined cans, or into hot, sterile, glass jars, hot crown-closure bottles, or hot screw-top glass jugs, taking care to remove foam and to fill each container completely.

Close each container immediately. Cool as directed above.

Raspberry Juice

This procedure may be followed for the making and preservation of red, black and purple raspberries, loganberries, blackberries, youngberries, boysenberries and dewberries.

Select soft, ripe fruit, preferably of a highly colored and flavored variety.

Wash carefully.

Mix with granulated sugar, using 10 parts fruit to 1 part sugar by weight.

Fill into heavily paraffined paperboard cartons or fruit-enamel-lined tin cans.

Close containers.

Freeze in locker plant or farm freezer, and store until needed.

Without opening containers thaw in a refrigerator or at room temperature before a fan.

Immediately after thawing, press in barrel-press or hydraulic press. (See alternate method.)

Strain juice through muslin or cheesecloth.

Heat juice promptly and rapidly to 170–172° F. in upper part of double boiler.

Without cooling, fill immediately into fruit-enamel-lined cans, or into hot, sterile, glass jars, hot crown-closure bottles, or screw-top glass jugs, taking care to remove foam and to fill each container completely.

Close each container immediately. Invert or turn containers on side. Air cool for three to five minutes, then place in warm water 150° F., and

Cool with running water.

Alternate Method for Pulpy Juice

If pulpy berry juice is desired, the thawed juice may be passed through a collander or a taper screw press to separate the seeds and more solid parts.

Heat juice promptly and rapidly to 180°–185° F., in the upper part of a double boiler and continue as directed for pressed juice.

Because of difficulty of removing the air from the juice, it will not retain its quality as well as pressed juice, but may be used satisfactorily in fruit juice blends.

Plum Juice

Select soft, ripe fruit, preferably of a highly colored and flavored variety. Press (undried) prunes may be used.

Wash carefully.

Pit and cut in halves.

Mix well with granulated sugar, using 10 parts of fruit to 1 part sugar by weight.

Fill into heavily paraffined paper-board cartons or fruit-enamel-lined tin cans.

Close containers.

Freeze in locker plant or farm freezer and store until needed.

Thaw, without opening containers, in a refrigerator or before a fan.

Immediately after thawing, crush in grape crusher.

Press in barrel-press or hydraulic press.

Strain juice through cheese cloth.

Heat juice promptly and rapidly to 170°–175° F. in upper part of double boiler.

Without cooling, fill immediately into fruit-enamel-lined cans, or into hot sterile glass jars, hot crown-closure bottles or screw-top glass jugs, taking care to remove foam and to fill each container completely.

Close each container immediately.

Invert or place container on side. Air cool three to five minutes.

Place in hot water at 160°–165° F. then cool as directed above.

Blended Fruit Juices

Apple Berry

Most berry juices are too strong to be palatable unless they are either sweetened and diluted with water or blended with a bland juice, such as that of the apple. The exact proportion of any berry juice to use will depend upon its acidity and amount of flavor. In the case of highly flavored berries, such as black raspberries, three to four parts of apple juice may be used with one of raspberry juice. If the berry juice is milder in flavor, two parts of berry juice to one part of apple juice may be a satisfactory proportion.

Blending of the preserved juices may be done just before use, or preserved (canned or frozen) berry juices may be blended with either freshly prepared or preserved apple juice and the product pasteurized as directed below. Apple juice may be added to the thawing berries (see section on berry juices) and pressed with the berries. The latter method will result in the greatest yield from the berries.

Blend juices to give pleasing flavor.

Sweeten with sugar, if a sweeter juice is desired.

Heat juice rapidly to 170°–180° F. in upper part of double boiler.

Without cooling, fill immediately into fruit-enamel-lined cans, or into hot, sterile, glass jars, hot crown-closure bottles, or hot screw top glass jugs taking care to remove foam and fill each container completely.

Close each container immediately. Invert or place container on its side. Air cool for three to five minutes. Place in water at 160°–165° F.

Cool with running water.

Apple Cherry

The juice of sour cherries (Montmorency, English Morello and Early Richmond) is too intense and tart for consumption without sweetening, and dilution with water or blending with apple juice. Two parts of apple juice should be blended with one part of hot pressed cherry juice. Cold pressed cherry juice should be blended with an equal amount of apple juice.

The blended juice should be pasteurized in the same manner as the apple-berry blend.

Apple Plum

The juice of most varieties of plums is too strong for consumption without dilution with water or blending with apple juice. As a rule, plum juices should be blended with an equal volume of apple juice. The blended juice should be pasteurized in the same manner as the apple-berry juice blend.

Tomato Juice

Select fully red ripe tomatoes.

Wash thoroughly and remove all green, yellow, black and rotted areas. Cut out stem end.

Quarter tomatoes, and heat to boiling.

Strain through a sieve, collander or tapered screw press or any other type of screening device. The finer screens give a smoother juice.

Add salt, approximately 1 teaspoon per pint.

Heat rapidly to 190°–200° F. in the upper part of a double boiler or in a kettle with continued stirring.

Without cooling, fill immediately into hot sterile glass jars, hot crown closure bottles or tin cans, taking care to remove foam and to fill each container completely.

Close each container immediately and place in hot water at 190° F. and hold for five minutes.

Remove containers, cool.

Tomato Juice Blends, or Cocktails

Tomato juice blends or cocktails of delightful flavor may be prepared by mixing with various vegetable juices. After blending they are preserved by the methods as described for tomato juice. A few satisfactory blends are as follows:

Four parts tomato juice and one part sauerkraut juice.

Four parts tomato juice and one part celery sauerkraut juice.

Tomato juice with five per cent turnip juice.

Tomato juice with two to three per cent beet juice.

The beet juice may also be added to the first combination to give a deeper red color to the cocktail.

If sauerkraut juice is not available, vinegar or lemon juice may be used but they do not impart the same flavor.

Sauerkraut Juice

Excess sauerkraut juice is nearly always present in vats or barrels of sauerkraut. This juice contains approximately the same nutritive qualities that are present in kraut including from one to two per cent of digestible organic acid.

Kraut juice may be obtained from the bottoms of barrels of kraut made in the home or from vats of kraut made by commercial packers.

Heat the juice to 160°–165° F. in the upper part of a double boiler or in a kettle stirring continuously.

Without cooling fill immediately into hot sterile crown closure bottles, glass jars or tin cans taking care to fill each container completely.

Close each container immediately and invert to air cool three to five minutes. Place in warm water, 150° F. and add running water to cool.

Rhubarb Juice

Select tender and juicy stalks of rhubarb. Red varieties give the best color.

Wash thoroughly and drain excess water.

Cut in pieces four to six inches long.

Cook until tender, but not until mushy.

Press in a barrel type press while hot, using muslin cloth or fine pore press cloth.

Sweeten if desired. One pound per gallon is usually sufficient.

Heat to 165°–170° F. in an aluminum or white enamel kettle, stirring continually.

Without cooling, fill immediately into fruit-enamel-lined tin cans, hot sterile glass jars or hot crown closure bottles, taking care to remove foam and to fill each container completely.

Close each container immediately, invert it or turn on its side and cool for three to five minutes.

Place in warm water 155° F. and add running water to cool.

Alternate Procedure

After washing and cutting the rhubarb as directed above it may be mixed with sugar ten to one, placed in fruit-enamel-lined tin cans or heavily paraffined paperboard container for freezing.

Place in a sharp freezer or in a cold storage locker and freeze.

At a later date, thaw.

Extract the juice, sweeten and preserve as directed above.

Carrot Juice

Select large, well formed carrots.

Wash, peel and remove green parts.

Shred and extract the juice in a centrifugal extractor.

Three different procedures may be used for preserving the juice.

First Method—Freezing

The extracted juice may be placed in heavily paraffined paperboard containers and frozen in a sharp freezer or locker cold storage plant.

Second Method—Acidification

The carrot juice may be mixed in equal quantities with high acid sauerkraut juice.

After blending, heat rapidly to 190°–195° F. in double boiler or kettle, stirring continuously.

Without cooling, fill immediately into hot sterile glass jars, hot crown closure bottles or tin cans, taking care to remove foam and to fill each container completely.

Close each container immediately and place in hot water at 190° F. and hold for five minutes.

Cool by running in cold water.

Third Method

The carrot juice may be sterilized in a pressure cooker.

Heat juice to 160°–170° F. Fill into glass jars or tin cans. Sterilize in a pressure cooker at 10 pounds pressure for 30 minutes. When pressure is released, glass jars should be tightly sealed.

Air cool.

Other Vegetable Juices—Celery, Beet, Turnips

Various vegetable juices are nutritious and flavorable and blend very well with each other as well as with tomato juice. The centrifugal extractor or similar device is required to extract fresh juice from the hard vegetables.

Wash, trim and extract juice.

Three different procedures may be used for preserving these juices.

First Method—Freezing

The extracted juice may be placed in heavily paraffined paperboard containers and frozen in a sharp freezer or locker cold storage plant.

Second Method—Acidification

The vegetable juice may be mixed with one-half the quantity of high acid sauerkraut juice.

After blending, heat rapidly to 190°–195° F. in a double boiler or kettle, with continued stirring.

Without cooling, fill immediately into hot sterile glass jars, hot crown closure bottles or tin cans, taking care to remove foam and to fill each container completely.

Close each container immediately and place in hot water at 190° F. and hold for five minutes.

Cool by running in cold water

Third Method

The vegetable juice may be sterilized in a pressure cooker.

Heat juice to 160°–170° F. Fill into glass jars or tin cans. Sterilize in a pressure cooker at ten pounds pressure for 30 minutes. When pressure is released, glass jars should be tightly sealed.

Air cool.

STORAGE OF FRUIT AND VEGETABLE JUICES

Regardless of the care taken in reparation and preservation of ruit and vegetable byproducts, they vill slowly deteriorate in quality ver a period of time particularly if ept in a fairly warm place. This oss of quality may be reduced to a ninimum by storage at low temerature, for example in a cold dry ellar.

Prepared Orange Beverage

Sugar	64 lb.
Salt	2 oz.
Powdered Orange Juice	2 lb. 8 oz.
Citric Acid	1 lb. 8 oz.
California Orange Oil	2 oz.
Orange Color, Pure Food	3 oz.

Place 48 lb. of sugar, salt, powdered orange juice and citric acid in machine bowl and blend thoroughly. Blend 16 lb. of sugar, orange oil and color carefully and add to first mixture. Use additional sugar as required.

Mix well and sift through fine sieve. One pound produces one gallon or more of beverage.

Chapter Four

COSMETICS AND DRUGS

Cosmetic Creams

	Formula No.	1	2	3	4
A.	1. Sorbitol Syrup	2.5	3.9	5.0	2.5
	2. Potassium Carbonate	—	—	—	—
	3. Potassium Hydroxide	0.9 **	0.8	0.9 **	—
	4. Triethanolamine	—	—	—	1.7
	5. Water	77.8	74.5	50.0	72.5
	6. Preservative *	0.15	0.15	0.15	0.15

* "Moldex".
** 0.7 to 1.0.

		1	2	3	4
B.	7. Stearic Acid	15.0	15.4	6.0	4.0
	8. Lanolin	0.5	1.2	0.5	2.0
	9. Mineral Oil (65/75)	2.0	1.5	35.3	15.0
	10. Sorbitan Mono-oleate	—	0.8	—	—
	11. Mannitan Monostearate	2.0	—	2.0	2.0
	12. Spermaceti	—	1.5	—	—
C.	13. Perfume	0.2	0.2	0.2	0.2

Warm (A) and (B) separately to 75–80° C.
Add (B) to (A) slowly and with thorough agitation.
Add (C) at 50° C.

Note: Stir No. 1 and No. 2 intermittently and when body has increased, turn over by hand. No. 1 cream may be poured directly into jars at 58° C.

No. 4 cream must be passed through colloid mill or homogenizer while still fluid.

No. 3 cream should be stirred until cold, set aside over night and remixed the following morning.

No. 1—Vanishing Cream
No. 2—Hand Cream
No. 3—All Purpose Cream
No. 4—Liquid Cream

Modifications

For modification of the above formulas, the following will serve as guides:

Brushless Shaving Cream: Formula No. 2. Increase the potassium hydroxide to 1. 3 parts.

Foundation Cream: Formula No. 2. Add 2% titanium dioxide.

Powder Cream: Formula No. 2. Add 5% titanium dioxide and color akes.

Emollient Skin Cream: Formula No. 3. Increase lanolin to 5% or as equired.

Liquid Cleanser: Formula No. 4.

Hand Lotion: Formula No. 4. Reduce mineral oil to not over 5%, in-reasing water by like percentage.

Liquid Powder Base: Formula No. 4. Reduce mineral oil to 5% as above.

Hair Dressing: Formula No. 4.

Simple Cold Cream

White Beeswax	1 oz.
Paraffin Wax	2 dr.
Light Mineral Oil	4 fl. oz.
Water	3 fl. oz.
Borax	25 gr.
Perfume	q. s.

Melt the first three ingredients together. Add the borax dissolved in the water at 70° C. Stir until thick. Perfume may be worked in before mixture becomes too thick.

Vitamin Cold Cream

Niacin	10 g.
Pantothenic Acid	10 g.
Distilled Water	20 g.
Lanolin (Anhydrous)	40 g.
Whipped White Petrolatum Jelly	30 g.

"Nourishing" Cream

Formula No. 1

Beeswax	18
Raisin Seed Oil	30
Olive Oil, Odorless	16
Lanolin	2
Cetyl Alcohol	2
Cholesterol	0.1
Water	30.9
Borax	1.0

No. 2

A special nourishing cream contains 7.5 cetyl stearate, 7.5 cetyl alcohol, 5 glycerin, 10 absorption base, 9.7 white vaseline, 0.5 vitamin F, 15 lauryl sulfonate, water 54.5.

Honey Cream

6.5 diethylene glycol myristate, 18 lanolin, 41 absorption base, 14 cucumber seed oil, 2 cetyl alcohol, 2 myristyl alcohol, 3 honey, 13.5 water or rose water.

Dermal Cream

Water, Distilled	74.0
Sesame Oil	20.0
Citrus Pectin	5.0
Citric Acid	1.0

Liquid Creams

Formula No. 1

Stearic Acid	3.0
Raisin Seed Oil	10.0
Liquid Paraffin	10.0
Triethanolamine	1.5
Water	75.5
Preservative	Sufficient

No. 2

Propylene Glycol Laurate	15.0
Raisin Seed Oil	30.0
Water	55.0
Preservative	Sufficient

No. 3

Glyceryl Monostearate	14

Avocado Pear Oil	12
Cetyl Alcohol	1
Water	70
Glycerin	3
Preservative	Sufficient

No. 4

Propylene Glycol Stearate	15
Avocado Pear Oil	5
Raisin Seed Oil	5
Water	70
Glycerin	5
Preservative	Sufficient

No. 5

Diethylene Glycol Monolaurate	16.0
Liquid Paraffin	8.0
Raisin Seed Oil	8.0
Water	68.0
Preservative	Sufficient

No. 6

29 mineral oil, 6 isopropanolamine oleate, 64.5 distilled water and 0.5 perfume.

No. 7

Falba Absorption Base	15.0
Liquid Paraffin	20.0
Raisin Seed Oil	23.0
Water	42.0

Salt Containing Cosmetic Cream

Polymerized Glycol Stearate	10
Stearic Acid	10
Paraffin Wax	15
Mineral Oil	15
Petrolatum	10
Sodium Chloride	6
Water	To make 100

Aluminum Chloride Cream

Polymerized Glycol Stearate	15
Spermaceti	5
Glycerin	3
Aluminum Chloride, Hydrated Crystals	12
Water	To make 100

Antiseptic Cream

Emulgor A (Polymerized Glycol Ester)	10.0
Stearic Acid	10.0
Paraffin Wax	15.0
Mineral Oil	15.0
Petrolatum	10.0
Oxy-Quinoline Sulfate	0.2
Water	To make 100

Foundation Cream

A	Lanolin Concentrate (Parachol)	3¼
	Lanolin	6
	Cocoa Butter	5
	Cetyl Alcohol	15
	Glyceryl Stearate (Mono)	30

Melt to 70° C.

B	Water	41
	Heat to 70° C.	

Add A to B and agitate until cool. Perfume to suit.

Vanishing Cream

Stearic Acid	25.0
Spermaceti	5.0
Aminomethylpropanediol	1.5
Glycerin	8.0
Water	60.5

The aminomethylpropanediol, glycerin, and water are heated together to about 75° C. (167° F.) while the stearic acid and spermaceti are heated in another container. When both mixtures are at the same temperature and homogeneous, the stearic acid-spermaceti mixture is slowly and thoroughly stirred into the aqueous solution. The temperature must be maintained at approximately 75° C. throughout this operation. After the acid-wax melt has all been added, heating is discontinued. However, vigorous stirring is maintained until the mixture thickens at which point the stirring should be slowed down and changed to a kneading action. Like all stearate creams, it should be allowed to stand overnight and then thoroughly remixed.

Beauty Mask

Bentonite	10.0
Alcohol	10.0
Purified Kaolin	35.0
Glycerin	13.0
Tragacanth (Mucilage)	20.0
Water	10.9
Perfume	1.0
Methyl-p-hydrobenzoate	0.1

Skin Protectives
Formula No. 1

Beeswax	5.0
Glyceryl Monostearate	12.5
Hydrous Wool Fat	5.0
Sodium Silicate	5.0
Ammonium Hydroxide (10% sol.)	0.5
Petrolatum	72.5

The ammonium hydroxide is incorporated to prevent the precipitation of the sodium silicate. The above formula with 5% latex gives a product which produces a rubber-like film on the skin. The base is melted, the latex added with stirring and the mixture stirred constantly until it congeals.

No. 2

Beeswax	10.0
Hydrous Wool Fat	5.0
Glyceryl Monostearate	12.5
Stearic Acid	2.0
Petrolatum	75.5

This cream has a pH of 5.4 and is recommended where there is prolonged contact of the hands with soapy water.

No. 3

Beeswax	10.0
Hydrous Wool Fat	5.0
Sulfonated Olive Oil (75%)	10.0
Petrolatum	75.0

This cream is useful to waterproof the skin when there is prolonged contact with water.

The following is a non-greasy preparation which dries on the skin and does not rub off. Its use is indicated for dry work as a protection against dust-borne irritants or where workers must guard against soiling the materials or objects with which they are working.

No. 4

Glyceryl Monostearate	12.0
Beeswax	12.0
Hydrous Wool Fat	6.0
Cholesterol	1.0
Sodium Silicate	5.0
Ammonium Hydroxide (10% sol.)	0.5
Water	63.5

Melt together the white wax, glyceryl monostearate, the wool fat and cholesterol. Heat the water to the same temperature as the wax mixture and add the sodium silicate and ammonium hydroxide solutions. Stir the aqueous solution into the wax mixture and continue stirring until it congeals.

No. 5

Glycosterin	40 g.
Diglycol Laurate, Neutral	90 g.
Triple Pressed Stearic Acid	180 g.
Glycerin	60 g.

Sulfatate (2% Solution)	100 cc.
Methyl Cellulose (5% Solution)	150 cc.
Water	350 cc.
Perfume	5 cc.

This gives a very smooth hand cream which is not greasy but softens the skin and minimizes the absorption of grease, oil or dirt.

No. 6

Stearic Acid	2 oz.
White Beeswax	1 oz.
Petrolatum	2½ oz.
Mineral Oil	1½ oz.
Melt and emulsify with heated Triethanolamine	4 dr.
Boiling Water, To make	24 oz.
Then incorporate Magnesium Stearate	2 oz.

	No. 7	No. 8
Mixture of Cetyl and Stearyl Alcohols Containing Sulfated Derivatives	12	2
Liquid Petrolatum	12	12
Lanolin, Anhydrous	6	6
Water, Distilled	70	70
Perfume, To suit		
Cetyl and Stearyl Alcohols	To suit	10

Heat together all constituents except perfume over a water bath until emulsified and stir until cold, working in the perfume. It is not necessary to run through a homogenizer but the latter procedure improves the product. A preservative should be incorporated unless the cream is to be used immediately. For various consistencies, paraffin wax may be substituted for lanolin and the proportions altered accordingly. A suitable mild organic acid might be added to the water phase to adjust the pH to that of normal skins.

Cover Paint for Skin Discolorations

So-called cover paints used to conceal and blend the white, pigment-free skin spots characteristic of vitigilo (acquired leukoderma) can also be used to temporarily cover other smooth skin blemishes, scars or marks. Moreover, such products, if properly made, can also be useful to soften and blend the often sharp sunburn lines so frequently seen at the end of the summer season.

Zinc Oxide	45 g.
Prepared Calamine	45 g.
Glycerin	4–16 cc.
Rose Water, To make	500 cc.

To this is added, drop by drop, sufficient ichthyol to cause the paint to match or blend with the surrounding skin. Usually from 10 to 60 drops are needed, and the addition should be made carefully, since success of the paint depends on the closeness of the match obtained. Face powder may be applied after using this concealing paint.

Hand Lotion

Formula No. 1

Stearic Acid	4.0
Cetyl Alcohol	1.0
Butyl Stearate	3.0
Amino-methyl-propanediol	0.8
Quince Seed	0.5
Water	90.7
Preservative	As required

A thick cream is made from all of the above ingredients except the quince seed, preservative and half the amount of water specified, using the technique described for the vanishing cream.

The quince seed is soaked overnight with the rest of the water and the preservative, strained, and the resulting "mucilage" is stirred into the cream. The lotion thickens considerably after standing several days and may be thinned with water as desired. Perfume is usually added as a highly concentrated solution in specially denatured alcohol, in the last stages of manufacture.

No. 2

Stearic Acid	1.75	g.
Glycerin	1.00	cc.
Cholesterol Base (Parachol)	0.50	g.
Cetyl Alcohol	0.50	g.
Triethanolamine	0.15	cc.
Alcohol (15%) To make	100.00	cc.

Allow the stearic acid and the triethanolamine to react until the reaction is complete (about 20 minutes). Add the cholestrol, cetyl alcohol and glycerin (which must have been brought to the same temperature as the other ingredients), and when this mixture is melted, gradually stir in the 15% alcohol (also warmed to the same temperature). Stir till cool. Perfume as desired.

The value of this lotion lies in the fact that the emulsion is permanent without any tendency to break down. In addition, the lotion will take a variety of dermatologically important medicaments such as calamine, zinc oxide, kaolin, phenol, salicylic acid, boric acid and any of the alkalies within reason.

Facial Astringent
German Patent 700,229

Gum Arabic	20	g.
Sodium Perborate	5	g.
Soap, Powdered	3½	g.
Dissolve before use in Water	15	l.

Skin Lotions
U. S. Patent 1,631,384
Formula No. 1

Pure Citrus Pectin	1 oz.
Tincture of Benzoin	1 fl. oz
Glycerin	4 oz.

No. 2

Pure Citrus Pectin	8
Tincture of Benzoin	20
Glycol	100
Boric Acid	4
Water	850

No. 3

Citrus Pectin	0.9
Diethylene Glycol	10.0
Tincture of Benzoin	2.0
Boric Acid	0.4
Water	86.7

No. 4

Zinc Oxide	5.0
Talc	5.0
Sodium Borate	5.0
Menthol	1.4
Alcohol	15.6
Bentonite (6% Lime water suspension)	100.0

Lotion for Blotched Skin

Orange Water (or Rose Water)	2	fl. oz.
Extract of Witch Hazel	1	fl. oz.
Magnesium Sulphate	½	oz.
Borax	¼	oz.

(Wash face with soap and hot water, then apply lotion at bedtime.)

Baby Skin Oils
Formula No. 1

Olive Oil, Best Deodorized	80
Almond Oil	20
Oil-Soluble Rose Perfume	To suit

No. 2

Raisin Seed Oil	60
Avocado Pear Oil	20
Olive Oil, Deodorized	20
Perfume	To suit

These products are prepared by simple mixing.

Sunburn Preventative

Salol	10
Cold Cream	90

Sun Protective Lotion

Butylcinnamoyl Pyruvate	0.5 cc.
Dihydropyrone	30.0 cc.
Alcohol	30.0 cc.
Water, To make	100.0 cc.

Glenn's Liquid Powder

Colloidal Kaolin	7
Titanium Dioxide	5
Bentonite	5
Water	78
Glycerin	5
Pigment	To suit

Face Powders

For the coloration of face powders the eye of the artist is necessary. The colors must not be glaring, but delicate. Various dyestuffs which may be used have already been mentioned and here is a general outline of the method of working. Either a stock of diluted color is made up in calcium carbonate (precipitated chalk) or talc, or the dyestuff is separately ground into part of the talc necessary for formulation, before adding it to the rest of the mixture. This permits much better mixing; it would be asking for trouble to add undiluted color directly to the mill.

Face Powder
(Medium Peach Color)

Talc	65.00
Purified Kaolin	20.00
Magnesium Carbonate	6.00
Titanium Dioxide	4.00
Magnesium Stearate	3.00
Perfume	0.50
Ochre	1.45
Brilliant Pink Lake (Eosine Type)	0.05

Here the ochre and the pink lake are separately mixed with a portion (known weight) of the talc and, when a sample from either grinding, rubbed out on a white paper card, shows no streakiness or color spots, they can be mixed together. The final color mixing is then rubbed out evenly on to a card which is retained for matching when dealing with later consignments of the same shade. The color is then added to complete the mixture, which should be milling during the time that the color is being prepared.

Certain colorists find it easier to work with stock solid "solutions." There is little doubt that this is the best method to employ when matching is being carried out. "Nude" for example may be made up as follows:

Nude Shade Powder

Precipitated Chalk	69.25
Golden Ochre	30.00
Tetrabromofluorescein Lake	0.75

and then 7.5 to 10 per cent, or whatever is required to give the right depth, is added to the white face powder batch which is separately prepared.

Color Stocks

Stocks of all the colorist's dyes may be prepared in 25 per cent mixtures with calcium carbonate. The above shade would then be made by using 120 parts of 25 per cent ochre "solution" and 3 parts of 25 per cent tetrabromofluorescein lake "solution." The powderman and the colorman must work together however on these points in order to control the amount of calcium carbonate in the final product.

Burnt sienna is often used in conjunction with eosine and ochre for the formation of brunette and sunburn shade powders. Baby powders are often very delicately tinted and perfumed, the perfume oil and color lake or pigment being rubbed out carefully in a portion of the talc before adding it to the bulk mixing.

Sun tans are made by numerous different procedures but the coloring matter, usually a pigment brown mixture, is ground up in petrolatum and then added to the prepared oil which may have something like a nut oil base. Oil soluble dyestuffs are often employed. Artificial sun tan liquids (not oils) are made up by using basic dyestuffs in very small proportions. Owing to the fact that dyestuffs used must be perfectly dissolved, the colorist works with very dilute solutions, usually 0.5 to 0.1 per cent in spirits or water. For example here is a kind of artificial tan used for theatrical purposes:

Theatrical Sun Tan Color

Bismarck Brown R (1% Solution in water)	80.0
Methyl Violet 4B (1% Solution in water)	1.5
Methylene Blue 2B (1% Solution in water)	1.0
Glycerin	6.0
Alcohol	11.0
Perfume	0.5

Fluorescent Rouge and Face Powder

Fluorescent rouge and face powder for theatrical work is made by using a base of barium sulfate or magnesium oxide and adding a phosphor to color the rouge. Phosphors may be zinc beryllium silicate or the fluorescent organic dyes.

Ordinary petrolatum when applied to the face gives a bright bluish fluorescence suited to theatrical application. Lanolin and other grease-like substances may also serve in this capacity.

Rouges for compacts require brilliant red lakes and carmine may be supplemented by an eosine lake, or the latter may be used alone. An example is:

Compact Rouge

Talc	36
Zinc Oxide	20
Kaolin	20
Zinc Stearate	5
Bromo Acid Lake	11
Petrolatum, Heavy	3
Gum Tragacanth (1% Solution)	3
Perfume	2

In this case the color lake may be rubbed out in the petrolatum and perfume oil, and then mixed with a portion of the talc before adding to the mass. The object of the tragacanth and liquid petrolatum, of course, is to enable the powder to set in a compact mass. It will be noticed that here much more color is required than has been mentioned in examples of other types of cosmetics; this may be explained by the fact that usually four or five times more color is required when using lake colors as compared to pure dyestuffs.

Liquid "Lip-Stick"
U. S. Patent 2,230,063
Formula No. 1

Ethyl Cellulose	3.1%
Ethyl Alcohol	68.4%
Petroleum Ether	20.0%
Hydrogenated Methyl Abietate	7.5%
Rhodamine	1.0%

No. 2

Ethyl Cellulose	3.06%
Bleached Wax-Free Shellac	4.94%
Ethyl Alcohol	65.00%
Petroleum Ether	14.66%
Hydrogenated Methyl Abietate	12.00%
Fuchsine	0.3 %
Saccharin	0.04%

No. 3

Ethyl Cellulose	3.5%
Ethyl Alcohol	90.7%
Castor Oil	5.0%
Tetra-bromo Eosine	0.8%

No. 4

Ethyl Cellulose	3.0%
Bleached Wax-Free Shellac	2.5%
Methyl Abietate	7.5%
Petroleum Ether	11.0%
Ethyl Alcohol	77.0%
Oil-Red O	1.0%

Accessory agents may be added, such as perfumes, preservatives, anti-oxidants, etc. The preparations are preferably clear and transparent and are of a viscosity ranging between 3 and 5 hundred centipoises, while best results are obtained in a range between 20 and 50 centipoises.

Fluorescent Lipstick

For most theatrical purposes the lipstick ordinarily used will fluoresce brightly under ultraviolet light.

Coco Butter or Lipstick Base	100
Fluorescent Dye	up to 5

Physiologically harmless dyes should be used. Various shades may be obtained by mixing the dyes.

Coloring Lipsticks

Lipsticks formulae vary widely, but the simplest way of explaining the procedure of coloring is by an example:

(*a*)	Castor Oil	45.5
(*b*)	Beeswax	25.0
(*c*)	Peanut Oil	15.0
(*d*)	Paraffin	7.5
(*e*)	Lanolin	1.5
(*f*)	Perfume	1.0
(*g*)	Oil Red	2.0
(*h*)	Tetrabromofluorescein	2.5

The castor oil is heated to 55° C. and the tetrabromofluorescein dissolved in it. The oil soluble red is then dissolved and the whole thoroughly incorporated; (*b*), (*d*) and (*e*) are then mixed and melted together. The peanut oil (*c*) is then added, followed by the color and castor oil, and finally the perfume. The whole is well milled and then run into cooled molds to set.

Another type of procedure is as follows:

Lanolin	11
White Carnauba Wax	12
White Beeswax	20
Spermaceti	1

are melted together and then the fluorescein derivative is dissolved in:

Castor Oil	37
Oleyl Alcohol	9
Propylene Glycol	8
Cetyl Alcohol	2

at 50° C., and sieved before adding

to the wax melt. Thorough agitation follows and the mass is then run into molds to set.

When hydrocarbons are used as a basis for a lipstick, which is a common practice, castor oil cannot be used in normal formulation. On this account another solvent for the dyestuff must be employed. Butyl stearate may then be used and the presence of a binder such as triethanolamine is then necessitated. Eosine reacts with the fatty acids present in the lipstick and produces a combined bromo acid and thus the color does not bleed in use. A lake usually is four or five times weaker than the dyestuff itself and consequently allowances must be made in such cases, and also the lake substratum naturally tends to harden the stick and this factor must also be regulated. The bright orange and yellow lipsticks which give a red color when applied to the lips are perhaps a puzzle to the uninitiated. This double color effect is accomplished by the use of the free acid of eosine.

It must be pointed out that when using artificial dyestuffs, the color, in every case should be full batch strength, *i.e.*, it must contain no salt or diluent.

Eye Lotions

Formula No. 1

Sodium Chloride	5 gr.
Borax	4 gr.
Sodium Bicarbonate	2 gr.
Distilled Water, To make	1 fl. oz.

No. 2

Mercuric Oxycyanide	1 gr.
Distilled Water	6 fl. oz.

Dilute with equal amount of warm water for use.

No. 3

Boric Acid	8 gr.
Zinc Sulphate	½ gr.
Distilled Water	1 fl. oz.

Dilute with equal amount of warm water for use.

Liquid Mascara

Liquid mascara consists of Tr. benzoin 20, ester gum 2, castor oil 1, lampblack 14 and alcohol 23 parts each.

Nail Lacquer

U. S. Patent 2,215,898

Formula No. 1

Solids

Nitro-Cellulose (½ sec.)	132
Ethyl Cellulose	16
Tricresylphosphate	24
Liquid Resin "Hercolyn" (Dihydro Methyl Abietate)	34
Alkyd Resin ("Glyptal")	50

Solvent

Denatured Alcohol	83.5
Xylol	21
Ethyl Acetate	260
Butyl Acetate	113
Amyl Acetate	112

No. 2

Solids

Nitro-Cellulose (½ sec.)	121
Nitro-Cellulose (20–30 sec.)	10
Ethyl Cellulose	15
Gum Camphor	20
"Santolite" (Toluene Sulfonamideformaldehyde)	39
"Hercolyn"	10
Dibutylphthalate	15
Sodium Lauryl Sulfate	7

Solvent

Butyl Acetate	190
Ethyl Acetate	140
Ethyl Alcohol	140
Monethyl Ether of Ethylene Glycol	80
Toluol	57

No. 3

Ethyl Acetate	50.0
Butyl Acetate	20.0
Diethyl Phthalate	15.0
Camphor	4.5
Nitrocellulose Lacquer	10.0
Color, *e.g.*, Eosine	0.5

Eosine is usually dissolved in the ethyl acetate, although many processes involve direct solution in a portion of the nitrocellulose lacquer. Matt types of varnishes are often produced by adding inert white or colored pigments in definite amounts; color lakes may be employed here. The fluorescein range is probably the only color group of importance in this section of the industry although various other colors are used for certain jobs.

No. 4

Rosin, WW	7
Alcohol	68
Butyl Acetate	68
Ethyl Lactate	1
Nitrocellulose Film Scrap (Washed)	15
Glycol Ricinoleate	1

Liquid Nail Wax

Petrolatum	5
Paraffin Wax	10
Ethyl Acetate	10
Chloroform or Deodorized Kerosene	175
Perfume, To suit	

Mix together with gentle heat until dissolved.

Fluorescent Nail Polish

Many of the lacquer polishes contain organic dyes which fluoresce brightly under ultraviolet light and are therefore suited to theatrical work.

By adding eosine, uranine, or other fluorescent dye to a lacquer base in concentration of from 0.2 to 0.5 per cent, a luminous lacquer may be made which can be applied to the fingernails.

Nail Varnish Removers
U. S. Patent 2,197,630

Emulsified compositions containing a blown or sulphonated vegetable oil, a nitrocellulose solvent and an activated clay are proposed as combined varnish removers, nail cleaners and cuticle softeners.

Formula No. 1

Sulphonated Olive Oil	10.0 g.
Caustic Soda (10%)	1.0 g.
Butyl Acetate	15.0 g.
Acetone	15.0 g.
Titanium Oxide	0.5 g.
Bentonite	2.0 g.
Water To make	100 cc.

No. 2

Linseed Oil	5.0 g.
Caustic Soda (10%)	8.0 g.
Butyl Acetate	15.0 g.
Acetone	15.0 g.
Titanium Oxide	1.0 g.
Tin Oxide	1.0 g.
Bentonite	2.0 g.
Water To make	100 cc.

The proportion of caustic soda in each case is intended to give a final *p*H of 7.5 to 10. In making up the emulsion it is advisable to leave the bentonite in contact with about 20 per cent of the total water for about twelve hours in order to become hydrated. Pigment and polishing ingredients (*e.g.*, tin oxide) are then added, followed by the oil, solvent, alkali and the balance of the water. It is claimed for these emollient emulsions that they prevent colored nail varnish from running under the nails or on to the cuticle in course of removal. They have a whitening effect on the underside of the nails. Properly prepared emulsions on these lines should not settle or break.

Bath Salts

Bath salts are made in a variety of colors and ways; a simple example is given here to show the lines on which they are prepared:

Sodium Sesquicarbonate	99.0
Perfume	0.5
Color	0.5

The color may be applied in two chief ways; either the crystals are immersed in strong solution of the dye (in water or spirits), or the color is sprayed on whilst mixing is taking place. In the first case the crystals are finally centrifuged and allowed to dry before being packed. The basic dyestuffs are usually employed and these are water or spirit soluble. An example showing how a water dye solution is prepared for this purpose is:

Tangerine

Chrysoidine R Crystals	2
Glycerin	150
Distilled Water	848

The dyestuff, of course, must be carefully dissolved in water not exceeding 60° C. and this is particularly important for auramine which is used for lemon shades. Other colors commonly employed are Acridine Orange R, Safranine GR, Methyl Violets and Basic Greens.

Besides basic colors the fast oil colors are used in spirit solutions. When using these it is a common practice to add a small amount of a resin or ester gum to the spirit solution. On evaporation of the spirit, the small amount of gum is left in a very thin coating all over the surface of the crystals. This holds the color without being sticky. A spraying solution is made up as follows:

Lemon

Auramine Base	3
Ester Gum or Spirit Soluble Resin	10
Industrial Alcohol	987
	1000

All the oil colors mentioned earlier are applicable by this process.

Perspiration Deodorant

Formula No. 1

Liquid

Aluminum Chloride	8.0
Aluminum Sulphate	5.0
Borax	0.5
Water	86.5

No. 2

Cream

Triple Pressed Stearic Acid	10.0
Diglycol Laurate, Neutral	10.0
Fir Balsam Needle Oil	5.0
Water	35.0
Glycerin	3.5

This cream neutralizes body and perspiration odors for men.

No. 3

Aluminum Sulphate	17
Urea	11
Acid Cream Base	72

No. 4

Greaseless (Vanishing) Cream	80
Aluminum Phenolsulfonate	20

No. 5

Aluminum Phenolsulfonate	20.00
Perfumed Spirit, N.F.	50.00
Lavender Oil	0.25
Ethyl Acetate	0.50
Distilled Water, To make	100.00

Hygienic Deodorant Powder
Australian Patent 109,904

Zinc Oxide	3.0
Boric Acid	24.8
Chalk, French	67.0
Zinc Stearate	4.0
Benzoic Acid	1.0
Aluminum Chloride	0.1
Sodium Bicarbonate	0.1
Perfume, To suit	

Foaming Bath Powder

Bicarbonate of Soda	100
Tartaric Acid	80
Starch	20
Saponin	3
Perfume and Color, To suit	

Baby Powders
Formula No. 1

Boric Acid	100
Starch	100
Purified Talc, To make	1,000

No. 2

Boric Acid	100
Zinc Oxide	100
Kaolin	500
Magnesium Stearate	50
Purified Talc	250

No. 3

Purified Talc	500
Kaolin	200
Magnesium Stearate	50
Boric Acid	100
Starch	150

No. 4

Calamine	250
Starch	500
Boric Acid	200
Zinc Stearate	50

These powders, or variations of them, are prepared by simple mixing.

AROMATIC WATERS

Cherry Laurel Water

True Fruit Extract of Wild Cherry	8 oz.
Bitter Almond Oil	1 dr.
Water to make	1 gal.
Filter through Magnesia	1 oz.

Orange Flower Water

Orange Flower Essence	8 oz.
Water, To make	1 gal.
Filter through Magnesia	1 oz.

Rose Water

Rose Essence	4 oz.
Water (Lukewarm), To make	1 gal.
Filter through Magnesia	1 oz.

Peppermint Water

Peppermint Oil	2 dr.
Water, To make	1 gal.
Filter through Magnesia	½ oz.

Essences for the Above Aromatic Waters

Orange Flower Essence

Neroli Oil	2 dr.
Alcohol	1 pt.

Rose Essence

Rose Oil *	2 dr.
Alcohol	1 pt.

* Rose geranium oil may be used, being cheaper than the natural rose oil.

Synthetic Lavender Oil

Ethyl Amyl Ketone	1
Benzylidene Acetone	1
Citronellal	4
Geranyl Acetate	5
Coumarin	4
Linalyl Acetate	30
Linalol	20
Linalyl Isobutyrate	5
Terpinyl Acetate	10
Rosemary Oil	5
Spike Oil	10
Trichlorphenylmethyl Carbinyl Acetate	2
Benzoin R	2
Benzyl Cinnamate	2
Civet	1

Synthetic Geranium Oil

Citronellol	15
Geraniol	27
Phenylethyl Alcohol	15
Diphenyl Oxide	5
Geranyl Isobutyrate	5
Rhodinol	15
Benzophenone	2
Geranyl Propionate	5
Geranyl Formate	5
Aldehyde C_9 (10%)	1
Palmarosa Oil	5

Synthetic Bergamot Oil

Linalyl Acetate	40
Terpinyl Acetate	15
Linalol	10
Terpinyl Isobutyrate	4
Geranyl Methyl Ether	4
Methyl Anthranilate	10
Citronellyl Acetate	10
Limonene	4
Aldehyde C_8 (10%)	1
Aldehyde C_{12} (1%)	2
Chlorophyll, To color	q.s.

Perfume for Deodorant Cream

Rose Oil	0.25 cc.
Clove Oil	0.06 cc.
Ethyl Acetate	0.6 cc.
Alcohol (95%)	50.00 cc.
Water, To make	100.00 cc.

Solid Perfumes (Non-Greasy)

Formula No. 1

Japan Wax	20 g.
White Beeswax	20 g.
Diethyl Phthalate	40 ml.
Perfume Oil	20 ml.

No. 2

Melt 4.5% of stearin, add 0.5 part of sodium carbonate and 95% Eau de Cologne and heat during 1 hour in an autoclave. When nearly cool, pour into molds.

Cosmetic Stockings

Formula No. 1

Methyl Cellulose	0.5
Gum Arabic	0.1
Titanium Dioxide	2.0
Zinc Oxide	0.1
Precipitated Chalk	10.0
Zinc Stearate	1.0
Talc	0.4
Bentonite	2.0
Alcohol	8.0
"Aguaresin"	4.0
"Sulfatate"	0.5
Water	68.0
Dye and Pigment, To suit	

No. 2

Precipitated Chalk	20
Zinc Stearate	2
Talc	8
Bentonite	20
Sodium Bicarbonate	1
Alcohol	10
"Glycopon 50"	10
"Protoflex"	6
Water	124
Dye and Pigment, To suit	

No. 3

Stearyl alcohol, 100 grams; Lanette wax SX, 20 grams; liquid paraffin, B.P., s.g. 0.89, 150 ml.; water, distilled, 500 ml.; powdered mixture, 50 grams; water, distilled, 180 ml.; preservative perfume, q.s. Another formula consists of: Diglycol stearate, 10 grams; titanium dioxide, 3 grams; colloidal kaolin, 5 grams; water, about 200 grams.

No. 4

Bentonite	9
Precipitated Chalk	11
Magnesium Silicate	5
Aquaresin (Glycol Bori-Borate)	6
Protoflex	8
Water	68
Diglycol Stearate	1
Dye or Pigment, To suit	

No. 5

Vanishing Cream	62
Gum Solution (Tragacanth, Karaya, etc.)	2
Titanium Dioxide	1
China Clay, Colloidal	3
Face Powder Base (Suntan, etc.)	12
Glycerin or Substitute	5
Water	15

The pigments need to be carefully and thoroughly dispersed. The final product is best run through a paint or similar mill. Naturally, much will depend upon the weight of the pigments used, the consistency of the cream, and other variable factors.

In so far as possible, earth colors are preferred because of their inert nature and freedom from bleeding. The selection of earths is a wide one, and practically any shade desired can now be obtained with them. Judicious use of organic lakes will give life and brightness to the tints.

No. 6

Liquid Powder

Zinc Oxide	6
Precipitated Chalk	8
Kaolin	3
Zinc Stearate	2
Glycerin	3
Witch Hazel Extract	10
Orange-Flower Water	68
Color and Perfume, as required	

No. 7

Cream

Paraffin Oil	1
Glyceryl Mono-Stearate	6
Distilled Water	48
Titanium Dioxide plus Earth Pigments	21
Talc	9
Perfume, To suit	

No. 8

Powder Cream

Vanishing Cream	70
Talc	24
Titanium Dioxide	6
Color and Perfume, as required	

No. 9

Powder

Titanium Dioxide	3.00
Zinc Oxide	20.00
Talc	65.06
Zinc Stearate	4.00
Precipitated Chalk	6.00
Color and Perfume, as required	

No. 10

Lotion

Bismarck Brown (1% Solution)	79.0
Methyl Violet (½% Solution)	3.0
Methylene Blue (½% Solution)	2.0
Glycerin	5.0
Alcohol	10.5
Perfume	0.5

Hair "Tonics"

Formula No. 1

Isopropyl Alcohol	70.0
Propylene Glycol	5.0
Eau de Cologne	5.0
Cholesterin	0.5
Perfume	0.5
Distilled Water	19.0

Dissolve all the ingredients except water in isopropyl alcohol, add the water last.

No. 2

Gum Benzoin	2 dr.
Castor Oil	4 oz.
Alcohol	1 qt.

Shake well together, then add

Lavender Oil	1 dr.
Bergamot Oil	1 dr.
Clove Oil	30 drops
Rosemary Oil	30 drops
Lemon Oil	30 drops
Neroli Oil	30 drops
Tincture of Cantharides	½ oz.

Shake well to cut the oils.

No. 3

Il biondo Dio (Italian "Hair Tonic")

Castor Oil	1 pt.
Jamaica Rum	½ pt.

Shake well; use immediately after hair has been shampooed and is perfectly dry. For many years this has been a favorite tonic in parts of Italy.

Hair Lacquer

Formula No. 1

Elastolac	20–35%
Water or Alcohol	80–65%
Perfume, To suit	

This product gives a soft sheen to the hair and does not become brittle. The more alcohol used in the formula, the faster the rate of drying.

No. 2

Ten parts of orange shellac is dissolved with 2 parts borax in 70 parts of water, to which 20–30 parts of eau de cologne is added. If a gold effect is desired, powdered aluminum bronze may be added.

Cheap Hair Cream

White Wax	10
Paraffin Oil	130
Distilled Water	15
Borax	1
Perfume, To suit	

Melt the wax in 50 parts of paraffin oil. Place in a mortar and stir in the remainder of the liquid paraffin. Add the distilled water in which the borax has been dissolved and stir the cream formed consistently until cold.

Hair Fixatives

Formula No. 1

Water	96.00
Glycerin	3.00
Pectin	1.00

No. 2

Water	98.50
Pectin	1.00
Citric Acid	0.50

No. 3

Water	96.00
Glycol Borate	2.00
Pectin	0.25
Alcohol	1.75

Permanent Wave Solutions

Formula No. 1

Sulphonated Castor Oil	8 oz.
Ammonia	20 oz.
Sodium Pyrophosphate	2 oz.
Sodium Bisulphite	8 oz.
Potassium Pyrosulphite	12 oz.
Sodium Sulphite	12 oz.
Water	428 oz.

No. 2

Sulphonated Castor Oil	8	oz.
Sodium Pyrophosphate	3	oz.
Potassium Pyrosulphite	16	oz.
Sodium Sulphite	16	oz.
Soda Ash	16	oz.
Monoethanolamine	5	oz.
Water	456	oz.

Permanent Wave Creams
(Cream Paste Type)
Formula No. 1

A.	Glycosterin (Glyceryl Monostearate)	9.40
	Lanolin	1.70
	Wetanol	0.10
	Water	37.30
B.	Sodium Sulfite	9.50
	Ammonium Carbonate	4.70
	Ammonia (28° Bé.)	2.00
	Water	30.30

No. 2

A.	Glycosterin	10.00
	Lanolin	2.00
	Glycerin	0.72
	Water	40.00
B.	Permosalt	7.85
	Ammonia (28° Bé.)	7.53
	Water	31.90

Melt the Glycosterin and lanolin together, heat to about 160–170° F. Add the water, or water and Wetanol (wetting agent) heated to the same temperature and stir to about 100° F. or a little lower. At the same time, mix Group B *cold* and add to Group A mixture at 100° F. or lower, slowly with thorough agitation. Both these materials will give a heavy fluid emulsion which will set overnight into cream form which can be packed in jars or tubes.

No. 3

Ammonia (28° Bé.)	20
Caustic Soda	1
Sodium Sulphite	10
Water	89

Dissolve above by mixing.

Melt together

Parachol	40
Lanolin	40

Add the water solution to the above, a little at a time, mixing each portion until well absorbed.

STABILIZATION OF FINGER WAVE CONCENTRATES

In finger wave concentrates using gum karaya, difficulty is experienced in keeping the gum from settling to the bottom and caking into a hard mass which is difficult to redissolve. This can be overcome by the addition of Diglycol Laurate to the gum-alcohol mixture, giving a uniform dispersion of the gum which will dilute easily in water. Diglycol Laurate gives added gloss to the hair and prevents flaking of the gum on drying. The following formula is suggested:

Gum Karaya	3
Diglycol Laurate	1–1½
Alcohol	1–2

Increasing the gum content gives a paste suitable for tubes.

Kinky or Wavy Hair Straightener
British Patent 543,066

Sodium Sulphide	20	g.
Caustic Soda	2	g.
Soda Ash	2	g.
Triethanolamine	4	g.
Water	1	l.

Hair Waving Pad Heat Generating Composition
Formula No. 1
British Patent 531,250

Mercurous Chloride	1
Sodium Nitrate	5
Aluminum, Powdered	10

Before use, moisten with 1% of ammonium or stannous chloride.

No. 2

A sheet of fibrous material is impregnated with a 40% copper chloride solution and dried and a sheet of aluminum foil is used with a solution of

Ammonium Chloride	6.7
Ammonium Nitrate	28.6
Sodium Salicylate	8.3
Water	56.4

No. 3

U. S. Patent 2,261,221

Aluminum	4.0
Potassium Chlorate	1.5
Cuprous Oxide	1.0
Benzene Sulfonic Acid	0.6

Hair Dyeing Soap

British Patent 524,293

Eighty grams of *para*-aminophenol is mixed with 75 grams of soda ash and 100 grams of the methyl tauride of oleic acid, and converted into a paste with 75 grams of water. This paste is milled with 675 grams of toilet soap chips with a fatty acid content of 75 to 80 per cent. The homogeneously milled product is plodded and cut to shape. The product dipped into a 2 per cent solution of hydrogen peroxide and applied to the hair—*e.g.*, by rubbing—is stated to give a reddish-brown shade.

Solid Hair Dye Block

British Patent 524,293

Methyl p-Phenylene Diamine Sulphate	120 g.
Ammonia (25%)	65 g.
Fullers' Earth	40 g.
Water	50 g.

Mix together until uniform; then mill with soap (75–80% fatty acid), 760 g.

The hair is first moistened with 2% hydrogen peroxide and treated with above is dyed brown-black.

Removing Hair Dye

U. S. Patent 2,149,319

A dye-removing composition is formed of 3% of 40° Baumé nitric acid, 1% hydrochloric acid, 1% oxalic acid, 1% of acetic ether, 0.02% of cholesterol, 3% of diethylene glycol, 1% of sodium formaldehyde-sulfoxylate and 89.98% of water.

Coal Tar Shampoo

A liquid tar shampoo may be made from 66 grams coconut oil, 78 cc. cottonseed oil, 36 grams stearic acid, 10 cc. rectified oil of tar, 42 grams potassium hydroxide, 9 grams potassium carbonate, 42 cc. alcohol, 10 grams purified talc and sufficient water to make 1000 cc. The fats are melted and heated to 82° C.; the alkalies are dissolved in 100 cc. distilled water and added. Finally, the alcohol containing the coal tar is added and the whole heated until saponification is complete. When cold, add distilled water to measure 1000 cc.

Blond Hair Rinse

Tincture of Rhubarb	40
Alcohol	70
Propylene Glycol	10
Perfume, To suit	
Water, To make	1000

Depilatories

Formula No. 1

Barium Sulfide	30 g.
Atropine	5 g.
Spermaceti	100 g.
Distilled Water	200 g.
White Petrolatum Jelly	300 g.

This cream can be used as a depilatory or can be applied every day for 20 minutes to stop the growth of unwanted hair.

No. 2

Strontium Sulphide	1
Chalk, Precipitated	2
Talc, Powdered	5
Zinc Oxide	5

Mix well and pack in moisture proof containers. Before use, make into a paste, with water, and leave on skin for about five minutes.

No. 3

U. S. Patent 2,202,829

Rosin	50
Mineral Oil	25
Beeswax	3

Brushless Shaving Cream

Formula No. 1

Stearic Acid (Triple Pressed)	900
Lanolin	160
Glycopon S	60
Mineral Oil	180
Triethanolamine	51
Borax	54
Water	2600
Perfume	2

Melt the stearic acid, add the lanolin, Glycopon S and mineral oil and bring the temperature to about 70° C. Heat the water, triethanolamine, and borax in a separate container to boiling. Now add the first mixture to the second with vigorous stirring until a smooth emulsion is obtained. Add the perfume and stir gently until cool.

No. 2

Stearic Acid	35
Cottonseed Oil	18
Anhydrous Ammonium Laurate	7.5
Glycerin	12
Water	100

This produces a very smooth cream, of good consistency. It spreads easily and washes off the razor blade readily.

No. 3

Stearic Acid	70 g.
Cottonseed Oil	30 g.
Diglycol Laurate	5 g.
Anhydrous Ammonium Stearate	15 g.
Glycerin	25 g.
Sulfatate (2% Solution)	55 cc.
Water	200 cc.

This produces a cream which spreads readily, washes off the razor blade easily and remains moist on the face.

No. 4

Sodium Stearate (Soap)	4.0
Stearic Acid	14.2
Spermaceti	3.8
Mineral Oil	7.5
Water	70.5

After Shaving Lotion

Formula No. 1

(especially for users of electric shavers: disinfectant, refreshing, improving resistance of skin.)

Glycerin	5 cc.
Tannic Acid	5 gr.
Menthol	1 gr.
Alcohol, Diluted To make	100 cc.

No. 2

Water, Distilled	85.0
Glycerin	10.0
Citrus Pectin	2.0
Alcohol	2.0
Boric Acid	0.5
Zinc Sulpho Phenolate	0.2
Rose Oil	0.3

Electric Razor Shave Lotion

Alcohol	70.0
Lecithin	0.5

Wetting Agent (Wetanol) 1.0
Witch Hazel Extract 28.0
Perfume Sufficient Quantity

Tooth Paste
Formula No. 1

Citrus Pectin	3.0%
Glycerin (28° Bé.)	71.5%
Water, Distilled	25.0%
Citric Acid	0.5%

Mix 58% of this jelly with 42% of titanium dioxide.

No. 2

Citrus Pectin	8.0%
Glycerin (28° Bé.)	10.0%
Water, Distilled	81.0%
Citric Acid	1.0%

Mix 50% of this jelly with 48% titanium dioxide and 2% peppermint oil.

No. 3

Tragacanth	1.0%
Citrus Pectin	5.0%
Glycerin (28° Bé.)	30.0%
Water, Distilled	63.0%
Citric Acid	1.0%

Mix 36% of the solution with 37% titanium dioxide, 22% Pepsin U.S.P., 8.3% invert syrup and 2% Menthol.

No. 4

Tragacanth	1.5%
Pectin	5.0%
Glycerin (28° Bé.)	30.0%
Water, Distilled	61.0%
Invert Syrup	2.5%

Mix 39% of this solution with 20% Pepsin U.S.P., 8.39% titanium dioxide and 2% Peppermint Oil.

No. 5

Mix 50% of the formula 2 with 30% titanium dioxide, 16% fine kaolin, 2% saponin-acetate and 2% Peppermint Oil.

Glycerin Free Tooth Pastes
Formula No. 1

Methyl Cellulose (4% Stock Solution)	20
Cold Water	34
Mineral Oil (Cosmetic Grade	3
Precipitated Chalk	32
Magnesium Carbonate	10
Perfume	1

No. 2

Precipitated Chalk	46.75 lb.
Bentonite	3.75 lb.
Gum Tragacanth Mucilage (5%)	3.75 lb.
Mineral Oil	1.50 lb.
Saccharine	0.50 oz.
Flavoring Oil	5.00 lb.
Water	45.00 lb.

The method of manufacture is to take about one-third of the water, then add the chalk mixed with the binder (absorbent clay or bentonite) slowly, so that the mass remains smooth. Then add the gum tragacanth mucilage and saccharin dissolved in a little of the water. Afterwards incorporate the flavor and the lubricant (mineral oil).

No. 3

Precipitated Chalk	552.00 g.
Hard Soap	60.00 g.
Soluble Saccharin	0.25 g.
Heavy Mineral Oil	25.00 cc.
Sorbitol Syrup (Yumidol)	348.50 g.

Inexpensive Tooth Powder

Precipitated Chalk (Very Fine)	200 g.
Orris Root (Powdered)	20 g.
Common Salt (Pulverized)	10 g.
Sodium Bicarbonate	20 g.
Menthol Crystals (Pulverized)	2 g.

Mix all the above ingredients and

flavor with 10–20 drops of oil of wintergreen or oil of peppermint. Keep this mixture in a dry place and use it on a dry toothbrush for best results.

Tooth Powder for Pyorrhea

Sodium Perborate	10.00	g.
Sodium Benzoate	2.50	g.
Sodium Bicarbonate	0.50	g.
Menthol, To suit		

Tooth Powder for False Teeth

Mix together 10 parts of calcium carbonate, 2 parts of sodium perborate, 2 parts of sodium bicarbonate, 2 parts of tricalcium phosphate and 2 cc. of oil of cinnamon.

Denture Cleaner
Canadian Patent 391,039

Citric Acid (Saturated Solution)	1
Isopropyl Alcohol	15

Mix with water for use.

Lotion for Massaging Gums

Zinc Chloride, Zinc Iodide	1.0	g.
Tincture of Iodine, Distilled Water	10.0	g.
Novocaine	0.5	g.

Mouth Lotion

Water, Distilled	84
Citrus Pectin	5
Citric Acid	1
Peppermint Oil	10

Mouth Wash

Citric Acid	0.1	g.
Tartaric Acid	0.1	g.
Benzoic Acid	0.1	g.
Boric Acid	2.0	g.
Glucose	0.5	g.
Glycerin	10.0	g.
Chlorothymol	0.033	g.
Eucalyptol	0.1	cc.
Thymol	0.07	g.
Menthol	0.045	g.
Alcohol	25.0	cc.
Distilled Water, To make	100.0	cc.

Soapy Mouth Wash

Powdered Neutral Soap	20.0
Glycerin	90.0
Peppermint Oil	6.0
Wintergreen Oil	2.5
Cinnamon Oil	1.0
Clove Oil	0.5
Alcohol	300.0
Distilled Water	580.0

Astringent Mouth Wash
Formula No. 1

Sodium Bicarbonate	12.5
Borax	12.5
Zinc Chloride	1.5
Menthol	0.25
Alcohol	25.0
Glycerin	50.0
Cinnamon Water	200.0

No. 2

Quinine Hydrochloride	½
Urea	4
Glycerin	25
Water, Distilled, To make	100

Flavor and color to suit. Adjust pH, with caustic soda to about 7.

Throat Gargle

To 75 cc. of 95% ethyl alcohol (not the denatured variety) add 45 cc. of water. Then, with stirring, add 5.4 grams of ground ferric chloride, hexahydrate. When solution is effected add 45 grams of potassium chlorate and 200 cc. of water. Next add 240 cc. of glycerin and enough water to make 1000 cc. total volume. Mix well.

Dosage: 2 tablespoonfuls in half a glass of water and gargle at intervals of one hour.

Ephedrine Spray

Ephedrine Hydrochloride	4 gr.
Sodium Chloride	2 gr.
Water, To make	1 fl. oz.

Buffered Ephedrine Nose Drops

Ephedrine Sulphate	0.5
Potassium Phosphate Monobasic	0.5
Sodium Phosphate Dibasic	0.5
Potassium Chloride	0.15
Sodium Chloride	0.15
Dextrose, Anhydrous	0.9969
Preserved Water, To make	100 cc.

If a one per cent solution of ephedrine sulphate is desired, the amount of dextrose in the above formula is reduced to 0.7867 g., and for a two per cent solution the dextrose is reduced to 0.3663 g. The quantities in the above formula must be weighed accurately on an analytical balance and made up to the indicated volume in a volumetric flask in order to obtain the desired tonicity and pH values. In addition the chemicals used must be of the highest purity.

Antiseptic Snuff

Sulfathiazole	10
Magnesium Carbonate	90

This is used to prevent secondary nasal infections.

Ear Drops

Formula No. 1

Salicylic Acid	4 gr.
Alcohol, Denatured	1 fl. oz.

Dilute with 10–20 parts water for use.

No. 2

Phenol, Liquefied	12 min.
Glycerin, To make	1 fl. oz.

Do not mix this water.

Eye Drops

Boric Acid	10 gr.
Zinc Sulphate	1 gr.
Adrenaline Solution (1–1000)	30 min.
Distilled Water, To make	1 fl. oz.

No. 2

Silver Protein	40 gr.
Distilled Water	1 fl. oz.

No. 3

Atropine Sulphate	2 gr.
Distilled Water	1 fl. oz.

No. 4

Soluble Fluorescein	2 gr.
Distilled Water	1 fl. oz.

Alkaline Eye Ointment

Sodium Borate	1
Sodium Bicarbonate	2
Wool Fat	10
Distilled Water	10
White Petrolatum	80

Eye-Wash

Boric Acid	16.2 g.
Zinc Sulphate	0.8 g.
Glycerin	15.4 cc.
Distilled Water	740.0 cc.

Toothache Drops

Clove Oil	12
Camphor	6
Chloroform	6
Phenol	2
Menthol	2
Cinnamon Oil	3

Apply to cavity.

Antiphlogistic Poultice
British Patent 521,215

Aluminum, Powdered	2–5
Copper Sulphate, Anhydrous	2–6
Sodium Chloride	1–4
Silica	0.2–13

Poison-Gas Ointment

Eucalyptol	5
Ichthyol	0.8
Olive Oil	45
Calcium Hydroxide (Saturated Solution)	50

Mustard Gas Itch Ointment

Benzyl Alcohol	50
Stearic Acid	30
Glycerin	10
Pontocaine	1
Menthol	1

Protective Against Mustard Gas

Glycerin	3
Gelatin	2
Water	3

This is applied to feet of animals in contaminated gregs. To render it waterproof, it is treated with formaldehyde after application.

Tear Gas Cleaner for Skin

Alcohol (50%)	96
Sodium Sulphite	4

Air Raid Abrasion Antiseptic

Potassium Iodide	640 gr.
Iodine	200 gr.
Distilled Water	1 fl. oz.
Glycerin, To make	20 fl. oz.

Grind the potassium iodide to a fine powder, add the iodine, mix, add the water and make up to volume with glycerin.

This solution may be used full strength or half strength. In the latter case it is diluted with glycerin, never with water.

The preparation is used to paint on abrasions. It may also be used to pour into wounds instead of sulfanilamide, or to swab out wounds which cannot be excised.

Ointment Base
Formula No. 1

Glyceryl Monostearate	15
Cetyl Alcohol	15
Glycerin	35
Diethylene Glycol	35

The ingredients are heated together until all of the solid particles are melted and it is then stirred slowly until cool. This formula can be modified by incorporating water into each ointment as indicated in the parentheses below. The amount of water added is governed by the consistency of the finished ointment.

This base is compatible with the following medication:

Phenol, 2 per cent (42% water in final ointment).

Rectified Oil of Birch Tar, 10 per cent (30% water in final ointment).

Sulfur, 15 per cent (37% water in final ointment).

Ammoniated Mercury, 10 per cent (45% water in final ointment).

Coal Tar, 5 per cent (32% water in final ointment).

Burrow's Solution, 10 per cent (no water added to final ointment).

Balsam of Peru, 10 per cent (12% water in final ointment).

No. 2

Glyceryl Monostearate	10 g.
Glycerin	25 g.
Bentonite	2 g.
Distilled Water, To make	100 g.

The bentonite is mixed with 50 cc.

of distilled water and stirred into a uniform magma. The glyceryl monostearate is melted in the glycerin on a water-bath and added to the magma, warmed to the same temperature and more distilled water added to make the product weigh 100 g. It is then stirred until cool.

Emulsified Ointment Base

Sodium Lauryl Sulphate	0.5
Cetyl Alcohol	8.0
Cocoa Butter	6.5
Petrolatum, White	20.0
Water	65.0

Water content has been retained at a fifty per cent ratio in ointment emulsions in which medication is added. There is no reason other than that this water ratio is a convenient amount to work with and gives satisfactory results. The following chemicals have been incorporated in this emulsion system, either separately, or combined in varying percentages:

Salicylic acid 3%/c Benzoic acid 6%.

Juniper tar (Oil of cade) 10%/c Salicylic acid 3%.

Coal tar 5%.

Kaolin 10%/c Sulfur precipitated 10%.

Ammoniated mercury 5% and 10%.

Sulfur precipitated 10%.

Phenol 2%.

Sulfathiazole 1%.

Balsam Peru 10%/c Sulfur precipitated 10%.

Salicylic acid 2%/c Sulfur precipitated 5%.

Calamine 8%.

Emulsion Ointment Base

Propylene Glycol	6.00
Water	1.92
Wetanol (Wetting Agent)	0.25
Petrolatum, White	91.83

Germicidal Emulsion Base
U. S. Patent 2,281,249

O-Phenylphenol	24
Sulfonated Castor Oil (Partially neutralized)	36–136

A 5% emulsion of this, with water, gives a pH above 7.

Ointment for Dry and Rough Skin

A compound ointment specified for use in the treatment of excessively dry skin conditions and for promoting exceptionally rapid healing of abraded skin contains the essential unsaturated fatty acids, together with lecithin, cholesterol and other synergistically acting ingredients.

Lecithin	7.0
Anhydrous Lanolin	35.0
"Vitamin F" Concentrate	11.5
Special Lard	46.0
Peppermint Oil	0.5

To this base, sun-bleached beeswax may be added to procure satisfactory stiffening of the product. About 10 per cent of beeswax seems to give the most desirable consistency.

Ringworm Treatment
Formula No. 1

Twenty (20) grains (1.3 g.) of salicylic acid is dissolved in one ounce (32 cc.) of alcohol, and this lotion is applied repeatedly. A dram (4 cc.) of glycerin will prevent too-rapid evaporation; a minim or two of rose water will impart a pleasing odor.

No. 2

Salicylic Acid	1.0 g.
Benzoic Acid	1.0 g.
Alcohol (90%)	30.0 cc.

Apply twice daily with a cotton-wrapped match.

For the dry, horny type, Whitfield's ointment should be used:

No. 3

Benzoic Acid	2.0 g.
Salicylic Acid	1.0 g.
Wool Fat	1.6 g.
White Petrolatum	30.0 g.

Both of these remedies should be used long after the last traces of the disease have disappeared.

A case of eczema of the hand, of 15 years' duration, cleared up in six weeks after the patient was instructed to scrape the ringworm-infected *toenails* with the edge of a glass slide, following a hot potassium permanganate foot bath. The allergic manifestations of mycotic infections of the feet are almost always on the hands, and are more resistant than the original focus. The great majority of hand lesions are dermatophytids (fungus-free) and should be treated with soothing remedies while the feet are undergoing active treatment.

The hands should be soaked in Burrow's solution, 1 to 10 in warm water, for one hour, three times daily. Overnight and between soakings, this combination should be applied to the hands: Burrow's solution, 10 parts; anhydrous lanolin, 20 parts; Lassar's paste, 30 parts. The feet should be soaked in a potassium permanganate solution (5 grains to a basin of hot water) for 30 minutes. Whitfield's ointment should be used between the toes overnight.

No. 4

Petrolatum 190

Melt and Add:

Benzoic Acid 32

Salicylic Acid 14

Mix till dissolved.

When nearly cool add:

Phenol 2

Thymol 2

Mix

Stir till cool

Pass through an ointment mill.

No. 5

Itching Feet (Whitfield's) Ointment

Salicylic Acid	15 gr.
Benzoic Acid	25 gr.
Soft Petrolatum	2 dr.
Cocoanut Oil q. s. to make	1 oz.

Apply at night for a week, then omit treatment for a week and repeat if necessary.

No. 6

Camphor	1–3 g.
Phenol U. S. P. (Melted)	3 g.

Rub together in a mortar until uniform. Do not apply to wet skin for then it is caustic.

No. 7

To:

Ethyl Alcohol	50%
Water	50%

Add: (Based on amount of above mixture)

Salicylic Acid	6%
Benzoic Acid	12%

Apply once or twice a day after washing.

Skin Coating to Prevent Dermatitis
U. S. Patent 2,249,523

Formula No. 1

Cellulose Tributyrate	1
Butyl Stearate	3

Heat to 160° C. while stirring slowly. On cooling a gel is formed.

No. 2

Cellulose Acetate Stearate	1
Butyl Stearate	2
Cottonseed Oil	2
Zinc Oxide	2

Impetigo Lotion

Boric Acid	4 gr.
Zinc Sulfate	1 gr.
Saturated solution of Sulfanilamide in freshly distilled water, To make	1 fl. oz.

Dermatological Cream

Glyceryl Monostearate	10
Glycerin	25
Bentonite	2
Distilled Water, To make	100

The suggested procedure is to sprinkle the bentonite on 50 parts of distilled water, and, after it has become thoroughly wetted, to stir the mixture until a uniform magma results. The glyceryl monostearate is melted in the glycerin on a water-bath, and the mixture is added to the magma, warmed to the same temperature, together with sufficient distilled water to bring to the required weight. The cream is then stirred until cold. The bentonite can be preserved, if desired, and used in the form of a 7 per cent magma. Creams containing boric acid, tannic acid, iodine, sulphur, benzoic acid, calamine, ichthammol, and potassium iodide, respectively, were prepared using this basis, and appeared to be satisfactory.

Poison-Ivy Protective

Ferrous Sulfate	30
Water	250
Alcohol	250
Glycerin	25

Poison Ivy Treatment
Formula No. 1

Tannic Acid U. S. P.	2 g.
Boric Acid U. S. P.	2 g.

The above is dissolved in one pint of water. Apply solution, when cold, to parts affected, with cupped hand once every two hours for about a half a day. The application of this solution will remove the itch instantly, and the postules will be removed overnight.

No. 2

This ointment should be thickly applied to exposed parts, such as the hands and face, if necessary, before contact with poison ivy. Clothes must be removed after exposure before the ointment is washed off, otherwise the unprotected skin may be exposed to clothes which have been contaminated. Tools and instruments or clothing which have been used in cutting poison ivy must be decontaminated before being used again. Decontamination can be effected by washing clothes or immersing tools for 15 to 20 minutes in a 1% solution of calcium hypochlorite.

No. 3

Castor Oil	21.5
Olive Oil	21.5
Lanolin, Anhydrous	21.5
Diglycol Stearate	12.9
Paraffin, Refined	8.6
Boric Acid	2.0
Sodium Perborate	10.0
Duponol WA, Pure	2.0

No. 4

Cetyl Alcohol	35.1
Stearyl Alcohol	5.3
Ceresin	3.5
Castor Oil	20.8
Mineral Oil	21.9
Duponol WA, Pure	1.7
Sodium Perborate	10.0
Boric Acid	1.7

Keep in closed containers.

Calamine Lotion

Prepared Calamine	150 g.
Bentonite	20 g.
Water, To make	1000 cc.

Mix the bentonite with 800 cc. of water, agitate well and frequently and allow to stand 12 hours or more. Thoroughly incorporate the calamine with about one-tenth of the bentonite sol and gradually incorporate the remainder of the sol in small portions at a time until the mixture is complete. Finally add enough water to make 1000 cc. and shake well.

Burn Treatments

Formula No. 1

Glycerin	30 g.
Distilled Water	70 cc.
Brilliant Green	0.1 g.

No. 2

Tannic Acid Jelly

Tannic Acid	10
Glycerin	10
Tragacanth	3
Euflavine	0.1
Water, To make	100

No. 3

Silver Nitrate	0.5
Glycerin	10
Tragacanth	3
Activated Charcoal, Powdered	15
Water, To make	100

No. 4

Aluminum Naphthol Disulphonate	2
Ichthyol	4
Liquefied Phenol	2
Wool Fat	62
Yellow Petrolatum	30

Burn Jelly

Formula No. 1

Tannic Acid	20.0
Proflavine Sulphate or Methyl Violet	0.1
Procaine	2.0
Glycerin-Tragacanth Base, To make	100.0

Glycerin-Tragacanth Base

Glycerin	10.1
Pulv. Tragacanth	2.0
Distilled Water, To make	100.0

No. 2

Tannic Acid (10% Solution)	32
Polymerized Glycol Oleate	2

Warm together and mix until dispersed.

Sulfanilamide Ointment

Formula No. 1

(a) Dissolve 10 parts of sulfanilamide in 25 parts of nearly boiling water; filter. (b) To 4 parts of sodium alginate add 75 parts of boiling water and then strain the resulting mucilage through fine gauze. Mix the sulfanilamide solution and the sodium alginate while hot, and stir until cool. (c) Add 16 parts of anhydrous wool fat, 1 part of sodium chloride dissolved in 4 parts of water, and 78 parts of white petrolatum base to the sulfanilamide-sodium alginate combination and mix until smooth.

No. 2

1. Dissolve 10 parts of sulfanilamide in 25 parts of almost boiling water and filter.
2. Add to 100 parts hot starch solution.
3. Cool and add 15 parts of hydrous wool fat and 50 parts of petrolatum.

Sulfanilamide Paste

To make 1000 g. of paste, 8 g. of sulfanilamide and 75 g. of medicinal pectin are rubbed with 180 g. of glycerin. To this is added, at once and with sufficient stirring, a hot solution of benzoic acid in 735 g. of Ringer's Solution. Emulsify with hand homogenizer or with mortar and pestle.

No. 2

Five per cent of the powdered drug in a base consisting either of equal parts of zinc oxide and liquid petrolatum or of equal parts of iodoform and liquid petrolatum.

Sulfanilamide Throat Spray

Sulfanilamide	1.7 g.
Alcohol (90%)	7.5 cc.
Acetone	7.5 cc.
Glycerine, To make	100.0 cc.

Sulfanilamide Emulsion

Powdered Sulfanilamide	175 gr.
Cod Liver Oil	4 fl. oz.
Oleic Acid	36 min.
Solution of Calcium Hydroxide, To make	8 fl. oz.

Sulfathiazole Emulsion

Sulfathiazole (Finely powdered)	5
Triethanolamine	2
Distilled Water	24
White Beeswax	5
Liquid Petrolatum	64

Melt the beeswax in the liquid petrolatum on a water bath. Mix the triethanolamine with the water and heat to the same temperature as the petrolatum-wax mixture. Mix the sulfathiazole with the triethanolamine solution and add the petrolatum-wax mixture, stir vigorously until a creamy emulsion is obtained.

Wounds and ulcers are packed with gauze which has been soaked in the emulsion. The preparation keeps the walls of the cavity oiled, preventing the adherence of clothing or lymph. Loose packing over the emulsion-soaked gauze permits constant drainage and the packing need not be removed for several days. It is readily removed with little discomfort to the patient.

Sulfathiazole Ointment

Formula No. 1

Cold Cream	70
Absorption Base (Parachol)	25
Sulfathiazole	5

No. 2

Lanolin, Anhydrous	47.5
Sulfathiazole, Finely divided	5.0
Vanishing Cream	47.5

Melt the lanolin and add the vanishing cream slowly with adequate beating, so as to produce a good, fluffy cream. Add the very finely divided sulfathiazole and continue beating until a uniform mixture is obtained. If the mixture is not readily emulsified, add three or four drops of ethanolamine and continue the beating.

This ointment is particularly adapted to various skin irritations, eczema, and similar skin disturbances.

No. 3

Stearic Acid	35
Triethanolamine	5
Glycerin	10
Lanolin, Anhydrous	250
Powdered Sulfathiazole	25
Distilled Water, To make	50

Melt the stearic acid and lanolin and use portion of water, add glycerin and triethanolamine and add to the melted stearic acid and lanolin. Allow to cool to about 35–40° C. Rub sulfathiazole in mortar with water and add to the ointment base under mechanical, constant stirring.

Foot Corn Ointment

Benzoic Acid	1.0
Lanolin	44.5
Salicylic Acid	5.0
Sulfathiazole	5.0
Vanishing Cream	44.5

Warm the lanolin and vanishing cream together and beat in a finely pulverized mixture of the benzoic acid, salicylic acid, and sulfathiazole. When all the ingredients have been added, continue to beat until a good, fluffy cream is obtained. This cream is particularly useful for soft corns between the toes and similar foot irritations.

Cod Liver Oil Ointment

Cod Liver Oil U.S.P.	45%
Paraffin Wax	7%
Stearic Acid U.S.P.	20%
Cetyl Alcohol	3%
Lanolin Hydrous, To make	100%

Weigh the cod liver oil and lanolin and heat, melt paraffin wax, stearic acid and cetyl alcohol and pour the molten waxes at a temperature of 70° C. into the cod liver oil and lanolin at 70° C. Stir until cooled to room temperature and put through ointment mill.

Dhobie Itch Ointments

Formula No. 1

Salicylic Acid	4
Bismuth Subnitrate	10
Mercury Salicylate	4
Eucalyptus Oil	10
Lanolin or Petrolatum	100

No. 2

Resorcin	2 dr.
Salicylic Acid	10 gr.
Lanolin	4 dr.
Petrolatum	4 dr.

Zinc Peroxide Ointment

Zinc Peroxide	10 g.
Glyceryl Monostearate	3 g.
Peanut Oil, To make	100 g.

The oil is heated at 150° C. in a hot-air oven and cooled to normal temperature. The glyceryl monostearate is added and the mixture heated to 130° C. for 1 hour. After cooling to normal temperature the peroxide is incorporated with the mixture in a sterile mortar by sifting through sterile muslin under aseptic conditions. The suspension is stored in sterile glass-stoppered bottles capped with sterile parchment paper.

Allantoin Ointment

Petrolatum	45
Chlorthymol	0.15
Paraffin Wax	6
Stearic Acid	6
Allantoin	2
Lanolin, Hydrous, To make	100

Analgesic Salve

Menthol	15
Methyl Salicylate	15
Hydrous Lanolin	70

Melt together and mix well.

Acne Lotion

Formula No. 1

Alcohol (90%)	50.00
Ether	50.00
Rhodinol (ex Geranium)	0.10
Methyl Cinnamate	0.02
Phenylethyl Alcohol	0.10

No. 2

Menthol	0.10
Ti-Tree Oil	2.50
Alcohol	60.00
Propylene Glycol	24.70
Sodium Hexametaphosphate	5.00
Sulfonated Olive Oil	7.70
Water	250.00
Caraway Water	50.00
Witch Hazel Extract	100.00

Scabies Ointments

Formula No. 1

Benzyl Benzoate	200 cc.
Lanette Wax S or Aminostearin	10 g.
Water	800 cc.

No. 2

Benzyl Benzoate	200 cc.
Glyceryl Monostearate	25 g.
Water	1000 cc.

Scabies Sulphur Soap

Sulphur, Sublimed	18
Soap, Powdered	82

Mix well and work in a little mucilage of acacia and water. Roll out the mass and cut into 60 gr. tablets and dry quickly at room temperature.

Scabies Treatments

For infants and young children the following formula may be used:

Formula No. 1

Precipitated Sulfur	½ dr.
Zinc Oxide Powder	½ dr.
Petrolatum	1 oz.

In older children, the strengths of the powders may be increased. Six applications for three nights and three mornings are usually found satisfactory.

In adults, the following formula may be applied for three nights only, its action being rapid and effective, but not to be used in excess:

No. 2

Betanaphthol	½ dr.
Sublimed Sulfur	1 dr.
Peruvian Balsam and Petrolatum, of equal parts sufficient to make	1 oz.

No. 3

Five to fifteen per cent alcoholic solutions of Peruvian balsam or a 40% aqueous solution of sodium thiosulfate may also be used. Following the latter by 15 minutes, however, a 4% solution of hydrochloric acid must be applied and both solutions reapplied one hour later. An 18% bland sulfur soap paste is effective.

No. 4

Another mixture to be painted over the body with a soft brush consists of:

Benzyl Benzoate	50
Liniment of Soft Soap	65
Alcohol (90%)	30
Distilled Water	5

No. 5

A lotion of benzyl benzoate is composed of equal parts of soft soap, isopropyl alcohol and benzyl benzoate. About 150 grams of the lotion are required for each patient.

Head Lice Treatment

Formula No. 1

Use 15 per cent of Lethane (n butyl-carbitol-thiocyanate) in deodorized purified kerosene. One treatment applied to the head without a towel covering kills nits and lice immediately. The head is not to be shampooed for a few days.

No. 2

Lauryl Thiocyanate	25
Mineral Oil, Refined	50

Scurf Pomade

Salicylic Acid	30 gr.
Borax	15 gr.
Soft Paraffin Wax	1 oz.
Balsam of Peru	30 gr.
Cinnamon Oil	3 drops
Bergamot Oil	10 drops

Mix.

Treatment for "Swimming Pool Ear"

Fungous infection of the external ear may be prevented by instilling a few drops of

Boric Acid	2 g.
Mercury Bichloride (1:1000)	8 cc.
Alcohol, To make	30 cc.

after swimming.

Suppositories

Formula No. 1

Extract of Belladonna	0.05 g.
Cocoa Butter	0.40 g.
Sugar of Milk	2.00 g.

No. 2

White Wax	2.50 g.
Hard Paraffin Wax	5.00 g.
Wool Fat	5.00 g.
Spermaceti	35.00 g.
Liquid Paraffin, To make	100.00 g.

No. 3

Cocoa Butter	65
Stearic Acid	35

No. 4

Cocoa Butter	2
Chloral Hydrate	1
Lactose	1

Soothing Massage Oil

Eucalyptol	5 g.
Camphor	5 g.
Spermaceti	20 g.
Olive Oil or Almond Oil	75 g.

Liniment of Soft Soap (Tincture of Green Soap)

Soft Soap U.S.P.	650 g.
Cedar Leaf Oil	20 cc.
Alcohol, To make	1000 cc.

Healing Rubbing Alcohol Compound

To a pint of any good grade of rubbing alcohol add:

Phenol (88%)	5 drops
Glycerin	5 drops
Castor Oil	5 drops

Shake thoroughly, and use for tired muscles of the feet and limbs, and hands. This is especially good for chapped and rough hands.

Aromatic Liniment

Formula No. 1

Camphor	10 oz.
Menthol	4 oz.
Thyme Oil	4 oz.
Sassafras Oil	4 oz.
Tincture of Myrrh	6 oz.
Tincture of Capsicum	6 oz.
Chloroform	5 oz.
Alcohol	2 to 2½ pt.

No. 2

Methyl Salicylate	5.0
Cedar Leaf Oil	1.0
Sassafras Oil	1.0
Hemlock Oil	1.0
Chloroform	5.0
Tincture of Opium	5.0
Spirit of Camphor	5.0
Rectified Turpentine	3.0
Tincture of Capsicum	3.0
Alcohol	71.0

No. 3

Ground Capsicum	8.0
Soft Soap	5.0
Camphor	3.0
Eugenol	0.5
Methyl Salicylate	2.0

Sassafras Oil	1.0
Soap Liniment	30.0
Alcohol	50.5

Macerate the capsicum, soft soap, camphor and soap liniment for ten days. Filter and add to the rest of the ingredients.

Anodyne Liniment

Tincture of Aconite	12.5
Tincture of Iodine	12.5
Chloroform	12.5
Ammonia Water	12.5
Soap Liniment, To make	100.0

Analgesic Balm Liniment

Methyl Salicylate	30.0
Menthol	10.0
Saponin	0.5
Lanolin	32.5
Chloroform	5.0
Water	100.0

Stimulating Liniment

Tincture of Capsicum	26.5
Ammonia Water	26.5
Soap Liniment, To make	100.0

Athletic Rub

Methyl Salicylate	2.0
Witch Hazel	50.0
Alcohol, To make	100.0

Acetone Liniment

Methyl Salicylate	4.0
Peppermint Oil	0.5
Sassafras Oil	0.4
Chloroform	10.0
Soap Liniment, To make	100.0

Hay Fever Treatment

Formula No. 1

Potassium Chloride	7.5 g.
Distilled Water, To make	100 cc.

Take 1 teaspoonful three times daily.

No. 2

After years of suffering with this malady, doubtless from rag weed pollen, the contributor of this note has tried numerous remedies without avail, but has discovered that, by taking into the system extra amounts of common salt the flow of mucus is greatly reduced if not entirely arrested.

It seems that the loss of salt to the system during the summer months, as well as during the hay fever period, reduces the resistance to pollen attack. For an adult weighing between 130 and 150 pounds, about 1 gram (15 grains) of salt in a glass of water morning and evening, produces a marked effect.

Note: The Editor would like to hear, from those who try the above remedy, as to its efficacy.

Asthma Treatment

Methyl Atropine Nitrate, or Bromide	0.14
Papaverine	0.08
Sodium Nitrate	0.08
Adrenalin	0.05
Lactic Acid	2.50
Glycerin	10.00
Distilled Water, To make	100.00

Treatment of Hiccup

Such simple measures as holding the breath, pressing on the abdomen, a series of deep breaths or rebreathing from a paper bag, traction on the tongue, pressure on the ensiform cartilage, or inhalation of aromatic spirits of ammonia, will often relieve hiccup. In a stout person, after a heavy meal, the tendency to hiccup is relieved by bending back in the chair so as to give more space in the upper abdomen.

A simple carminative, such as:

Menthol	8 gr. (0.5 g.)
Compound Spirit Ammonia Spirit of Chloroform Tincture of Ginger	1 fl. oz. (32 cc.) of each

Two teaspoonfuls in water, taken as strong as possible, is of value, or a few drops of oil of cloves or cajuput on sugar.

If hiccup is due to gastric irritation and does not respond to simple measures, washing out the stomach and the use of an alkaline powder will be effective.

Therapeutical Pectin Dressing

Dissolve 2% flake pectin in distilled water and shake the solution occasionally for 24 hours. Then autoclave during 15 minutes at 20 lb. and keep sealed until used. Float the pectin-solution with a bulb syringe onto gauze until saturation. Cover the infected burn or wound with this dressing and remoisten every 4–8 hours. Burns are usually healed within 8 days.

Uses and Treatment of Pectin

Pectin has many valuable properties indicating its use for various therapeutic and cosmetic compositions. In every treatment of this material, special care has to be taken of clotting. Pectin should therefore never be contacted with water in dry state. It must preliminarily be moistened with small quantities of alcohol and then carefully comminuted in a mortar whilst gradually increasing the additions of water. The swollen mass is heated on a water bath. If an addition of glycerin is provided for, it should be comminuted with the alcohol moistened pectin in the mortar before adding water. Glycerin may be replaced by glycol, diethylene-glycol, etc.

If aqueous pectin solutions are added with alcohol, a jelly is formed when the amount of alcohol attains 40%; a much smaller quantity of alcohol is required to bring about gelation if the pectin solution already contains glycerin.

Pectin has the property of stimulating the papillary glands thus increasing the salivary excretion and is therefore suggested as an addition to toothpastes.

Pectin Jellies

Pectin jellies are prepared by moistening the required quantity of pectin with a small amount of alcohol, adding acidulated water and stirring the mixture in the cold. The rest of the water is then slowly added and the whole mass is heated on a water bath, whilst pouring in the rest of the alcohol. The mixture congeals after cooling. Water may separate out from the softer jellies on standing. This "bleeding" is prevented by the addition of small amounts of tragacanth or Galagum.

Formulas for pectin jellies, suitable for carriers of pharmaceutical and cosmetic compositions:

Formula No. 1

Stiff Cutting Jelly

Pectin	1.5
Water	65.0
Glycerin	2.5
Alcohol (92%)	30.0
Citric Acid	1.0

No. 2

Pectin	4
Glycerin	30
Water	80
Citric Acid	1

No. 3
Semi-Fluid Jelly

Pectin	0.9
Glycerin	18.0
Water	53.6
Alcohol (92%)	27.0
Citric Acid	0.5

No. 4

Pectin	1.3
Tincture of Benzoin	1.0
Glycerin	96.4
Citric Acid	0.5
Lime Oil	0.8

Sugarless Sweetener

Tragacanth, Powdered	2.50 g.
Alcohol, To suit	
Chloroform	0.25 cc.
Soluble Saccharin	0.25 g.
Distilled Water, To make	100.00 cc.

Enteric Coating Base

Monostearin is being used as a replacement for shellac for enteric coatings. Monostearin is insoluble in water, but readily emulsifies and breaks down in the system in contact with alkalies.

Enteric Coatings (Pill and Tablet)
Formula No. 1

Salol	20 g.
Shellac (White, Arsenic-Free)	30 g.
Ether	30 cc.
Absolute Alcohol, To make	100 cc.

No. 2

Salol	22.5 g.
Stearic Acid	2.5 g.
Alcoholic Solution of Shellac (10%)	10.0 cc.

Use at temperature at which shellac is fluid.

No. 3

Shellac	25 %
Ammonia Water	37½%
Alcohol	37½%

No. 4

Peptonized Keratin	1 g.
Spirit of Ammonia (with a few drops of water)	6 cc.

Apply warm.

No. 5

Mastic	25%
Methylpropylketone	75%

Sprinkle dried coating with magnesium stearate.

No. 6

Cetyl Alcohol	10 g.
Mastic	10 g.
Acetone	100 cc.

No. 7

Myristic acid, 68%; "Opal" wax, 25%; castor oil, 2%; cholesterol, 1% and sodium taurocholate, 4%. This coating prevents tablets from disintegrating in the stomach for at least 6 hours but is a coating which will dissolve in the intestines in from ½ to 2 hours. The fatty acids in the mixture are not digested by stomach juices but form diffusible suspensions with the bile salts in the intestines contrary to the old theory.

Binder for Yeast and Other Tablets

Yumidol (Sorbitol syrup) is being used very successfully as a binder for yeast and similar tablets. The necessary amount of Yumidol is weighed out and mixed with 50% of yeast. It is then screened through 40 mesh, and the remaining yeast added and screened again. It is aged 16–20 hours in air tight containers and screened again after aging prior to compression. No heating or drying operation is necessary. The specific amount of Yumidol required as binder can only be determined by specific formulation but, in general, about 2½% is used, based on the weight of the yeast or other material to be made into tablet form.

Polishing or Pharmaceutical Tablets, Pills, Etc.

Pharmaceutical tablets are generally polished by tumbling with a mixture of natural waxes in a suitable solvent such as acetone. This mixture of natural waxes can be replaced by glyceryl tristearate which will not spot or discolor colored or black coatings. In this connection anhydrous ethyl acetate or deodorized naphtha are suggested as solvents.

Tablet Making Lubricant

Add 1–4 oz. glyceryl tristearate per 100 lb. of granulation to prevent sticking to dies.

Medical Prophylactic (Preventive) Capsules

Mercuric Cyanide	0.075
Levigated Calomel	25.000
Anhydrous Lanolin	15.000
Yellow Soft Paraffin Wax	27.500
Mineral Oil, Refined	27.500
Acrawax	5.000

All ingredients and mixing apparatus must be dry. Mill between stone rollers. Pack into capsules of 36 grains.

Activated Carbon (Medicinal) Tablets

Activated Carbon	200
Tragacanth, Powdered	8
Sugar	195
Water	68

Synthetic (Medicinal) Coal Tar

Anthracene	1.10
Naphthalene	10.90
Phenanthrene	4.00
Carbazole	2.30
Picolene	0.58
Pyridin	0.58
Quinolin	0.58
Phenol	0.70
Cresol	0.75

Universal Poison Antidote

Activated Carbon	2
Magnesium Oxide	1
Tannic Acid	1

Use in teaspoonful doses in small glass of water.

Improved Aspirin
U. S. Patent 2,134,714

Aspirin	20
Saccharin	½
Starch	4
Tincture of Vanilla	4

This composition has an improved taste and increased rate of absorption when taken.

Anti Acid Stomach Tablets

Sugar	1 lb.
Calcium Carbonate	½ lb.
Peppermint Essence	

Mix sugar and carbonate with one and one half pints of water and boil until mass will set on a cold spoon. Cool and add one fourth ounce of essence of peppermint while stirring is still possible. Spread thin over a flat pan and cut into ¼ in. cubes.

Anthelmintic Tablet
U. S. Patent 2,282,290

Phenothiazine	80
Starch	8
Sodium Bicarbonate	5
Tartaric Acid	4
Sodium Choleate	2
Phenolphthalein	1

Cod Liver Oil Emulsion

Cod Liver Oil	50
Gum Arabic	3
Gum Tragacanth	¾
Irish Moss	1
Calcium Glycerophosphate	1
Glycerin	6
Water	38

Dissolve arabic in 9 parts of water while heating to 145° F. When dissolved, cool to 90° F. Disperse tragacanth in 9 parts of water while heating to 180° F. When well dispersed, cool to below 90° F., add to arabic solution. Make up a 10% Irish moss solution, boil thoroughly, then pass through homogenizer after straining and cool. Take aliquot equivalent to above percentage and add to gum solution. With stirring, add glycerin, then glycerophosphate and, finally, the oil in a small stream. Do not add oil so that "lakes" are formed on the surface; add just sufficient for the body of the mixture to take up.

When thoroughly mixed, homogenize at a total pressure of about 2500 lb., with the greater pressure, about 2000 lb., on the first valve.

After homogenization let set for a few hours to effect readjustment; then stir slightly and bottle.

Mineral Oil Emulsion (Laxative)

Heavy Mineral Oil	45.0
Acacia, Gum	2

Mix together

Water	51.9
Sodium Alginate	0.5
Sodium Benzoate	0.1
Citric Acid	0.5

Mix and dissolve.

Add to oil—acacia mixture. Mix thoroughly and pass through homogenizer at 2500 lb.

Pine Needle Oil Emulsion

Pine Needle Oil	12 oz.
Methyl Cellulose	2½ oz.
Water	6 qt.
Preservative (Moldex), To suit	
Color, To suit	

Cacao Butter Substitutes
Formula No. 1

Lanolin	6
Spermaceti	3
Olive Oil	1

No. 2

Paraffin Wax	1
Lanolin	1
Olive Oil	1

No. 3

Lanolin	9
Stearin	1
Triethanolamine Stearate	3

No. 4

Beeswax, Sunbleached	2.5
Paraffin Wax, Hard	5.0
Lanolin	5.0
Spermaceti	35.0
Mineral Oil U.S.P.	100.0

Nux Vomica Extract, Substitute for

Strychnine Hydrochloride	8 gr.
Burnt Sugar	12 min.
Alcohol (90%)	192 min.
Distilled Water, To make	1 fl. oz.

The preparation has the same

strychnine and alcoholic content as the official liquid extract and if necessary it can be diluted satisfactorily with alcohol to produce a substitute for the tincture. If it is prepared in bulk and kept for any length of time prior to use, filtration is usually necessary, although this does not have any appreciable effect upon its color.

Barium X-Ray Meal

Two types of x-ray films are required to rule out cancer or ulcer of the stomach: (1) A thick barium suspension, to completely fill out the stomach and outline the greater and lesser curvatures; and (2) a thin suspension to delineate the gastric rugae. By use of this new, inexpensive preparation (which can be made anywhere), both views are shown on one film, thus saving time and films.

Formula No. 1

Seven ounces of water are stirred thoroughly with 20 grams of a mixture containing 4 parts of barium sulfate, 1 part of gum acacia, 1 part cocoa, and 1 part granulated sugar, by volume. The meal is taken and films are made as usual; no compression is needed.

No. 2

Methyl Cellulose	350 gr.
Water	16 fl. oz.

Shake well and let stand overnight.

Rub down

Barium Sulphate	2.5 lb.
Water	5–10 lb.

Then add the above mucilage and

Sodium Benzoate	40 gr.

Add flavor to suit.

Fast Setting Plaster of Paris

Use a solution of:

Sodium Chloride	10 g.
Potassium Chloride	10 g.
Water (50° C.)	500 cc.

Quick Painless Remover for Surgical Dressings

Solution No. 1

Pure White (72%) Olive Oil Soap Flakes	100 g.
Alcohol (95%)	300 g.
Distilled Water, To make	1000 cc.

No. 2

Same diluted as follows: 1 part in 3 parts distilled water.

First soak the adhering dressing by gently pouring a small quantity of Solution 1. Leave on a few minutes. Rinse by pouring on Solution 2. The dressing comes off easily without pain.

Acné Salve

Cade Oil	5 g.
Balsam Peru	8 g.
White Petrolatum Jelly	87 g.

Mix and whip to make a light "mousse."

Athlete's Foot Salve

Colloidal Sulfur, Boric Acid	@ 12.5 g.
Almond Oil, Whipped Anhydrous Lanolin	@ 65.0 g.
Rose Water	10.0 g.

Sore Throat Tablets

Potassium Chlorate	10.000 g.
Sugar	80.000 g.
Borax	8.000 g.
Eucalyptol	0.125 g.
Thymol	0.500 g.
Menthol	0.025 g.

Macerate the capsicum, soft soap, amphor and soap liniment for ten ays. Filter and add to the rest of he ingredients.

Add glycerin and distilled water to make a paste. Mold and dry. For 200 tablets.

Waterproof Dressings and Bandages
British Patent 448,742

Dissolve 1 polymerized acrylic acid ethylester in 4 acetone, methylene chloride mixture. Add to the solution a small quantity of dibutylphthalate (up to 5%). Pour the solution on a hard surface bearing a layer of fibrous material and allow to evaporate while in contact with the fibrous material.

Adhesive Tape Remover

A composition for removing plasters and adhesive tape from the skin comprises a mixture of preferably pure carbon tetrachloride with not more than 40%, preferably not more than 20%, of white oil. The mixture may contain up to 10% of alcohol.

Hemostatic Preparation
British Patent 490,432

Mix 300 cc. of a 10% aqueous solution of a polyvinyl-alcohol with 50 cc. of an aqueous solution of iron chloride containing 80% $FeCl_3 \cdot 6H_2O$. Add 1000 cc. of a mixture of equal volumes of sterile 96% alcohol and water. Allow the solution thus obtained to flow slowly into 2700 cc. of 96% sterile alcohol whilst stirring vigorously. The yellowish precipitated powder is filtered off, washed with ethyl acetate and dried in vacuo at ordinary temperature.

Moth Proofing Solution

Magnesium Silico Fluoride	1.5
Butyl "Carbitol"	5.0
Santomerse S	0.5
Water, To make	100.0
Adjust to pH	6.5

Use 3% by weight on woolen goods.

Mosquito Repellent
Formula No. 1

Castor Oil	10 oz.
Thyme Oil	1 oz.
Pyrethrum Powder	1 oz.

No. 2

Citronella Oil	5 g.
Clove Oil	5 g.
Eucalyptus Oil	5 g.
Alcohol	255 g.

No. 3

Dihydropyrone	5 cc.
Alcohol	95 cc.

No. 4

Phenol	2
Pennyroyal Oil	4
Spirits of Camphor	6
Tar Oil	6
Lard Oil	12

No. 5

Citronella Oil	1 fl. oz.
Spirits of Camphor	1 fl. oz.
Cedar Oil	½ fl. oz.

No. 6
U. S. Patent 2,293,255

Butyl "Carbitol" Acetate	50
Ethyl "Carbitol"	15
Corn Oil	7
Alcohol	28

Remedy for Insect Stings
Formula No. 1

Dilute ammonia water or a paste made with Sodium Bicarbonate, followed with Zinc Oxide, Lime Water, or Calamine Lotion.

No. 2

Rub bites briskly with absorbent cotton saturated with

Chloroform	100
Diglycol Laurate	2

Marked relief is gotten, even many hours after the bites have been received.

Silverfish Insecticide

Oatmeal (Ground to Flour)	100
White Arsenic	8
Granulated Sugar	5
Salt	2½

Mix together dry, the oatmeal, white arsenic, sugar, and salt. Moisten the mass and mix thoroughly to bind the substances together. Then thoroughly dry the bait to prevent mold, and work it into small bits so that it can be scattered easily.

One level tablespoon of sodium fluoride powder can be substituted for the white arsenic in the above formula. If the substitution is made, simply mix the materials thoroughly, but do not add moisture.

Anti-Midge Preparation

To avoid discomfort from midges or gnats when working or walking in the open, the following glycerin-containing preparation may be applied to the exposed parts of the body as a preventive measure:

Glycerin	1 dr.
Tincture of Wormwood	3 dr.
Thymol	1 gr.
Cologne Water, To make	2 oz.

Those who have been bitten by these or other insects, will find that the preparation given below offers a means of soothing the bites:

Phenol	15 gr.
Glycerin	½ oz.
Rose Water, To make	4 oz.

Perfume for Hypochlorite Sprays

Formula No. 1

Synthetic Neroli	1.0 g.
Cetyl Alcohol	1.5 g.
Carbon Tetrachloride	10.0 cc.
Sulphonated Castor Oil	15.0 cc.
Water, To make	100.0 cc.

No. 2

Synthetic Musk	1.0 g.
Cetyl Alcohol	1.5 g.
Carbon Tetrachloride	10.0 cc.
Sulphonated Castor Oil	15.0 cc.
Water, To make	100.0 cc.

One cc. of either 1 or 2 is added to 100 cc. of 10% Sodium Hypochlorite Solution.

Cesspool Deodorant and Germicide

Formula No. 1

Dowacide A (Antiseptic)	17
Aluminum Chloride	10
Paraformaldehyde	2
Hydrated Lime	66
Pine Oil	5

No. 2

Aluminum Chloride	50	A
Formaldehyde	100	A
Water	300	A
Pine Oil	50	B
Sorbitol Monolaurate	50	B

Mix together with vigorous stirring.

Germicidal Oil

Mineral Oil	100
Pine Oil	5
Parachlormetaxylenol	2
National Oil Brown (Dye)	2

Saponified Cresol Disinfectant

Water	200 cc.
Cresol	500 cc.
Sodium Oleate	240 g.

Heat with stirring to 65° C. until dissolved. Cool and add

Water, To make	1 l.

Mix until uniform.

Disinfectant and Bleach Powder
U. S. Patent 2,121,501

Calcium Hypochlorite	15–20
Tartaric Acid	3–10
Sodium Chloride	45–80
Sodium Acetate, Anhydrous	2–10
Sodium Chromate	0.1–1
Borax	0–22

Disinfectant Emulsion
U. S. Patent 2,289,476

Orthophenyl Phenol	10–40%
Terpene Ether	15%
Sulphonated Castor Oil, To make	100%

Disinfecting Infected Shoes

Fungi-infected shoes can be inexpensively and effectively sterilized by spraying a 37% solution of formaldehyde into the shoes on three nights in succession.

Non-Alcoholic Iodine Antiseptic

Iodine	1.000
Calcium Iodide	0.044
Potassium Iodide	0.048
Sodium Iodide	1.104

This is more penetrating and less irritating than tincture of iodine. A 3% solution of the above in water is recommended.

Mild Tincture of Iodine U.S.P.

Iodine	20 g.
Sodium Iodide	24 g.
Diluted Alcohol, To make	1000 cc.

Antiformin Type Antiseptic

Anhydrous Soda Ash	100 g.
Distilled Water, To yield	1000 cc.

Saturate this solution with chlorine gas, under the pressure available with the compressed gas in commercial cylinders.

Antiseptic

Alcohol (95%)	525.00 cc.
Acetone	100.00 cc.
Cresol, U.S.P.	5.00 cc.
Mercuric Chloride	0.70 g.
Eosin Y	0.60 g.
Acid Fuchsin	0.08 g.
Water, To make	1000.00 cc.

Alcohol and acetone alone will kill 96 per cent of bacteria on the skin surface, and acetone is also a fat solvent. Cresol (tricresol) and mercuric chloride kill 98 per cent of such bacteria. The eosin and fuchsin are added as dyes, so that the antisepticized surface will be plainly visible. Such an antiseptic is quick-drying, quick-acting, is not injurious to the skin (unless the patient is allowed to lie in a puddle of it), to the operating room personnel, or to linens, and is capable of sustained action. If tax-free alcohol cannot be obtained, ordinary rubbing alcohol may be used (the pharmacist will recompute the formula to compensate for the weaker strength of the alcohol). It is an interesting fact that *50 to 70 per cent alcohol will kill organisms more quickly than will the 90 per cent strength.*

Embalming Fluid

Sodium Nitrate	1 lb.	A
Methanol	1 pt.	A
Water	5 pt.	A
Sodium Acetate	2 lb.	B
Methanol	1 pt.	B
Glycerin	5 pt.	B
Water	13 pt.	B

Formaldehyde Counter-Odor

Patchouli Oil	10
Sassafras Oil	50
Methyl Salicylate	50
Eucalyptus Oil	50

Ointment for Embalmers' Eczema

Salicylic Acid	2.0 dr.
Phenol	1.5 dr.
Zinc Ointment	4.0 oz.

Maggot Repellent for Embalming

Kerosene	4
Turpentine	2
Carbon Tetrachloride	1
Methyl Salicylate	1

Use on cotton and plug the nostrils.

Iodized Oils

This method is applicable to all vegetable and animal oils, which contain unsaturated fatty acids.

At first the free fatty acids are removed, as they are very detrimental. This is accomplished by taking 1000 grams of the oil, vegetable or animal oil with about 2000 cc. of denatured alcohol and then N/2 sodium-hydroxide and a few drops of phenolphthalein. The addition of the N/2 sodium hydroxide is continued until it imparts a pink color to the alcohol. This is now mechanically stirred. Should this pink color disappear, more N/2 sodium hydroxide is added. The amount of sodium hydroxide will depend on the degree of acidity of the vegetable or animal oil.

The oil is then allowed to settle, the alcohol is syphoned off and two liters of water is added and stirred for a few minutes. This frees the oil from the alcohol. The oil is allowed to separate and floats on top. The water is now syphoned off by means of suction. Another two liter portion is now added to the oil and stirred again for a few minutes. It is allowed to settle, and the water is again syphoned off. To this purified oil, any oil solvent, such as chloroform, ethylenedichloride, carbon tetrachloride, or ether may be added.

The second step comprises the preparation of the iodizing and chlorinating agent in aqueous solution. This is prepared in the following way:

In 720 cc. of distilled water 357 grams of potassium iodide is dissolved and 239 grams of potassium iodate is added to this cold solution; 720 cc. of concentrated hydrochloric acid U.S.P. is added with mechanical, vigorous stirring in one rapid stream. A dark precipitate of iodine monochloride is first formed, which on continued stirring will go in solution. The stirring should be continued until the solution has a dark reddish color.

This iodizing and chlorinating solution is now cooled to about 15 degrees C. and is now ready for use.

The third step comprises the actual introduction of iodine and chlorine into the purified oil, by means of this aqueous solution of iodine monochloride. The iodizing solution is mechanically stirred and well cooled with ice and water. Now the purified oil, dissolved in an organic solvent is added in a thin stream to the iodizing solution with vigorous stirring. Care should be taken that the temperature remains between 20–25° C. After all of the oil has been added, stirring for about ten minutes will completely iodize the oil. The remaining iodine monochloride solution has lost all of its color and is faintly yellow or almost water white.

Upon allowing the iodized oil to stand, it will settle to the bottom, as the specific gravity is greater than that of water, and the supernatant aqueous portion is syphoned off and at this stage, some more solvent may be added to facilitate the washing of the oil. This oil solution is now washed in a wide mouth bottle with running water, until all of the hydrochloric acid is removed. This can be established by the addition of a few drops of bromthymol blue, which will give a yellow color if mineral acid is still present, but a blue color if all of the acid has been removed. This usually takes several hours. The fourth step:

Now the oil solution is separated from water as much as possible, then 750 cc. of denatured alcohol is added, and stirred mechanically. To this is then added alcoholic sodium hydroxide, until a pink color remains with phenolphthalein. When this is the case, it is allowed to settle. The alcohol is syphoned off. Then the iodized oil is washed with water, until the pink color to phenolphthalein disappears and the wash water has lost its soapy appearance.

Then the oily layer is separated from the water, and is dried with sodium sulfate, treated with activated charcoal, stirred and filtered. The solution should be brilliantly clear and of a light yellow color. Then it is placed in a round flask and the organic solvent removed by distillation in a vacuum, at first without heat, then later it is placed in a controlled water bath and the temperature is maintained at about 55 degrees C. After about six to eight hours distillation, all of the solvent is removed and a nice yellow viscous iodized oil, remains behind, which contains about 27.5% of iodine and 7.5% chlorine.

CALCIUM IODOXYBENZOIC ACID

4800 grams of anthranilic acid and 13,200 cc. of conc. hydrochloric acid to which 6000 grams of chopped ice has been added, are mixed, then 6000 grams of water is added and the entire mixture is stirred mechanically and cooled very thoroughly with ice and salt. When the temperature has gone to about 5 degrees C., a solution of 2540 grams of sodium nitrite dissolved in 4200 cc. of water, is added slowly. Care should be taken that the temperature does not rise too much, so that no nitric oxide is formed. The solution is tested with potassium iodide starch paper, which indicates the presence of free sodium nitrite by a blue iodine starch reaction. So long as this test is negative, the addition of the sodium nitrite solution is continued. When a blue iodine starch color appears on the paper, which on continued stirring of the solution still is present, the diazotization is finished. About 150 grams of talcum is added and the solution filtered. To the clear diazo solution, 7270 grams of potassium iodide dissolved in 9600 grams of water, is added in a slow stream to the diazo solution under stirring. When all the iodide solution is added, careful and slow heating is started by means of steam, heating the water outside the container of the diazo and iodide solution. The reaction, which has already begun in the cold, proceeds and evolution of nitrogen gas occurs and at first the benzoic diazo iodide is formed as a black mass, which as the heating proceeds goes over into the ortho iodobenzoic acid. The heating is continued until the outside water is boiling and kept boiling for several hours. Then it is allowed to cool overnight, filtered and washed with water and then emulsified with a solution of 500 grams of sodium bisulfite in 5 liters of water, which removes all the free iodine. Then the iodobenzoic acid is placed on a tray and dried. Approximate yield 6085 grams.

6085 grams of iodobenzoic acid, finely powdered, is suspended in 7,100 cc. of conc. sulphuric acid and 50,700 cc. of water, heated to about 85 degrees C. and now 7100 grams of potassium bromate is added in portions of 300-500 grams under constant stirring. This process is continued for several hours and the bromine vapors, which are formed are carried off by means of an exhaust fan. After about six hours of heating it is allowed to cool overnight and filtered and washed with cold water and allowed to dry. It is important that the crude iodoxybenzoic acid should be completely dry. Yield about 5500–6000 of crude iodoxybenzoic acid.

These 6000 grams of iodoxybenzoic acid are extracted with 5000 cc. of denatured alcohol, which removes the coloring matter and the unoxidized iodobenzoic acid, filtered off, which will give about 5000 grams of fairly pure iodoxybenzoic acid, which is dried and finely powdered then suspended in 14,000 cc. of water, heated to about 40 degrees C. Twenty per cent ammonia is added slowly, about 2400 cc. of it are needed, then 200 grams of Darco (decolorizing carbon) is added and the solution is filtered.

To the clear, yellow solution of ammonium-iodoxybenzoate, a solution of 1332 grams of calcium chloride U.S.P. in 2000 cc. of water is rapidly added. All of a sudden the calcium iodoxybenzoate comes out as a white solid cake, which is allowed to cool, by standing overnight, then filtered off and emulsified with about 5,000–6,000 cc. of ice cold water, filtered and placed on trays and allowed to dry on exposure to air. White hard pieces of calcium iodoxybenzoate, yield about 5 kilos.

GOLD SODIUM THIOSULFATE
(*Pharmaceutical*)

Thirty grams of gold chloride (light) are dissolved in 25 cc. of water and put into a dropping funnel. 105 grams of sodium thiosulfate are dissolved in 120 cc. of water. The solution is filtered and agitated by means of a small electric stirrer. The gold chloride is added gradually a few drops at a time, then stirred until the yellow color disappears, when a few more drops of gold chloride are added. The solution assumes a milky white appearance, due to the precipitated sulphur and colloidal sulphur and gives a strong odor of sulphur dioxide. The addition of gold chloride lasts for about half an hour and the stirring is continued for some time. Then this milky liquid is filtered once to free it from the precipitated sulphur. After filtration however, it still possesses a milky white appearance due to the colloidal finely suspended sulphur. This colloidal sulphur is removed later on. This milky white fluid is then stirred and while being stirred, receives an addition of two and half its volume of 95% of alcohol. A white precipitate is obtained after a little stirring. The addition of 95% alcohol causes a precipitation of the gold sodium thiosulfate, which is very readily soluble in water, but insoluble in alcohol. At the same time the colloidal sulfur is precipitated by the alcohol. This white precipitate is then dissolved in as little water as possible, the solution filtered (and in this way freed from the colloidal sulfur) and the clear filtrate reprecipitated with 95% alcohol; the white crystalline precipitate can be redissolved and reprecipitated as before to obtain a perfectly pure compound.

Preparation of Fatty Acids of Cod Liver Oil

One gallon of U.S.P. Cod Liver Oil, is placed in a 5 gallon Pyrex flask. To this is added one gallon and a half of denatured alcohol No. 1, and to this is added a solution of potassium hydroxide in water made in the following way: 2 lb. of potassium hydroxide is dissolved in 2 pints of water, and this saturated solution is added, while still hot, to the mixture of cod liver oil and alcohol. This is then placed in a boiling water bath and heated on a reflux condenser until saponification is completed. This can be tested by adding a small amount to water and if a clear solution occurs, saponification is complete.

To this is now added hydrochloric acid (concentrated) until Congo paper turns red, then one gallon of ethylene dichloride is added and water is added. The ethylene dichloride brings the liberated acids in solution, and as the specific gravity of ethylene dichloride is high, the fatty acids in ethylene dichloride will settle down and it then can be washed by a constant stream of tap water, until all of the hydrochloric acid is removed. This is checked with silver nitrate solution, allowing for the chloride test as contained in ordinary tap water as control.

The ethylene dichloride solution of the fatty acids are then separated from the water by syphoning, and dried with anhydrous sodium sulfate and filtered. The sparkling clear solution of fatty acids in ethylene dichloride is now placed in a round bottom flask and distilled, in vacuo, until no ethylene dichloride distills over. The residue consists of very pure fatty acids of cod liver oil absolutely clear.

Chapter Five

EMULSIONS

Santowax Emulsions

Formula No. 1

Santowax M	50.00 g.
Stearic Acid	3.10 g.
Turpentine	50.00 g.
Triethanolamine	0.94 g.
Water	400.00 g.

Dissolve Santowax and stearic acid in turpentine, heating to about 85° C. Dissolve triethanolamine in 100 grams of soft water and heat to 85° C. Add the solution of triethanolamine to the hot turpentine solution with rapid mixing. Mix for about two minutes, then add 300 grams of soft water at room temperature. Continue to mix for ten minutes or more. This gives a smooth, fairly fluid, creamy-white suspension. Follow the same procedure with the formula given below.

No. 2

Santowax M or Santowax 9523	50.00 g.
Turpentine	50.00 g.
Stearic Acid	2.00 g.
Triethanolamine	1.23 g.
Water	400.00 g.

Synthetic Wax Emulsion

Acrawax	10
Ethanolamine Oleate	1
Water	84

Heat to 95–100° C. with vigorous mixing. Continue mixing until temperature drops to 75° C. Then add

Methyl Ethyl Ketone	5

Paraffin Wax Emulsion

Diglycol Stearate	10	A
Paraffin Wax	40	A
Water	250	B
Granulated Sugar	0.5	B
Ammonia (26° Bé.)	1	B

Melt A and add B (warmed to 40° C.) with vigorous stirring.

Candelilla Wax Emulsions

Formula No. 1

Candelilla Wax	60
Myristic Acid	15
Emulsifier S489	19
Water, To make	375

This gives a very translucent, almost clear emulsion which dries to a transparent film.

No. 2

Candelilla Wax	60
Myristic Acid	13
Morpholine	10
Water, To make	375

This produces a very thin emulsion which is translucent and dries to a transparent film.

No. 3

Refined Candelilla Wax	60
Myristic Acid	13
Emulsifier S489	10
Water, To make	375

This gives a dark but very translucent, thin emulsion with high wax content, which dries to a transparent film.

No. 4

Candelilla Wax	60
Oleic Acid	16
Morpholine	8
Water, To make	375

Very thin, translucent and dries to a transparent film.

No. 5

Candelilla Wax	40
Abietic	8
Isopropanol	10
Ammonia	15
Borax	1
Water, To make	400

This gives a thin, amber emulsion, translucent, which dries to a transparent film.

No. 6

Candelilla Wax	30
Flexo Wax C	10
Oleic Acid	8
Ammonia	15
Butanol	10
Water, To make	400

This gives a translucent, thin emulsion which dries to a transparent film.

No. 7

Candelilla Wax	20
Flexo Wax C	10
Oleic Acid	11
Morpholine	4
Water, To make	225

Carnauba Wax Emulsions

Formula No. 1

Carnauba Wax	60
Myristic Acid	16
Oleic Acid	4
Emulsifier S489	19
Water, To make	375

This gives a very clear, light colored emulsion, almost transparent and which dries to a transparent film.

No. 2

Carnauba Wax	60
Oleic Acid	16
Morpholine	11.5
Water, To make	375

This makes a very light colored, translucent emulsion of low viscosity which dries to a transparent film with good gloss.

No. 3

Carnauba Wax	60
Linseed Oil Fatty Acids	13
Emulsifier S489	17
Water, To make	375

This gives a very thin, almost transparent emulsion which dries to a transparent film with good brightness.

Fruit Coating Wax Emulsions

Formula No. 1

Paraffin Wax	168.0
Beeswax	42.0
Oleic Acid	22.0
Sodium Bicarbonate	6.6
Sodium Chloride	2.2
Water	599.2

It is necessary that the soap be formed simultaneously with the first stage of emulsification by the addition of the alkali, in a minimum amount of water, to a wax. These emulsions will readily coat fruits and vegetables with nonwaxy surfaces; but spreading agents are necessary when they are used for waxy fruits. Alkyl naphthalenesulfonates are suitable spreading agents and also inhibit breaking of the emulsions.

No. 2

U. S. Patent 2,153,487

Caustic Soda	6
Triethanolamine	20
Stearic Acid	42
Paraffin Wax	165

Carnauba Wax	55
Shellac	100
Water	2000

Dilute above before use.

No. 3

U. S. Patent 2,019,758

Paraffin Wax	553
Carnauba Wax	68
Cottonseed Oil	98
Oleic Acid	183
Triethanolamine	98
Water (Containing Soda Ash)	to suit

It is necessary that the soap be formed simultaneously with the first stage of emulsification by the addition of the alkali, in a minimum amount of water, to a mixture of fatty materials.

Agricultural Emulsion Sprays

U. S. Patent 2,258,833

Formula No. 1

Glyceryl Oleate	2
Oil Soluble Nicotine Compound	5
White Mineral Oil	93
Water, To suit	

No. 2

Glyceryl Oleate	1.5
Aluminum Oleate	2.0
Nicotine	2.0
Naphthenic Acid	3.1
White Mineral Oil	91.4
Water, To suit	

Plant Growth Regulation Emulsion

Indoleacetic Acid	0.3 mg.
Lanolin	5 0 g.
Soap Flakes	5.0 g.
Water	100 cc.

Melt lanolin and mix in indoleacetic acid; dissolve soap in water and mix vigorously.

Agar	0.25 g.

may be dissolved in the water (used for soap solution) if desired.

Dilute the above to 1000 cc. with water for use.

Acid Wax Emulsions

Polymerized Glycol Distearate	5
Candelilla Wax	20
Phosphoric Acid (85%)	5
Water (Boiling)	70

Another acid emulsion is made ot the following materials, in the proportions shown:

Polymerized Glycol Distearate	15
Monostearate of Glycerin	15
Paraffin Wax	6
Hydrochloric Acid	9
Water	55

Acid Oil Emulsions

Polymerized Glycol Dilaurate	15
Oleic Acid	5
Mineral Oil	30
Hydrochloric Acid	6
Water	44

Another mineral oil emulsion is made of the following formula:

Polymerized Glycol Dilaurate	15
Oleic Acid	5
Mineral Oil	30
Sodium Chloride (10% Solution)	50

In the above formula the sodium chloride solution may be replaced by a solution of tartaric acid, say, of concentration 10%. Also the mineral oil may be replaced by other oily materials as, for example, by toluene, pine oil, a fatty vegetable oil, kerosene or naphtha, the proportion of water being decreased somewhat if necessary to give desired stability.

Emulsions With Acids, Salts and Hard Water

These emulsions are made by dissolving the EMULGOR-A (Glaurin and Oleic Acid where necessary) in the oil to be emulsified, by heating to about 50° C. The salt, acid, etc. are dissolved in the water which is added at about 50° C. to the oil, EMULGOR-A mixture and stirred rapidly. All the emulsions are fluid with the exception of the paraffin wax emulsion, which is a paste.

Material to Be Emulsified		Electrolytes		Water	Emulgor-A	Oleic Acid	Glaurin
Amyl Acetate	50.00	Boric Acid	1.8	38.2	10		
Cottonseed Oil	41.15	300 parts per million calcium carbonate contained in water used.		41.15	11.8	5.9	
Mineral Oil	30.00	Acetic Acid (Glacial)	5	45	15	5	
Mineral Oil	30.00	Aluminum Chloride (14% solution)	5	45	15	5	
Mineral Oil	40.00	Citric Acid	4	36	10	10	
Mineral Oil	31.6	Formic Acid	2.3	39.85	10.5	5.25	10.5
Mineral Oil	30.0	Hydrochloric Acid (22° Bé)	6	44	10		10
Mineral Oil	30.0	Oxyquinoline Sulphate	0.2	49.8	15	5	
Mineral Oil	31.6	Phosphoric Acid (85%)	2.4	39.75	10.5	5.25	10.5
Mineral Oil	30.0	Sodium Chloride	2.5	47.5	20		
Mineral Oil	31.6	Sulphuric Acid (Conc.)	2.15	40	10.5	5.25	10.5
Mineral Oil	40.0	Tartaric Acid	4.5	40.5	10	5	5
Paraffin Wax	30.0	Aluminum Acetate (28% solution)	30	20	10	5	5
Pine Oil	32.0	Sodium Chloride	2.35	44.45	21.2		
Pine Oil	31.3	Hydrochloric Acid (22° Bé)	6.25	41.6	18.75	2.1	
Toluol	33.3	Sodium Chloride	2.22	42.25	17.75	4.45	
Toluol	30.0	Hydrochloric Acid (22° Bé)	6	44	20		

Emulsion Containing Lime

Polymerized Glycol Stearo-Oleate	10
Mineral Oil	20
Paraffin Wax	20
Lime	5
Stearic Acid	12
Water, To make	100

The acid is melted in the oil at a temperature of about 65° C. The oil-acid mixture is poured into the water solution with agitation and the emulsion is stirred occasionally while it cools.

Sulfathiazole Emulsion

Sulfathiazole 5%, triethanolamine 2%, distilled water 24%, white beeswax 5% and paraffin 64%. The sulfathiazole is stirred into a heated mixture of the triethanolamine and water, to which is added the melted beeswax-paraffin mixture. Vigorous stirring produces the proper product.

Anise Oil Emulsion

Anise Oil	10
Tincture of Quillaia	5
Water	100

Castor Oil Emulsion

Sodium Hydroxide	9.8 gr.
Water	25 oz.
Castor Oil	6 lb. 5 oz.
Sodium Benzoate	100.5 gr.
Triethanolamine	288 min.

Run through colloid mill after flavoring mix with lemon and vanilla and sweetening with saccharine to suit.

Kerosene-Pine Oil Emulsion

Kerosene	40–60 %
Pine Oil	0–20 %
Aerosol OT	5– 6.5%
Oleic Acid	6– 6.5%
Water	35–50 %

Mineral Oil Emulsion

White Mineral Oil	43.0
Stearic Acid	5.1
Aminomethylpropanediol	1.9
Water	50.0

Pectin in Emulsions
(*Sesame Oil*)

Various emulsions may be prepared with pectin. For oil emulsions a pectin concentration of 3–5% is required. Vigorous stirring of an acidified 5% pectin solution with 20% sesame-oil yields a stable viscous emulsion which readily disperses when poured into water. If such emulsions are homogenized and prevented from desiccation, they will remain stable for years. Dispersed systems consisting of pectin solutions and paraffin-oils are excellent cosmetics and skin protectors. Pectin solutions are not sticky, quick drying and form a protective layer on the skin.

Copper Naphthenate Emulsion

1. Copper Naphthenate	10
2. Ammonium Linoleate (B585)	20
3. Water	70

Melt 1 and stir in hot mixture of 2 and 3.

Benzyl Benzoate Emulsion

Methyl Cellulose (Medium Viscosity)	1
Benzyl Benzoate	25
Water	74

Toluol Emulsions

Polymerized Glycol Dilaurate	20
Toluol	30
Hydrochloric Acid	6
Water	44

Another toluol formula is the following:

Polymerized Glycol Dilaurate	10.2
Diglycol Monolaurate	6.7
Toluol	44.5
Water	38.6

Acid Pine Oil Emulsion

Polymerized Glycol Dilaurate	10
Pine Oil	40
Diglycol Monoleate	7
Oxalic Acid	2
Water	41

Benzene Emulsion

Glyceryl Mono Oleo Laurate	14
Benzene	88
Water	50

Flexalyn (Synthetic Resin) Emulsions

	Formula No. 1	No. 2	No. 3
Flexalyn (80% solids in Xylene)	41.25	47.5	47.22
Duponol ME Dry	0.33		
Triethanolamine		0.67	
Potassium Hydroxide			0.27
Oleic Acid		1.33	1.13
Sulfated Castor Oil			0.88
Water	58.42	50.5	50.5

Formula No. 1 requires the use of a colloid mill. The Duponol ME is added with stirring to warm water (40–50° C.). When solution is complete, Flexalyn solution is added with continued stirring. The spontaneous semi-emulsion thus formed is passed through a colloid mill, forming a stable low-viscosity emulsion with an average particle size of 85% to 100% under 5 microns, and a pH of 6.7 to 7.5.

Spontaneous emulsion of Flexalyn solution may be prepared with either Formula No. 2 or No. 3 above, providing the following procedure is followed. With either formula, the oleic acid required is carefully stirred into the Flexalyn solution; ½ the amount of water required is mixed with either the triethanolamine in Formula No. 2, or potassium hydroxide and sulfated castor oil in Formula No. 3. The oleic acid-Flexalyn solution is then added slowly to this water phase with vigorous agitation (highspeed propeller-type stirrers have been found satisfactory.) This produces a spontaneous emulsion, which is allowed to stir for approximately 30–45 minutes, and the remaining water slowly added. After addition of the water is completed, the emulsion is allowed to stir for about ½ hour before using.

Vinyl Resin Emulsion
U. S. Patent 2,238,956

Copolymerized Vinyl Chloride and Acetate	25
Methyl n-Amyl Ketone	45
Toluene	30
Water	33 1/3
Alkanol B	1 1/5

Dissolve resin in solvents and mix vigorously with water solution of Alkanol B. Then pass through a colloid mill.

Cumar Emulsion

Cumar	44 lb.
Water	36 lb.
Soap "A"	2 lb.

Soap "A"

Rosin	4 lb.
Caustic Potash (31° Bé.)	200 cc.
Water	500 cc.
Alcohol	200 cc.

To make Soap "A," melt rosin, heat to 180° F. Then combine caustic potash, water and alcohol, heat to 160° F. and add to the melted rosin. Stir.

To make cumar emulsion, combine Cumar, water and Soap "A," heat mixture to 200° and run through colloid mill or high speed agitator.

Natural Resin Emulsion

Elemi	70.3 lb.
Batavia Dammar	7.7 lb.
Oleic Acid	11.0 lb.
Triethanolamine	4.0 lb.
Water	210.0 lb.

Melt the Elemi and add the Dammar. Keep the mixture at about 125° C. until it is homogeneous. Put the mixture in a boiling water bath. Add the oleic acid and then the triethanolamine, making use of vigorous stirring as additions are made. Heat the water to boiling and add it slowly to the mixture, still using vigorous agitation. Continue the agitation until the emulsion has cooled.

A milky white emulsion is formed which yields a clear, tacky film.

Resin-Wax Emulsions
Formula No. 1

Batavia Dammar	12.5 lb.
Beeswax	12.5 lb.
Morpholine	1.0 lb.
Stearic Acid	1.5 lb.
Water	100 lb.

Melt the beeswax and add the Dammar, stirring until the mixture is homogeneous. Stir into this mixture the morpholine and the stearic acid. Add this mixture slowly with stirring to the water which has been heated to about 85–95° C.

This emulsion remains stable and is capable of infinite dilution with water without breaking.

No. 2

Batavia Dammar	25 lb.
Paraffin Wax	25 lb.
Oleic Acid	5 lb.
Morpholine	2.5 lb.
Water	250 lb.

Melt the wax and sift in the Dammar. When all is melted, add the oleic acid and stir until the mixture is homogeneous. Do the same with the morpholine. Keeping the mixture hot and continuing the stirring, add the water slowly. The water should previously be heated to about 85–95° C.

No. 3

Batavia Dammar	37.5 lb.
Beeswax	37.5 lb.
Oleic Acid	8 lb.
Morpholine	4 lb.
Water	375 lb.

Melt the wax. Sift in the Dammar. When all is melted, add the oleic acid and stir until the mixture is homogeneous. Do the same with the morpholine. Keeping the mixture hot and continuing the stirring, add the water slowly. The water should be previously heated to 85–95° C.

Plasticized Rosin Emulsion

Rosin	17
Theop (Plasticizer)	3
Caustic Potash (45%)	2
Water	78

Cleaning Emulsion Bases

Duponol LS Paste or Duponol ES Paste	47.6
Hexalin	4.8
Tetralin	47.6

Prepare by adding the Duponol LS Paste to the previously mixed solvents.

Duponol LS Paste	15.0
Hexalin	2.0
Pine Oil	3.0
Tetralin	70.0
Water	10.0

Prepare by mixing together at room temperature the Duponol, Hexalin and Pine Oil. Stir until a clear mixture is formed. Continue stirring and add the Tetralin slowly. After the Tetralin has been added and the mixture thoroughly agitated, add the water.

Die Lubricating Emulsion

Mineral Oil	4
Aerosol OT (100%)	1
Glycostearin	1

Heat together to 100–120° C., then cool to 80° C. and then add slowly with stirring to 100 water (heated to 80° C.).

Waterproofing Emulsion
British Patent 540,650

Aqueous basic aluminum formate (aluminum oxide 20, formic acid 28%) (20) is diluted with hot water (40), melted wax (acid value 19, 10 kg.) added at 70°, and the mixture stirred to give a stable emulsion.

Emulsifying Base

Aerosol OT (100%)	40
Oleic Acid	42
Triethanolamine	18

The Aerosol OT is dissolved in the oleic acid by heating to 50° to 60° C. The mixture is cooled to room temperature and the triethanolamine added. 5% of this emulsifier can be mixed with 95% of oil and to it added 30–50% water, whereupon a thick creamy paste will be obtained, which can be further diluted with additional water.

Soluble Oils
Formula No. 1

Mineral Oil	82.9
Caustic Potash (45%)	0.1
Diethylene Glycol	2.0
Ultranate, Refined	15.0

No. 2
U. S. Patent 2,230,556

Potash Soap	30
Olein	7
Cyclohexanol	2
Paraffin Oil	32

Lacquer Emulsions

Combination of a lacquer with an emulsifying mixture gives some degree of emulsification and can be realized by any customary method. Most useful concentration ratio is 2½ lacquer to 1 of water phase.

Stability generally varies with particle size—all other variables being constant.

Poorest emulsions are made by rapid stirring with a high speed mixer. They are greatly improved by a stator-rotor type sheering action, of a colloid mill. The most stable emulsions are usually produced by high pressure extrusion, i.e. in an homogenizer. For easily dispersed systems, an homogenizer alone is sufficient. More difficultly dispersable systems must be predispersed by other types of equipment (colloid mill) before passing through homogenizer.

An emulsion with many large

agglomerate particles can in general be expected to have a short life. The dispersed droplets should be small and uniform.

Almost any good emulsifying agent, and this includes soap will form some type of an emulsion with a lacquer. Useful stabilities and this means three months or better, can only be realized by a careful selection of the emulsifier. One of the best found contains 0.5% Dupanol, 1.0% sulfonated castor oil eithe alone or mixed with other emulsifier or lacquer solvents.

Lacquers rich in resins are mor easily dispersed than those that ar nitrocellulose rich, and as a rul have a longer life. (Table A No. and 5.)

Lacquer emulsions have hig flash points and are usually broke by freezing. Lacquers of very hig viscosity are difficult to disperse.

Naphthenic Miscible Oils (U. S. Patent 2,289,536)

Formula No.	1	2	3	4	5	6	7	8	9	10
1. Potassium naphthenate from—										
Heavy gas oil	65	48								
Light Lube oil			37	37	37	23	20	30		
Heavy lube oil									13	30
2. Light lube oil	20	35	50	45	50	58	62	55	69	64
3. Common solvents										
Group 1—Water soluble										
Ethylene glycol										3
Diethylene Glycol	5	5	5	5	5	5	5	5	5	
Group 2—Oil soluble										
Sub-group A										
Pine Oil		10		10						
Alpha terpineol	5		5			5	5	10	5	
Octyl alcohol					5					
Sub-group B										
Rosin	5					5			5	3
Cresylic acid							5			
4. Water		2	3	3	3	4	3		3	

Nitrocellulose Emulsions

Table A—Lacquer Base Formulae (clear)

Formula No.	1	2	3	4	5
Nitro Cellulose (Dry Weight)	30	25	30	30	40.0
Castor Oil	30	—	—	—	—
Triethyl Citrate	—	20	—	—	5.0
Tricresyl Phosphate	—	—	12	—	—
Dibutyl Phthalate	—	—	—	13	—
Resin (Alkyd)	—	10	15	15	—
Butyl Acetate	25	25	25	22	30.0
Xylene	—	5	—	—	7.5
Butanol	15	15	18	20	17.5

No. 6

Pigmented Nitrocellulose (Dry Weight)	20%
Resin (Alkyd)	8
Castor Oil (Blown)	6
Diamyl Phthalate	6
Carbon Black	2
Xylene	14.4
Octyl Acetate	10
Butyl Acetate	25
Alcohol (Present in Nitro Cellulose)	8.6

Table A

Any of these formulae can be emulsified without much difficulty using any of the emulsifiers of Table B. The difference lies in the properties of the dried films.

They are all formulated electrolyte free—small amounts can be included and dispersed. The presence of gelatin or casein in the water phase will help disperse a certain concentration of electrolytes.

Lacquer Emulsifiers

Table B

Formula No. 1

Duponol	0.5%
Sulfonated Castor Oil	1.0
Water	98.5

No. 2

Duponol	0.5%
Sulfonated Castor Oil	1.0
Acetone	10.0
Water	88.5

No. 3

Duponol	0.5%
Sulfonated Castor Oil	1.0
Casein	0.5
Water	98.0

No. 4

Duponol	0.5%
Sulfonated Castor Oil	1.0
Gelatin	2.0
Water	96.5

No. 5

Duponol	0.5%
Sulfonated Castor Oil	1.0
Casein	0.5
Acetone	10.0
Water	88.0

No. 6

Sodium Oleate	6.0%
Water	94.0

No. 7

Duponol	0.5%
Sulfonated Castor Oil	1.0
Methyl Cellulose	0.3
Water	98.2

Table B illustrates many of the various possible combinations of emulsifiers.

No. 1.—The emulsion stability varies with the composition of the lacquer phase. The presence of resins in lacquer usually aids the dispersion, both as to the ease of forming the emulsion and the uniformity of droplet size.

No. 2.—The presence of acetone gives (1) increased emulsifying power for n/c rich lacquers, (2) depresses freezing temperature of the emulsion, (3) acetone replaced by a water miscible high boiling solvent decreases the tendency of film blush.

Nos. 3, 4, 5, 8 are all of the same general type, i.e., they contain an auxiliary emulsifier or stabilizer. This additional material helps produce emulsions from lacquers of very high viscosity and those that are relatively rich in n/c.

No. 6.—While the stability of the

emulsion using this emulsifying mixture may be satisfactory, too high a concentration of the emulsifiers that are incompatible with the lacquer results in a decrease of gloss upon the surface of the dried films.

No. 7.—Will produce a W/O emulsion. If the concentration is decreased to get an O/W emulsion a poor emulsion results (only partially emulsified).

Lacquer emulsions with casein emulsions are O/W; gelatin emulsions are W/O; 0.03% is the upper limit of methyl cellulose that can be included without producing "flatting" of the film. Methyl cellulose is an exceptionally good emulsifying aid.

Demulsifier for Crude Petroleum Emulsions

U. S. Patent 2,127,140

Formula No. 1

Ammonium Ricinoleate	H9.78
Cyclohexanol	12.05
Ammonia	1.91
Furfuraldehyde Potassium Bisulphite	3.20
Castor Oil	20.45
Water	12.09

No. 2

Ammonium Ricinoleate	46.00
Ammonia	0.86
Furfuraldehyde Potassium Bisulphite	8.93
Hydro Furamide	0.29
Castor Oil	30.00
Water	13.69

CHAPTER SIX

FARM AND GARDEN SPECIALTIES

Weed Killers

Formula No. 1

Dissolve 20–50 grams of sodium chlorate in 1 liter of water. Apply 1–1.5 liter solution per square meter for all common weeds. For weeds with tubers or bulbs apply the same solution twice or three times in intervals of one to two weeks or use

No. 2

100–200 grams sodium chlorate per liter of water.

Caution: These concentrated solutions when dried on inflammable material like cloths, paper, etc., cause a fire hazard.

Less dangerous in these cases is:

No. 3

100–200 grams sodium chlorate, 200–400 grams calcium chloride dissolved in 1 liter of water. Apply, by spraying or by a watering can, 1–1.5 liter solution per square meter.

No. 4

Less effective, but without fire hazard and not poisonous to livestock: 25–100 grams ammonium thiocyanate in 1 liter of water. Apply 2 liters per square meter by spraying or by watering can.

For arable land which will not be cultivated for 4–6 months, formulas 1, 2 or 3 may be used at a rate of 100 gallons per acre. When formula 4 is used at a rate of 200 gallons per acre, the field may be cultivated after 1–3 months.

No. 5

Eradication of weeds in cereal crops: Dissolve 1 volume concentrated sulphuric acid (66° Bé.) in 10–20 volumes of water. Apply 50–200 gallons per acre by spraying.

The effect of these herbicides and all other mixtures depends also on the nature of the weeds, on the weather (very dry and very wet times are unfavorable), on the properties of the soil (sandy soils often need less chemicals than clay), the pH, the organic matter and the nitrates in the soil, etc.

Furfural is very toxic to dandelions and other annual weeds, but is too expensive to be used alone as a weed killer. This has led to the preparation of mixtures usually containing kerosene or light petroleum distillates which themselves have some lethal properties toward vegetative materials.

The advantage of the furfural weed killers is in their ease of handling, effectiveness, and in their non-persistency in the soil which facilitates reseeding.

No. 6

One recommended formula for a furfural herbicide comprises 45 parts of kerosene, 45 parts xylol, and 10 parts of furfural. The xylol is used as a mutual solvent for the kerosene and furfural. In making this preparation the kerosene and xylol are first mixed and the 10 parts of furfural are added with stirring.

No. 7

An emulsion of the herbicide may also be prepared comprising 89 parts of kerosene or light petroleum distillate and 1 part of crude oil, such as is commonly used in oiling roads, and 10 parts of furfural. The emulsion is not permanent and, consequently, it is necessary to agitate the material thoroughly from time to time during the period of use; otherwise, the constituents will separate and lose their maximum effectiveness. The solution form is generally recommended in preference to the emulsion form.

Two methods of application are used. The first comprises individual treatment of dandelions in lawns, golf courses, and the like. The application is made by means of a so-called "dandelion gun," which comprises a tube with a valve in the bottom so arranged that when the gun is placed on the crown of a weed and pressed lightly a few cubic centimeters of the liquid is discharged on the crown of the dandelion. This causes it to turn brown, dry, and disintegrate. Care must be exercised to see that the weed killer does not contact grass because it has a considerable lethal action on the grass as well as on the weed.

A second method of treatment comprises the so-called broadcast procedure in which the weed killer is sprayed on thickly infested areas resulting in the killing of practically all the vegetation in the area, and this must be followed by reseeding after about a week or ten days. Usually about 500 gallons per acre will suffice for a substantially complete kill of all annual weeds and weakening of many of the deep-rooted perennials.

Where the individual treatment is used, usually the grass surrounding the treated plant fills in the bare spots. However, if desired the treated areas may be reseeded in order to speed up growth of new grass.

No. 8

German Patent 696,671

Copper Chloride	½
Potassium Chloride	½
Potassium Nitrate	½
Water	98½

No. 9

German Patent 683,877

Ferric Sulphate (10% Solution)	75
Copper Sulphate	5
Magnesium Sulphate	10
Sulphuric Acid (60° Bé.)	10

No. 10

Canadian Patent 391,459

Sodium Chlorate	12.5 lb.
Borax	15–25 lb.
Caustic Soda	3.2–5.2 lb.
Water	10 gal.

No. 11

Ammonium Sulfamate	0.5–1 lb.
Water	1 gal.

Use 1–1½ gal./100 sq. ft.

No. 12

Arsenic Trioxide	4 oz.
Sodium Hydroxide	4 oz.
Water	1 gal.

Boil until clear. Use 1 cupful of this solution per gallon of water and apply the resulting solution to the weeds.

Citrus Potato Leaf Hopper Wash

Damage to oranges is prevented by following wash:

Lime, Slaked	100 lb.
Zinc Sulphate	25 lb.
Blood Albumin	12 oz.
Water	300 gal.

Citrus Thrip Spray

Tartar Emetic	1½ lb.
Sugar	2 lb.
Water	100 gal.

Use 3–5 gal. per tree. This is toxic to thrips and does not injure foliage.

Combating Thrips on Flax

Dust with

Naphthalene, Crude	30–40 kg.
Lime, Slaked	10–13 kg.

per hectare.

First dust before buds appear and repeat as often as necessary.

Fruit Fly Lure

Orange juice is used as the attractant.

Banana Leaf Spot Control Dust

Copper Sulphate	20
Sulphur	40
Lime, Slaked	30
Kaolin	10

Peach-Borer Spray

p-Dichlorbenzene	3 lb.
Cottonseed Oil	6 qt.
Water	18 qt.
Fish Oil Soap	3 oz.

Mix well.

Apply ½ pt. per tree.

Alfalfa Snout Beetle Poison

Soyabean Meal	100 lb.
Sugar	15 lb.
Sodium Silicofluoride	6 lb.
Water	12 gal.

Mix well; press into pellets.

Spray for Leaf Roller Moth on Plums

Formula No. 1

Quassia Extract	2
Water	98

No. 2

Nicotine (2% Solution)

Crop Dusting Insecticide

British Patent 527,054

Cuprous Oxide	4	10
Magnesium Oxide	1	
Sulphite-Cellulose Residue	2	
Flour		10
Talc		80

Insecticidal Dust

4% Derris or Cube (Ground)	25.0
Talc or Clay	74.5
Santomerse D. (Wetting Agent)	0.5

European Corn Borer Spray

Formula No. 1

Water	25.0 gal.
Pure Ground Derris Root (4% Rotenone)	1.0 lb.
Areskap 100	1.5 oz.

No. 2

Water	25.0 gal.
Liquid Tannin	4.0 oz.
50% Free Nicotine	4 oz.
Areskap 100	1.5 oz.

Gypsy Moth Spray

Lead Arsenate	30.0 lb.
Fish Oil	6.0 lb.
Motor Oil	1.2 lb.
Aresket 240	1.2 lb.
Water	45.0 lb.

Dilute to suit.

Stinking Smut on Wheat-Spray (Stock Solution)

Formaldehyde (40%)	2.7 lb.
Santomerse No. 3	1.6 lb.
Water, To make	1.0 gal.

One pint of stock solution added to 40 gallons of water for use.

Insecticide for Gherkins

Water	50 gal.
Copper Sulphate	2 lb.
Quicklime	3 lb.
Lead Arsenate	1 lb.
Nicotine Sulphate	3/8 pt.
Soap, Liquid	1/2 pt.

This is effective against downy mildew, mosaic, and aphids.

Grasshopper Poisons

Per 100 lb. of bran, any of the following may be used:

Sodium Arsenite (30–32% Solution)	1 qt.
or	
Paris Green	2 lb.
or	
Arsenic Trioxide	2–3 lb.
or	
Sodium Arsenate	2 lb.

Colloidal Insecticides

Lead Arsenate	10.0 lb.
Gum Arabic	3.5 lb.
Water	100.0 oz.

Mill on colloid mill or paste mill.

Red Copper Oxide	730
Cup Grease	200
Butyl Stearate	30
Lubricating Oil (No. 20 SAE)	220

Mill on colloid mill and dilute to spraying consistency with light oil. (Heat and mix before milling.)

Red Copper Oxide	1200
Colloidal Kaolin	200
Gum Arabic	120
Water	1050

Mill on colloid mill and dilute to spraying consistency with water.

Colloidal Sulfur

Flowers of Sulfur	30 lb.
Gum Arabic	6 lb.
Water, To make	54 lb.

Mill on a colloid mill; three runs at zero setting will produce colloidal sulfur.

HOME-MIXED INSECTICIDE OIL EMULSIONS

The principle involved in the preparation of an oil emulsion is the bringing together of oil and water in the presence of an emulsifying agent and under conditions of violent agitation during which operation the oil is finely dispersed throughout the water medium and remains so for a considerable length of time. There are two types of materials commonly used to emulsify oils for spraying purposes: (1) Dry materials which are dissolved in a small amount of water previous to mixing with the oil; (2) oil soluble emulsifiers which are added directly to the oil.

Methods of preparing summer oil emulsions are explained in the following examples:

Blood Albumin.—Two ounces of actual water soluble blood albumin or eight ounces of the commercial blood albumin-filler mixture are dissolved in about one quart of water in a large pail (16 quart). The oil required for 100 gallons of spray is then added. Emulsification is produced by turning water from the spray gun under high pressure into the mixture of water, blood albumin and oil. By the time the pail is filled with the creamy emulsion it is sufficiently mixed to pour into the spray tank where it is diluted to make 100 gallons. Agitation with both the spray gun and tank agitator is continued until the full amount of water is added. The tank agitator is kept in action until the spray is applied. When preparing emulsions using more oil, a larger container is desirable for the original emulsification. Summer oil sprays are not usually prepared directly in the spray tank as is the customary practice when making dormant oil sprays for the reason that the small amount of oil required for foliage sprays could very easily be lost by adhering to the sides of the tank or by floating on the water.

One part of most oil soluble emulsifiers (glycerylokate) after being thoroughly mixed into 99 parts of oil will produce an emulsion when water is added under high pressure from the spray gun. The important consideration is to have the emulsifier thoroughly mixed into the oil. For illustration, 150 cc. of emulsifier is added to 4 gallons of oil in a 5-gallon can. Violent shaking of the container will thoroughly mix the emulsifier in the oil. To prepare the emulsion, measure out the required amount of oil in a 16-quart pail or larger container if more than 200 gallons of spray are to be prepared. (No allowance is made for the small amount of emulsifier when measuring out the oil.) The water is then added under high pressure from the spray gun to produce the emulsion after which the mixture is poured into the spray tank and diluted to the required concentration similar to the method described under blood albumin.

A weak Bordo mixture (¾–3–100) is not only an emulsifier for summer oil sprays but also affords some protection against apple scab and acts as a corrective for arsenical injury. To emulsify with Bordo, after preparing the mixture in a small amount of water, follow the method described for blood albumin. To avoid russeting the fruit, it is advisable to use only the weak Bordo in the summer and not to apply copper sprays before July first.

Locust Poison
British Patent 535,339

M-Dinitrobenzene	½
Sodium Sulphate (Anhydrous)	½
Grind together and mix with	
Bran	28
Molasses	1
Finally add	
Water	28

Gladiolus Thrip Spray

Tartar Emetic	4 lb.
Brown Sugar	16 lb.
Water	100 gal.

Apply frequently in a fine mist.

Chinch Bug Control

Mineral Oil Emulsion, Refined	2.500 oz.
Nicotine Sulfate	0.125 oz.
or	
Derris Extract	0.125 oz.
Water	1.000 gal.

Apply 70 gal. per acre of corn at 2–4 ff. stage.

Insecticide Spreader and Sticker
U. S. Patent 2,109,961

Fish Oil	500
Caustic Potash	85

Water	83
Alcohol	85

The above is mixed and then used with powdered insecticides.

Chigger Control
Formula No. 1

Spray 1% kerosene emulsion on bushes and grass.

No. 2

Dust every 10 days with sulphur (1 lb. to 1000 sq. ft.).

Insect Repellent Spray for Animals
Australian Patent 109,014

Carbon Tetrachloride	46
Kerosene	46
Castor Oil	3
Mineral Oil	3
Eucalyptus Oil	2

Nicotine Spray Activator

Efficiency is trebled by adding a 1–500 dispersion of agar-agar or gum karaya.

Snail and Slug Bait

Molasses, Black Strap	1 pt.
Calcium Arsenate	1 lb.
Metaldehyde	½ lb.
Bran	16 lb.
Water	2 gal.

Walnut Husk Fly Bait

Glycerin, Technical	2%
Caustic Soda	3%
Water	95%

Nematode (Fungi) Destroyer
Canadian Patent 391,483

Use 0.01–0.1% solution of sodium chlorite.

Walnut Caterpillar Spray

Lead Arsenate	3 lb.
Lime	6 lb.
Fish Oil	3½ lb.
Water	100 gal.

Red Spider Spray
U. S. Patent 2,105,727

Selenium	302
Lime-Sulphur Solution	1 gal.

Codling Moth Spray for Pears

Copper Sulfate	2 lb.
Hydrated Lime	10 lb.
Spreader	
Lead Arsenate	3 lb.
Water, To make	100 gal.

Japanese Beetle Spray
(Patented)

Rosin	10 lb.
Lye	1 lb.
Fish Oil Soap	1 lb.
Corrosive Sublimate	1 lb.
Nicotine Sulfate	1 oz.

The first three ingredients are boiled in 10 gallons of water until the mixture turns black; the corrosive sublimate and nicotine sulfate are then added in 20 gallons of water. For spraying fruit trees, hardy shrubs, grass and weeds, the mixture should be diluted with two parts of water. For beans and other delicate plants, dilute with three or four parts of water.

Japanese Beetle Lure

Geraniol	10
Eugenol	8

Mosquito Repellents

Diethylene Glycol Monobutylether Acetate	33
Diethylene Glycol Monoethylether	32
Alcohol	28
Corn Oil	7

This is far better than citronella oil.

Mosquito Spray

Tobacco Dust	5 lb.
Water	1 gal.

Steep for 1–2 days. Filter and add

Soft Soap	9.65 oz.

Elimination of Silverfish

Powdered Arsenious Oxide	8 g.
Cracker Crumbs, Fine	100 g.

Mix well, moisten slightly, roll into small loose pellets and distribute back of books, etc., where infestation occurs.

Silverfish cannot climb a clean vertical glass wall. A trap may then be devised by covering the outside of a vaseline jar up to the mouth with rough adhesive paper or cloth. Leave inner wall untouched, and place a teaspoonful of flour in the jar. Silverfish climb up and fall in, but cannot get out. Occasionally shake out remains and put in fresh flour.

Roach Destruction

Sprinkle sodium fluoride freely and constantly where roaches frequent, until they all disappear. Then as needed.

To Destroy Ant Hills or Nests

Pour a sufficient amount of carbon bisulphide over the hill or nest to follow the holes or trails down into the ground. Then set fire to it. The fumes and fire tend to exterminate the ants. Use when nest is occupied.

Ant Repellents and Poisons

Homes, gardens, fences and other places infested with ants can usually be freed of the pests by the use of a mixture made of:

Arsenic Trioxide	½ oz.
Glycerin	5 oz.
Honey	1 oz.
Water	1 pt.

Heat the arsenic with the glycerin (to about 150° C.) till clear. When cold add the water in which the honey has been dissolved.

A recommended procedure is to place a piece of wet cloth or blotting paper on a plate (to be kept wet by additions of water as necessary), sprinkle a few grains of sugar on the surface and add about 10 to 20 drops of the above solution. This solution is effective because it attracts the ants to their death.

Sodium fluosilicate, mixed with glycerin and honey, also makes a good ant poison, but not quite as effective as the arsenic solution. An old fashioned yet useful ant poison can be made from equal parts of tartar emetic and sugar made into a paste with glycerin. This poison paste is spread around the places frequented by the ants.

Fumigating Stored Food with Carbon Disulfide

Where large quantities of beans or cereal products are to be fumigated, it is more convenient to use carbon disulfide to kill the insects in them. Carbon disulfide is a very effective fumigant where infested materials can be confined in a small, relatively air-tight space such as a barrel or box. It is advised for treating both food and clothing to rid them of insects.

Carbon disulfide is a heavy clear liquid with a disagreeable odor. It is highly inflammable, and its fumes are explosive when mixed with air. For this reason it must never be used around an open flame or by a person who is smoking. It is used at the rate of 1 pound to each 100 cubic feet of space or about 1½ ounces, liquid measure, of carbon disulfide to each 10 cubic feet of space in the container.

Place the chemical in shallow pans or pour it on some rags on the top of the material to be fumigated. The gas is heavier than air; as the material evaporates it will sink through the articles to be fumigated. The container should be kept tightly closed or sealed for 24 to 36 hours at a temperature between 65 and 80 degrees F. Best results are secured when the temperature is at or above 70 degrees F. It is not effective when the temperature is below 60 degrees. When used according to the above directions fumigation will not harm grains or beans for plating, or cereals and dried fruits for food.

Colchicine Plant Treating Emulsion

Water	20 cc.
Glyceryl Monostearate	2 g.
Lanolin	8 g.
Colchicine	⅛ g.

Add first two items together. Mix until melted. Stir to creamy soap solution. Add lanolin. Heat till melted and just below boiling point. Add colchicine. Stir till thick creamy emulsion results. Stir intermittently until it has cooled to room temperature.

Seedless-Fruit Forming Hormont Spray

Dichlorophenoxyacetic Acid	10–25
Water	1,000,000

Seed Disinfectant
French Patent 847,302

Methyl Mercury Iodide	1.2
Ethyl Mercury Iodide	1.5
Calcium Arsenate	10.0
Quartz Sand	87.3

Pea Seed Treatment

Spergon, Semesan, and 2% Ceresan are used for treating seed of all pea varieties except Alaska and Rogers Winner which do not appear to be quite so susceptible to injury from decay organisms. Red copper oxide is an excellent treatment for the early varieties but may injure some of the late-maturing sorts. Treatment is simple. The seed and chemical are placed in a rotary mixer, such as a barrel churn, and tumbled until each seed is thoroughly coated. This usually takes four to five minutes with 25 to 30 revolutions a minute. The gardener can accomplish the same thing by shaking the seed and chemical in a tightly covered fruit jar.

Spergon is applied at the rate of 1½ ounces per bushel of seed, while Semesan and 2% Ceresan are applied at the rate of 2½ ounces per bushel. If the seed is to be sown by a grain drill, graphite should be added to Semesan and 2% Ceresan at the rate of 1¼ ounces per bushel. Spergon is a natural lubricant and does not need any added graphite.

Treating Oat Seed

In treating seed oats with formaldehyde, a mixture of 1 pint of formaldehyde and 2 pints of water may be carefully and evenly sprayed over each 75 bushels of seed, applying the spray to each scoopful of oats as it is turned over. The seed hould then be shoveled from one ile to another several times and hen covered tightly with burlap ags or blankets for two to five ays. The temperature at all times hould be above 50° F. The oats hould be planted from two to five ays after treatment or else passed hrough a fanning mill to drive out he formaldehyde gas. The seed can hen be safely stored and planted at ny time and any excess seed safely d to livestock.

Tree Canker Salve
Swiss Patent 209,590

Rosin	80
Coal Tar	20

Melt together, mix and apply hile fluid.

Cannibalism (Poultry) Remedy

Cannibalism among hens may be medied by dissolving permanganate of potash in the drinking water ntil it is the color of cherry juice nd then give them no other drinkng water. A little salt should be laced in the water which will make em drink more frequently and ill give them more of this mediated water. If one will use motor l, lamp black, a little kerosene and rpentine and creosote in a mixture nd bathe the injured parts to take way the bloody appearance, canibalism will be stopped. Chickens ill not pick feathers where there is n odor of creosote or turpentine.

Control of Chicken Chiggers or Harvest Mites

If an abscess has not yet formed, e inflamed area may be treated ith sulfur ointment, Peruvian balam, or a mixture of one part of kerosene with three parts of lard. If suppuration has occurred the scab should be removed and the area washed with 4 per cent carbolic acid solution. Frequent light dusting with flowers of sulfur will keep the chickens from becoming infested.

Treatment of Poultry Favus

Favus is a disease of the skin of chickens characterized by the formation of white area on the comb and wattles. Occasionally this infection, caused by a fungus, may extend to the feathered parts of the body. Since this disease is transmitted from fowl to fowl by direct or indirect contact, affected animals should be promptly isolated and treated with an iodine-glycerin mixture. This mixture is composed of:

Tincture of Iodine	1
Glycerin	6

This combination should be applied two or three times per week to the infected areas. Since it will probably not be effective when the feathered skin is involved, animals so infected should be destroyed to prevent their becoming centers of infection for the whole flock.

Control of Chicken Feather Mite

Feather mites may be destroyed by dipping affected fowls in a tub containing a mixture of water, 1 gallon; flowers of sulfur, 2 ounces; and soap, 1 ounce.

The feathers should be thoroughly wet to the skin. The head should be submerged for an instant. During the dipping process the mixture should be stirred so as to keep the sulfur in suspension. Dipping should be done only on warm, sunny days, or in a heated building. If treatment is found to be necessary during the winter or early spring months, thorough dusting of the fowls with flowers of sulfur should take the place of dipping.

The nesting material should be removed and burned, and nest boxes, roosts, walls, and floor should be sprayed or painted with anthracene oil as recommended for use against the common chicken mite.

A simpler method than the one just mentioned, reported as satisfactory in some cases but not so successful in others, consists in painting the perches with a 40 per cent nicotine sulfate solution shortly before the fowls go to roost, and dusting the nest with sulfur. Nests of English sparrows in the immediate vicinity of the poultry house should be destroyed.

Chicken Lice Powders and Their Application

A very satisfactory way of eliminating lice from poultry is to treat each fowl separately with sodium fluoride. If applied to all fowls as directed, one treatment is sufficient to kill all lice and their eggs. It is applied either by the so-called pinch method, or by means of a duster made by punching small nail holes in the bottom of a can having a tight-fitting cover, or by dipping. If the pinch method is used, the bird is held on a table while sodium fluoride is applied next to the skin under the feathers, as follows: One pinch on the head, one on the neck, two on the back, one on the breast, one below the vent, one on the tail, one on each thigh, and one on the under side of each wing. The feathers should be ruffled to allow the powder to get next to the skin. If th bird is held in a large shallow par the small quantity of powder whicl falls off will be saved. If the powde is dusted on by means of a shake the quantity of sodium fluoride use may be reduced by using three part of road dust or flour to one part o the chemical. This method require the services of a second person t hold and turn the fowl.

When considerable numbers o birds are to be treated the sodium fluoride should be used in the forn of a dip, a rounded tablespoonful (ounce) of commercial sodium fluo ride being used to each gallon o water. The birds should be held b the wings and plunged into a tul filled with the solution, the head be ing left out, while the feathers ar ruffled with the hand to allow th solution to penetrate to the skin The head is then ducked once o twice and the bird held for a few sec onds to drain, and then released Dipping is just as effective as th other methods and is quicker an more economical of material. Dip ping should be done only on warm sunny days.

To sick fowls or to very youn chickens or turkeys, sodium fluorid should be applied only by the pincl method, and especially in the cas of young birds it should be used cautiously, in very small pinches.

One pound of sodium fluoride wil treat 100 birds by the pinch method

Another good lice powder is flowers of sulfur, which should be applied with a duster. Although sulfu is considerably cheaper than sodium fluoride, it is less effective agains lice and hence must be applied mor liberally, so that its use is in realit nore expensive than that of sodium luoride. Many other powders, some f which contain pyrethrum (insect owder), are commonly used, but hey have no advantage over sodium uoride.

Dust baths containing a mixture f tobacco dust or other insecticide nd ordinary road dust are often ecommended to destroy lice. While is a good plan to let the birds dust hemselves when they wish, no nethod which allows the bird to reat itself for lice can be expected o eradicate them all, since fowls annot get the dusting powder on all arts of the body where lice are, and nany lousy birds will not use the ust baths.

A simple procedure, which is effective in controlling lice but will ot eradicate them, consists in the pplication of undiluted 40 per cent olution of nicotine sulfate to the op surface of the roosts by means f a paint brush. This is done a short ime (from 15 to 20 minutes) before he fowls go to roost. The fumes of he nicotine kill the lice during the rst, second, and third nights after pplication. The head lice, natually, are least affected. As some of he lice are not killed and the eggs re not destroyed, it is necessary to epeat the treatment frequently.

Disinfection of Poultry Houses

In case of an actual outbreak of irulent disease it is advisable to use or disinfecting purposes a whiteash made by dissolving 1 pound of ommercial lye (containing 94 per ent of sodium hydroxide) and 2½ ounds of water-slaked lime in 5½ allons of water. If the solution is ot used at once, it should be tightly covered to prevent deterioration. This solution is cheap, odorless, and destructive to almost all kinds of disease germs. On prolonged contact, however, it may be injurious to painted or varnished surfaces, and to some fabrics. It is corrosive to aluminum, but relatively harmless to the metallic fixtures ordinarily found about chicken houses, and to wooden construction or equipment. The poultry yard may be disinfected by wetting it thoroughly; use from ½ to 1 gallon of the solution per square yard of soil surface, depending upon the absorbent quality of the ground. It is essential, however, that all refuse matter be removed from the surface and burned or buried before poultry runs are disinfected.

The above-mentioned disinfectant solution has been found through experimentation to be ineffective against the germs of tuberculosis. For combating that infection, the chicken houses, enclosed runs, and all eating, drinking, and other utensils should be thoroughly cleaned and disinfected with a strong solution of such germicides as carbolic acid or compound solution of cresol. Carbolic acid may be used in 5 per cent solution and compound solution of cresol in 3 per cent solution.

Kerosene emulsion is frequently used to destroy mites. To make the emulsion, shave half a pound of hard laundry soap into half a gallon of soft water and boil the mixture until all the soap is dissolved; then remove it to a safe distance from the fire and stir into it at once, while still hot, 2 gallons of kerosene. This makes a thick, creamy emulsion or stock mixture. When it is to be used for killing mites in the houses, 1 quart of this emulsion is mixed with 10 quarts of water.

CHICKEN FEED

Starting Mash

No. 2 Yellow Corn Meal	28	lb.
Standard Wheat Middlings	20	lb.
No. 2 Heavy Oats, Finely pulverized	10	lb.
Standard Wheat Bran	16	lb.
Fish Meal (55% Protein)	8	lb.
Meat Scraps (55% Protein)	8	lb.
Dried Skim Milk (or Buttermilk)	5	lb.
Ground Limestone (39% calcium and not over 2% magnesium)	3	lb.
Salt Mixture (Made by mixing 12½ lb. common salt with 3.4 oz. anhydrous manganous sulfate)	1	lb.
Cod Liver Oil	1½	lb.

Starting Scratch Grain

Cracked Wheat	40 lb.
No. 2 Yellow Cracked Corn	60 lb.

Starting and Growing Mash

Formula No. 1

Ground Yellow Corn	17.5
Wheat Middlings	30.0
Wheat Bran	10.0
Dried Skim Milk (or Dried Buttermilk)	10.0
Meat Scrap	5.0
Fish Meal	4.0
Soybean Meal	4.0
Corn-Gluten Meal	4.0
Alfalfa-Leaf Meal	10.0
Ground Limestone (or Oyster-shell)	2.0
Steamed Bonemeal	1.5
Salt Mixture	1.0
Cod Liver Oil	1.0

No. 2

Ground Yellow Corn	21.0
Wheat Middlings	30.0
Wheat Bran	10.0
Dried Skim Milk (or Dried Buttermilk)	10.0
Meat Scrap	15.0
Alfalfa-Leaf Meal	10.0
Ground Limestone (or Oyster-shell)	2.0
Salt Mixture	1.0
Cod Liver Oil	1.0

Laying Scratch Grain

No. 2 Cracked Yellow Corn	60 lb.
Whole Wheat	40 lb.

During winter 100 lb. No. 2 yellow corn can be used for scratch.

Laying Mash

Formula No. 1

No. 2 Yellow Corn Meal	24	lb.
Standard Wheat Middlings	20	lb.
No. 2 Heavy Oats, Finely pulverized	20	lb.
Standard Wheat Bran	10	lb.
Fish Meal (55% Protein)	6½	lb.
Meat Scrap (55% Protein)	6½	lb.
Dried Skim Milk (or Buttermilk)	7	lb.
Steamed Bone Meal	1	lb.
Ground Limestone (39% Calcium and not over 2% magnesium) or oyster shell	3	lb.

Salt Mixture (Made by mixing 12½ lb. common salt with 3.4 oz. anhydrous manganous sulfate)	1	lb.
Cod Liver Oil	1½	lb.

No. 2

Wheat Middlings	31.5
Wheat Bran	20.0
Dried Skim Milk (or Dried Buttermilk)	10.0
Meat Scrap	4.5
Fish Meal	5.0
Soybean Meal	2.3
Corn-Gluten Meal	2.4
Alfalfa-Leaf Meal	12.0
Ground Limestone (or oyster shell)	6.8
Steamed Bonemeal	1.5
Common Salt (or Salt mixture)	1.2
Cod Liver Oil	2.8

No. 3

Finely Ground Oats	19.0
Wheat Middlings	15.0
Wheat Bran	10.0
Dried Skim Milk (or Dried Buttermilk)	10.0
Meat Scrap	10.0
Soybean Meal	5.0
Linseed Meal	4.0
Alfalfa-Leaf Meal	15.0
Ground Limestone (or oyster shell)	5.6
Steamed Bonemeal	2.4
Salt Mixture	1.2
Cod Liver Oil	2.8

Mixed Protein-Vitamin Concentrate for Chickens

Alfalfa-Leaf Meal	25
Dried Skim Milk or Dried Buttermilk	20
Fish Meal or Meat Scrap	20
Soybean Meal	10
Corn Gluten	5
Linseed Meal (Old process)	5
Steamed Bonemeal	10
Ground Limestone	1
Salt Mixture	2
Cod Liver Oil	2

A mixture of 100 pounds of common salt and 1.7 pounds of anhydrous manganous sulfate (or 2.5 pounds of manganous sulfate tetrahydrate).

Or 0.4 part, by weight, of fortified cod liver oil that contains at least 400 A. O. A. C. chick units of Vitamin D and 3,000 International Units of Vitamin A per gram.

Fowl Reviver Pills

As most poultry men know, medication for fowls is best given in pill or tablet form. In making these items glycerin or glycerin-containing preparations known as glycerites are usually employed as the excipient of massing material. An instance of such usage is given in the following recipe, modified from an English formula, for making fowl tonic pills:

Ferrous Sulfate	2	dr.
Extract of Gentian	½	dr.
Calcium Phosphate	1	dr.
Glycerite of Tragacanth	sufficient to make a mass	

Divide into 5 grain pills.

Tonic pills for pigeons are made from a similar formula. They consist of:

Ferrous Sulfate	1	dr.
Powdered Capsicum	20	gr.
Extract of Nux Vomica	6	gr.

Powdered Gentian	$\frac{1}{2}$ dr.
Glycerite of Tragacanth	sufficient

Mix the ingredients and divide into sixty pills. Coat with sugar. Allow each pigeon six of these pills per day.

Glycerite of tragacanth is made from:

Powdered Tragacanth	125 g.
Glycerin	775 cc.
Water	185 cc.

Heat together in a double boiler, mixing until a smooth paste is formed.

Pigeon Malaria Control

A thorough cleaning of the pigeon nests at intervals not to exceed 25 days, using a light spray of kerosene extract of pyrethrum in the interior of buildings, and one of the following procedures: Dusting the squabs and pigeons with fresh pyrethrum powder, derris powder, or tobacco powder containing about 6 per cent of nicotine; dipping them with either an aqueous extract of pyrethrum with soap or derris extract with soap and water; or using the kerosene extract of pyrethrum spray on the birds.

Removing Oil Gloss from Eggs

U. S. Patent 2,109,575

Dip in following solution for 2–3 minutes:

Sodium Dihydrogen Phosphate	5– 10%
Sodium Lauryl Sulphate	0.1–0.25%

Bedbug Extermination

German Patent 701,761

The walls and especially the habitat of the insects is first sprayed with an irritant. After 15–30 minutes the disinfectant proper is atomized within the room. The irritant is made as follows: 1 liter of a 15% pyrethrum extract is mixed with 150 liters of highly refined mineral oil; to this mixture is added wintergreen oil 2 kgms., citronella oil 1 kgm., and paradichlorobenzene 3 kgms. The disinfecting mixture is made up of: 15% pyrethrum extract 40%, highly refined mineral oil 35%, petroleum ether or carbon tetrachloride 20%, and Turkey red oil (50%) 5%. Neither of these preparations is injurious to men or animals.

House-Fly Preparations

(1) Pyrethrum $\frac{1}{2}$ lb., kerosene 1 gal. and methyl salicylate 3 fl. oz. (2) soap $\frac{1}{4}$ lb., water $\frac{1}{2}$ gal. and kerosene 1 gal. A safe and effective poison bait consists of 3 teaspoonfuls of formalin to 1 pint of milk or water with a little sugar added. Fly-papers or wires can be prepared by use of rosin 2 lb. and castor oil 1 pint.

Lures for Rats

The following materials attract rats in the order given, with bread having a standard rating of 100:

Oatmeal	80
Tallow	70
Bread and Milk	60
Banana	60
Flour	50
Barley	50
Bloater Paste	30
Fish (Smoked)	20
Drippings	20
Lard	10

Rat Extermination Compositions

Formula No. 1

Red-Squill (Fine Powdered)	1
Oatmeal	4

Mix well and add sufficient oil of aniseed to give a faint smell to the bait.

No. 2

Red-Squill (Fine Powder)	1
Oat or Maize Meal	2¾
Dripping or Tallow	1¼

Melt the fat and quickly pour on to the dry ingredients, which have been previously well mixed. Stir well until a stiff paste is produced, then add oil of aniseed to give a faint smell to the bait.

No. 3

Red-Squill (Fine Powder)	4
Grated Cheese	5
Fat	5
Oatmeal or Flour	6

Mix the ingredients into a thick paste and cut into pieces about the size of a walnut and bake lightly in a quick oven.

Red-squill powder, when mixed with flour, is very suitable for making up into small biscuits, which can be used as baits in hedgerows, refuse dumps, barns, stables, and even dwelling-houses. One biscuit is sufficient to kill a rat. It is advisable to use the biscuits freely so as to poison as many rodents as possible with the first application. The biscuits should be laid in the evening and pushed well down the rat holes. They keep well and retain their toxicity.

No. 4

Red-Squill (Fine Powder)	1
Flour	2¾
Fat	1¼

Mix the squill and flour well together and knead in the fat, adding sufficient water to make a fairly stiff dough. Roll out into a sheet ¼ inch thick and cut into about 1,400 biscuits for each pound of squill powder used.

No. 5

Red-Squill (Fine Powder)	6
Breadcrumbs	15
Kipper (Minced)	4
Fat	5

Make a paste. One dessertspoonful is a toxic dose. This preparation should be made up as required for use, as it retains its toxic value only for three or four days.

Red-Squill Liquid Extracts

No. 6

Red-Squill Bulbs	1
Water	2
Salicylic Acid 2 oz. to every 10 gallons.	

Chop up the bulbs as finely as possible and place in an earthenware vessel. Boil the water and pour it on to the bulbs, stir in the salicylic acid and macerate for several hours. When cold, strain through fine muslin, pressing out as much as possible. Pour into amber-colored glass-stoppered bottles and store in a cool, dark place.

The bait is made as follows: The liquid is mixed with an equal quantity of fresh milk, which has been scalded and allowed to cool. Bread (finely divided) may be used in place of oatmeal, making a thin bread-and-milk mixture. A quantity of the bait is poured into a saucer or similar receptacle, which is placed near the rat holes and runs at night. Another method is to soak small squares of stale bread in the liquid squill. These can then be laid in the rat holes, using a wooden spoon for the purpose. Vessels containing water should be placed in the vicinity for the rodents to drink.

No. 7

Red-Squill (Bulbs, Chopped)	60	lb.
Distilled Water	10	gal.
Hydrochloric Acid	3½	fl. oz.
Salicylic Acid	7	oz.

Macerate bulbs with the acid liquid for twenty-four hours, stirring occasionally. Pour off the liquid, press the bulbs thoroughly, adding this to the decanted liquid. Pour in amber-colored stoppered bottles, preferably covering the surface with a layer of sweet oil or nut oil. To use: Place the extract in saucers with an equal quantity of milk, containing oatmeal, gruel or bread. 20 minims, equivalent to about 10 gr. of bulb, should be fatal. The extract does not keep well and should be used as soon as possible.

No. 8

A new rat-bait formula with powdered red-squill is prepared as follows:

The basic food is dried bread-crumbs. Obtain dried but not moldy white bread in whole loaves (not sliced); a non-milk bread is best. Break up the bread with the hands and put through grinder with ⅜-inch holes, then press through a sieve with around 40 holes to the square inch.

Ground Dried Bread-crumbs	52
Ground Fresh Pork Fat (Back Strips Are Best)	4
Ground Fresh Halibut or Haddock or Cod	16
Powdered Red-Squill	8

Mix the bread, pork fat and halibut well, and then add the squill and thoroughly mix.

No. 9

Ground Dried Bread-crumbs	68
Glycerin, Nut Oil or Mazola Oil	4
Powdered Red-Squill	8

Mix the bread and glycerin and then add the squill and thoroughly mix.

No. 10

Ground Dried Bread-crumbs	30
Glycerin, Nut Oil or Mazola Oil	2½
Fresh Bait—Hamburger or fish or ground sweet-potatoes or ground apples or ground bananas	40
Powdered Red-Squill	8

Mix the bread, glycerin and squill. Then add the fresh bait. Mix thoroughly.

No. 11

Ground Hamburger	5	lb.
Powdered Red-Squill	8	oz.

Thoroughly mix the hamburger and squill, adding water to make quite moist.

Bait Exposure

Place bait in half-teaspoonful quantities at intervals along runways and walls and in all places where rats may feed. Use plenty of baits and cover the entire premises at one time. If using formula No. 11 or 10, place the baits in the late afternoon or evening, before dark. Uneaten baits should be collected and destroyed after three days. If using formula No. 8 or 9, the baits can be placed at any time of day. These baits will not deteriorate, so they can be collected at a later date if uneaten and exposed again after four weeks. Do not use the fresh or moist baits during freezing temperatures. It is best to keep the bait out of reach of animals, children and irresponsible persons.

Most of the baits can be prepared by the manufacturers of the poison. The liquid extract of squill should be placed in dark-colored bottles and well sealed. The baked squill biscuits will keep quite well and retain their toxicity. The squill powder must necessarily be sold in sealed containers which will keep the contents perfectly dry. Complete instructions for use, for the information of purchasers, should be enclosed with all packages.

Experiment Rat Food

Whole Wheat (Finely ground)	1600 g.
Whole Milk (Powdered)	800 g.
Salt	32 g.

Sheep Dips

Formula No. 1

Soft Soap	5 lb.
Liquid Phenol (Minimum 97% Tar Acid)	3 qt.
Water q. s. to	100 gal.

No. 2

Sulphur	18 lb.
Quicklime	9 lb.
Water	21 gal.

No. 3

Tobacco Waste	35 lb.
Water	21 gal.
Sulphur	10 lb.

The first of these is quite simply made up by dissolving the soap, with gentle warming, in the liquefied phenol, the solution being subsequently admixed with the water.

In the second formula, the sulphur and lime are made into a thick paste with water. The whole is then placed in a strong cloth, the ends tied, and placed in a boiler containing 10 gallons of water, so that the contents are completely covered, avoiding burning. After boiling for two hours the solids are thrown away and the liquor again made up to 10 gallons and stored in tightly-closed drums. The quantity is sufficient for 100 gallons.

The third product is made by steeping the tobacco "offal" in the water for a period of four days, after which the sulphur is added to the strained liquor, the resulting mixture being sufficient to make 100 gallons of dip.

No. 4

White Arsenic	2½ lb.
Sodium Carbonate	1¼ lb.
or	
Caustic Soda	½ lb.

These proportions are finally made up to 100 gallons; 4 to 8 lb. of sulphur may be added to this formula, thus giving a sulphur-arsenic-washing soda type of dip

No. 5

Tar Acids	29.0
Paraffin Oil	36.0
Lanolin	8.0
Soft Soap	17.5
Water	9.5

One gallon of this to be used in 100 gallons of water.

No. 6

Soap	1 lb.
Crude Carbolic Acid	1 pt.
Water	50 gal.

No. 7

Copper Sulphate	8 lb.
White Arsenic	2 lb.
Hydrochloric Acid	1.6 pt.
Water To make	100 gal.

Sheep Blow-Fly Wound Dressings

Formula No. 1

Zinc Sulphate	5 lb.
Water	95 lb.

No. 2

Phenol	4 lb.
Whale Oil	96 lb.

No. 3

Boric Acid	3 lb.
Glycerin	1 gal.

Blow-Fly Destroyer

Formula No. 1

Carbon Disulfide	2 fl. oz.
Sulphonated Castor Oil	2 fl. oz.
Water	2 gal.

Mix vigorously before use. Apply 2 gal. per sq. yard of soil surface.

No. 2

Australian Patent 110,141

Anthracene Oil	100
Rosin	20
Phenol	5
Zinc chloride	5
Petroleum Ether	130

No. 3

Boric Acid	30
Tar Oil (B.p. 170–210° C.)	2
Bentonite	2
Sulfatate (Wetting Agent)	½
Water	65½

Tapeworms in Sheep

Nicotine sulphate, commonly known under the trade name of Black Leaf 40, is effective against tapeworms and hookworms in sheep. The dose should be one-half ounce or two teaspoons mixed with one quart of water. Lambs weighing 40 to 60 pounds should be given one ounce as a dose. Sheep 100 pounds or over should have from two to four ounces. Sheep should be kept off feed for 24 hours before drenching. Use a long-necked bottle and be careful not to hold the head too high. Repeat in three weeks and change pasture for sheep every year.

Cow Milk Stimulation Powder

Potassium Nitrate	1
Alum	1
Sublimed Sulphur	1
Prepared Chalk	1
White Bole	2
Red Clover	5
Anise	10
Fennel	10
Salt	10

Give 1 or 2 handfuls in the morning feed, making sure that the powdered composition is thoroughly mixed.

Molasses Cattle Feed

British Patent 521,332

Hay, Finely Ground	47
Molasses	53

Mix together and dry to 12% moisture content.

Calf Scour Treatment

A very good treatment for calf scours is to reduce the feed approximately one half. A 6 gram dose of bismuth sub-nitrate will help to check the profuse diarrhea. One level teaspoonful of bismuth sub-nitrate will weigh about 6 grams. Care should be taken to get the calves back on full feed. A second dose may be given if needed.

Livestock Screwworm Remedy

Diphenylamine	3.5
Benzol	3.5
Turkey-Red Oil	1.0
Lampblack	2.0

Apply thoroughly to wounds.

Treatment of Cattle Ringworm

The ringworm is treated by scraping off the scab with a knife blade and applying tincture of iodine or a mixture of equal parts of iodine and glycerin. This should also be put around the margin of the infected area. Immediate treatment of the first spot seen in order to prevent spread to other animals through indirect contact is necessary.

Cattle Salt Sickness Cure

Prevention: Add one pound of either the sulfate or chloride of Cobalt to each ton of the regular salt sick mixture. This material must be thoroughly mixed, to secure results.

For smaller amounts, use this formula:

Salt	100 lb.
Red Oxide of Iron	25 lb.
Powdered Copper Sulphate	1 lb.
Cobalt Salt	22 g.

This, too, *must be mixed thoroughly* to secure proper results.

Perhaps the best manner in which to make the mixture, is to mix the salt, copper and iron, and then dissolve the cobalt, 22 grams, in a small amount of water, and then spray this water over the mineral mixture, using a fly sprayer for this purpose.

For Advanced Cases: Dissolve 10 grams of cobalt chloride or cobalt sulfate in one gallon of water. Give each animal six ounces, as a drench one weekly for three or four weeks. Calves should have 3 fluid ounces, and other animals in like proportion.

For Animal Ear Aches

What to do when one of the farm animals suffers from ear pains is often a problem. This can usually be solved, however, by the use of the following preparation which is very similar to that employed in human cases of ear ache:

Phenol	3
Glycerin	97

Add boric acid until the glycerin will not dissolve any more. Let stand overnight and strain.

To use—put half an eyedropperful into the animal's ear and remove the excess.

Animal Dandruff Treatment

When the skin is dry, treatment consists of washing the affected areas with soap and water or gasoline and applying antiseptics. A recommended solution is the following:

Salicylic Acid	1
Glycerin	3
Alcohol	60

When the skin is moist, the treatment is slightly different. Cleanse with gasoline, clip the hair and then apply the antiseptic.

Glycerin Treatment for Ear Wax in Dogs

Treatment consists essentially of removing the wax and cleaning the ear. Applying drugs which tend to lessen the secretion are helpful. The following preparation is useful in controlling this condition:

Salicylic Acid	1.5
Glycerin	5.0
Alcohol, To make	100.0

A great deal of itching is associated with the excess wax production, and in some cases the lining of the dog's ear will become thickened and red. Recovery is slow in many cases and the condition may last for years, with intermittent periods of relief. Hence the need for prompt care and treatment.

Animal Eczema Treatment

In cases of chronic and seborrhic eczema of animals iodized glycerin is one of the most useful remedies. As the first step in the treatment, all crusts must be removed. In most cases the skin may be washed with soap and water. Iodized glycerin to be used after these preliminary procedures is readily prepared from:

Tincture of Iodine	1
Glycerin	4

Treating Ear Mange in Animals

Ear mange is not difficult to control in farm and household animals, if the suggestions given below are followed carefully. Use a mixture of 1 per cent cresol, 1 per cent carbolic acid or creosote, added to glycerin. Some animals require slightly different mixtures, however. For foxes, equal parts of tincture of iodine and glycerin is advocated and for ear mites in goats, an ointment of flowers of sulfur in sweet oil has proven both economical and effective.

Whichever preparation is used, however, the inside of the ear is swabbed with a cotton pledget held with forceps and moistened with the prescribed solution, taking particular care not to injure the eardrum. The accumulations and detritus in the ear should then be carefully cleaned out with a swab and the solution applied again. This treatment should be repeated after a week to kill any mites that may have hatched in the interval since the first treatment. To prevent reinfection the pens and premises should be thoroughly cleaned and washed with a suitable insecticide.

Dog Coat Dressing

Mineral Oil	3
Pine Oil	1

Mix and add trace of oil soluble blue dye (0.1 g. per gal.).

Spray on dog's coat or moisten cloth with dressing and rub lightly to achieve desired gloss. This will improve appearance and assist in controlling fleas.

Hair Restorer for Dogs

Tincture of Cantharidin	1/4	oz.
Almond Oil	1	oz.
Glycerin	1	oz.
Lime Water, To make	4	oz.

Dress the skin where the hair is thin with daily applications of the above lotion.

Dog Wash

For washing dogs, to maintain the healthy condition of their skin and fur, the following solution has been recommended:

Soft Soap	8	oz.
Glycerin	2 1/2	oz.
Alcohol	2	oz.
Phenol	3/8	oz.
Eucalyptus Oil	1/4	oz.
Water, To make	35	oz.

Mouth Wash for Dogs

When a mildly astringent, soothing mouth wash is needed for dogs, a suitable preparation can be made from the following:

Tincture of Ferric Chloride	1 oz.
Potassium Chlorate	2 oz.
Glycerin	4 oz.
Water, To make	1 gal.

For Animal Ringworm

In the treatment of animal ringworm, a mixture of equal parts of glycerin, tincture of iodine, and chloral hydrate appears to be more efficient than the tincture of iodine alone, which is generally used. Others have found that a mixture of equal parts of tincture of iodine and glycerin is also very efficient in clearing up cattle ringworm.

Treatment generally consists of clipping whatever hair may be left on the diseased area. The scabs and crusts should then be softened with soap and warm, soft water. In some cases, it may be necessary to scrape the diseased area with a knife blade or even to sandpaper down the surface of the affected part. The diseased parts are then painted with the glycerin, tincture of iodine, chloral hydrate mixture.

Dog Mange Treatment

A glycerin wash is an important feature of a recommended treatment for sarcoptic mange in dogs.

The treatment calls first for the application of the following preparation:

Oil of Tar	1
Olive Oil	1
Turpentine	1

Two good dressings with this preparation, with an interval of three days, will destroy sarcoptic mange mites. After this treatment the dog is washed with a neutral soap and rinsed with a quart or two of warm water containing 5 per cent of glycerin.

This glycerin wash apparently serves to soothe the irritated skin, promotes healing and helps to restore the natural sheen of the animal's fur or hair.

(N.B. There are two forms of mange: sarcoptic mange and follicular mange. A preparation that will cure one form is practically useless in the other type. Follicular mange is very hard to cure and sometimes it is necessary to destroy the affected animal.)

Breeding Kennel Dog Foods

Ingredient	Auburn* Ration	Formula No. 1	No. 2	No. 3	No. 4	No. 5	No. 6†	No. 7	No. 8	No. 9†	No. 10†	No. 11	No. 12
Yellow Corn	35	86	66	44	56.5	56	58	58	60	46	55	43	47
Wheat Shorts	20		20	20	20	20	20	20	20	20	20	20	20
Wheat Bran	10												
Meat Scrap	10					18	20				10		5
Fish Meal	10				17			20					5
Tankage									18				
Peanut Meal										29	12		
Cottonseed Meal												27	15
Soybean Meal				27									
Skimmilk Powder	10	10	10	5	5	5						5	5
Bone Meal	2	3	3	2.5						2.5		1	
Alfalfa Meal	2												
Salt	1	1	1	1	1	1	1	1	1	1	1	1	1
Limestone				0.5	0.5					0.5	1	2	1
Sardine Oil							1	1	1	1	1	1	1

* Found to be complete for growth, maintenance and reproduction.
† Found to be complete for growth and maintenance.

Intestinal Parisite Remedy for Horses

Phenothiazine	40 g.
Carbon Disulphide	24 cc.

Administer after a 36 hour fast.

Glycerin-Phenol Treatment for Horse Galls

Of practical value in the treatment of horse galls is the simple glycerin-phenol mixture given below:

Phenol	1
Glycerin	15

This preparation has properties intermediate between those of ointments and lotions, both of which are used in the treatment of horse galls.

The animal should be permitted to rest for several days to allow healing to progress.

Horse Liniment

Camphor	5
Tincture of Cantharides	5
Tincture of Capsicum	5
Tincture of Arnica	50
Alcohol	35
Diglycol Oleate	5

Treating Horse Cracked Heels and Leg Irritations

Tincture of Iodine	1
Carbolic Acid	3
Glycerin, To make	32

A piece of cotton large enough to cover the injured area is saturated with the mixture, covered with a sheet of cotton and bandaged on. This should be repeated every day or two.

This is an excellent preparation for chronic scratches. It was used when horses' legs became blistered or irritated from the salt thrown on the street in cold weather. It is helpful in almost any leg condition.

Garden Fertilizers

Formula No.	1	2	3	4	5
Sulphate of Ammonia	3¾	—	1	5	2
Hoof-and-Horn Meal	—	5¼	1½	—	5
Superphosphate (16% P_2O_5)	10	—	10	10	8½
Steamed Bone Flour	2½	5½	4½	2½	2½
Bone Meal	—	5½	—	—	—
Sulphate of Potash	3¾	3¾	3	2½	2

Reinforced Manures

A ton of average mixed farm manure, in the moist condition, will contain about ten pounds of nitrogen, six pounds of phosphoric acid and ten pounds of potash. It is thus apparent that it is important to increase the phosphoric acid content of manure. It is well also to increase to some extent the amount of available nitrogen and potash, especially if it is to be used in place of commercial fertilizer. A suitable mixture could be made as follows:

Formula No. 1

	Containing		
	N lb.	P_2O_5 lb.	K_2O lb.
1400 lb. Farm Manure	7.0	4.2	7.0
400 lb. (16%) Superphosphate		64.0	

	Containing		
	N lb.	P_2O_5 lb.	K_2O lb.
100 lb. Sulfate of Ammonia	20.5		
100 lb. (50%) Muriate of Potash			50.0

In plant food content this would approximate 800 pounds of a 4–8–7 fertilizer, which would be suitable for a variety of crops.

No. 2

	Containing		
	N ib.	P_2O_5 lb.	K_2O lb.
1500 lb. Hen Manure	15.0	12.0	6.0
300 lb .(20%) Superphosphate		60.0	
100 lb. Sulfate of Ammonia	20.5		
100 lb. (50%) Muriate of Potash			50.0

One ton of this mixture would be approximately equivalent, in plant food content, to 800 pounds of a 4–9–7 fertilizer.

No. 3

	Containing		
	N lb.	P_2O_5 lb.	K_2O lb.
1600 lb. Dried Poultry Manure	51.2	51.2	32.0
300 lb. (16%) Superphosphate		48.0	
50 lb. Sulfate of Ammonia	10.3		
50 lb. (50%) Muriate of Potash			25.0

One ton of this mixture would approximate in plant food content 1000 pounds of a 5–10–5 fertilizer, which would be suitable for a variety of vegetable and other crops.

It is frequently advisable to use superphosphate on the dropping boards in poultry houses. This reinforces the manure at its weakest point—phosphoric acid—and likewise aids in retaining ammonia which might otherwise be lost. The ammonia in the manure, which is more or less volatile, and thus easily lost, unites with the gypsum of the superphosphate to form a stable compound, ammonium sulfate.

It is not advisable to use lime with manure. This reacts, with compounds in the manure, in a way that causes the loss of ammonia.

Manure in piles should not be exposed to leaching rains, which would naturally cause a great loss of soluble constituents, but on the other hand should be kept moderately moist to prevent loss of ammonia and destruction of organic matter through undue heating. Unless manure can be kept moist and well packed, without exposure to heavy rains, it is best, where possible, to spread on the fields as produced.

Manure is also reinforced and saved by the proper use of bedding. The liquid portion of the manure is the most valuable, and the greater the absorbing properties of the bedding the greater its value in this respect. There are a number of materials which may be used for bedding, as for example, the different straws, salt marsh hay, peanut shells, wood shavings, sawdust, and peat moss. Of these, peat moss has the greatest absorbing power but contains comparatively little plant food. Straws, salt marsh hay, and peanut shells contain an appreciable amount of plant food constituents.

Poultry Lime Granules
British Patent 539,298

Calcium Carbonate	25,000
Calcium Phosphate	500
Magnesium Carbonate	200
Iron Sulphate	200
Sulphur, Flowers of	10
Potassium Permanganate	5

The above is mixed and tumbled in a rotary drum while being sprayed with 18–20% water. Dextrin or skimmed milk may be dissolved in the water if a smooth finish is wanted.

Substitute for Farm Manure

Because manure is not available on many farms now, a fairly good substitute can be made up as follows: To one ton of straw or similar material, including garden refuse and leaves, add a mixture of about 75 pounds ammonium sulfate, 50 pounds superphosphate, and 100 pounds of powdered limestone. This is mixed with the straw. Keep the whole moist by frequent watering, and turn it over occasionally. After about three months, the straw will become well rotted and will be quite similar in appearance, composition, and effects to the ordinary farm manure.

Inexpensive Fertilizer

Wood Ashes	1	lb.
Ammonium Phosphate	2	oz.
Sodium Nitrate	2	oz.
Ferrous Sulphate	¼	oz.

Mix all together and use 1 teaspoonful placed around the plant, *not closer than 10 inches,* else the mixture may burn the tender roots.

Conservatory Fertilizer

Potassium Nitrate	2	oz.
Disodium Phosphate	5	oz.
Ammonium Nitrate	1	oz.
Ferrous Sulphate	⅛	oz.
Calcium Chloride	⅛	oz.

Use ¼ oz. to a gallon of water. Avoid sprinkling foliage with the mixture.

Plant Growth Stimulant
German Patent 682,023

Fluorescein	600	mg.
Iron Sulphate	500	mg.
Copper Sulphate	40	mg.
Peat, Powdered	100	kg.

Mix the above well and make weakly alkaline with sodium bicarbonate solution.

Chapter Seven

FOOD PRODUCTS

Uncooked Dried Fruit Candies

Whole, Seeded Muscat Raisins	2.5 lb.
Prunes, Pitted, Finely Ground	2.5 lb.
Coarsely Chopped, Walnuts or Almonds	2.5 lb.

Warm and mix the fruits in a jacketed kettle until well softened, add the nuts and mix. Press into rectangular metal or wooden forms, or roll out on an oiled slab or on oiled paper. Let harden, cut to size if on a slab. Coat with ground nuts or in other suitable manner. Wrap as 5¢ bars. The prunes and raisins are rather sticky and therefore bind the mass together. Figs may be substituted for the prunes.

The dried fruits should contain about 22% moisture and should neither be too dry and hard, nor too moist and soft. Dried apricots may be substituted in part for the prunes, equal to about 50% of the prunes.

A better bar is obtained if a binder of fondant or fudge is used to give a firmer texture to the candy.

Raisins (Seeded Muscat or Seedless)	2.5 lb.
Pitted Prunes, Finely Ground	2.5 lb.
Coarsely Chopped Walnuts or Almonds	2.5 lb.
Freshly Made Warm Fondant Preferably Made with Invert	2.5 lb.

Mix in a mechanical mixer, or warm and mix in a jacketed kettle. Spread on oiled slab or paper to harden. Cut to desired size.

The flavor of the bar in the second formula is improved if about 10%, by weight, of orange peel jam is added. It may be made by taking dried whole orange peel not previously used for oil recovery, adding to it 6–7 parts of water, boiling 2–3 minutes, soaking overnight, cooking soft, grinding finely, adding one half its weight of sugar and cooking to a heavy jam.

"Arctic Ice" Dried Fruit Candy

High Melting Fat (90° F. or Above)	7 lb.
Powdered Sugar	10 lb.
Powdered Milk	3½ lb.
Chopped or Ground Dried Fruit	20 lb.
(Figs, Apricots, Prunes or Raisins or a Mixture of Any Two or More)	
Vanilla Flavoring, To suit	
Chopped Walnuts or Almonds	3–5 lb.

Melt the Fat. Stir in the powdered milk and sugar. Allow to cool until it begins to thicken. Add the fruits and nuts. Mix well. Spread on slab or paper to harden. Cut to desired size.

Fruit Rolls: Another uncooked fruit candy can be made by mixing ground dried fruits with about 20% their weight of chopped nuts and a little honey (about 20%), or orange concentrate, invert syrup, or other heavy syrup as a binder, and extruding through a sausage casing filling machine, cutting off the "sausage" in short lengths as it is extruded, and coating these pieces with chopped nuts, ground cocoanut, or powdered sugar. All this is done mechanically in making date rolls. These are coated with cocoanut.

A mixture of a sticky dried fruit such as ground raisins with a less sticky one, such as ground prunes or apricots, may be used, and the syrup binder omitted. Figs need little syrup binder and no added raisins if they are previously processed in a packing plant or by the confectioner to about 23% moisture, that is until they are pliable. They keep much better if syrup is added.

Fruit Jelly Candies

If made with fruit pulp or fruit juice, the candy is more flavorful and contains the dietary value of the fruit used. Fresh fruits are cooked with a little water until soft and are then sieved; dried fruits are cooked in sufficient water to cover well and when soft are sieved to give a juice or pulp. Canned pie grade fruits in No. 10 cans may be sieved without additional cooking. As a general rule, the dried fruits will be more suitable since they are inexpensive and give pulps that are "meaty" and not too thin or watery. Canned or bottled fruit juices may be used, such as pineapple, grape, orange or grapefruit.

Fruit Pulp or Juice	40	lb.
Cane Sugar	30	lb.
Invert Syrup or Glucose Syrup	30	lb.
Powdered Confectioner's Pectin (High Jellying Power)	1¼	lb.
Citric Acid, 8 Ounces Dissolved in 2 Pints of Water		

Mix the dry pectin with half the sugar. Heat about 3 gallons of water to boiling and sprinkle the mixture of pectin and sugar with it, stirring rapidly to avoid lumping. Several minutes heating will be necessary to dissolve it.

Mix the fruit pulp (or juice), invert and syrup or glucose syrup, remaining sugar and then the pectin solution. Cook in a jacketed kettle to 223° F., add the citric acid solution, and then boil to 223½°–224° F., that is 11½–12° F. above the boiling point of water. Cast in starch molds, or pour on to an oiled slab or oiled paper to harden. It is advisable to follow the latter procedure as it is sometimes difficult to cast the hot jelly. Coat with coarse sugar or in finely cut cocoanut.

Other Suggestions

Dried fruits can be used in various standard candies by adding them after the batch has been cooked to the finishing point. For example it can be whipped into marshmallow, or added to fondant after creaming, or to fudge before pouring. The writer does not pretend to be, or does not pose as, an expert confectioner; he is primarily a fruit products chemist and technologist. Therefore, you as experienced candy makers, will see other possible uses of fruits in candy that do not occur to me, and will be able to improve greatly upon the formulae I have offered in this article.

The principal fact to remember concerning dried fruits is that they contain from 50 to 65% of sugar and hence are excellent substitutes for sugar in certain kinds of candy to which they are adapted. Also they produce candies of exceptional dietary value.

Maple Bonbon Coating

Water	3½	gal.
Condensed Maple Syrup (2½ Fold)	10	lb.
Sugar (Sucrose)	75	lb.
Salt	¾	oz.
Egg White (Beaten)	½	pint
Sheet Gelatin (Melted in 2 oz. Hot Water)	1	oz.

Cook sugar, condensed maple flavor, salt and water together to 240–242° F. in steam-jacketed kettle. Turn off steam and add melted gelatin.

Pour batch on Ball beater and allow to cool to blood heat (97° F.) before heating. When half beaten, add egg whites which have been previously whipped to a light fluff.

Let set overnight before using. Produces a fine bonbon coating for highest-class goods.

Praline Crunch

Granulated Sugar	1 lb.
Butter	1 lb.
Nut Meats, Roasted	8 oz.

Nut meats should be chopped, not too fine. Melt the sugar to a golden brown, 340 to 350° F. Remove from the fire and thoroughly mix in the nut meats and butter. Pour onto a buttered bun pan. When cool, crush.

This can be used in icings or to dress the top and sides of a cake.

Retention of Flavor in Hard Candies

Flavor retention in hard candies can be obtained by the use of glyceryl oleate (edible). It acts as a solvent fixative for extract flavors used in hard candies such as peppermint, cherry, lemon, etc., and prevents the flavor from volatilizing on addition to the hot melt (300° F.).

Billowy Marshmallow

Gelatin	1 lb.
Water (Cold)	8 lb.
Eggs (Whites)	15 lb.
Sugar (Icing)	65 lb.
Standardized Invert Sugar	10 lb.
Flavor and Color, As desired	

Soak the gelatin in the cold water. Heat the soaked gelatin carefully until fluid (about 140° F.). Add the fluid gelatin to the egg whites, 35 lb. of sugar and the standardized invert sugar. Whip this mixture until it is fairly stiff. Just before finishing the whipping, add the remaining 30 lb. of sugar along with the flavor and color. Beat until of the consistency of marshmallow.

For banana flavor use, for each 10-lb. batch, about 4 oz. of ripe banana powder containing dry milk solids. Incorporate it in any way convenient, preferably first blended with about an equal weight of sugar.

Marshmallow Filler

Sugar XXXX	4	lb.
Glucose	1	lb.
Egg Whites	½	pt.
Invert Syrup	4	lb.
Vanilla Gelatin (220 Bloom), High Test	2¾	oz.
Water	1	pt.
Salt	⅛	oz.

Soak gelatin in the water for one hour and then heat to 180° F. Place in the mixing bowl with the other ingredients and beat for 5 minutes at medium speed. Change to high speed and beat until stiff. Add flavor to taste.

Marshmallow for Chocolate Dipping

Granulated Sugar	40 lb.
Standardized Invert Sugar	10 lb.
Corn Syrup	50 lb.
Gelatin	3 lb.
Water	14 lb.

Dissolve the gelatin in the water until tender, then heat until fully dissolved. Add the granulated and standardized invert sugar and heat until they are dissolved. Place the batch into the beater, add the corn syrup and beat until the marshmallow is sufficiently light. Flavor and color can be added as desired. The proportion of water must be increased or decreased according to the conditions under which the goods are manufactured.

Caramel Base

A

Water	1 lb. 8 oz.
Brown Sugar	5 lb.
Butter	1 lb.

B

Sugar (Granulated Sucrose)	3 lb. 5 oz.
Powdered Skim Milk	1 lb. 11 oz.
Water	2 lb. 8 oz.

Place ingredients *A* (water, brown sugar and butter) in a copper kettle and stir until the sugar is dissolved. Place on fire and boil to 320° F. If no thermometer is available, boil to a hard crack. Wash sides of kettle with a wet brush while boiling, but do not stir.

When temperature reaches 300° F., turn flame down low and boil slowly until the temperature reaches 320° F.

Of the ingredients marked *B*, mix together the sugar and dry milk. Pour this mixture into the water, and stir with a wire whip until all ingredients are dissolved.

Remove kettle containing *A* from fire when temperature has reached 320° F. and add gradually the mixture *B* (sugar, milk, water) while stirring *A* briskly with a wire whip. Continue this stirring until base is smooth.

If a darker color is desired, add 2 oz. of burnt sugar color. Cool base before using. Store in a covered container in a cool, dry place.

This caramel base has excellent keeping properties, and may be used as required.

Caramel Sauce

White Sugar	5
Water	5
Cream	1
Butter	0.3
Flour	0.1

Brown the white sugar, add water and boil until it forms a soft ball in cold water. Stir quickly. Let come just to a boil, turn off heat and add butter.

"Homemade" Fudge Icing

Granulated Sugar	4 lb. 8 oz.
Corn Syrup	12 oz.
Milk	1½ pt.
Concentrated Maple Syrup (2½ Fold)	8 oz.
Butter	6 oz.

Boil the granulated sugar, corn syrup and milk together to 234° F., and at the finish of the cook, add the concentrated maple syrup. Cook to 130° F. Finally, add, and mix until smooth, 6 oz. of butter and sufficient icing sugar to give the consistency desired. Apply warm or at about 110° F.

Chewing Gum Base
U. S. Patent 2,137,746
Formula No. 1

Jelutong (Dry)	92%
Paraffin Wax (M. P. 168° F.)	8

No. 2

Ester Gum	30%
Coumarone Resin	45
Latex (Dry)	15
Paraffin Wax (M. P. 180° F.)	10

No. 3

Jelutong (Dry)	80%
Gutta Siak	15
Paraffin Wax (M. P. 176° F.)	5

Chewing Gum

A chewing gum mix is made employing about 20% of the chewing gum base, about 60% of pulverized sugar (sucrose), about 19% commercial corn syrup, and about 1% of a desired flavor.

Activated Charcoal Chewing Gum
British Patent 523,122

Gum Base	20 lb.
Glucose	20 lb.
Sugar	60 lb.
Activated Charcoal	3¼ lb.
Flavoring	1 lb.

Pumpernickel Bread

Rye Meal	4 lb.
Rye Flour	4 lb.
Strong Clear Flour	5–6 lb.
Water	1 gal.
Yeast	3 oz.
Salt	4 oz.
Malt Extract	2 oz.

Temperature of stove should be approximately 82° F. to obtain the best results. Approximate time for first rising, 1 hour, 30 minutes; second rising, 30 minutes; and take-in, 30 minutes.

Scale and round out the dough. Make up and place close together on boards or boxes dusted with cornmeal. Grease loaves when placing them together.

Soy Noodles

Flour, Spring Wheat	117 lb
Soy Flour, Full Fat	13 lb.
Whole Eggs	24 lb.
Water	24 lb.

Mix soy flour with wheat flour, then proceed as with regular noodles.

Silver Doughnut Glaze

Glucose	1 lb. 4 oz.
Salt	1 oz.
Water	4 lb. 8 oz.

Add glucose and salt to water. Heat to boiling and remove from fire.

Gelatin	1 oz.
Water, Cold	8 oz.

Stir gelatin into the cold water. Let stand about 5 minutes, or while glucose-water syrup is heating to a boil.

Add gelatin solution to the hot syrup immediately after removing from the fire. Stir until gelatin is thoroughly dissolved in the hot syrup.

Sugar, XXXX	20 lb.
Vanilla Extract	1 oz.

Sift sugar to remove all lumps.

Stir while slowly pouring the *hot* gelatin-glucose solution on the sugar. Mix thoroughly. Cool to about 120° F. to use.

Doughnuts give off considerable moisture while cooling. A good glaze can be ruined by dipping doughnuts while too warm. For best results, cool doughnuts to about 110° F. and have the glaze at about 120° F. Dip the doughnuts and lift them out immediately to drain on wire screens or racks.

If the glazed doughnuts are packaged, they must not be placed in airtight packages. For this will promote sweating and soaking the glaze.

Prepared Doughnut Glace

Sugar, Confectioner's	100 lb.
Pure or Imitation Vanilla (16 Concentrate)	1 oz.
Algin	1 oz.
Powdered Gum Karaya	1 oz.
Gelatin (Fine Powdered)	8 oz.
Tapioca Flour	3 lb.

Mix ingredients together (add vanilla mixed with part of the sugar). Add boiling water to produce consistency and stiffness required to coat doughnuts. Keep over hot water bath while using.

Macaroon Cup Cakes

Egg Whites	5 lb.
Granulated Sugar	9 lb.
Salt	¾ oz.
Cream of Tartar	1 oz.
Vanilla Extract	1 oz.
Macaroon Cocoanut	5 lb. 10 oz.
White Flour	5 oz.

Whip egg whites to a wet peak, but not quite as stiff as for angel food. Mix 4 lb. granulated sugar with the salt, cream of tartar, and vanilla. Add mixture to the beaten whites. Do not overbeat. Mix remaining 5 lb. of granulated sugar with the macaroon cocoanut and white flour. Fold in gently, either by hand or by machine.

Use crinkled parchment cup cake liners in standard cup cake pans. Measure 15 to 16 oz. per doz. cups. Bake approximately 20 to 25 minutes at 350° F. For variety place pieces of candied fruit or nuts on top of the batter before baking.

For commercial production the egg whites should be 75° F. Hand finishing of mixing is recommended.

Homemade Pie Dough for Fruit Pies

Flour (Ordinary Pastry Flour)	10 lb.
Shortening	7 lb.
Ice Water	2 lb. 12 oz.
Salt	6 oz.
Corn Sugar	5 oz.

Rub the shortening and flour together until the flour is partly coated with fat. Do not mix to a paste but, on the other hand, do not leave the shortening lumps too large.

Dissolve the salt and corn sugar in the water, add to the flour and shortening mixture until the mass is of uniform consistency.

Washing of pies made with this dough is not necessary. If correctly mixed, crusts will color uniformly on both top and bottom. If the flour is too soft, the crust will not remain crisp. If bread flour is used, the crust will likely shrink. Therefore, a medium-quality pastry flour is recommended for this formula.

Pointers for preparation:

1. The best results are obtained by using cold ingredients.
2. Store the flour to be used for pies in a cool place.
3. Water should be below 45° F.
4. Shortening should be kept in a cool place.
5. Hot flour or shortening will ruin the crust.

The shortening used in pie dough should be a product having a mild lard flavor; a better body than is normally obtained in lard; and a wide plastic range.

Waffle Mix

Sugar	36	lb.
Salt	15	lb.
Totally Hydrogenated Shortening	120	lb.
Soy Flour	18	lb.
Milk Powder	36	lb.
Soda	12	lb.
Cream of Tartar	12	lb.
Egg Yolk	40	lb.
Malted Milk	9	lb.
Cake Flour	360	lb.
Bread Flour	240	lb.
Powdered Turmeric	2.5	oz.

Mix ingredients together. Add a sufficient amount of liquid milk or water to produce a batter of medium consistency. Usually approximately 1 lb. 4 oz. of liquid milk to 1 lb. of flour. Batter should not, however, be excessively soft.

Cinnamon Meringue

Egg Whites	1	lb.	8	oz.
Sugar	2	lb.	10	oz.
Vanilla Extract			½	oz.
Cinnamon			¼	oz.

Beat egg whites until firm. Add sugar and vanilla slowly. Add cinnamon and stir in until smooth.

Marshmallow Cake Icing

Albumen	1¼	lb.
Gelatin (220 Bloom), High Test	9	oz.
Granulated Sugar	10½	oz.
Water (140° F.)	11½	lb.
Vanilla	2½	oz.
Granulated Sugar	21	lb.
Salt	5¼	oz.

Sift the albumen, gelatin and sugar three times. Place water in a mixing bowl and add the sifted albumen, gelatin and sugar and whip by hand until dissolved. Place bowl in machine and whip at high speed over a period of 6 minutes. Then whip at low speed for a few minutes. Add vanilla and mix 30 seconds. Add sugar and salt gradually over a period of 30 seconds. Remove whip and fold a few times by hand.

Cream-Type Icing Stock (Non-Sticky)

Butter	2	lb.	8	oz.
Hydrogenated Shortening	5	lb.	4	oz.
Water	10	lb.	6	oz.
Sugar (Powdered 4X)	50	lb.	0	oz.
Salt			8	oz.
Dry Milk Solids *	10	lb.	6	oz.

Blend butter and shortening thoroughly at 75° F. Add water carefully with slow-speed beating to make an emulsion. Blend sugar, salt and dry milk solids thoroughly and add gradually to emulsion. Cream well.

Desired variations may be had by adding various flavorings, colorings, fruits or nuts to this icing stock.

* Not over 1½% fat content.

Butter Cream Icing

Butter	12 oz.
Hydrogenated Shortening	2 lb. 6 oz.
Water	2 lb. 4 oz.
Sugar (Powdered 4X)	12 lb. 8 oz.
Dry Milk Solids *	3 lb. 12 oz.
Vanilla Extract, To suit	
Egg Whites	12 oz.

Blend fats well at 75° F. Add water carefully with slow-speed beating to make an emulsion. Sift sugar and milk solids several times and add gradually to mix. Beat smooth. Add vanilla flavor. Beat egg whites stiff and fold in. Mix until smooth.

If chocolate flavor and color are desired, add 1 oz. natural or "domestic" cocoa powder for each 15 oz. of butter cream icing stock used. Mix on low speed to get uniformity in mix.

Caramel Custard Pie

Whole Milk	8 lb.
Brown Sugar	3 lb.
Salt	1/4 oz.
Vanilla	1 1/2 oz.
Frozen Eggs	2 lb. 12 oz.
Maple Flavor	1/8 oz.
Caramel Color	1/2 oz.

Heat the milk to 140° F. Add the sugar, salt, flavors and color.

Stir until the sugar is melted. Add the eggs and stir. Do not whip.

Strain through a small sieve to remove any insoluble matter. Skim off the foam from the surface.

Pour into unbaked shells. Bake at 450° F.

Improved Yeast

U. S. Patent 2,223,465

After yeast has been "filter-pressed" 0.12–1% of any of the fol-lowing is incorporated:

Propylene Glycol Laurate
Glyceryl Mono Laurate
Sorbitol Laurate

A whiter color; improved cuttin[g] and increase in water binding ca-pacity results.

* Not over 1 1/2% fat content.

Fruit Purées for Ice Cream

The preparation of raspberry[,] boysenberry, loganberry, young-berry, blackberry and blueberr[y] pectinized purées is a relativel[y] simple matter, e.g.

Pulpy Fruit Juice or Purée	100 lb.
Granulated Sugar	50 lb.
Enzyme-Converted Corn Syrup (43° Bé)	60 lb.
Pectin * (110 Grade)	2 lb.
Citric Acid †	1 1/2 lb.
Water	20 lb.

The pectin is stirred into 15 lb. o[f] the enzyme-converted corn syrup[,] which has previously been warme[d] to about 160° F. This suspension o[f] pectin in corn syrup is then slowl[y] stirred into the water which ha[s] been brought just to the boilin[g] point. The solution is agitated rap-idly and kept hot for 5 to 10 min-utes, or until a smooth syrupy solu-tion is obtained. Heating is then dis-continued, and while the agitatio[n] is continued the remainder of th[e] enzyme-converted corn syrup i[s] added. The fruit purée, to which ha[s] been previously added the granu-lated sugar and finely powdered cit-

* Less pectin is required for black rasp-berry purée. One pound per 100 lb. of fruit will be sufficient.

† The more tart the fruit, the less citric acid should be added. Thus little, if any, citric acid needs to be used in making loganberry purée.

c acid, is then stirred with rapid ;itation into the solution of the ctin in the corn syrup.

The product may be heated to 70° F. and then canned without oling, or it may be cooled for im- ediate use or for freezing preser- ation. If the purée is canned, it ould be filled into cans lined with ecial fruit enamel, taking care to ll each can completely. The cans re closed immediately, then they re inverted or turned on the side order to sterilize the covers. After tanding for five minutes, they are ooled in running water.

If the purée is to be used immedi- tely or frozen for future use it hould be cooled, under slow agita- on, by running water in the jacket f the kettle in which it was made. f the purée is to be preserved by reezing, it should be cooled as cold s possible in the kettle and then un into 30- or 50-lb. enamel-lined lip-cover cans and frozen at 0° F. r lower.

Peach Purée

Ground Peaches	100	lb.
Granulated Sugar	50	lb.
Enzyme-Converted Corn Syrup (43° Bé)	60	lb.
Pectin	2½	lb.
Citric Acid	1¼	lb.
Water	20	lb.

The solution of the pectin in the enzyme-converted corn syrup is obtained in exactly the same manner as described above under pectinized berry purée. In this case the solution is raised to and maintained at 175° F. The peaches are peeled either by scalding in boiling water or steam for about 90 seconds or by immersion in hot lye solution.

After the peels have been rubbed off, the peaches are halved and the pits removed. The halved peaches are placed in the hopper of a food chopper, especially arranged to grind soft fruits, or in a similar grinding device, sprinkled with some of the sugar and some citric acid solution prepared by dissolving the citric acid in a small amount of water. As fast as the peaches are ground, they are stirred directly into the pectin-corn syrup solution, which is maintained steadily at 175° F. When the proper weight of ground peaches has been added, any remaining sugar and citric acid is stirred into the batch. Then the pectinized purée is cooled by circulating cold water in the jacket of the kettle.

Currant Purée

Pulpy Currant Juice	100	lb.
Granulated Sugar	50	lb.
Enzyme-Converted Corn Syrup	60	lb.
Pectin (100 Grade)	1	lb.
Water	20	lb.

Currant juice is so tart that no citric acid need be used.

Pineapple Purée

Canned Crushed Pineapple	100	lb.
Granulated Sugar	45	lb.
Enzyme-Converted Corn Syrup	60	lb.
Pectin (100 Grade)	2½	lb.
Citric Acid	1½	lb.

First the crushed pineapple should be disintegrated. The pulpy product is combined with the sugar, syrup and pectin by the procedure described under berry purées. A suitable color should be added, as otherwise the product is scarcely noticeable in the ice cream.

Grape Purée

Concord Grape Juice	100	lb.
Granulated Sugar	50	lb.
Enzyme-Converted Corn Syrup	60	lb.
Pectin (100 Grade)	2	lb.
Citric Acid	1½	lb.
Water	20	lb.

The method of making this product is the same as that described for berry purées, the Concord grape juice being substituted for the pulpy fruit juice.

Damson Plum Purée

Damson Plums Frozen with Sugar (3 + 1)	100	lb.
Granulated Sugar	12½	lb.
Enzyme-Converted Corn Syrup	45	lb.
Pectin (100 Grade)	1	lb.
Water	15	lb.

The thawed plums should be converted into a purée in a tomato juice extractor of the tapered screw type. The resultant purée is combined with the sugar, corn syrup, and pectin according to the manner described under berry purées. Damson plums are so tart that no acid need be added.

English Morello Cherry Purée

English Morello Cherry Purée	100	lb.
Granulated Sugar	50	lb.
Enyzme-Converted Corn Syrup	60	lb.
Pectin (100 Grade)	2½	lb.
Water	25	lb.

English Morello cherries usually are so acid that no citric acid need be used. Either fresh or frozen cherries can be used.

Maraschino Cherry Purée

The Maraschino cherries shoul be pulped in a suitable machir which will comminute even tl skins. The pulp should then l mixed with the syrup and the cherr juice. From then on the procedu is the same as that described fo berry purées.

Maraschino Cherries in Syrup (Pulped)	35	lb.
Hot Pressed Montmorency Cherry Juice	65	lb.
Granulated Sugar	40	lb.
Enzyme-Converted Corn Syrup	60	lb.
Pectin (100 Grade)	2½	lb.
Citric Acid	8	oz.
Water	20	lb.

Orange Marmalade Purée

An orange marmalade is prepare by cooking the following ingredient for about 15 minutes.

Orange Juice	5 qt.
Lemon Juice	1 qt.
Finely Sliced and Chopped Cooked Orange Peel	2 lb.
Granulated Sugar	7 lb.
Enzyme-Converted Corn Syrup	8 lb.
Pectin	5 oz.
Water	2 lb.

The pectin is dissolved in corn syrup and water according to the procedure described under berry purées. The finely sliced and chopped cooked orange peel is heated to boiling in the mixture of juices. Then the pectin solution is stirred in. Next the granulated sugar is added and finally the remainder of the corn syrup. The whole is oiled for about 15 minutes but the otal solids content should not be ermitted to rise above about 55%. he product may be used in making ither a variegated ice cream or herbet.

Milk Chocolate Ice Cream

Heavy Sweet Cream *	30
Condensed Skim Milk	30
Sugar, Granulated	10
Corn Syrup Solids	5
Malt Syrup	2.5
Cocoa Powder	2.6
Gelatin	0.3
Soybean Lecithin	0.1
Water	19.5

Combine cocoa powder with equal arts of granulated cane sugar and nix dry, then add the water. Prepare the chocolate syrup by adding the lecithin and malt syrup with the aid of an agitator until smooth.

Place cream and condensed skim milk in the pasteurizing tank and apply heat. When the temperature on the recording chart of the pasteurizer reaches 120° F., add the remainder of the granulated sugar, the gelatin, and corn syrup solids. Finally add the prepared chocolate syrup and proceed with the pasteurization process by heating the complete chocolate ice cream mix to 155° F. for 30 minutes. Homogenize at 2500 lb. pressure. Cool to about 45° F. Place in storage tank until ready to freeze.

The preparation of the chocolate syrup is best accomplished in a steam-jacketed receptacle or double boiler.

* Use sweet cream with 40% butterfat ontent.

Ice Cream Mixes

(Using all butter in place of cream.)

40 quarts — 12% fat — 37½% Total Solids

		Using Whole Milk	
Sweet Butter	13 lb.	Sweet Butter.	11 lb.
Sugar	13 lb. 8 oz.	Skim Milk Powder	4 lb.
Skim Milk Powder	9 lb. 8 oz.	Whole Milk	30½ qt.
Gelatin	6 oz.	Sugar	13 lb. 8 oz.
Egg Yolk Powder	6 oz.	Gelatin	6 oz.
Water	54 lb.	Egg Yolk Powder	6 oz.

14% fat — 38½% Total Solids

		Using Whole Milk	
Butter	15 lb.	Butter	13 lb.
Skim Milk Powder	2 lb. 5 oz.	Skim Milk Powder	3 lb.
Sugar	13 lb. 8 oz.	Sugar	13 lb. 8 oz.
Gelatin	6 oz.	Gelatin	6 oz.
Egg Yolk Powder	6 oz.	Egg Yolk Powder	6 oz.
Milk	29½ qt.	Milk	30 qt.

16% fat		$39\frac{1}{2}\%$ Total Solids Using Whole Mil	
Butter	17 lb.	Butter	15 lb.
Skim Milk Powder	7 lb. 3 oz.	Skim Milk Powder	8 lb. 6 oz.
Sugar	13 lb. 8 oz.	Sugar	13 lb. 8 oz.
Gelatin	6 oz.	Gelatin	6 oz.
Egg Yolk Powder	6 oz.	Egg Yolk Powder	6 oz.
Water	52 lb.	Water	$26\frac{1}{2}$ qt.

Procedure

1. Dissolve milk powder in water. Add butter. Heat to 110° F.
2. Mix thoroughly sugar, dry egg yolk and gelatin powder.
3. Add to milk while stirring continually.
4. Heat to 145–150° F. Hold for 10 to 20 minutes.
5. Stir vigorously and pour into bowl.
6. Homogenize with nozzle nuts slightly loosened.
7. Cool immediately and put into ice box for 24 to 48 hours aging.

Artificial Whipping Cream
(Quantity 1 Gallon)

Heavy Cream

Sweet Cream Butter	32 oz.
Shortening	20 oz.
Whole Milk Powder	12 oz.
Cold Water	72 oz.

Coffee Flavored Cream

Butter	32 oz.
Shortening	20 oz.
Cold Water	72 oz.
Whole Milk Powder	12 oz.
Liquid Coffee Concentrate	4 oz.

(Do not use coffee extract.)

Banana Cream

Butter	24 oz.
Shortening	24 oz.
Whole Milk Powder	6 oz.
Prepared Banana Powder	8 oz.
Cold Water	72 oz.

Chocolate Cream

Butter	24 oz.
Shortening	20 oz.
Chocolate Liquor (Best Grade Bitter Chocolate)	12 oz.
Whole Milk Powder	10 oz.
Cold Water	72 oz.

Mocha Cream

Whip together chocolate and coffee cream—50:50.

Extra Heavy Cream

Butter	28 oz.
Shortening	24 oz.
Whole Milk Powder	16 oz.
Cold Water	72 oz.

(Use in hot weather and with fresh fruits.)

Dissolve milk powder in cold water using wire whip. Add the fat and heat to 145° F. Don't heat over 150°! Stir thoroughly before pouring mix into the machine bowl. Homogenize with nozzle nuts slightly loosened. Cool finished cream immediately with stirring. Store in cold ice box not higher than 40° for at least 48 hours before using. Whip the same as dairy cream. Sugar and flavor to taste.

The butter used in these mixes must be a pure sweet cream butter hich has not been neutralized. The avor depends upon the butter used. reat the finished product the way ou treat a regular dairy cream.

Heavy Coffee Cream
(Cut this with milk as usual)
Formula No. 1

Milk	4 lb. 12 oz.
Butter	3 lb. 6 oz.

No. 2

Butter	3 lb. 10 oz.
Whole Milk Powder	7 oz.
Water	5 lb.

Break butter into small pieces. 'ut into milk. Heat in double boiler o 140–145° F. Adjust valves to lowedium pressure and homogenize.

To dissolve milk powder in water, se wire whip agitating briskly. Then follow procedure above. Alvays use cold water to dissolve milk owder.

Fudge Milk (10 Gallons)
(3% Milk)

Cane Sugar	3 lb. 12 oz.
Corn Sugar	3 lb.
Chocolate Liquor (Bitter Chocolate)	1 lb. 6 oz.
Gelatin	3 oz.
Cold Water	34 qt.
Sweet Butter	3½ lb.
Skim Milk Powder	6 lb.

If desired, 1½ lb. of malted milk powder can be added to the mix and then only 5 lb. of skim milk powder should be used.

The mix may be made up with one pound of butter in place of 3½ lb. The mix then cannot be called a milk, it must be called a drink, e.g. *Chocolate Fudge Drink Fudge Malted Drink, etc.*

Whip milk powder into cold water until dissolved. Break up butter into pieces and put into milk. Add sugar, gelatin, broken chocolate and heat to 140–150° F. Stir for ten minutes until everything has dissolved. Strain while pouring into machine. Homogenize at high pressure with valves one turn from the top.

Cream Neutralizers
(For Butter Manufacture)
Formula No. 1

Sodium Sesquicarbonate	100

No. 2

Sodium Carbonate	65
Sodium Bicarbonate	35

No. 3

Sodium Carbonate	80
Sodium Bicarbonate	20

No. 4

Sodium Carbonate	55
Sodium Bicarbonate	45

Increasing Melting Point of Butter

Butter, for use in hot climates, is mixed with melted glyceryl tristearate (an edible product), in various amounts, to increase its melting point.

Smoked Cheese in Sausage Casing

Use only processed cheese. Heat cheese until an emulsion forms as in processing for loaf cheese and then, while hot, force through standard sausage stuffer into the Visking No. 2x20-in. red fibrous casing. Chill until solid. Place in smokehouse and subject to a hickory smoke for 48 hours at as low temperature as possible to obtain in smokehouse.

Remove from smokehouse and cool for 12 hours at room temperature.

If desired, the finished product may be wrapped in a decorative or trade-marked transparent sheet.

Also, a sharp, snappy natural cheese may be used instead of the processed cheese. In that case the cheese is cut up, ground through a 1/8-in. plate, and worked in the same machine as meat packers use for sausage or in a large vertical bowl type mixer and at the speed as used in creaming butter.

The natural cheeses are not heated as in the case of processed cheese but transferred from the sausage machine or mixer bowl into the stuffer. Stuffing, smoking and cooling procedure is same as when processed cheese is used.

The mechanically worked natural cheese may be made into 2- or 5-lb. loaves, held for 48 hours, cut into four pieces, smoked and wrapped. This product has good demand in eastern markets at present.

PROCESS CHEESE SPREADS

Pimento Cream Spread

Cream Curd from 6% Fat Milk	252	lb.
Pimento Through 5/16-in. Grinder Plate	62½	lb.
Vinegar-Sugar Solution	40	lb.
Gum Paste	31	lb.
Salt	5	lb.
Cheese Color	6	oz.
Cayenne Pepper	1⅓	oz.
Sugar (Granulated)	4	lb.

Mix thoroughly before putting into cooking kettle. Heat to 170° F. Run through filling machine into sterilized glass containers of desired capacity. Use of emulsifier optional. If used, emulsifiers are: Sodium citrate, 1%, or di-sodium phosphat 2%, calculated on weight of batc mix.

Relish Spread

Cream Curd from 6% Fat Milk	250 lb.
Sweet Pickle Relish	60 lb.
Vinegar-Sugar Solution	40 lb.
Pimento (Through an 1/8-in. Grinder Plate)	27 lb.
Condensed Whey or Whey Powder Plus Water	30 lb.
Gum Paste	40 lb.

An emulsifier may be used; som operators prefer its use, some do no

Mix constituents thoroughly, hea to 165° F. or slightly higher, depend ing upon flavor desired, and ru through filling machine into steri lized glass containers of desired ca pacity.

Limburger Cheese in Glass

Limburger Cheese	180 lb.
Cream Curd from 6% Fat Milk	60 lb.
Water	10 lb.
Disodium Phosphate	10 lb.
Salt	2 lb.

Mix thoroughly and heat tc 174° F. Run through filling machine into sterilized glass containers of desired capacity.

Preferred way to cook limburger cheese to assure good spreading consistency is to use 80 lb. of soft smooth limburger, 40 lb. of a firm cheese but with a sharp limburger flavor and 60 lb. of limburger with a "clean" but slightly "off" flavor.

Put 70 lb. of this limburger mixture in cooking kettle with steam turned on and heat with stirring until fat begins to separate sufficiently to give a slight glistening appearance to the cheese which pulls together to form a mass.

Add 5 lb. of the phosphate and 1 lb. of the salt, with continued heating and stirring. Add another 40 lb. of the limburger mixture and stir until mass is smoothed out. Add 30 lb. of the cream curd and stir until smoothed out. Add remainder of limburger mixture. Heat and stir until smooth before adding the remaining 5 lb. of phosphate, 1 lb. of salt and 30 lb. of cream curd.

Caution: After the first charge of cheese is put into kettle, additions of more than 60 lb. should not be made. Also, subsequent additions should not be made until the charge in the kettle has been thoroughly smoothed out by gradual heating and constant stirring by the mechanical agitator in the cooking kettle. The procedure followed in building up the batch in the kettle may be varied according to the operator's experience and the particular properties desired in the finished limburger cheese mixture.

Special "Sharp" Spread

American Cheddar Cheese (Smooth)	340	lb.
Salt	2½	lb.
Disodium Phosphate	6	lb.
Condensed Whey	84	lb.
Water	40	lb.
Cheese Color, To suit		

If needed to firm up the batch, 1 to 2 lb. of sodium citrate or of sodium metaphosphate may be added.

The American cheddar cheese should be partly broken down. One-half should be of clean flavor and aged.

Add cheese to heated kettle slowly as in making the limburger product. Add 15 lb. of the water at start. Do not add the condensed whey until after the batch has been heated to about 135° F.

Finish by heating to 160° F. and packaging in containers other than glass. Make into packages similar to those of processed cheeses.

"Supply" Type Cheese

Aged cheese is used. It should have a pleasing nutty flavor, be not too firm nor too soft in texture. Color is of no consequence unless finished product is to be designated as a "white" cheese.

Cheese	491	lb.
Salt	5	lb.
Sugar	4	lb.

If a high-acid cheese is used, molds will grow more rapidly than if a low-acid or neutral cheese is used, particularly in the case of the 4-oz. ground-cheese package. Also, the low-acid cheese permits whipping in of as much as 3% water. Development of mold growth can be checked largely through the use of propionic acid or its salts.

The cheese used is first cleaned and ground, as in the making of processed cheese, through a 3/32- or ⅛-in. plate. It is then placed in a sausage machine or an upright bowl and beater-type mixer, and salt, sugar, color, other seasoning and water are added, if used. Mix thoroughly preparatory to second passage through grinder in which the perforated plate has been replaced with a funnel of predetermined shape and size to give desired extrusion cross section.

As the helical screw in the grinder forces the ground mixture through the funnel, it is caught on a strip of paper of required width which reels off a roll as the extruded mixture forces the paper strip over a roller conveyor about 6 ft. long.

The paper supports the extruded cheese until it is cut into desired length and put on trays for transfer to cooler or chilling room.

Caciocavallo Cheese

The manufacture of this cheese is peculiar to southern and central Italy; but today it is manufactured in the northern part and on the island as well.

The quality of the taste and its convenient shape make this cheese popular with both Italians and foreign consumers. It is made by a mixture of skimmed evening milk and skimmed morning milk. Sometimes cow's milk mixed with sheep milk is used.

Coagulation: The milk is allowed to coagulate at 35° C. (95° F.) in a tub or pail using as much rennet as may be necessary to complete the coagulation in about half an hour.

Breaking: When the mass has reached the proper consistency, it is broken with a sort of ladle and then with a buckler (harp) into particles, about the size of a hazel nut, which are allowed to settle at the bottom of the container.

In the meantime, the whey is gradually separated until the mass, having been exposed to the air, collects itself and is compressed somewhat with a stirrer. It is then covered with hot whey and left until matured.

Maturation is complete when a piece of the mass, having been immersed in boiling water, stretches out into a smooth elastic thread, an the milky whey from which the ma came has a value of 6.5 to 8 (So hlet), .14 to .18 acidity.

At this point the milky mass sliced and placed into a tub ov which boiling water is poured. I this way, the slices merge into a elastic mass which is made unifor by continued pressing and stretch ing by hand through the use c wooden spatula. Finally it is pulle into one long cord which, in tur is subdivided, into pieces about 3 k (6½ pounds).

A very convenient kneader ha been devised to prepare the curd fc this cheese. Every piece is reduce to a sort of clew * which is left i hot water and is then worked into solid sphere without folds.

Shaping the form: This operatio which can be carried out only b one who has had a long experienc in this type of work, is called "clos ing of the curd." This closing give the cheese its characteristic lon fusiform appearance with a sort o "neck" at one end. Then, placing i in a tin form † or on a short piec of linen, so that it does not lose it shape, it is cooled in cold water.

Salting: As soon as the cheese be comes cold, it is transferred to th brine basin where it stays for thre days. After this, the cheese is take from the brine and each piece o cheese is tied individually by a vege table string at the head. They ar later hung up in pairs on loome poles ‡ near each other in a dr

* This has reference to a ball, as o string.

† Refers to a piece of iron which ha been tinned over.

‡ Poles with linen forms hanging fron them.

amber at 15 to 18° C. (59 to 64°). During this seasoning, this eese receives no care other than ing it and cleansing it of greenish olds which cover it. Then the eese will become yellowed by a ıoking process.

Ripening: It matures in from 10 12 months and weighs from 1.5 2 kg. (3.3 to 4.4 pounds).

Yield: One hundred liters (about 0 pounds) of whole milk yield kg. (17.6 pounds) of fresh cheese ıd 7 kg. (15.4 pounds) of ripened eese.

From this work a very rich whey left which furnishes a good qualy to the skim milk. Caciocavallo a product which is indicated parcularly for grating purposes havg the quality of resisting the heat Summer which is so destructive most cheeses.

This type of cheese rarely swells ut when it does, it is so violent it oes so far as to crack the entire orm.*

Such extraordinary action is atibuted to butyric fermentation hich occurs when the fermenting hey of the curd is substituted for ıe whey heated to 80° C. (176° .), which is intended to soften the urd especially when it is made with ilk that has been skimmed too uch.

Under these latter conditions, the egetative bacterial cells are killed, aving the spore-forming bacteria hich cause butyric fermentation.

It is easy to discover the existence f this fermentation when having ierced the form of cheese with a eedle and bringing a lighted match close to the hole, the escaping hydrogen gas lights.

To avoid this, the temperature of the whey, which is used to hold the curd in a hot bath, must never exceed 50° C. (122° F.).

* A complete piece of cheese is referred o as a "form."

Provolone

Provolone, like the Caciocavallo, originated in southern Italy and Sicily and its manufacture is identical with that which has been described for Caciocavallo. The only difference lies in the fact that the forms are spherical or oval and that whole milk is used. It is a table product (used for eating and not grating purposes) being highly esteemed for its tenderness and butyric quality.

Yield: 10 kg. (22 pounds) of this cheese are obtained from a corresponding 100 kg. (220 pounds) of fresh milk. This 10 kg. (22 pounds), when it is ripened, is then reduced to around 8.7 kg. (19.1 pounds). Each form weighs from 2 to 4 kg. (4.4 to 8.8 pounds).

Provole

Originated in Calabria and Sicily, these cheeses are manufactured from wild ox or buffalo milk by means of a process similar to Mozzarella.

As soon as the milk is drawn, it is coagulated with lamb or goat rennet. The coagulated mass is broken coarsely and let set till the particles become settled to the bottom of the container, then the container is covered or sealed. Afterwards the cheese mass is raised, sliced, and thrown into a bucket containing nearly boiling water or whey. In short time, the curd becomes tender and soft and at this point it is manipulated, reducing it into small spherical pieces of 1 kg. (2.2 pounds) weight which are then transferred to brine where they are kept for about a day, after which they are ready for consumption.

Pecorino Romano

This is the only unique type of sheep milk cheese which belongs distinctly to Italy. It gets its name from the Roman fields where it is produced on a large scale and from where it is exported. Today, its manufacture is extended through southern Italy and into Sardinia.

It is manufactured in a sort of cottage or barn built with poles and straw, using the following utensils:

1. A copper kettle or boiler.
2. "La chiova"—a long ruler-like piece of wood used for stirring.
3. "Cascine"—woven forms for soft cheese—made by weaving willows or twigs.
4. Cheese table or platform for working cheese.

The milk, as soon as it is drawn, is taken into the barn or hut and strained through a linen cloth stretched across the kettle. Then it is heated to about 38° C. (100° F.) and a paste of rennet is added, which is made from dried baby lamb stomachs cut up and kneaded in brine.

Coagulation: The coagulation should take place in 15 to 20 minutes.

It is left to harden further for 10 to 15 minutes, after which it is broken with the "chiova."

It is cooked for 15 or 20 minutes, constantly stirring the particles (using the same device) and raising the temperature to 50° C. (122° F.). The kettle is removed from the fire and is left to allow the curd to sett for 10 to 15 minutes until it is gat ered at the bottom in a mass.

In the meantime, some pails fresh water are poured into the ke tle to lower the temperature of t whey.

Then the cheesemaker very car fully compresses the cheese at t bottom of the kettle, turns it ove and continues the pressing on a sides. In doing this, always worki under the whey, with a string h cuts the mass in one or more piec and turns them over to his assis ants, to be kneaded in the forms c the table.

Molding the cheese: The chees is pressed in the forms, to overflow ing, in such a fashion as to make great part of it rise over the rim c woven mold. This becomes manipu lated, by hand, into a sort of con shape. This is worked until the ve tex of the cone is flush with the ri of the form. The whey, meanwhile is being squeezed out of the chees and trickles out through the open ings on the side of the form. From these same holes, punctures ar made in various directions into th mass to let out as much whey a possible. In the meantime, th cheese is still under hand pressure This operation is called "friction" and lasts around one-quarter hour The pressing is continued for 2 minutes with much care and skill i order to eliminate all perforation and make the mass entirely solid.

Salting: After two days, the salt ing is begun, in the salting room or chambers, by applying dry sal abundantly. During the salting operation, the cheese is removed from the form and deposited on a rt of stone trough and then, after few days, they are transported to wooden table, situated near the me place. Every other day, they e turned over and rubbed with sh salt and placed in piles of twos threes. The salting period lasts ound two months.

From the salting chambers, the eeses pass into the ripening rooms here they are immediately scraped d cleaned.

The only cure which they then re- ive is turning over and a rubbing hand (comparable to massag- g).

The cheeses are properly ripened one year.

Yield: 100 l. (200 pounds) of ilk yield 20 kg. (44 pounds) of esh cheese which, during ripening, come reduced by 18 to 30% or to out 16 to 17 kg. (35.2 to 37.4 unds). Each form is from 12 to 14 n. (4.8 to 5.6 inches) high with a ameter of 24 to 26 cm. (9.6 to 10.4 ches), and weighs on the average kg. (15.2 pounds).

The whey, remaining after the rocess, is of milky appearance and highly esteemed in the Roman untry for its high residual of al- umin and fat content which makes ossible a 10 to 12 kg. (22 to 26.4 ounds) yield of ricotta of a fine nd delicate quality.

---

merican Formulae for Provolone Type

A liquid smoke can be used to roduce a fine smoked cheese of the merican or Provolone type. It is ssential to add the liquid smoke irectly to the milk. It cannot be dded to the curds; and there is no pecial advantage in adding it to the whey directly after cutting the curd. There is no way of determining the traces of liquid smoke which remain in the curd. The practice in no way affects the healthfulness of the cheese. However, any one contemplating using it should determine its legal status as used in cheese that is to be consumed in the state, or that enters the interstate trade channel.

0.07% of liquid smoke, three-fourths of a pound per 1000 pounds of milk, added to the milk is sufficient.

The milk for the Provolone type is prepared by mixing three parts of whole milk with two parts of skim milk by weight. 1% of culture, and 1½ ounces of color, and six ounces of rennet are used per 1000 pounds of milk. The setting temperature is 86° F.

After cutting, the curd is cooked to 94° F. in 40 minutes, and to 108° F. in 10 more minutes. The curd is then cooled from 108 to 94° F. in about 45 minutes and allowed to mat over night. The following morning the curd is soaked in 120° F. water for 30 minutes. The salting rate is 2.5 on the curd weight basis. The cheese is rubbed freely after making. The cheese is made in regular Young American hoops to give a cheese weighing eight pounds.

American Romano Type

Romano type can be made from whole milk with 2½% of starter. When the acidity reaches .20 to .22 and the temperature 86° F., 2½ ounces of rennet per 1000 pounds of milk are added. After 25 minutes the curd is cut as in American cheesemaking.

The curd is cooked from 86 to 110° F. for 40 minutes, and held until properly firmed. This can be guided by a whey acidity of .3.

The drained curd is salted at the rate of 1½ pounds per 1000 pounds of milk. The cheese is hooped and pressed with cloths in the hoops for 24 hours. After three days, and again after seven, the cheeses are rubbed with salt. As the cheeses tend to dry they are also rubbed with a neutral oil.

American Style Process Cheese

Green (Young) Cheddars	5½	Cheddars
Sharp (Aged) Cheddars	1	Cheddar
Skim Milk Cheddars	¼	Cheddar
Salt	3	lb.
Sugar	2	lb.
Disodium Phosphate (Anhy.)	13	lb.
Water	4	gal.

Dissolve the salt, sugar, and disodium phosphate in the water. Slowly add the cheese, ground medium fine, to the above solution at 170° F. with constant stirring until uniform and smooth. Pour into box molds.

Romano Type Process Cheese

Green (Young) Cheddars	1½	lb.
Sharp (Aged) Cheddars	2	lb.
Sbrinza Cheese	2	lb.
Monterey Cheese	½	lb.
Salt	1	oz.
Disodium Phosphate (Anhy.)	2	oz.
Water	20	fl. oz.

Prepare in the same manner a American Style Process Cheese.

Quick Curing Cheddar Cheese

By following the method d scribed, a good marketable chee can be made from milk which co tains a low, medium or high pe centage of acidity. Low acidi ranges from 0.17 to 0.21%; mediu from 0.22 to 0.23%; overripe, 0. to 0.24%, and high, from 0.24 0.30%.

The procedure in the case of mi whether low acidity or high, is t same up to a certain point in a cases. After the milk has been p into the cheese vat, 5% of a good a tive starter is added, followed in mediately by the addition of t usual amount of rennet.

The coagulation of the milk o curs within a short or long perio depending on the acidity of the mi at the time of adding the rennet. T curd is cut into cubes, about ⅜ an inch, immediately after the mi has coagulated firmly.

The acid test of the first whe that separates out, in the case of t low acid milk .2, would be approx mately .12, and in the case of mi with an acidity of .23, the first whe would be .15, and, in every case, t whey would test approximately points less than the milk.

The first whey from milk contai ing 0.3% acid would test 0.22.

Whey must be removed when i acidity is approximately 0.175% Slightly higher acidity (for instanc .18) would make little differenc

ut .18 would be about the limit of
ıe whey acidity that could be al-
wed.

The addition of water at the time
f cutting the curd should be in such
uantity as to lower the acid con-
ent of the whey in all cases to near
.2. The addition of water later on
hould be with the idea of having
he whey acidity approximately
.175 when drawn. After drawing
he whey the curd is piled and
ıilled, etc., not different from the
rocedure in the case of cheese made
y the process now in common use.

In making cheese for the market
y the procedure outlined above, all
teps in processing the milk are
hortened. The average time taken
or the milk to coagulate is little if
ny longer than half the time usu-
lly required. The curd is cut as
oon as it is firm enough, and this is
eft to the judgment of the cheese
naker.

Improved Cream Cheese
Cream Cheese
35% Fat

Standardize cream to contain 12%
at.

Pasteurize at 150° F.; homogenize
at 1500 single stage.

Cool to 70° F.; add 1% starter.

Let set overnight to develop .7%
acidity.

Cut with cheese knives and cook
to 110–115° F.

Drain on cloth racks for about
an hour.

Test cheese curd for fat and moisture.

Calculate amount of cream and
water to add to give 35% fat and
55% water.

Put water and cream in vat.

Calculate amount of salt using
.75%.

Calculate amount of gum using
.25%. The gum should consist of
mixture of equal parts of locust bear
gum, Irish moss, and gum Karaya.

Mix salt and gum together and
sieve onto the cream while agitated
(gum and salt may be mixed in part
of the cheese).

Start heating and add cheese curd.

Pasteurize to 160° F.; homogenize
at 2500 single stage.

Package while hot.

Precautions: It is best to have the
cheese curd tested for fat percentage
desired in the finished cheese so that
excessive quantities of cream or
water need not be added. The cheese
curd is made most fluid by added
water. Too high a water content
makes a sticky, soft cheese. Also
low fat and high gum content makes
a sticky cheese.

It is best to add no extra skim-
milk solids to the original milk,
skimmilk, and cream from which
the cheese curd is made.

The lower the cooking temperature of the milk the easier the cheese
curd melts down. Low homogenization pressures of the milk promote
fat drainage but too low a pressure
increases losses of fat in the whey.
The whey should test below .2%.

Any tendency toward moisture
drainage in the cheese can be prevented by increased amounts of
gum.

The boxes should be inverted
when the cheese is cooling. If the
cheese has shrunk too far from
boxes, the hot cheese mixture may
be cooled to 140° F. before homogenization to reduce shrinkage in
volume due to cooling.

Formula for 250 Gallon Batch
Milk test 3.5%; cream test 55%.

Cream	354 lb.
Milk	210 gal.

Cheese curd yield	700–720 lb.
Water and cream	175
	875 lb. cheese

Salt 6½ lb.

Gums 2.2 lb. or 35.2 oz. made up of

Irish Moss	12 oz.
Karaya Gum	12 oz.
Locust Bean Gum	12 oz.

Cream Cheese
25% Fat

Standardize milk to contain 6% fat.

Then proceed as given for 35% fat cheese except that cheese is to test 25% fat and 60% water and .5% gum is to be used. The milk should be cooked about 5° lower.

Formula for 250 Gallon Batch
Milk test 3.5%; cream test 55%.

Cream	104 lb.
Milk	238 gal.

Cheese curd yield	630–650 lb.
Water and cream	155
	785 lb. cheese

Salt 6.0 lb.

Gums 3.9 lb. or 62.4 oz. made up of

Irish Moss	21 oz.
Karaya Gum	21 oz.
Locust Bean Gum	21 oz.

Mold Control in Cheese Making

Simple means of mold control are now in use in the cheese industry. Washing of walls, ceilings and shelves with a 10% copper sulphate solution, or a 5% borax solution, ha been of great aid. Likewise, chedda cheeses can be dipped in a 5% bora solution before being paraffined. O if the paraffin is agitated durin dipping, the borax can be added di rectly to the paraffin. This not onl controls molds, but also prevent fungus or rind rot from developin

It has also proved possible, i mixing water paints or cement coat ings for cheese rooms, to substitut 10% copper sulphate solution fo the water. Whether such a paint i used or not, walls and ceiling should be well made and kept wel painted. Every possible effort shoul be made to keep foreign matter an mold from walls and ceilings.

If condensation on the ceiling i hard to control, then the ceilin should be washed more frequently with the 5% borax solution. In an case, walls and ceilings should b washed at regular, frequent intervals with borax solution, after being first scrubbed with soap and water

Borax should be spread on floors that are washed and kept dry.

SMOKED POULTRY

Birds for production of smoked poultry should be killed by piercing the brain and bleeding, followed by any method of picking which will cause a minimum of injury to the skin surface of the bird. In drawing, great care should be taken to avoid rupturing the intestines and subsequent contamination of the body cavity with the intestinal contents. The deposition of large numbers of microorganisms in the body cavity, which may later cause off-flavors and difficulty in the pickling process, is thus prevented.

Following evisceration, it is well to wipe out the body cavity with a paper towel. This tends to absorb any leakage or excess moisture which may remain. A better cure in the pickle as well as a more thorough smoke is obtained if the bird is split down the back before immersing it in the pickle. From a sales standpoint the halved bird is an added advantage, since if it seems desirable to sell either a whole or a half bird, the quantity in each part can be controlled.

Poultry Pickling Solution

Water	5	gal.
Salt	4	lb.
Sugar	30	oz.
Celery Oil	8	cc.
Black Pepper Oil	8	cc.
Parsley Leaves Oil	8	cc.
Sage Oil	5	cc.
Thyme Oil	5	cc.
Marjoram Oil	5	cc.
Bay Leaves Oil	6	cc.
Sweet Basil Oil	6	cc.
Coriander Oil	5	cc.
Cardamon Oil	5	cc.

These oils are dissolved together in 200 cc. of ethyl alcohol. To prepare 5 gal. of pickle, 4 lb. of salt and 30 oz. of sugar are used. Twenty-two cubic centimeters of the alcoholic solution of essential oils is added to a small amount of the sugar, together with about one gram of gum tragacanth, and the whole thoroughly mixed with a mortar and pestle. The remainder of the sugar and salt is dissolved in 5 gal. of water to which the essential oil-sugar-tragacanth material is added slowly with constant stirring. Properly prepared, the pickle should have a cloudy appearance.

Meat Curing Brine

Sodium Chloride	3	lb.
Granulated Sugar	4	oz.
Sodium Nitrate	2	oz.
Sodium Nitrite	1/4	oz.

Enough for 100 pounds of meat.

Meat and Vegetable Stew
Army Ration C

Beef	50	lb.
Potatoes	15	lb.
Or Dehydrated Potato Cubes	3	lb.
Carrots	15	lb.
Dry Beans	8	lb.
Tomato Juice or Pulp	12	lb.
Salt	14.5	oz.
Pepper, Black	0.5	oz.
Celery Salt	5	oz.

The beef component shall be based on the weight of the raw beef before braising. If fresh (unfrozen) beef is used, it is to be ground in a meat grinder fitted with a plate having holes approximately 3/4 in. in diameter. If beef is frozen the holes will be approximately 1 in. in diameter. In lieu of grinding, the beef may be diced in cubes approximately 1/2 in. in diameter. Beef shall be braised until medium well done, with only enough added moisture to prevent scorching. All juices from the meat shall be included in the product.

The potato component shall be based on the weight of the peeled, trimmed and diced potatoes. These shall be cut into cubes of approximately 3/8 in. The dehydrated potato cubes, when used, shall be based on the weight of the dehydrated potato cubes. Dehydrated potato cubes shall be reconstituted in water before use.

The carrot component shall be based on the weight of the cleaned and diced carrots. These shall be diced into cubes of approximately 3/8 in.

The bean component shall be based on the dried beans before soaking.

Beans in the finished product shall be whole, separate and tender, shall not be mashed or pasty, but the skins may be split.

The braised beef, diced carrots and potatoes, soaked beans, tomato juice or pulp, and spices shall be thoroughly mixed. Water may be added, if necessary, to bring the moisture of the finished product within the prescribed range. After thorough mixing, the product shall be filled into tin cans, size 300 x 308. If solids and juices are separated for convenience in canning, they shall be placed in cans in proper proportion. The filled cans shall be exhausted, sealed and processed. Moisture content of the finished product shall be not less than 72% and not more than 76%. Net contents of the cans shall not be less than 12 oz.

Genuine Dill Pickles

Remove head from a thoroughly cleaned and washed barrel.

Throw in handful of dill weed, put in from 5 to 5½ bu. of cucumbers, and spread about 1 lb. of dry spices on top of cucumbers.

Replace head of barrel.

Lay barrel on its side, remove the bung and fill with 32 to 35 salometer brine. The brine will likely vary in salt content in different parts of the country.

Permit fermentation to proceed and replace lost brine daily. Fermentation will usually be complete in about 10 days in regions having warm weather.

After fermentation ceases, the bungs are replaced and the barrel of pickles stored in refrigerated warehouses if they are to be kept in stock. Sometimes they are held for a year.

During the elapsed time the brine slowly dissolves some of the oil from the dill weed and brings it in contact with the outside of the pickles from where it may eventually penetrate into the pickle.

Use of Spice Oil to Replace Dill Weed

By using an emulsion of dill weed oil, the flavoring of the pickles will take place in a considerably shorter length of time.

From the beginning, the emulsion will impart a dill flavor to the brine. As the brine penetrates into the cucumber during fermentation, the dill flavor will also be carried in and with probably deeper penetration than is the case when the dill weed itself is used.

At one time genuine dills did not contain any flavor other than that of dill weed. Today many packers use spices to get variations in flavor. In such instances spice oil emulsions can quite readily be used instead of the dry spices that may be added. Care must be taken not to overflavor. After the amount of spice oil needed has been calculated on the basis of dry spices used, the quantity used should be half of the calculated amount. The spice oils are more quickly available to the pickles than when dry spices are used.

Mixed Dill Pickle Spices

Mix well the following materials: cinnamon saigon broken and sifted 30 lb., whole clove 27 lb., whole allspice 27 lb., small red chili peppers 8 lb., and unbroken bay leaves 8 lb.

Dill Pickle Flavor

A satisfactory emulsion or colloidal suspension is made by grinding 3½ oz. of dill weed oil with 1 oz. of a gum emulsifier in a mortar with a pestle and by making up to 1 pt. with water and vigorous stirring. Beating for a few minutes with an egg beater helps. Two ounces of the emulsion flavors one barrel of pickles carrying 15 to 18 gal. of brine.

Other methods used to get dill weed oil into brine include: (a) using ethyl alcohol as a solvent, (b) using salt or sugar as a carrier One pound of dill weed oil made up to 1 gal. with the alcohol may be used on the basis of 2 to 4 oz. of the solution to 15 to 18 gal. of brine; enough for 1 bbl. of dill pickles. Or ½ oz. of the dill weed oil mixed thoroughly with 1 lb. of salt or sugar will flavor a barrel of dills.

Keeping Pickles Green

Several factors seem to govern the retention of a fresh green color in pickles. One is to allow plenty of time for the preliminary soaking in salt water, as it seems that "short-cut" methods have a tendency to spoil the color. Soft water should be used in making pickles when possible, as the substance in hard water seems to darken the color.

Some recipes call for alum, and this is to make them crisp and green —but while this is not absolutely injurious, it is not recommended as a general practice.

Hydrated or powdered lime may be added to the solution used in pickling to increase the green color.

Free Flowing Salt
U. S. Patent 2,288,409

Magnesium Carbonate	40
Calcium Stearate	1
Salt	359

Vinegar-Sugar Solution

Cider Vinegar	3 lb.
Sugar, Granulated	2 lb.

Warm and stir till dissolved.

Artificial Cinnamon

Cinnamaldehyde (96%)	3–4%
Eugenol	4%
Powdered Almond Shells, To make	100%

Curry Powder, Indian Type

Coriander	4 oz.
Tumeric	4 oz.
Cinnamon	2 lb.
Cayenne	8 oz.
Mustard	1 lb.
Ginger	1 lb.
Allspice	8 oz.
Fenugreek	2 lb.

All ingredients should be dry. Grind to fine powder and sift through a fine-mesh sieve. Mix thoroughly. Package in airtight, moisture-tight container to avoid loss of flavor.

Spice Essential Oil Equivalents *

Quantity of Dry Spice	*Spice*	*Equivalent quantity of spice essential oil*
100 lb.	Allspice	3 ½ lb.
100 lb.	Almond, bitter	½ lb.
100 lb.	Bay leaves	2 lb.
100 lb.	Black pepper	1 ½ lb.
100 lb.	Black pepper	Oleoresin 6 lb.

*When compared with best grade of dry spice.

Quantity of Dry Spice	*Spice*	*Equivalent quantity of spice essential oil*
100 lb.	Angelica root	¾ lb.
100 lb.	Angelica seed	1 lb.
100 lb.	Anise star (Chinese)	3 lb.
100 lb.	Basil, sweet	1 2/5 oz.
100 lb.	Calamus root	2 ½ lb.
100 lb.	Caraway seed	5 lb.
100 lb.	Cardamom seed	5 lb.
100 lb.	Carrot seed	2 lb.
100 lb.	Cassia cinnamon	1 lb.
100 lb.	Capsicum	Oleoresin 8 lb.
100 lb.	Celery seed	3 lb.
100 lb.	Cinnamon, Ceylon	1 lb.
100 lb.	Cloves	17 lb.
100 lb.	Coriander seed	½ lb.
100 lb.	Cumin seed	3 lb.
100 lb.	Dill seed	3 ½ lb.
100 lb.	Dill weed	½ lb.
100 lb.	Estragon	½ lb.
100 lb.	Fennel seed	5 lb.
100 lb.	Garlic	Imit. flavor 4 oz.
100 lb.	Ginger	Oleoresin 10 lb.
100 lb.	Horse radish	Imit. flavor 1 oz.
100 lb.	Laurel leaves (distilled) (bay leaves)	2 lb.
100 lb.	Lavage root	½ lb.
100 lb.	Mace	12 ½ lb.
100 lb.	Marjoram, sweet	½ lb.
100 lb.	Marjoram, wild	2 lb.
100 lb.	Mustard seed	¾ lb.
100 lb.	Nutmeg	12 ½ lb.
100 lb.	Onion	Imit. flavor 4 oz.
100 lb.	Paprika	Oleoresin 8 lb.
100 lb.	Parsley seed	3 lb.
100 lb.	Pimento berries	3 ½ lb.
100 lb.	Sage	2 lb.
100 lb.	Thyme	2 lb.
100 lb.	Valerian root	1 lb.

Mayonnaise

Formula No. 1

Egg Yolk (Fresh or Frozen)	15	lb.
Sugar	3	lb.
Mustard Flour	1¼	lb.
Salt	1	lb.
Paprika	¾	lb.
Pepper (White)	1	oz.
Oil (Salad)	12	gal.
Vinegar (100 Grain, White)	4	qt.
Water	2	qt.

Beat the yolks at high speed until they are thoroughly broken, which may require about one minute.

Now add the sugar, salt and spices and beat for another 2 to 5 minutes until the mixture is well creamed.

Now start adding the oil slowly, at the rate of a gallon per minute, until 8 gallons have been added. Change the speed of the beater to first speed while adding one quart of vinegar and one quart of water; the mixture should be thoroughly agitated. Turn the beater on high speed and add the remainder of the oil at the rate of one gallon every two minutes. Then shift the beater to slow speed and add the remainder of the vinegar and water.

The beater should now be shut off to scrape down the sides of the bowl. Then mix the entire batch at slow speed for at least another two minutes taking care not to overbeat which tends to make the mayonnaise thin.

No. 2

Egg Yolk	5	lb.
Oil	2¼	gal.
Vinegar (45 Grain)	1¼	qt.
Sugar	10	oz.
Mustard Flour	4	oz.
Salt	6	oz.
White Pepper	½	oz.
Paprika	½	oz.

No. 3

U. S. Patent 2,264,593

Egg Yolk	120
Cottonseed Oil	330
Spice and Salt Mixture	24

Cider Vinegar (45 Grain)	160
Monochloracetic Acid	¼

Salad Dressings

Formula No. 1

Whole Egg	340 g.
Vinegar	320 cc.
Water	80 cc.
Salt	44 g.
Sugar	40 g.
Mustard	16 g.
Oil	3160 cc.

No. 2

Oil	1420 cc.
Egg Yolk	200 g.
Vinegar (6% Solution)	174 cc.
Mustard	20 g.
Salt	15 g.
Sugar	50 g.
Water	121 cc.

No. 3

Corn Starch	90.0 g.
Tapioca Flour	20.0 g.
Sugar	152.0 g.
Vinegar	152.0 g.
Salt	31.0 g.
Mustard	12.5 g.
Pepper	3.0 g.
Egg Yolk	100.0 g.
Water	455.0 cc.
Cottonseed Oil	830.0 cc.

Add cold water to starch, heat to 184° F., mix until it cools to 120° F.; add spices, vinegar and sugar; cool to 90°; then add egg. Pour oil in slowly and homogenize the mixture.

No. 4

Corn Starch	110.0 g.
Tapioca Flour	25.0 g.
Sugar	152.0 g.
Vinegar	152.0 g.
Salt	31.0 g.
Mustard	12.5 g.
Pepper	3.0 g.
Egg Yolk	125.0 g.
Water *	540.0 cc.
Cottonseed Oil	700.0 cc.

Salad Dressing (Non-Fattening)

Water (1080 cc. for Agar, 1800 cc. for Starch)	2880.0 cc.
Agar	48.5 g.
Starch	142.5 g.
Pepper	.5 g.
Saccharine (10% Solution)	7.2 cc.
Mustard	28.5 g.
Vinegar (5% Solution)	513.0 cc.
Paprika	5.0 g.
Gum Tragacanth	28.0 g.
Mineral Oil (Viscosity 175–185)	1995.0 cc.
Turmeric	2.0 g.

Combine the ingredients in the order given and homogenize.

Dried Fruit Bars

Formula No. 1

Fig Paste	3
Prunes, Pitted	3
Apricots, Dried	3
Honey or Heavy Invert Syrup	1

No. 2

Fig Paste	3
Apricots, Dried	6
Honey or Invert Syrup	1

No. 3

Fig Paste	3
Apricots, Dried	3
Raisins, Ground †	2

* 200 more cc. of water are added during homogenizing; aside from this, procedure is as above.

† Preferably seeded Muscat, these may be replaced by dates of high invert sugar content.

The average moisture content must be below 20% in the finished product.

Grind the fruit, mix, then regrind. Extrude in a sausage machine fitted with a rectangular opening. Cut to desired size. Wrap in moistureproof cellophane, then again in heavy cellophane. Heat in air until the centers of the bars are at 140° F. Cool in air. For Army use pack in insect-proof containers such as 5-gal. cans.

Dried-Fruit Cereal

Ground Dried Fruit	100 lb.
Corn Sugar	100 lb.
Wheat Bran	50 lb.
Whole-Wheat Flour	100 lb.
Salt	3 lb.
Baking Powder	3 lb.

Mix all ingredients except baking powder with water to give the desired consistency. Then mix the baking powder in quickly and make the dough into loaves. Dust with whole-wheat flour and place in pans greased with nonrancidifying fat and bake to about 410° F. The loaves are then sliced and the slices are dried bone dry on trays in a dehydrator. These are then coarsely crushed, screened to uniform size, toasted a short time at 300° F. and packed in attachment line, paraffin wrapped cartons.

Fruit Preserve Thickener

Corn Sugar	35
Cane Sugar	25
Vegetable Gum (Galagum)	40

Fining or Clarifying Powder for Fruit Juices

Tannic Acid	1¼–1½ oz.
Gelatin	1½–6 oz.

Use the above amount for 120 gal. fruit juice.

Sulphured Fruits

To produce the solution of sulphur dioxide used in preserving the fruit, the following method may be followed: Place on a scale a barrel containing 378 lb. of water. Run sulphur dioxide gas, which must be free from arsenic, into the water through a laboratory tube. This is continued until enough gas has been dissolved to bring the weight of the contents of the barrel to 400 lb. Specific gravity of the solution should be 1.025 or upwards. The water should be as cold as possible, and the gas must be fed in gradually. Six to 10 hours are usually required to obtain a solution of the required strength.

Generally speaking, the fruit is prepared for preservation by this process by the same methods as for canning or cold packing.

Weighed quantities of this prepared fruit are placed in the pulping tank. (If a jacketed kettle or cooker with closed coils is used, 15% of water based on the weight of the fruit should be run in first and brought to a boil.) Steam is turned on gently as soon as the open steam discharge (or jacket or coil) is covered and the balance of the fruit is added as quickly as possible. The steam is then turned on full and the batch is boiled vigorously for from 3 to 5 minutes in order to sterilize it.

With certain soft fruits, such as raspberries, no further cooking is necessary. But, generally speaking, additional boiling is required to make the fruit sufficiently soft and fluid. No hard portions should remain and the heat should penetrate thoroughly to all portions of the fruit. With plums and similar fruits, stones and pits should be reasonably free from adhering fruit structure. The fruit should not be cooked to a mush, but should be kept as whole as possible. In the case of large fruit, some breaking up is necessary, but as distinct pieces as possible are desirable. The general texture of the product should be that of "stewed" fruit.

Of course, if the fruit is too firm, the user can render it softer. But, if it is too soft, there are no means of rendering it more firm. However, it is essential that the fruit be softened sufficiently to permit penetration by the sulphur dioxide solution, and also to permit easy passage through discharge valve and through the bung of the barrel. The additional cooking required to reach this condition usually takes from 5 to 20 minutes, depending on the degree of firmness of the raw fruit and the efficiency of the equipment. With soft fruit, it may be necessary to stir with a paddle to obtain uniform cooking.

When cooking is completed, the batch should be discharged to a mixing tank or receiver, where it is standardized with water to the calculated volume, if a definite "fruit strength" is specified. About 10% dilution usually occurs owing to condensation where "live" steam is injected into the fruit. Or, if a jacket or closed heating coil is used, the loss by evaporation from the 15% added water usually results in about the same final dilution. With some fruits, additional dilution may be necessary if the pulp proves too thick to flow and to allow thorough mixing of the sulphur dioxide solution.

Adding the Solution

The pulp is then run into barrels, with as little delay as possible so as to avoid cooling. The barrels are filled to within 2 or 3 in. of the bung hole, by means of a funnel. The consistency should be such that the pulp will not run out of the receiver entirely by gravity, but must be assisted by a flat paddle, "squeegee" or equivalent.

The cooperage used for this product should be new oak or fir barrels of 50 U. S. gallons capacity. *Waxed barrels must not be used* for this product.

Before filling, 5 lb. of the sulphur dioxide solution is placed in each barrel. After filling, 5 lb. more of the solution is placed in each barrel. The batch should be stirred constantly while filling, to avoid floating of the fruit and to assure uniformity between barrels. After the final sulphur dioxide solution addition is made, the barrels are closed. They are then rocked and next rolled over a number of times, to mix the sulphur dioxide solution thoroughly through the pulp. Poor mixing may result in spoilage.

Samples of each batch should be taken in sealed glass jars, stored in a warm place and examined frequently. These samples should be taken from the barrels the day following their filling and will serve as a check on the efficiency of the preservation process. If the pulp has been properly prepared, the color will bleach and there will be no fermentation.

In carrying out this method of fruit preservation, it should be remembered that sulphur dioxide is volatile and has an irritating effect on the membranes of the nose and throat, although it should be harmless in any concentrations normally met with. The solution had best be made up in a separate room or out of doors. Also, the barreling is best carried out in a separate department.

Removing Skins from Nuts
U. S. Patent 2,273,183

Immerse meats, with skins, in a solution of approximately 3% of one of the alkalies of the group which consists of sodium hydroxide, sodium carbonate, sodium bicarbonate and borax for 1–4 minutes. Remove meats; rinse in water. Immerse meats in a 2–4% acid bath of the group which consists of hydrochloric, acetic and citric acids for 1–4 minutes. Remove meats; subject to a forcible action of water to remove the skins.

Cracking Walnuts (American Black)

Hold the walnut on end, and continue cracking after the first breaks until the quarters, which usually separate at first, are each cracked further. Any quarter in which the meat is not well exposed can be cracked again. When this procedure is successful, the meat can be removed in whole quarters.

Improving Firmness of Canned Tomatoes

Peeled tomatoes are dipped in 2% calcium chloride solution for 2–3 minutes.

Improved French Fried Potatoes
U. S. Patent 2,212,461

Sliced potatoes are soaked in vinegar, diluted (0.4–0.7% acetic), for 5 minutes to 5 hours prior to frying.

Removing Egg Shells

Shells will come off hard-boiled eggs easily if the shells are slightly cracked and then dropped quickly into cold water.

Sauerkraut

Kraut is prepared in the home from large, firm, well-ripened heads of cabbage. These should be allowed to stand at room temperature for a day to wilt, the wilting causing the leaves to become less brittle and thus not so likely to break in cutting. The outer leaves are trimmed down to the white leaves and the heads are then washed. With ordinary home equipment, the heads must be cut in halves or quarters with a large knife and the core removed or cut fine. The kraut is cut with an ordinary kraut cutting board with blades set to cut shreds about the thickness of a dime. The setting of the blades varies, some preferring to cut the cabbage very fine, while others use a coarse cut.

The cabbage is packed in clean, paraffined barrels or jars with a light sprinkling of salt. If it is necessary to paraffin the barrels, a high-grade paraffin wax should be melted and applied to the dry container with a brush.

Salt is used in the proportion of 1 pound to 40 to 45 pounds of cut cabbage. If the salt is allowed to remain on the shredded cabbage a short time before packing, less breaking of the shreds is obtained.

Packing is quite often the cause of much unnecessary bruising and tearing of shreds and results in a oftening of the kraut. A large vooden tamper should be used and vith it the kraut should be firmly pressed or pushed down to force out the air rather than pounded until juice is produced. Ordinarily, pounding is not necessary to draw out the juice for if the salt is added as noted above, it will draw out more than enough juice to cover the cabbage by the time the container is filled. When the container is filled the juice should come to the surface.

The kraut should be covered with a clean white muslin cloth and then with a round paraffined cover of such size that it just fits within the container. A weight is placed upon the cover of such size that the juice comes to the bottom of the cover, but not over it. This will keep the cloth moist but juice will not cover the cloth. The weight necessary for this purpose varies, especially during the first few days of fermentation and with changes in temperature, and therefore should be watched carefully. For smaller containers a weight consisting of a jar to which water can be added or taken from serves very well. The placing of weights on the kraut is very important in producing good quality sauerkraut. Commercial packers are very particular about this.

Fermentation starts within a day after packing as is usually evidenced by the formation of gas bubbles on the surface. Although fermentation is more rapid at higher temperatures, more spoilage is also likely to occur. The best quality kraut is produced at 70° F. or lower. It requires a month to six weeks for the kraut to cure properly at these temperatures.

If it is used quite frequently and intended to be entirely consumed in fall and winter, kraut may be left in the container in a cold room; otherwise, it should be canned. Canning is simple and insures a good supply of kraut throughout the year. The cold pack method is more commonly used, although kraut may be successfully packed by hot pack methods. In the cold pack method the kraut is warmed to between 110° and 130° F. in its own juice, packed into sterilized jars, covered, and cooked in a kettle of boiling water for 20 to 25 minutes. The jar is sealed when removed and placed in a cold place so that it may cool as rapidly as possible. If allowed to remain hot for a long period, the kraut softens and darkens in color. When canned and cooled properly the kraut is very much like the raw product in texture and flavor.

Although many people have made kraut for years, few realize what happens in the typical curing process. When cabbage is cut there are a great many bacteria, both good and bad, yeast and molds upon the cut shreds. The salt sprinkled upon the cabbage draws out the sugar which is used by certain of the bacteria. They change the sugar to acids and other by-products. A typical mellowing of the cabbage takes place with these changes, resulting in the product we term sauerkraut (acid cabbage).

When kraut is properly packed, with the correct amount of salt, and at a reasonable temperature, only certain good types of bacteria are able to grow. The various spoilage types of bacteria, yeast, and molds are stopped from growing by the absence of air and the acid which is rapidly formed by these desirable bacteria. At times insufficient amounts of salt are used, or the kraut is not thoroughly packed, or the juice has drained off, or the kraut may not be covered properly. Under such conditions spoilage may occur.

Tagging Foods and Beverages

The source of beverages may be traced by their appearance under ultraviolet light. Many soft drinks and some foods contain natural fluorescent dyestuffs.

A small amount of neutral acriflavine, one part in several million or one part in 50 million of fluorescein or uranine causes a distinct fluorescence in liquids.

Chapter Eight

HIDES, LEATHER AND FUR

Rapid Vegetable Tanning

Vegetable tanning can be done in a very short period of time when the hide is first treated with sodium hexametaphosphate. The tannage may be performed in a drum, paddle or vat. The strength of the vegetable tanning solution may vary from 30° Barkometer to 80°. The time, depending upon the equipment, will vary from a few hours to seven days for heavy leather.

The unhaired hide should be thoroughly cleansed, i.e., it should be washed as free as possible from lime salts after bating or it may be neutralized with ammonium chloride or sulphate and then washed with cold water to free it of soluble salts. The extent of bating apparently will not effect subsequent plumpness. After having washed the stock, it is drained and placed into the tan drum with the following amounts of materials, based on the white weight:

Water	125 %
Sodium Hexametaphosphate	5 %
Sulphuric Acid	1.5%

After milling 15 minutes, the pH is determined; this should be adjusted to 2.2 to 2.4 with sulphuric acid. The stock is milled for a period of three to four hours, then permitted to rest in the drum overnight. The next morning the stock is washed with cold water for about one hour. It is then ready for vegetable tanning.

The vegetable tanning material may consist of a mixture of hemlock, chestnut and quebracho or any other suitable mixture. Very good results are obtained with quebracho alone. It is, however, very important that the pH of the vegetable tanning material be adjusted to a pH of 3.0 to 3.2. This may be done with an organic acid such as formic, acetic or lactic, or with a synthetic tanning agent such as mentioned previously. Inorganic acids are not satisfactory excepting phosphoric. We have found that phosphoric acid is the most economical to use. The 85% syrupy type is recommended.

The strength of the vegetable tanning solution may vary from 30° to 80° Barkometer. The acid or syntan is added directly to it, stirred well.

Tanning may be done in one of two ways:

1. *In the drum:* The drum may be the ordinary tanning drum, but one of the following dimensions is preferable: 8 feet in diameter, 8 to 10 feet face. The speed should not exceed 14 r.p.m., the total load to be no more than 100 sides of light cows or the equivalent in smaller stock. The total amount of tanning material need not exceed 18% on a pure tannin basis. Time of milling will vary from four to six hours, depend-

ing upon the plumpness of the stock. After the stock is tanned, it is horsed up for 24 hours, then pressed or put through the wringer and split. It may then be retanned and shaved, fatliquored or stuffed. The spent tan liquor is refreshed and used for a subsequent pack.

2. *In the rocker vat:* For this purpose it is not necessary to have a rocker section in which the liquor is varied in strength from a weak tail to a strong head, but each vat may be of the same strength. The liquor may vary from 30° to 60° Barkometer. The stock is hung into the rocker vat in the conventional manner and rocking proceeded with until the stock is tanned through. Steer hides for belt or sole will be tanned through in from three to five days' time. Sole leather stock may then be placed into the layaways and the balance of the process continued as usual. The spent or sapped liquor in the rocker vats is refreshed and the next lot entered.

In drum tanning some drawn grain is obtained, which is, however, easily set out. Drawn grain is directly blamed to the speed of the drum. In the rocker vat no drawn grain is obtained, nor is it necessary always to retan after rocker tanning, especially for bag, case, and strap leather.

Two outstanding properties result from metaphosphate treatment before vegetable tanning, namely, plumper leather and stronger leather.

Chrome Tanning

Soaking

After the hides are trimmed and split into sides, lots are made consisting of 3000 lb., one side more or less. This is called a "pack" and will be so distinguished throughout. The pack is placed in a paddle vat with 1450 U. S. gallons of water at a temperature of no higher than 65° F., preferably between 60 and 65° F. The pack is milled for a half hour, then allowed to rest. The total time for soaking should not exceed 18 hours, and during this period the pack should be milled for a 15 minute period each three hours.

After the 18 hour period and while the pack is being milled, it is washed with water at a temperature of 60 to 65° F. for a period of one hour. The amount of water consumed for washing should total 3000 gal., flowing into the paddle vat at the surface and in a far corner and discharging from an opening in the bottom of the paddle in a position opposite to the point where the stream enters the paddle vat.

After the washing, the pack is pulled from the paddle, placed flat on trucks, drawn to the fleshing machine, where the sides are "green" fleshed as clean as possible, i.e., no fleshy tissue should remain on the sides after passing through the machine.

Liming

Liming is preferably done in a paddle vat which is prepared as follows:

The paddle vat should contain approximately 400 gal. of an old or mellow lime.

To this is added 1000 gal. of water (65 to 70° F.) and

Hydrated Lime	100 lb.
Sodium Sulphide	30 lb.

The pack is thrown into the paddle vat while the wheel is in motion. When all of the pack has been entered, milling should continue for 15 minutes. It is assumed that at this stage it is about mid-day. During the afternoon the stock is milled for 15 minutes each two hours. It is then permitted to rest. The next morning the stock is again milled for a period of 15 minutes, then after three hours the following is added to the paddle vat:

Hydrated Lime	100 lb.
Sodium Sulphide	20 lb.

During the next 24 hours the pack is milled 15 minutes every three hours, excepting at night time, when it is permitted to rest.

On the third day nothing is added to the paddle vat, but every three hours the pack is milled for a period of 15 minutes. During the last milling period of the third day the contents of the vat is warmed to 75° F., then allowed to rest overnight.

On the morning of the fourth day the pack is milled for 15 minutes. The sides are then removed from the vat, placed on trucks and taken to the unhairing machine. The sides are then unhaired, head split, refleshed, scudded, and examined for fine hair.

Acid Wash

From the fine hair stage the pack is placed in a paddle vat which is filled with water. The pack is then washed for a period of a half hour, the water being at a temperature of 65 to 70° F., the amount being approximately 1500 gal. The operation is carried on in a similar manner as the wash after soaking. After the sides are washed and the paddle vat filled with water, the acid wash is applied as follows:

Have available approximately 20 lb. of muriatic acid in a small tub to which is affixed a hard rubber pet cock or syphon. The acid is run into the paddle vat in a small stream at one side of the paddle; the operator constantly takes a small portion of the liquor and tests it for acidity with methyl orange indicator. At this stage the operator must use judgment, gained by experience, and retard the flow of the acid at intervals so that the fluid contents of the vat remains acid to methyl orange indicator for a period of two minutes. After this period, hydrated lime is added to a degree to obtain a constant pH = 8.5. The sides are now ready for the bating operation.

Bating

The temperature of the contents of the paddle vat is now raised to 85° F. After the temperature remains constant for five minutes the bating material is added. The amount used is ¾% on the white weight of the pack.

It is quite impossible to describe the bating operation excepting to state that during the time of bating the temperature should be held at 85° F. and the pH at a point not to exceed 9.5. The time required for bating may vary from 15 minutes to 45 minutes, depending upon the extent to which the whole process is kept in balance.

The extent of bating is a predetermined stage which only the operator can judge by feel and appearance, and then only after a prolonged period of experience.

When the operator has decided the end point of the bate, the pack is washed with a stream of cold water, 65° F. or below, for a half hour, requiring approximately 1500 gal. of water. The sides are then removed from the paddle and transferred to the pickle paddle.

Pickle

The pickle is prepared in the following manner:

To 1450 gal. of water add sufficient common salt to make a 10% by volume solution. Sufficient sulphuric acid is added to make a 0.15% by volume solution. To this solution the sides are added.

On the white weight of the stock the following is added to the paddle wheel while the stock is being milled:

Sodium Chloride	10 %
Concentrated Sulphuric Acid (66° Bé.)	1.4%

The stock is milled for a period of four hours, then allowed to rest overnight. Next morning the stock is milled for one hour. The sides are then removed from the paddle wheel and "horsed" up to drain for a period of 24 hours.

The spent pickle is readjusted with water, salt and acid, if need be, to its original strength of 10% salt and 0.15% acid. It is now in order for the second pack. After every tenth pack the contents of the paddle is discarded.

The liquor may be used for a longer period if provision is made to permit it to settle for several days, syphon off the clear supernatant and discard the calcium sulphate.

Tanning

Tanning is done in drums.

A one bath chrome liquor is used.

After the pickled stock has drained for 24 hours it is made into lots of 2000 lb. each, that amount being chosen as a charge for a tan drum.

The drum is prepared as follows:

Water 75%, on the drained pickle weight of stock.

Salt 4%, on the drained pickle weight of stock.

Place the stock in the drum containing the above and mill for five minutes, then add the following amounts of chrome liquor:

10 gal. Chrome Liquor, mill 20 minutes, then
10 gal. Chrome Liquor, mill 20 minutes, then
10 gal. Chrome Liquor, mill 30 minutes, then
10 gal. Chrome Liquor, mill 3 hours

Then add 0.5% or 10 lb. of sodium bicarbonate, which is first dissolved in 25 gal. of water and divided into three equal portions, each of which is added at intervals of 15 minutes.

After the last portion of bicarbonate is added, milling is continued for 3½ hours. The drum is then stopped, the door removed; note that all stock is submerged, and let stand overnight. The next morning the drum is closed and the stock milled for a period of one hour. A boil test is then made. If the boil test is satisfactory, a pH determination is also made on the stock. If this is satisfactory, the stock is removed from the drum and piled flat on trucks to drain overnight.

Mechanical Operations

The following morning the stock is put through a wringer, then set out by machine. The sides are then sorted for splitting. Three weights are generally sorted, namely, heavy, medium and light. Sorting is regulated so that after splitting, the thickness of the whole side will correspond to the thickness of the belly, which should not be split. It is often found that cow sides vary considerably in thickness between belly and back. For this reason, judgment must be exercised by the sorter not to permit too heavy a split to be removed.

After the stock is split (the belt knife machine is generally used), it is sorted for thickness, by gauging, and the type of leather it is best suited for.

Kinds of Side Leather

Side leather tanners usually limit the weights of raw material (green salted weight) to from 25 to 60 lb. hides. Within these weights will be found hides which when split into sides will produce leathers known to the American trade as:

Buck sides, 25 to 35 lb. hides; for misses' shoes.

Veal sides, 25 to 35 lb. hides; for children's and misses' shoes.

Sport Elk, 25 to 43 lb. hides; for misses', girls' and boys' shoes.

Gunmetal sides, 35 to 55 lb. hides. The heavier hides are usually "spready," being thin cow sides, usually called "extremes," used largely for men's semi-dress shoes.

Elk sides, "43 lb. and up" hides, used largely for work shoes.

Waterproof sides, "43 lb. and up" hides. A choice leather for "high cuts," waterproof boots and shoes.

Retan sides, "43 lb. and up" hides. A side on which are found barbed wire scratches and other grain defects, and which are buffed before they are finished. "Retan" denotes a chrome tanned leather which is later retanned with vegetable tanning material, then heavily stuffed with greases. It is used principally for work shoes and "high-cut" boots.

There are, of course, specialties made by many side leather tanners. These special leathers are usually designated by some trade name and used for children's, misses' and girls' shoes. In addition, large quantities of sides are used for the purpose of making white leather for women's, girls' and children's shoes. Practically all white leathers are chrome tanned, excepting for a small percentage, which are tanned with alum, formaldehyde, and synthetic tanning materials.

Coloring

After the sides have been split, shaved and sorted, lots of 500 lb. are formed for each weight. This is the standard amount per color drum.

The stock is placed in the drum with water at a temperature of 70° F. and rinsed for five minutes. The drum is then well drained, after which 150 gal. of water at a temperature of 70° F. is added. While the drum is in motion, 25 lb. (5%) of synthetic tanning material, first diluted to 20 gal. with water, is added. The stock is milled for 30 minutes, then is added 7.5 lb. sodium bicarbonate in 20 gal. of water in three equal portions at intervals of 10 minutes. After the last portion is added, the stock is then milled for 30 minutes longer. It is then washed with running water at 70° F. for 30 minutes. The temperature of the water is then gradually raised, at such a rate as to reach a temperature of 120° F. in the next 20 minutes. Washing is continued at this temperature for a 10 minute period. Total washing time is one hour, and 2500 gal. of water should be consumed for this washing operation. The stock at this stage should show a pH = 4.8 to 5.2.

The stock is now in proper condition for coloring and fatliquoring.

Preparing Sheepskins for Tanning

After cleansing the wool and removing all burrs, etc., pickle with acid and salt and then tan straight away. This method tends, however, to produce a somewhat hard leather and it is better to go to a little expense and trouble to soften the skins.

The following bath:

Water at 65° F.	300 gal.
Ammonia (0.910)	14 pt.
Hydrogen Peroxide (100 Vol.)	6 pt.

should be used and goods paddled in it for 30 minutes and then left overnight. Next morning the skins should be drained, lightly hydro-extracted and neutralized by use of a weak organic acid bath using 1% of lactic acid on fleshed weight. The pH of the cut surface should be 4.5–5.0.

The most suitable pickle is an alum one and a useful formula is:

Water at 65° F.	100 gal.
Potash Alum	50 lb.
Salt	50 lb.

Paddle the skins for 2½ hours, then drain overnight and tan next morning in the usual way. The potash alum pickle is particularly recommended in the case of chrome tanning.

Hide Dehairing Powder

U. S. Patent 2,226,883

Sodium Hyposulphite	50%
Sodium Sulphite Anhydrous	40–45%
Lime	5–10%

Fat Liquor for Hides

British Patent 534,244

Egg Yolk	1.42
Sulphonated Neats-Foot Oil	6.40
Fig Soap	0.54
Oxidized Cod-Liver Oil	1.34
Sulphonated Cod-Liver Oil	0.75
Pine Oil	0.62

Deglazing Leather

Trisodium Phosphate	4 oz.
Borax	2 oz.
Diglycol Stearate	½ oz.
Sulfatate (Wetting Agent)	¼ oz.
Water, To make	1 gal.

Warm and mix until dissolved. Apply, warm, with sponge or soft brush.

Leather Degreaser

French Patent 846,569

Trisodium Phosphate	70
Soda Ash	20
Castile Soap	10
Water	3333

Waterproofing Leather

British Patent 532,306

Leather is immersed at 50–55° C. for ½–1 min. and brushed or sprayed with

Paraffin Wax	1
Stearin	1
Benzine	8

Sole Leather Waterproofing
Japanese Patent 128,561

Asphalt	8%
Rosin	7%
Boiled Oil	13%
Heat together until uniform; add	
Copal Varnish	47%
Cool and add	
Turpentine	10%
Benzol	15%

Dyeing Glove Leather

An important item in the glove leather industry is the production of wash fast shades on the various combination tannages—chrome and formaldehyde—formaldehyde and alum and variations of these basic treatments on sheep and kid skins.

The dyeing procedure consists of giving the skins a preparatory drumming for 30–40 minutes with 4% ammonia at 130° F. The dyestuff may be then added to the same bath or applied from a fresh bath if desired and drumming continued at 130° F. for 30 minutes. The color is then exhausted and fixed with from 3–8% formic acid, depending on depth of shade, the acid being fed on in three portions at intervals to insure leveling and penetration. Fat-liquor is then added and drummed 10 minutes longer. Percentages are based on dry weight of the leather.

Out of some 250 distinct dyestuffs checked for maximum wash fastness, only 14 were selected as giving most satisfaction. These are listed:

Chrome Leather Yellow 2G Conc
Chrome Leather Yellow C Conc
Amanil Fast Orange PRZ
Amanil Leather Orange DB
Benzo Fast Orange S
Amanil Leather Orange PR
Amacid Leather Orange PGS
Leather Brown SR
Leather Brown RN
Amanil Leather Brown SG
Amanil Leather Brown BR
Amanil Leather Green LT
Amanil Leather Brown RLH or BR

Ten minutes agitation at 100° F. with 2 grams neutral soap chips per quart of water is the wash test used.

Belt Edge Coloring
Formula No. 1

Nelgin	6.15 g.
Titanox A (Water Dispersible)	17.00 g.
Water	29.50 g.
Moldex (Preservative)	0.09 g.

Dissolve Moldex in boiling water, then add Nelgin and let soak for ½ hour. Stir vigorously while adding Titanox.

No. 2

1. Diglycol Stearate	12 g.
2. Water	388 g.
3. Titanox A (Water Dispersible)	100 g.

Boil 1 and 2 together, cool and while mixing vigorously add 3 and mix until uniform.

To get different colored materials replace the Titanox by any other inert pigment or color.

Black Leather Dye (Stain)

Alcohol-Soluble Black Dye	50 g.
Benzyl Alcohol	100 cc.
Alcohol	550 cc.
Water	300 cc.
Glycerin	50 cc.

Antique Leather Finish

Antique leather is employed for furniture purposes. Usually sheepskin is employed, or, for better qualities, vachette leather. The so-called antique grain is produced either by embossing or by tanning with concentrated liquor. Such leather shows dark and light parts. If the raised parts are to be light and the deeper parts dark proceed as follows:

The skin is dyed in the usual manner in the shade desired for the raised parts. After drying lightly rub a protective wax mixture on the raised parts, leaving the deeper parts free. Then apply the solution of a black or other darker colored dye by means of a sponge.

The following mixture may be used as a wax reserve: Melt together: 75 parts stearine, 5 parts Japan wax, 5 parts Carnauba wax, 75 parts petrolatum.

Sometimes the deeper parts are required to be lighter than the raised parts. In this case dye the leather in the shade desired for the deeper parts, and then cover the raised parts with pyroxylin lacquer previously colored in the required darker shade and mixed with a little castor oil.

Fluorescent Leather

Chrome tanned leather is dyed with a solution of one of the following fluorescent dyes:

Fluorescent Blue G
Rhodamine 3B
Eosine G
Brilliant Flavine S
Fluorescent Violet G
Diazo Scarlet PRD

Mordanted leather can be made fluorescent by dyeing with yellow OX, Red BX, and Red 6G. Shellac or casein may be used as binders and barium sulphate added to obtain good coverage and an increase in brilliancy.

Artificial Leather
U. S. Patent 2,273,973

A flexible webbing is coated with

Nitrocellulose	40
Butylacetylricinoleate	30
Castor Oil	15
Polymerized Rapeseed Oil	15

The above can be colored by suitable pigments.

Numida Effect on Feathers

Gum Arabic Solution (25–50% Solution)	1 qt.
Cold Water	2 qt.
Glycerin	1 qt.

Strain thoroughly to remove all particles of dirt, etc.

Take the dry feathers and work in this solution until thoroughly saturated; wring through the ordinary wash wringer, and squeeze out as much of the solution as possible, after which rub through the hands thoroughly for about five minutes in order to evenly distribute the remaining portion of the liquid that is in the feathers. Then string the feathers and beat them out on a wooden board for several minutes until the fine stems separate, after which hang up and dry over night.

Self Sterilizing Bristles and Sponges
British Patent 543,948

Bristles, hair or sponges (100 g.) are immersed, at room temperature, for several hours in silver nitrate (3–5% solution), then drained and washed.

Bleaching and Dyeing of Horsehair and Pigs' Bristles

The most valuable horsehair for commercial purposes is that from the mane and tail. Tail hair fetches the highest prices when it is over

67½ centimeters long. And sometimes it is from 75 to 85 centimeters in length. Mane hair, on the other hand, is seldom longer than 45 cm. True white hair is in best demand, and next comes the coal black. The market value of sorrel, brown or gray horsehair is not anywhere near so great. Skin hair is also used for various purposes, as for instance, upholstery and cushions, and is also spun into coarse yarn to be used for weaving haircloths, such as that used for lining the lapels of men's overcoats to make them keep their shape. Skin horsehair, together with cow hair, frequently is mixed in the mortar used for ceilings to make it hold better. The shorter mane hair is very much in demand for high grade upholstering. Such horsehair seldom is dyed before use, but simply is freed from impurities by means of soap and alkali.

Now the chief consumers of all horsehair fabrics are the manufacturers of sieves and filters to be used in making flour, gun powder, oils, and so forth. Also this hair used for filters and filter goods once it is cleaned is seldom dyed. Horsehair of the better quality often is taken instead of cow or donkey hair, for making high grade driving belts. Hair goods often are used for military cravats, a kind of pliable collar or stock worn by military men in some countries. This fabric, used in cravats, has a cotton or linen warp and a horsehair filling. Tufts of white or colored horsehair frequently are used in the accoutrement of military uniforms in some countries. The principal use for high grade white horsehair, however, is for violin and cello bows. Only the best material when used in this way can withstand the friction from the constant action of the bow across the strings of the musical instrument.

A short description of the use of horsehair for violin bows may be of interest. The animals are especially chosen as suitable for the supply of raw material. The mane hair is cut off from the horse close to the root, the shortest length of the hair sorted out being about 75 centimeters. When it has been cut out the hair is tied in convenient bunches and sent to the finisher. The work of the finisher is usually not on a very large scale and therefore his outfit is simple consisting of little more than a few large vats to contain the cleaning fluid, an iron boiler of about 450 liters content, heated with wood or coal, for warming up the water, and a few other simple articles of equipment.

In the preparation of the hair for cleaning, the ends are cut evenly and each end is dipped into a flat tub, containing a melted mass that consists of paraffin, shellac and colophony in their proper proportion. When cooled the mixture becomes firm like ordinary sealing wax and the hair ends are fastened in a bunch. Sometimes instead of the shellac preparation melted pitch is used, some finishers even prefering pitch since it neither saponifies nor becomes sticky in a strong cleaning bath as the other agents do.

For cleaning bow hair, soap and ammonia, or—if the hair is very dirty—soap and pearl ash are used. In the cleaning the hair bunches are stretched crosswise over a portable wooden frame and the ends attached lightly to the side of it. The whole is then dipped a few minutes into the cleaning bath, whereupon the frame is taken out and laid on a wooden bench on which a row of parallel grooves or rills of about 2½ centimeters' width are carved out for the reception of each strand of hair. The worker then takes a flat brush of white pigs' bristles and brushes carefully with soap lye each of the strands lying in its groove. After this process the frame with the strands again is entered into the wash tub and treated in it for a short time. It then is rinsed in another tub several times in clear warm water in order to remove any remaining soap. The frame is then laid aside for the strands of hair to dry somewhat.

The goods are now ready for bleaching with sulphur fumes. The equipment used for this purpose consists of a gas-tight wooden box that holds about six of the abovementioned strand filled frames. A number of wooden holders are adjusted at the sides of the bleaching box, forming spaces between the frames. In the double bottom of the bleaching box there are a number of holes to permit the passage of the sulphur fumes produced by burning lump sulphur in an iron pan in the space formed by the double bottom. The goods are allowed to remain in the bleaching box over night, and on the following day the frames with their contents are removed from the apparatus and rinsed repeatedly with water, in which process a little washing blue is added to the last bath to remove any yellow tinge from the hair. The frames then are placed on a wooden stand for dripping. This usually takes place in some well-aired room apart from the cleaning plant, both to dry the hair and at the same time to remove any odor of the sulphur fumes. Finally the strands are cut from the frame and sorted for shipment.

Horsehair from the mane and tail, when it is to be used for weaving haircloth is always cleaned until it is perfectly free from all impurities. Before beginning the cleaning process, the hair is sorted out into bunches of suitable lengths, the fastening of which is left loose enough for the cleaning fluid to pass through it easily. The equipment used for cleaning the hair consists of a wooden tub of about 1.2 square meters surface, fitted with a double bottom with perforations, to provide for circulation of the fluid. A steam pipe to heat the bath is fitted into the space formed by the double bottom. The liquid is circulated by means of a centrifugal pump. The bunches of hair are put into the dry tub, and care should be taken that a firm package is formed that does not leave openings for the liquid to rush through. When the packing is completed, a wooden lattice is laid over the bunches of hair. This device consists of a number of hardwood staves, fixed crosswise as in a garden lattice. The railing is attached by metal clamps to prevent its rising up when the tub is filled with the cleaning fluid.

The cleaning is carried out at a temperature of about 36° C. using soap and soda or soap and ammonia. Generally, after a fifteen minute treatment of a lot, the liquid is let off into a reserve container, and the bunches taken out of the apparatus and then laid in again but in a different position, the purpose of the change being to effect a thorough cleansing. After a treatment of an hour all in all in the cleansing bath, the exhausted liquid is let off and the tub again filled with fresh warm water to rinse off the residual soap. After several such rinsings the bunches are removed from the apparatus and centrifuged.

If the cleaned goods are to be dyed, about the same kind of apparatus is used as is usually used for cleaning purposes. But almost any kind of package dyeing machine can be used for dyeing the bunches, it being unnecessary that the circulation of the water be effected by centrifugal pumps, since direct steam also can be applied for this purpose. Black and brown afterchroming dyestuffs usually are used for the purpose of fast dyeing of the horsehair for upholstering fabrics and for brush bristles. Also acid black is used for dyeing the hair, a black of excellent fastness to light and washing being selected. Acid blue black is used only when special fastness to light and rubbing is desired, but when fastness to washing is not so important.

In order to insure a thorough penetration of the color, the process is carried out at a boiling point for from one and a half to two hours. Experience shows that dyed horsehair turns out very much lighter if it is taken out of the bath sooner. By using a more strongly acidified bath, it is possible to apply any well known acid dyestuff, that goes on the fiber evenly, for the dyeing of horsehair. In cases where it is necessary to bleach out the natural color of the hair strands to some extent, a short treatment in a boiling, weakly acidified bath with an approximate 3% content of sodium sulphoxylate formaldehyde will serve the purpose.

Pigs' Bristles

High grade white bristles are used by the makers of tooth brushes and other toilet brushes. The natural colored goods are usually converted into cheap curled hair for filling mattresses and other similar goods. The bristles are clipped off from the slaughtered pig when it is taken out of the scalding-vat. The best material is from along the spine towards the end, and from the flanks. The various kinds are sorted out one from the other and arranged according to length and color—black, white and brown being the ordinary shades.

After the bristles are scraped off from the pig, they are dried in flat boxes in the open air, the containers being covered over with some light fabric when wind is blowing, to prevent loss. When the bristles are dried they are either sprinkled with salt or dipped in salt water, to prevent decay of the bits of skin and flesh still clinging to the roots. The material is then packed in barrels or bags for shipment to the consumer.

After the goods have been received at the plant of the manufacturer of brushes or of other articles, the bristles are sorted out by hand and the bits of skin or dirt on the goods removed by rubbing them across the ridges of the metal bottom of a flat trough. Thereupon the bristles are transported to another part of the equipment which is fitted with a bottom similar to a fine mesh sieve. The entire lot of bristles is then worked for a while on this sieve, when most of the dust falls through the mesh, while the worker removes by hand any particles of dirt which are too large to fall through. After the bristles have been treated in the above described manner they are packed for wet cleaning in sacks of five pound lots.

In small plants the cleaning apparatus consists of a number of large barrels, the upper part of which has been removed. The goods are treated in these. The sacks in the cleaning fluid are kneaded and pressed by a workman, a pounder being used, like that sometimes used by women in the hand laundering of wearing apparel. In better equipped plants there is a machine which is similar to the fulling apparatus for the manufacture of felt hats. This machine is fitted with a number of wooden pistons, to which are attached rubber suction caps, and the movement to and fro of these pistons, operated by the propelling force of the driving mechanism connected with them, removes all impurities from the bristles by suction and pressure. For the sake of convenience in handling, the goods are packed in 5 lb. lots in sacks.

When the bristles have been soaped, the bags are pressed into a contrivance which consists of two boards of heavy wood slung parallel, the lower of which is fastened to a portable stand, while the upper board is movable and is fitted with a long handle to obtain the necessary leverage for the desired pressure. Since the whole is portable, it can be used at the side of any tub in the plant wherever it is needed. The soap water which is squeezed out of the sacks that are thus manipulated is collected in a trough fastened underneath the apparatus. After the soap water is pressed out the sacks containing the bristles are rinsed in warm water until they are quite free from soap and then pressed out again, when they are ready for bleaching.

Before peroxide bleaching came to be generally adopted, pigs' bristles were cleaned with sodium bisulphite but the latter is now being used less and less since results are much better with the newer process. However, where a cheap process is necessary, bisulphite is still applied. The bath for bisulphite bleaching contains one liter of sodium bisulphite of 75° Tw. to 100 liters of water. The bags containing the bristles are soaked for one hour in the cold liquid, being turned from time to time, and are then entered in a tub of fresh cold water for careful rinsing.

For peroxide bleaching, wooden tubs are used, fitted with lead steam pipes for heating the liquid. For a bath of 135 liters, 9 liters of bleaching agent are used, consisting of hydrogen peroxide (12% by volume), in conjunction with a sufficient quantity of ammonia to make the bath slightly alkaline. The sacks containing the bristles are entered into the cold bath, and when they have been moved about by hand for a time, the temperature of the bath is slowly increased to 50° C. The bath is then permitted to cool from six to eight hours, when the bags are removed from the liquid, squeezed off and rinsed. The goods re then taken out of the sacks and aid in flat cloth-covered containers or drying.

The pigs' bristles are dyed in suit- ble sacks in a package dyeing ma- hine, where the circulation of the iquid is effected either by a pump or by direct steam pressure. In smaller establishments the sacks are reated in open tubs, in which case hey are hung on wooden poles laced across the top of the tub— ach sack containing about two ounds of bristles. Afterchroming lack and logwood black are the nain dyes for bristles which are in- ended for brushmaking. During his process care should be taken hat the bristles are thoroughly soaked through, as the ends of in- ufficiently soaked bristles, when nade into brushes and trimmed, vill have a gray or purple appear- nce.

Generally the black goods are given a light soap bath after the lyeing, in order to free the bristles rom any loose dyestuff particles. For this treatment olive oil soap is ised, a small quantity of pure olive oil being generally added to the bath also to give the goods a better hand and luster.

Fur Bleach
U. S. Patent 2,086,123
Formula No. 1

Hydrogen Peroxide (30%)	25 cc.
Iron Sulphate	5 g.
Ammonium Bifluoride	2 g.
Water, To make	1 l.

No. 2
U. S. Patent 2,092,746

Hydrogen Peroxide (100 Vol.)	3–5 gal.
Sodium Silicate	12 lb.
Sodium Oxalate	2 lb.
Water	100 gal.

Fur Softener and Lubricant
U. S. Patent 2,262,611

Lanolin	20–35%
Cedar Wood Oil	3–10%
Benzine, To make	100%

Softening and Flexibilizing Dyed Furs

Diglycol Stearate	10 lb.
Water	12 gal.

Boil together and mix until uniform. Cool to room temperature and add to it 107 gal. of dye-bath solution. Dye in the usual way. If salts are present in the dye-bath, the addition of a wetting agent, such as Wetanol is necessary.

Fur and Leather Conditioner

Diglycol Laurate	14.0
Neatsfoot Oil	36.0
Mineral Oil	49.5
Santomerse OS (Oil Soluble)	0.5

This gives a considerable improvement in the behavior and appearance over formulae made with sulphonated oils which frequently react unfavorably with fur pelts.

Lustering Furs

Immerse the degreased and dried fur in the following solution for about three hours:

Water	19 l.
Salt	800 g.
Potassium Iodide	150 g.
Iodine	90 g.

The iodine-impregnated fur is then subjected to a lustering operation with a suitable oil. The process is particularly effective on lamb furs, which have a tendency to curl and readily spot with water.

Glazing Solution for Furs

Mineral Oil	10
Chloroform	90

Spray the fur using an atomizer.

Fur Gloss

Dissolve 3 to 6 ounces of paraffin wax in 1 gallon petroleum cleaning solvent.

Approved cleaning solvent is preferable because of its safety during ordinary handling.

Precaution: Paraffin separates from the petroleum solvent at temperatures below 70° F. At —15° F. it is completely chilled out of the solvent.

This finish is used for the saturation of dry cleaned furs to replace any oils removed and to make them water repellent. It is also sponged or sprayed on materials that are lifeless or lusterless after cleaning and drying to produce high gloss.

Non-Curling Fur
U. S. Patent 2,234,138

The fur is treated for three hours with

Aniline or Phenylene Diaminc	60
Acetic Acid	50

and then for three hours with an aqueous solution containing

Formaldehyde	50

Then wash and dry.

Fur Carroting Solution
British Patent 532,370

Formula	No. 1	No. 2
Nitric Acid (63%)	15	12
Hydrogen Peroxide (100 Vol.)	20	17
Urea	—	0.05
Citric Acid	—	0.05

Chapter Nine

INKS AND MARKING MATERIALS

CARBON PAPER INKS

Black Carbon
(Certified Public Accountant)

Carnauba Wax	40
Mineral Oil	40
Carbon Black	10
Iron Blue	10

Black Carbon

Carnauba Wax	30
Methyl Violet Base	2
Oleic Acid	8
Beeswax	3
Carbon Black	15
Lard Oil	42

Black Universal Carbon

Extra Heavy Mineral Oil	40
Oleic Acid	5
Carnauba Wax	37
Crystal Violet	3
Carbon Black	15

Black Pencil Carbon

Mineral Oil	18
Oleic Acid	23
Rosin	7
Montan Wax	30
Lampblack	12
Milori Blue	10

Blue Pencil Carbon

Carnauba Wax	30
Paraffin Wax	10
Milori Blue	25
Petrolatum	35

Blue Pencil (Soluble Type)

Victoria Blue	10
Oleic Acid	30
Ceresine Wax	5
Carnauba Wax	25
Mineral Oil	30

Hectograph Carbon

Crystal Violet	60
Carnauba Wax	9
Opal Wax or Acrawax B	10
Mineral Oil	20
Lecithin	1

One-Time Ink

Montan Wax	30
Paraffin Wax	5
Methyl Violet	1
Lampblack	15
Mineral Oil	49

Manifolding Paper Coating Emulsion
U. S. Patent 2,299,694

Toluene	150
Gum Dammar	150

Dissolve the above by mixing and then add to the following until emulsified:

Glycerin	80
Diethylene Glycol	14
Carbon Black	6

TYPEWRITER RIBBON INKS

Black
Formula No. 1

Nigrosine Base	5
Oleic Acid	25

Carbon Black	15
Mineral Oil	55

Grind thoroughly before applying to ribbon fabric.

No. 2

Victoria Blue Base	1
Nigrosine Base	4
Oleic Acid	25
Black Toner	15–25
Mineral Oil	55

Black Two-Color

Formula No. 1

Carbon Black	20
Lard Oil	40
Castor Oil	40

No. 2

Milori Blue	5
Peerless Black	15
Sperm Oil	80

Red Record

Red Lithol Toner	30
Mineral Oil	70

Blue Record

Blue Toner	20
Lard Oil	40
Castor Oil	40

Hectograph Ribbon Ink

Crystal Violet	50
Mineral Oil	45
Petrolatum	5

Purple Checkwriter

Methyl Violet Base	10
Violet Toner	5
Oleic Acid	85

Multigraph

Nigrosine Base	10
Oleic Acid	75
Milori Blue	5
Drop Black	10

STAMP-PAD INKS

Green

Light Green SFA	0.40 g.
Diethylene Glycol	50.00 cc.

Heat till completely dissolved.

Blue

Formula No. 1

Methylene Blue A (ex conc.)	0.40 g.
Crystal Violet A	0.04 g.
Diethylene Glycol	50.00 cc.

No. 2

Victoria Pure Blue BO	0.50 g.
Diethylene Glycol	50.00 cc.

Heat till dissolved. Filter.

No. 3

Soluble Blue	16
Glycerin	64

Rub fine in mortar.

Glucose	32
Water	16
* Syrup de Gomme	16

Mix thoroughly and combine.

Red Stamp Ink

Formula No. 1

Crocein Scarlet MOO	37 g.
Diethylene Glycol	50 cc.

Heat till dissolved.

No. 2

AE Fuchsine	1½
Alcohol	12
Glycerin	64

Mix thoroughly.

Glucose	32
Water	6
* Syrup de Gomme	16

Gradually add to glycerin solution. Stir well.

* Syrup de Gomme

Gum Arabic Nubs	1
Water	5

Stir until substantially dissolved and then strain out dirt and lumps.

Black

Carbon Black	5
Glycerin	88
Rub down in mortar.	
Glucose	16
Water	4
Syrup de Gomme	20

Stamp Ink for Eggs

Gum Arabic	0.10
Water	4.00
Glycerin	4.37
Formaldehyde (40%)	0.03
Ultramarine-Blue	1.50

Meat Stamping Ink—Blue

Blue Dye (pure food)	30
Dextrin	20
Glycerin	82
Water	70

Aqueous Printing Ink
Canadian Patent 405,509

Rosin Powdered	25	lb.
Water	3	gal.
Ammonia	6½	pt.

Dissolve the above and add the ollowing solution:

Zein	50 lb.
Water	50 qt.
Ammonia	2 qt.
Color	To suit

Colloidal Carbon Black

Carbon Black	31
Water	59
Glycerin	5
Soda Ash	1
Quebracho	4

Mill on colloid mill; this formula makes a supercolloidal black paste.

* Syrup de Gomme

Gum Arabic Nubs	1
Water	5

Stir until substantially dissolved and then strain out dirt and lumps.

Thermo-Setting Printing Ink
British Patent 543,393

Cumar Resin	56
Blown Soyabean Oil	20
Carnauba Wax	10
Carbon Black	10
Violet Toner	2
Blue Toner	2

Printing Ink
German Patent 704,951

Varnish	28
Glycerin	36
Dye	18
Ammonium Nitrate	4
Ester Gum	14

Rotary Heliogravure Ink
French Patent 844,153
Formula No. 1

Alcohol	300 cc.
Butanol	100 cc.
Zein	8 g.
Water	80 cc.
Nigrosine NB	16 g.

No. 2

Alcohol	320 cc.
Zein	8 g.
Water	80 cc.
Nigrosine NB	8 g.

No. 3

Alcohol	400 cc.
Zein	50 g.
Water	100 cc.
Titanium Oxide	300 g.

Rotogravure Ink for Cellulose Foils
French Patent 848,697

Acetone	500 cc.
Benzene	500 cc.
Alcohol	2000 cc.
Shellac	1500 kg.
Victoria Blue	3 %
Cellulose Acetate	1–3 %

Intaglio Printing Ink
British Patent 513,247

Gilsonite	15–$33\frac{1}{3}$%
Naphtha, Petroleum	85–95%
Methyl Ethyl Ketone	5–15%

Raised Printing Resin (Thermoplastic) "Ink" Base
U. S. Patent 2,272,706

Lewisol 2L or Amberol 801 or Rauzene X-135	70
Ethyl Cellulose	$17\frac{1}{2}$
Hercolyn	$17\frac{1}{2}$

Raised Printing "Ink"
U. S. Patent 2,272,706

Formula No. 1 (Satin Gloss Finish)

Raised Printing Resin	$173\frac{3}{4}$
Zinc Stearate	$21\frac{3}{4}$
Surfex (Calcium Carbonate)	$14\frac{1}{2}$

No. 2 (Luster White Finish)

Satin Gloss Finish Compound (Above)	$21\frac{3}{4}$
Titanium Dioxide	$13\frac{1}{2}$
Zinc Oxide	$4\frac{1}{2}$
Ultramarine Blue	$\frac{1}{2}$

Ink for Printing on Rayon Ribbon

Ethyl Cellulose (High Viscosity)	$\frac{1}{2}$ oz.
Denatured Alcohol	$\frac{1}{2}$ pt.

Allow to stand for about 20 to 30 minutes until a jell forms. Add:

Tributyl Phosphate	1/5 oz.
Surfex	$\frac{3}{4}$ oz.
Titanox	$1\frac{1}{2}$ oz.
Magnesium Carbonate	1 oz.

For other colors substitute other pigments for the Titanox.

Use alcohol for thinning.

Textile Printing Ink
U. S. Patent 2,213,006

Nitrocellulose (viscosity 80 sec. 10, a pigment 15, a mixture of monc alkyl ethers of ethylene glycol an diethylene glycol (32), and (petrc leum naphtha) 27–29 parts is mixe to form a non-stringy, gel-like in capable of plastic flow under pres sure.

White Ink for Photostats

Black photostats can be writte on with white figures by using 50% solution of syrupy phosphori acid. Write on the black photostat with a clean gold or glass pen, leav to react for 3–4 minutes, then was the character with plain water an dry with a blotter.

Blue Indelible Ink

For an ink that will resist no only water and oil, but alcohol oxalic acid, alkalies, the chlorides etc., mix the following: Dissolve parts of shellac in 36 parts of boil ing water, carrying 2 parts of borax Filter and set aside. Now dissolve 2 parts of gum arabic in 4 parts o water and add the solution to the filtrate. Finally, after the solution is quite cold, add 2 parts of pow dered indigo and dissolve by agita tion. Let stand for several hours then decant and put in small bottles

Removable Marking Ink for Leather and Textiles

Ammonia (0.5%)	2 kg.
Dimethylglyoxime (0.5%)	2 kg.
Ferrous Ammonium Sulphate (1%)	700 g.

This ink is readily removed by bleaching with oxalic acid (1–3% solution).

Invisible Laundry Inks

(1) An invisible laundry ink hich fluoresces a bright blue in traviolet light and which cannot seen in daylight, is made by ssolving beta-naphthol in water id alkalinizing with sodium hyoxide. The ink is good for rayon, otton, and other fibers.

(2) Another ink is made by dislving oxynaphthionic acid in dite sodium hydroxide solution. nder ultraviolet light this ink fluosces a yellowish-green color. Inks ith sodium hydroxide should not used on animal fibers such as ool and silk.

Invisible Ink

Formula No. 1

Water	50 cc.
Uranyl Nitrate	1–2 g.

When applied to paper and simir objects and allowed to dry, this ik fluoresces a bright yellowishreen under ultraviolet light. It iould not be used on foods or other roducts intended for human conimption.

No. 2

Water	50 cc.
Cinchonine	0.2 g.
Sulphuric Acid	To make acid

This ink may be used on wood nd durable cellulose products and uoresces a bright blue under ultraiolet light.

Sympathetic (Invisible) Inks

Formula No. 1

The writing fluid is the juice of ie onion, and it may be obtained y plunging the pen into a peeled nion. On drying the writing is abolutely colorless, but on applying eat, such as from a lighted candle or match, the writing develops to a deep red brown.

No. 2

British Patent 235,968

A mixture of phenolphthalein, glycerin, alcohol, and carbon tetrachloride is the writing fluid. On drying it is colorless, but on the application of ammonia it turns reddish.

Invisible Printing Ink

(Pyrographic Ink)

Potassium Nitrate	90
Rosin	5
Sulphur	5

Mix the above and then add sodium silicate to bring to any consistency desired.

When the above ink is used on paper and the latter is ignited, it burns along impression.

Invisible Writing Carbon Paper

A fluorescent substance, which may be either organic such as anthracene, or inorganic such as zinc sulphide phosphor, is coated on paper in a base of several per cent drying wax and an organic solvent which also acts as a vehicle. This paper may be used in the typewriter instead of carbon copying paper for typing invisible messages which may only be read under the proper wavelength of ultra-violet light. Dark paper which has been padded to prevent impressions is preferred for the final message.

Hectographing (Copying) Invisible Writing

(1) One of the inks described in this section may be made up as hectograph ink with the proper colorless vehicle, such as glycerin, and used in the usual manner.

(2) A fluorescent dye can be added to ordinary hectograph ink for conveying an additional meaning to a message. For example, a red fluorescing dye, which would be invisible in white light, would mean that the reader should ignore that part of the message containing the red fluorescent dye whereas that containing the green fluorescent dye should be read.

Invisible Pass-Out Ink

(1) A saturated aqueous-alcohol solution of beta-methyl umbelliferone may be used as an invisible fluorescent stamping ink. A little glycerin or other suspending or thickening agent may be added. When used with a clean rubber stamp marks may be applied to the skin which are not visible in ordinary light but, under ultra-violet light, they glow a brilliant blue color.

An alcoholic solution of tumeric may be used to harmlessly stain skin for tagging and marking purposes. Under ultraviolet light a yellowish fluorescence is noted.

Parachute Marking Ink

Mix 837 parts by weight of boiled linseed oil with 17 parts by weight of aluminum stearate. Heat to 115.6° C. until stearate is in solution. Cool the solution. Place it in a ball mill and add 525 parts by weight of ultramarine blue. Grind for at least eight hours. Add 450 parts by weight of spar varnish and continue to grind for at least eight hours more.

Silk Screen Inks

There are two types of inks used in the silk screen processes: Those that undergo chemical change o drying such as linseed oil inks, an those that dry by evaporation onl The following formulae are for thos that dry by evaporation and un dergo no chemical change.

For mixing the inks a 1/8 or 1/ H.P. motor having fastened to it shaft a metal or wooden propeller installed in a vertical position s that it can be used to mix the con tents of a one gallon can or larger.

It is only necessary to add a much color as is found to be neces sary to achieve the desired shad With a half dozen different flushe colors on hand it is possible to ob tain many colored inks of a variet of shades.

The process consists in the prepa ration of a gum solution, mixing i the fillers and dryers, and additio of the flushed color.

Gum Solution

Solvesso No. 1 or 2	35
Varnoline	35
Gum Batu, Finely Powdered	50

The solvents are first mixed an then the gum is added slowly an then stirred for about 15 minute A thick oily solution and gum sus pension results.

The following is added slowly un der constant stirring:

Magnesium Carbonate	10
Surfex	10
Petrolatum	3
Titanox	5

It is only necessary to add to thi any desired flushed color. To mak a blue ink for example, add to th above

Flushed Ultramarine Blue	25

When agitation has resulted in a uniform mixture, add

Water	30

slowly and stir constantly. This will cause the ink to swell up and become very thick, and will impart to the ink its fine drying properties. All of the above proportions can be altered to suit the occasion and any flushed color can be employed. This ink cannot be used with any silk screen using a glue background. If it is desired to use a glue screen it is then only necessary to omit the water. For thinning the following solvent is prepared:

Ink Thinner

Kerosene	5
Varnoline	95

The thinner is added slowly and the ink mixed very thoroughly each time before any more thinner is added.

For diluting the ink color a good base is needed. This is made exactly the same way as the ink. Use the following proportions:

Base Formula

Varnoline	35
Solvesso No. 1 or No. 2	35
Batu Gum	50
Magnesium Carbonate	10

When thoroughly mixed add slowly:

Water	35

For use dilute the ink with the base in any proportion desired. If too thick, add thinner and stir slowly to an even consistency.

Silk Screen Black

Solvesso No. 1 or 2	35
Varnoline	35
Gum Batu	50
Petrolatum	2
Kerosene	2
Carbon Black	20
Magnesium Carbonate	10

Marking Paper Money

Paper money, such as ransom money, may be invisibly tagged by dusting with anthracene. Under ultra-violet light a bright greenish fluorescence is observed.

By wiping a paper note with a cloth moistened with lubricating oil or petrolatum an invisible mark will be made which shows, under ultra-violet light, as a bright bluish region.

Cattle Marker

Paraffin Wax	40
Tallow	20
Woolfat	20
Ceresin (58/60° C.)	15
Oil Soluble Color	1–2
Pigment (To Match Color)	As desired

Lantern Slide Ink

An ink consisting of a 3% solution of celluloid in "Cellosolve" or other similar solvent, and colored sufficiently with crystal violet, or other dye, will write on clean glass, using an ordinary steel pen. Slides can be very easily prepared. It is necessary that the glass be unusually clean and free from any oil film. The dry glass, after washing with soap and water, should be sponged with acetone, and the surface must not be touched with the hand during the writing. The handle of a toothbrush is a satisfactory source of celluloid. The ink dries rapidly and corrections can be easily made, after erasing by scraping with a razor blade.

Ink for Glassware

Inks for printing on glass are useful for identifying or otherwise marking laboratory apparatus. The following may be applied by means of an ordinary rubber stamp and fixed by heating up slowly in a torch almost to the softening point of the glass.

Formula No. 1
British Patent 212,938

A mixture of 1 part of cupric oxide and 1.5 parts of boric acid is fused. A mixture of 1 part of litharge, 1.5 parts of zinc oxide, 0.4 parts of boric acid, and 0.1 parts of cryolite is also fused. The two are ground together in the proportion of two parts of the former to one part of the latter and this is ground in well with one-fifth its weight of castor oil.

No. 2
U. S. Patent 1,538,890

A mixture of seven parts of silver oxide and three parts of lead borate is well ground in with glycerin or linseed oil and gum dammar until a somewhat tacky suspension is obtained.

No. 3

Grind together well one part cobalt oxide, one part manganese oxide, one part chromic oxide, and one part lead borate and grind this mixture well with glycerin until a tacky paste is obtained.

Glass Marking Ink

Formula No. 1

Barium Sulphate	15
Ammonium Bifluoride	15
Ammonium Sulphate	10
Oxalic Acid	8
Glycerin	40
Water	12

Package in wax or hard rubber bottles. Apply in ventilated hood. If too thick, dilute with water. The addition of 1–5% sodium fluoride speeds up the action of the above mixture.

If a black marking is wanted, replace the barium sulphate by

Lampblack	8–10

No. 2
U. S. Patent 2,254,865

Lampblack	20.9
Titanium Dioxide	4.2
Silver Oxide	1.2
Glycerin	73.7

Ink for Ruling on Glass

Use India ink to which has been added 1% Lepage's Mucilage.

Glass Crayon

In the laboratory a worker often has need to write on glass vessels, flasks, and the like. If no glass pencil is available a good substitute can be made as follows:

Mix while it is melted,

Paraffin Wax	98 g.
Stearin	2 g.

Then add 10–15 g. of an oil soluble dye in either red, blue, black or yellow. Let the liquid cool slightly and before it sets to a hard mass, pour it into large paper soda straws, the ends of which have been plugged with a small cork. Next, set these filled tubes in ice water or better still in a refrigerator. When they have gotten hard, tear off one end and use them to mark on glass.

Etching of Designs and Lettering on Metals

For uniformity the formulae of solutions prepared by dissolving a dry salt in a liquid have been exressed in terms of grams per liter g./l.) and of avoirdupois ounces er gallon (oz./gal.) of the resultant olution, while the formulae of soluions prepared from a mixture of iquids have been expressed in terms f milliliters (often called cubic entimeters) per liter (ml./l.) and f fluid ounces (1/16 pint) per al. (fl. oz./gal.). Unless otherwise tated aqueous solutions are reerred to, that is, the substances are issolved in sufficient water to proluce the specified final volume.

The etching of designs on metals s usually accomplished in three teps, namely, (a) application of a rotective coating, (b) cutting the lesign through the coating, (c) etchng the design. In the first step, the netal is coated with a thin layer of substance known as a "resist," so alled because it is resistant to the ction of the etching solution. It hould also have the property of dhering firmly to the metal so as o confine the etching action to the lesired areas or lines, and should espond to the cutting tool readily vithout being disturbed adjacent to he cut.

The second step in the process is he cutting of the design. This may e done either mechanically or hemically, provided the resist is ompletely removed from the areas vhich it is desired to etch.

The final step is the etching with solution that will dissolve the netal. Usually acids are used, alhough acid salts, neutral salts of a nore noble metal than that being tched, or even alkaline solutions, nay be used for certain metals or lloys.

The particular process to be used for a specific purpose depends on a number of factors, such as the nature of the metal to be etched, the number of pieces to be etched with the same design, the complexity of the design, and the desired sharpness of the etched lines.

The surface to be etched should be smooth and free from scratches. It should also be clean so as not to interfere with the uniform etching of the materials. Alkali cleaners or organic solvents are generally used to remove grease from the metal pieces.

Since the application of the resist and the cutting of the design are to a large extent interdependent, they will be discussed together.

Waxes such as paraffin, ceresin, beeswax and ozokerite, are often used as resists. They may be used either singly or blended together in mixtures. For example, a mixture of 3 parts of paraffin and 1 part of beeswax has been found very suitable. The melted wax is applied to the surface to be etched, by dipping, brushing or flowing. When the wax is solidified, the design may be cut through it with a sharp tool by hand or by means of a pantograph. Some pantograph devices are constructed so that a number of tools can be operated simultaneously, with considerable saving in time.

In another method, the metal surface is coated with a solution of gum guaiacum in alcohol. After the coating has dried, the pattern to be etched is stamped on the surface by means of a rubber stamp wet with a concentrated solution of sodium hydroxide (for example, 300 g./l. or 40 oz./gal.). The alkali causes the gum to become soluble in water so that when the surface is washed the bare metal is exposed. This washing should be done quickly with a large quantity of water, so that the alkali will not spread over the surface and dissolve the resist from areas it is desired to protect.

A third pretreatment consists in rolling a thin film of ink on the surface to be etched. The type of ink is designated in German publications as "Umdruckfarbe" ("transfer ink"). The following mixture gives satisfactory results.

Printer's Ink	100
India Ink	10
Castile Soap	4
Beeswax	4
Animal Fat	2
Rosin	1

The components of this mixture are blended by heating and rubbing to produce a smooth paste.

The ink film, after being applied to the surface, is dusted with finely powdered asphalt, and the two are blended together by gentle heating. If powdered asphalt is not available, powdered rosin, although less satisfactory, may be substituted for it. If distortion of the metal article by heating is feared, the blending may be done by suspending the treated article for a few seconds in the vapor of alcohol or trichloroethylene.

The cutting of the design is readily carried out in the following manner: The assembled letters or characters in the form of steel stencils are clamped in a suitable holder. A sheet of tissue paper or fine sandpaper (No. 00) is placed on top of the resist coating and the design is impressed therein with light pressure, such as can be obtained with a small press. On releasing the pressure, the resist adheres to the paper and the desired design is exposed as bare metal.

Another procedure for applying the resist is as follows: The desired design is first printed on the surface with an ink which contains a considerable amount of fats. The inked surface is then coated with a rapidly drying spirit varnish, which dries readily over the bare metal but not over the greasy ink. The resist is then removed from the inked areas by a solvent that does not affect the dried varnish film. Petroleum oil has been recommended for this purpose. Before etching, the surface must finally be degreased by light rubbing with some material such as tripoli powder or by a suitable solvent. This method, though time-consuming, is specially useful on a curved surface or in locations where it is difficult to use the steel stencils described in the previous method.

Still another process in the preparation of the surface for etching consists in first printing the background, or that part of the surface to be protected from the etching, with heavy printing ink, which is then dusted with powdered "dragon's blood" or with a mixture of powdered asphalt and rosin. After the powder has been brushed or blown off from the dry, bare metal the ink and acid-resisting powder are fused together by baking, if heating does not affect the metal, or by the cold fusion process with alcohol vapor described previously. The contrivance for printing the design on the surface may be a simple rubber or linoleum stamp prepared with the design depressed instead of being in relief as in the ordinary rubber stamp. By another rather rough method, the design may be printed in reverse on tissue paper with heavy printer's ink. While the ink is still tacky, the tissue is applied to the metal surface and allowed to dry. The tissue is then removed after moistening it with water, leaving the ink film as the resist on the metal.

For quantity production, the printing of the design on the surface is most easily accomplished by first making a master plate in the manner used in photoengraving. A large drawing in black and white is made of the desired design, which then is photographed and reduced to the desired size. The design is transferred to a sensitized zinc plate by clamping the photographic negative over the zinc plate and exposing to the rays from an arc light. Light renders the exposed areas insoluble in the etching reagent, whereas unexposed areas (portions covered by the opaque film) are dissolved when the zinc plate is etched with dilute nitric acid. The resulting master plate with the exposed areas or background of the design raised in relief is used for transferring the design onto the article to be etched, which is often done in a flat-bed printing press. The master plate, after being inked with an asphalt-base ink, is rolled with a clean printer's roller which picks up the ink and transfers it to the article to be etched. This is followed by dusting with asphalt powder, as has already been described. Sometimes the etching of the master plate is omitted, since the areas of the sensitized zinc plate exposed to the light acquire the property of being wetted by and holding the ink, whereas ink will not stick to the unexposed areas.

Etching

It should be remembered that the solutions used in etching are corrosive to the skin and clothing and should be handled with care. If any of the acids should get on the skin, wash immediately with a large amount of water and neutralize any remaining acid with baking soda.

After a satisfactory resist has been applied to the metal surface and the design cut into it, the surface is ready for the etching. It is important that the etching action should take place uniformly over the entire exposed area and not be localized at certain points. It should also progress down into the metal and not undercut the areas covered by the resist and thereby spread over the surface.

The choice of the etching reagent is dependent on the nature of the metal being etched. It is often advisable to perform a few preliminary tests to determine the type and concentration of etching solution which gives the most satisfactory results. The formulae suggested here will serve as a starting point for experiments to enable one to decide upon the best reagent.

Reagents for Iron and Steel

The usual etching solution for iron and plain carbon steel is dilute nitric acid. One part of concentrated nitric acid diluted with three parts of water has been found to give satisfactory results in most cases. The following variation of the nitric acid solution has certain advantages, although it cannot be used where alcohol would dissolve the resist, as is the case with gum guaiacum, or with resists produced on the metal by photographic processes.

	ml./l.	fl. oz./gal.
Nitric Acid (sp. g. 1.42)	220	28
Hydrochloric Acid (sp. g. 1.18)	20.5	3
Ethyl Alcohol (95%)	110	14
Water, To make	1 l.	1 gal.

A concentrated aqueous solution of ferric chloride is often used to etch steel.

Resists such as the spirit-varnish film in the fourth resist described above will not withstand the action of strong acids. For such purposes a less corrosive reagent, for instance, a copper chloride solution slightly acidified with nitric acid, or one prepared according to the following formula will be useful.

	g./l.	avoir. oz./gal.
Copper Sulphate	200	27
Zinc Sulphate	8	1
Sodium Chloride	165	22

Stainless steel is etched with much more difficulty than ordinary low-carbon steel. Ferric chloride (sometimes called iron perchloride), usually with the addition of hydrochloric acid, is the most widely used reagent for this type of steel. The time required for the etching to be completed is much longer (about 30 minutes) than that for the plain carbon steel. After the etching has been completed and the resist removed, it is advisable to repassivate the surface by immersion in concentrated nitric acid, if practicable. This treatment restores the original property of corrosion resistance to the stainless steel.

Reagents for Nonferrous Metals and Alloys

Copper and its alloys may be conveniently etched with ferric chloride solutions. A formula which has been recommended for etching brass is

	g./l.	avoir. oz./gal.
Ferric Chloride	45	6
Hydrochloric Acid, (sp. g. 1.18)	55 (47 ml.)	7.3 (6 fl. oz.)

Some other formulas for etching copper and brass are:

		g./l.	avoir. oz./gal.
(a)	Potassium Chlorate	27	3.6
	Ferric Chloride	60	8
	Nitric Acid, (sp. g. 1.42)	46 (32 ml.)	6.1 (4.1 fl. oz.)
(b)	Ferric Chloride	390	52
	Hydrochloric Acid, (sp. g. 1.18)	83 (70 ml.)	11 (9 fl. oz.)
	Alcohol (95%), To make	1 l.	1 gal.
(c)	Potassium Chlorate	20	2.7
	Hydrochloric Acid, (sp. g. 1.18)	80 (68 ml.)	10.7 (8.7 fl. oz.)

Zinc may be etched with dilute nitric acid. A solution containing sulphuric acid and sodium dichromate has also been used for this purpose, as well as a solution of the following composition.

	ml./l.	fl. oz./gal.
Glacial acetic acid, $HC_2H_3O_2$ (sp. g. 1.05)	400	51.2
Nitric acid, HNO_3 (sp. g. 1.42)	100	12.8
Alcohol (95%), To make	1 l.	1 gal.

Aluminum may be etched with dilute hydrochloric or hydrofluoric acid. (If the latter acid is used, more than usual precautions must be taken to keep from breathing the fumes or getting any in contact with the skin.) A 5% solution of sodium hydroxide may also be used for etching aluminum, provided the resist is unaffected by this solution.

Nitric acid is the common etching reagent for silver, and may be used as a 20 percent solution in water.

After the design has been cut in the resist on the article and a suitable etching reagent has been selected, the next step is the actual etching. In working with small articles, the back and sides are first usually given an acid-resisting coating, such as asphalt varnish, which is allowed to dry. They are then completely immersed in the etching solution. Best results are obtained if the bubbles formed in the action are removed from the surface which is being etched. This can be accomplished by gently swabbing the surface. After a sufficient depth of cut has been obtained, the articles are removed from the etching bath, washed with water, and any acid remaining on the pieces is neutralized with a weak sodium carbonate solution. They are again washed, dried, freed from the resist coating with a suitable solvent and finally coated with a thin film of oil, varnish, or lacquer to prevent corrosion. In the etching of small designs on large surfaces, immersion of the whole article in the bath is not feasible. Here, the etching solution may be swabbed on the design with a rag held on the end of a stick, or a dam may be built around the design and the solution poured into the enclosed area. A rubber ring with a cross-section of about ½ inch square has been found suitable for a dam. A coating of petrolatum on the lower side of the ring prevents leakage of the solution between the ring and the resist. Dams may also be built up out of such materials as clay, plaster, molding wax or asphalt.

After the etching has been completed, the solution may be poured off if convenient, or most of it may be pipetted off by suction into a suitable container and the last traces removed with a damp sponge

The etch produced with acids is often not readily visible under all lighting conditions. Coloring of the design or of the unetched areas is often resorted to in order to produce a pleasing appearance and to attract attention to the design, trade mark or lettering. An almost unlimited field is available to the etcher for producing various effects in color. A coloring process may be substituted for the etching process by using a solution which colors as well as etches, or the etching may be completed first and followed by the coloring. Sometimes the colors are produced by spraying colored enamels on the etched design. After the enamel has dried, a solvent is used which dissolves the resist but does not affect the enamel, which therefore remains in the etched lines and areas. Two-color effects may be produced by the use of masks or stencils to block off parts of the design.

Coloring of the metal by chemical solutions is very often done. If the highlights are to be colored or plated, the appropriate treatment is applied to the whole article before the resist is applied. The design is later etched through the colored or plated coating in the usual manner. If it is desired to color the design, this is done before the resist is removed. A large number of solutions have been developed for obtaining various colors on metals. The following are typical:

Black Coloring of Iron

The objects are first copper-plated by immersing for about 10 seconds in a solution made by mixing solutions A and B diluting to 1 liter (*or 1 gallon*) with water.

Solution A—10 g. (*1.3 avoir. oz.*) Copper Sulphate, in 250 ml. (*32 fl. oz.*) water.

Solution B—15 g. (*2 avoir. oz.*) Stannous Chloride, in 17 ml. (*2.2 fl. oz.*) Hydrochloric Acid (sp. g. 1.18) and 100 ml. (*13 fl. oz.*) water.

After the plating is completed, the objects are rinsed and blackened by immersion for two or three minutes in a solution prepared as follows:

Dissolve by heating 1.5 kg. (*12.5 lb.*) of sodium thiosulphate in 1 liter (*or 1 gallon*) of water. Allow to cool and add 65 ml. (*8.3 fl.oz.*) of concentrated hydrochloric acid. Small amounts of hydrochloric acid are necessary from time to time to reactivate this solution.

Black Coloring of Brass

Immerse for about 20 minutes in the following solution:

	g./l.	avoir. oz./gal.
Basic Copper Carbonate	35	4.6
Aqua Ammonia (sp. g. 0.90)	103 (115 ml.)	13.8 (14.7 fl. oz.)

Black nickel plating is often used to color zinc, aluminum and other metals.

If it is desired simply to mark a steel specimen with some sort of an identification symbol or to write on the surface in large bold letters in which blurring of the edges is of no consequence, it is possible to apply the etching solution directly to the steel surface without the aid of a resist. The solution is best applied by a pen which is resistant to the acid, such as a wooden stylus or quill pen. A stainless steel pen is also satisfactory for this purpose. A rubber stamp with a pad of blotting paper or asbestos is sometimes used. The following "ink" has been used for work of this kind:

	g./l.	avoir. oz./gal.
Nitric Acid (sp. g. 1.42)	750 (530 ml.)	100 (68 fl. oz.)
Silver Nitrate	25	3.3

Electrical etching devices and tools are available with which lines can be drawn on metal by a rapidly vibrating point connected to a source of low voltage current. A small arc is formed by the make and break circuit between the vibrating point and the piece being etched, thereby leaving a permanent cut in the metal. This machine may be constructed as a multiple pantograph so that a large number of pieces may be marked at one time from a single master plate.

Electrolytic Etching

Etching of nearly all metals may be accomplished by making them anodic in a suitable solution. (This is just the opposite of electroplating). In general a solution of a salt of the same metal as that which is being etched may be used, or a neutral salt of sodium or potassium. For example, copper may be etched anodically in a solution containing 200 g./l. or 27 oz./gal. of copper sulphate. Steel may be etched by making it the anode in a solution containing 60 g./l. or 8 oz./gal. of sodium chloride. The cathode should be of carbon or of any metal that is not attacked by the solution. A direct current is used with a low voltage, usually from 2 to 4 volts. The voltage should be regulated so as to produce a current that dissolves the metal, but does not evolve much oxygen, which may lift off the resist. One advantage of electrolytic etching is that the metal is not attacked except when the current is passing, and hence the depth of the lines can be readily controlled.

A design may also be produced on metal sheet by mechanical means. The areas which are to remain bright and shiny are covered with masking tape or a stencil. The article may then be sandblasted to dull the surface and produce a pleasing contrast.

Ink for Brass

Rub 2 oz. of copper carbonate with a little water and add sufficient solution of ammonia (10%) to dissolve, then add 10% of glycerin.

Fluorescent Chalk

Chalk suitable for making blackboard writing, which is luminous, may be made by soaking ordinary calcium carbonate blocks, used to cut chalk sticks from, in a concentrated solution of calcium salicylate.

Many pastels are ordinarily fluorescent and these may be employed where a fluorescent chalk is required.

Invisible writing with chalk may be made by impregnating charcoal sticks or black chalk with theobromine or sodium salicylate. When this chalk is used, on a slate blackboard, the marks will not be visible at a distance but when illuminated with ultra-violet light they will show up with a bright blue color.

Fluorescent Inks

Fluorescent inks are very useful for marking documents, papers, books, etc. Various substances may be used.

Formula No. 1

Dissolve quinine sulphate in a small quantity of water to which a few drops of sulphuric acid have been added. This solution is applied to cloth or paper and when dry is invisible in daylight and shows a blue fluorescence under ultra-violet light.

No. 2

Dissolve anthracene (purified by recrystallizing from pyridine) in benzene. Apply to paper or cloth and dry. It is invisible in daylight and shows a brilliant blue fluorescence under ultra-violet light. Crude anthracene when applied in the same way is almost colorless in daylight and green under ultra-violet light.

No. 3

Luminous zinc sulphide is mixed with dammar varnish and applied to cloth or paper. When dry, the material fluoresces under ultra-violet light and then phosphoresces when removed.

Fluorescent Crayon

Fluorescent crayon is made by impregnating ordinary wax base used in crayon manufacture with a fluorescent dyestuff.

For phosphorescent crayons, calcium sulphide phosphor may be added to the crayon base and markings made with the crayon will glow for some time after excitation with white light or, preferably ultra-violet light.

Zinc orthosilicate or theobromine may be added to the crayon base to make its writing fluorescent.

Magic Writing Pad

Cardboard is coated with the following composition and covered with a sheet of waxed or oiled paper or cellophane. When the latter is written on with a stylus or pencil the writing appears. When this sheet is lifted away from the coated cardboard the writing disappears.

Beeswax	4
Venice Turpentine	9
Lard	4
China Clay	3½
Carbon Black	1
Mineral Oil	2

The consistency may be varied by varying the proportions of liquid in the formula.

Tailor's Chalk

White

French Chalk	20
Pipe Clay	20
White Curd Soap	6
Water, To suit	

Make into a stiff paste with sufficient water. From this form slabs of desired size and press into the oiled wooden or metal moulds. After moulding dry the pieces in a moderately heated place.

Yellow

Chalk Powder	28
Soapstone	18
Pipe Clay	10
Yellow Ochre	7
Lemon Chrome Yellow	1½
Water, To suit	

Proceed as above.

Blue

Chalk	20
Pipe Clay	20
Soapstone	15
Ultramarine Blue	10
Water, To suit	

Proceed as above.

Black

Soapstone	56
Bone Black	8

Yellow Soap	6
Gum Arabic	2
Glycerin	1

Dissolve the gum in a small quantity of water, add glycerin, mix in pigments. Then grind to a smooth paste with water and proceed as above.

Erasable Crayon
U. S. Patent 2,215,902

Charcoal, Powdered	130
Kaolin	50
Calcium Carbonate	20
Black, Greaseless	20
Soap	2
Gum Tragacanth	5

Soluble Crayon for Marking Textiles

Paraffin Wax	649
Stearic Acid	125
Ultramarine Blue HM	24
Emulsifier S-489	91

Melt paraffin and stearic acid to about 158° stir in the ultramarine blue. Turn off heat and add S-489, while stirring thoroughly. Continue stirring until it begins to solidify. Add to molds and chill as soon as possible.

Heat Resisting Stencil Sheet
U. S. Patent 2,242,313

Flexible "Cellophane" is coated with the following mixture:

Mineral Black	5
Castor or Cottonseed Oil	12
Carnauba Wax	49
Aluminum Powder	34

Ink Eradicator

Solution A—Dilute 4 cc. of concentrated hydrochloric acid to 100 cc. with water. Label—Solution A.

Solution B—Dilute 50 cc. of commercial "Chlorox" or "Javelle Water" with 50 cc. of water. Label—Solution B.

Application: Moisten the ink blot or mark with a drop or two of Solution A. Blot. Then moisten with a drop or so of Solution B. Blot and then let dry before writing over the bleached area.

Chapter Ten

LUBRICANTS AND OILS

Wood and Fiber Lubricant

Wood and nonmetallic surfaces of the type of synthetic fibers are difficultly lubricated, since the lubricant usually causes swelling or remains entirely on the surface. This is true even when the friction is against metal. A suitable lubricant for this purpose can be prepared from 1 oz. camphor, sufficient castor oil to make a thin paste, and 2.5 lb. of finely shaved paraffin wax. The camphor is first dissolved in the castor oil, with slight heating if necessary; the paraffin wax is then added and the whole well mixed until solution is complete (eventually with further heating). The mixture is cooled and stored in air-tight containers.

Driving Journal Lubricant
U. S. Patent 2,229,030

Sodium Laurate	25
Sodium Arachidate	20
Refined Petroleum	45
Heavy Black Petroleum	10

Diesel Engine Lubricant
U. S. Patent 2,234,005

Lubricating Oil	95
Aluminum Petroleum Sulphonate	½–5

Chassis Lubricant
U. S. Patent 2,264,353

Aluminum Stearate	3–8 %
Aluminum Naphthenate	0.5%
Polyisobutylene	0.1%
Viscous Lubricating Oil, To make	100 %

Gyroscope Bearing Lubricant

Sperm Oil	50
Mineral Oil, Refined	50

Dental Lubricant
(For removing inlay pattern)

Castor Oil	1
Diglycol Dilaurate	1

Lubricant for Glass Molds
Formula No. 1

Varnish	1 l.
Linseed Oil	½ l.
(Boiled with a little rosin)	
Minium (Finely Sifted)	5 kg.
Turpentine	40 cc.

After applying above to mold, dust lightly with powdered charcoal and dry.

No. 2

Varnish	1 l.
Linseed Oil	1 l.
Rosin	300 g.
Chestnut Flour	250 g.

Boil together.

Die Member Lubricant
U. S. Patent 2,276,453

Stearic Acid	20–30
Spermaceti	4–6
Lanolin	4–6
Borax	1–3
Water, To make a good emulsion	

Drawing Die Lubricant
U. S. Patent 2,121,606

Kaolin or Bentonite	17.00
Water	100.00
Boric Acid	5.75
Magnesium Oxide	0.45

Nickel Working Lubricant

Nickel alloys can now be drawn, stamped or formed and then annealed without cleaning, by using a dispersion of Diglycol stearate in water.

Diglycol stearate S is a white, wax-like solid, dispersible in water, and has a melting point of 51–54° C. It will fire off completely at any annealing temperature (400° F. min.) in an atmosphere of Oxygen or Hydrogen.

As an example, an alloy composed of 78% nickel, 4% manganese, copper and iron is drawn with a 2% dispersion of Diglycol stearate, as follows:

Flat plate approximately 5 inches square—

1st draw—2″ deep, 1½″ diameter (anneal).

2nd draw—3″ deep, 1″ diameter (anneal).

For drawing and stamping, usually a 1% concentration is sprayed on, and is sufficient as a lubricant. Even lesser concentrations can generally be used to replace sulphonated oils which are used as coolants in stamping. In all cases, there was no residue on firing.

Lubricant for Metal Working
U. S. Patent 2,238,738

A. For forming metals.

Soap	56
Iron Oxide	4
Lime Water	40

B. For drilling or cutting steel.

Soap	28
Iron Oxide	2
Lime Water	70

Metal Drawing Lubricant
British Patent 543,253

Mineral Oil	100
Bentonite	25–33
Curd Soap (1% Solution)	9900

Wire-Drawing Lubricant
U. S. Patent 2,126,128

Stearic Acid	65
Mineral Oil	35

Melt together, mix well and allow to cool. Apply melted and wash off with hot dilute caustic soda solution.

Non-Gumming Lubricating Oil
U. S. Patent 2,229,858

Lubricating Oil	99.5
Cyclohexylamine	0.5

Coolant and Lubricant for Rolling Strip Steel
U. S. Patent 2,246,549

Palm Oil	1 –50 %
Water	50 –99 %
Lactic Acid	0.3–33.3%

Pipe Thread Lubricant
U. S. Patent 2,205,990

Cup Grease	40 –45%
Zinc Dust	60 –55%
Nitrobenzol	0.1– 1%

Anti-Seize Compound for Threaded Fittings
Formula No. 1

Asphalt	9
Gilsonite	1
Lubricating Oil	10
Graphite (200 Mesh)	10

Petrolatum	10
White Lead	60

No. 2

Petroleum Oil, Refined	60
Mica (100 Mesh), Flake	40

Rustproofing Lubricant
U. S. Patent 2,222,487

Lubricating Oil, Light	97.5–87%
Lard Oil	2 –10%
Lecithin	0.5– 3%

Corrosion Preventive Compound
(For motors and engines)

Lard Oil	60
Methylaminophenol	2
Peanut Oil	30

Oil-Insoluble Stopcock Lubricant

Soluble starch	10 g.
Glycerin	25 g.

Stir the starch thoroughly into the glycerin, heat to 140° C., and allow to stand for a short time until any sediment has settled. Decant the clear liquid and allow to remain overnight. The resulting product is a thick, greasy preparation, soluble in water but not in oil.

Stop-cock Lubricant
(Ether Insoluble)

A gel is prepared by suspending 9 grams of soluble starch in 22 grams of glycerol and heating to 140° C. After standing for a short time, the clear solution is decanted from the sediment and allowed to cool. The mixture is then allowed to stand overnight, when it takes on the consistency of a thick grease and can be employed as a lubricant.

Hydraulic Fluids

	Formula No. 1	No. 2	No. 3	No. 4	
Castor Oil	45	25	40	20	cc.
Propylene Glycol	15	15	25	25	cc.
Hydroxyethyl Butyl Ether	10–20	10–20	—	—	
Alcohol	17½	38	32½	53	cc.
Phosphoric Acid	¼	¼	¼	¼	cc.
Cresylic Acid	¼	¼	¼	¼	cc.

Diethanolamine, Enough to give a neutralization number of 4–8.

Anti-Static Belt Dressing

Fish Glue	9 oz.
Glycerin	7 oz.
Sulphonated Castor Oil	7 oz.
Water	12 oz.
Lampblack	6 oz.
Ammonia (2%)	2 oz.

The glue and glycerin are first heated together, with frequent stirring, for two hours, at a temperature at which the mixture barely boils. The other ingredients are then stirred into this warm mixture in the order named. The two ounces of approximately 2% ammonia can be made by adding ¼ oz. of ordinary household ammonia to 1–¾ oz. of water.

These quantities are sufficient to make about a quart of dressing, which should be kept in a closed container. This dressing should be used regularly and may be applied to the side that bears against the pulleys while the belt is running.

For use on fan belts of automobiles and tractors, and for similar equipment, a simpler, very easily made preparation can be obtained by mixing graphite and glycerin in the desired proportions. Applied to the areas of friction this simple product is said to be very helpful in reducing static electricity.

Cutting Oil Bases

Formula No. 1

Cottonseed Oil	71.2
Rosin	28.8
Oleic Acid	2.0
Triethanolamine	1.2

No. 2

Talloil	32
Caustic Potash (45%)	8
Tertiary Butyl Alcohol	8
Mineral Oil	192

Rust Inhibitor for Cutting Oils

German Patent 680,884

Use 1/40–1/2 parts of sodium nitrate to each part of soluble oil in finished emulsion.

"Soluble" Oils

Formula No. 1

Sulphonated Fish Oil	75
Pale Oil No. 100	25

No. 2

Sulphonated Neat's Foot Oil	60
Pale Oil No. 100	40

No. 3

U. S. Patent 2,231,214

Rosin Soap	9%
Naphthenic Soap	9%
Rosin	2%
Naphthenic Acid	2%
Sulphonated Castor Oil	3%
Butyl Carbitol	1%
Water	4%
Mineral Oil, To make	100%

Non-Foaming Soluble Oil

U. S. Patent 2,265,799

Mineral Oil	85.10 %
Mahogany Sulphonates	9.50 %
Naphthenic Acid	2.50 %
Sodium Hydroxide	0.40 %
Water	1.60 %
Isopropyl Alcohol	0.90 %
Sodium Aluminate	0.2 –1%

Engine Crankcase Flushing Oil

U. S. Patent 2,259,872

Mineral Oil (28° or Higher A.P.I. Gravity: Saybolt Viscosity 80–140 at 100° F.)	60–80%
Dichlorbenzene	5–20%
Isopropyl Alcohol	5–20%

Slushing Oil for Protecting Steel

U. S. Patent 2,124,446

Magnesium Naphthenate	5–10%
Light Lubricating Oil, To make	100%

Sticky Viscous Oil

Where an extremely sticky oil is needed, use Standard Viscous Oil (Standard Oil Co. of Calif., Richmond, Calif.) which is a mixture of isobutene polymers. Commercial grades run from No. 8 (color and viscosity like honey in the summer) through grades of increasing viscosity, but same color, numbered as 16, 32, to No. 64, which is so stiff that it will scarcely run out of a bottle at room temperature. Stickiness can be increased by dissolving resins therein.

Drying Oil from Corn Oil

A kettle is filled ¾ full with corn oil and zinc shavings (6% of the oil). The oil is slowly heated to 200–290° C. and kept at that temperature until its viscosity at 20° is 300–400 seconds. The oil is then allowed to cool to 170–180° C. It is then poured into another kettle and mixed with 3% zinc resinate. When the latter dissolves, drier is added followed by 92% of mineral spirits. After standing 5 days, the drying oil has a viscosity of 6–9 seconds. The zinc shavings can be used over again 5 or 6 times.

Black Harness Oil

Ceresin	2
Ozokerite	2
Cod Oil	32
Herring Oil	26
Whale Oil	14
Pale Oil No. 100	22
Black Stain *	2

Mix all components, except dye, until melted and dissolved. Add dye; mix thoroughly and let it cool. This oil is good if it leaves a black spot on paper. If spot is not black, some more dye should be used.

* *Preparation of Black Stain*

Oil soluble (black) nigrosine base should be heated in red oil or stearic acid for 4–6 hours at a temperature above 100° C. but not exceeding 105° C. Proportion of dye to fatty acid is 1:3.

Transparent Lubricant Grease
French Patent 849,105

Spindle (Mineral Oil)	100 kg.
Aluminum Stearate	6 kg.
Vegetable Fat	4 kg.

Penetrating Liquid Lubricant Grease
U. S. Patent 2,247,577

Aluminum Stearate	1 –10%
Morpholine	0.5– 3%
Lubricating Oil, To make	100%

Gasket Grease

Potash Soap	22
Glycerin	6
Castor Oil, Heavy Bodied	72

Heat together, with stirring to 121° C. Turn off heat and stir till temperature falls to 38° C. Pour into containers.

Wool Grease Substitute

Petrolatum	50–100
Ester Gum	50

Heat to 100° C. and mix until uniform.

Drilling Fluids, Well
Formula No. 1
U. S. Patent 2,258,202

Graphite	2
Bentonite	98
Water, To suit	

No. 2
Dutch Patent 50,258

Crude Petroleum (Boiling Above 200° C.)	62.5
Hematite, Finely Ground	126.0
Mexican Asphalt	41.0

No. 3
U. S. Patent 2,109,337

Clay	1
Silica, Amorphous (325 Mesh)	99

No. 4
U. S. Patent 2,122,236

Iron Blast Furnace Flue Dust (Finer Portions)	30
Beidellite	10
Water	60

Dubbing, Leather

Ceresin	9
Aluminum Stearate	1
Neatsfoot Oil	40
Tallow	50

Petroleum Jelly

Formula No. 1

Petrolatum (White)	20
Ceresin	3
Paraffin Wax	4
White Light Mineral Oil	73

No. 2

Petrolatum (White)	10
Paraffin Wax	6
Ceresin	3
Ozokerite	2
Cetyl Alcohol	6
White Heavy Mineral Oil	73

Mix all solid components. Heat ntil melted and dissolved. Heat oil n separate container 5–10° above emperature needed to melt the solid omponents. Mix all components nd stir well. Pour in containers and llow to solidify without stirring.

Bleaching Oils and Fats

Melt the fat at 60–65° C. and stir vith 2% of its weight of hydrogen eroxide 30% strength. After two ours, during which a foam is produced, the whole is covered with a loth and left overnight. The oil nay be heated for a further hour, he temperature being raised to 0° C. An unmistakable improvenent in color and odor results from he above treatment, but rebleachng with a further 2% peroxide reults in very considerable lightening n color and diminution in odor. Bleaching with hydrogen peroxide is largely dependent on the efficiency of the agitation. The most efficient method appears to be the use of compressed air which gives an excellent emulsion. The stronger the strength of peroxide, added to the fat, the higher the temperature required for adequate bleaching. It is preferable to use a 30% rather than a 60% strength.

Bleaching soap fats with organic peroxides, such as benzoyl peroxide, has certain advantages as these chemicals are fairly readily soluble in oils and can be easily and quickly processed. Benzoyl peroxide is a white, almost odorless powder, insoluble in water, but completely soluble in oils and fats at 80° C. When warmed it decomposes readily into benzoic anhydride and free oxygen, finally forming benzoic acid, which is quite harmless to the oil but may, if desired, be volatilized away by blowing with steam. Only 0.1 to 0.2% by weight of oil is necessary to effect bleaching action. The usual method is to stir the requisite quantity of benzoyl peroxide into the oil at 70 to 80° C. until it is completely dissolved, when the whole is heated to 95 to 100° C. Bleaching is usually complete in 15 to 30 minutes or even less, leaving the fat quite clear from sediment.

Chapter Eleven

MATERIALS OF CONSTRUCTION

Fireproofing Solution for Wood
Formula No. 1

The solution is made by dissolving separately 1¼ oz. of sodium bichromate and 7 oz. zinc chloride and then combining in a total of 1 gal. of water.

Color may be added for decorative, identification or other purposes. However, the solution exerts a powerfully destructive action on most coal-tar colors. Of many so checked only the following three are unattacked:

Amacid Blue V
Kiton Red S
Amacid Yellow T Ex.

These may be combined to produce a number of shades.

The wood is brushed generously or allowed to soak in the solution and then dried.

No. 2

Ammonium Chloride	12
Zinc Chloride	8
Water	80

Wood Fence Post Preservation

Air dry for 24 hours after cutting and scrape off exuded pitch. Impregnate for 3 days in

Copper Sulphate or Zinc Chloride	2 lb.
Water	1 gal.

Then air dry for at least a month before setting.

Fluorescent Wood

Ordinary locust wood fluoresce a bright yellow under ultra-viole light and this material may be fab ricated into a wide variety of ob jects. Locust wood can be turned o the lathe and still retain its fluores cence.

An alcoholic solution of turmeri dyes wood a yellow color and unde ultra-violet light a bright yellowis fluorescence is noted. When tur meric dyed wood is treated wit acids or alkalies the fluorescence i brightened and changed in hue.

Uranine, eosine, acriflavine, an other fluorescent dyes in 0.5% alco holic or aqueous solution can b used to make wood fluorescent.

When wood is rubbed briskly wit lubricating oil or petrolatum a fluo rescent coating is imparted whic remains for a considerable time. Th fluorescence with these substance is bluish.

An alcoholic solution of anthra cene colors wood greenish and unde ultra-violet light a brilliant greenis fluorescence is seen.

Acid Resisting Glass Flux
U. S. Patent 2,225,159

Lead Oxide	50–75%
Silica	12–25%
Boric Oxide	1–10%
Titanium Dioxide	1– 8%

Infra Red Absorbing Glass
U. S. Patent 2,226,418

Phosphoric Acid	81
Alumina	7½
Barium Carbonate	4½
Ferric Phosphate, Hydrated	1–10

Fuse together.

Decorative Glassware Glaze
U. S. Patent 2,225,161

Lead Oxide	55–80%
Silica	15–35%
Titanium Dioxide	2– 7%

Transparent Red Glaze
U. S. Patent 2,130,215

Lead Oxide	686
Boric Acid	249
Flint	120
Cadmium Carbonate	94
Cadmium Sulphide, Red	30

Melt below 1000° under nitrogen.

Improving Adhesion of Vitreous Enamel
U. S. Patent 2,271,706

Dissolve 0.6 lb. antimony tetraoxide in a gallon of cold 18° Bé. muriatic acid. Then add 0.0006 lb. of arsenic trioxide. Then add enough solution of stannous chloride in strong muriatic acid to turn the solution colorless or pale green.

After the usual acid pickle, the article is dipped in the above bonding solution, then rinsed in cold water for 1–5 minutes and neutralized in a solution containing ¼ to 1 oz./gal. of a mixture of 10%–90% borax plus 90%–10% soda ash at 140°–180° F. for 2 to 8 minutes.

Cement Floor Tile
British Patent 524,227

Portland Cement	610 g.
Aluminous Cement	200 g.
Lime Cement	90 g.
Latex (60%)	1000 cc.
Casein (10% Solution)	200 cc.
Silver Sand	2500 g.
Washed, Sharp Sand	1250 g.
Color (Chrome Green)	100 g.

Wall Tile Composition
British Patent 541,695

Ball Clay	25
Flint	30
Talc	40
Felspar	5
Barium Carbonate	½

Glazed Finish for Bricks

The first thing to do is to build up a good surface on which to apply the glaze. Go over the brick with a stiff bristle brush or wire brush to remove all loose mortar, dust, etc. Then put on a priming coat mixed on this basis:

White Lead	100 lb.
Raw Linseed Oil	5 gal.
Turpentine	¾ gal.
Liquid Drier	1 pt.

This mix makes about 9⅛ gal. of paint that should cover about 200 sq. ft. of surface per gallon.

After the priming coat dries, apply a second coat mixed on this basis:

White Lead	100 lb.
Raw Linseed Oil	3 gal.
Turpentine	1 gal.
Liquid Drier	1 pt.

This mix makes about 7⅜ gal. of paint that will cover at the rate of 400 sq. ft. per gallon.

A final paint coat that dries fairly flat, on which to apply the glaze, is made on this basis:

White Lead	100 lb.
Lead Mixing Oil	4 gal.

This mix makes 7¼ gal. of paint that will cover about 500 sq. ft. of surface per gallon. The paint should be tinted with colors-in-oil to match the desired color.

When this coat is dry and hard, brush on the glaze coat. The glaze may be mixed as follows:

Lead Mixing Oil	3 gal.
Spar Varnish	1 gal.

or

Flatting Oil	2 gal.
Spar Varnish	2 gal.

To tint the glaze, add to it small amounts of the selected colors-in-oil. Colors often used are raw and burnt umber, and raw or burnt sienna.

Heat Insulation
U. S. Patent 2,130,091

Asbestos	7
Magnesium Oxide, Magnesium Sulphate (Addition Compound)	32
Vermiculite	61

Add the above to water, to make a thick slurry. Mix well and pour into molds. Set at 21° C. for 48 hours and dry at 150° C.

Insulating Firebrick
U. S. Patent 2,242,434

Calcium Sulphate ½ H_2O	37
Sawdust	37
Georgia Kaolin	100

Form to shape and burn before much moisture is lost.

Sound and Heat Insulating Cement
Canadian Patent 402,601

Kaolin	50 g.
Grog	50 g.

Mix the above and to it add 150–175 cc. of the following which has been beaten vigorously:

Water	190 g.
Bentonite	6 g.
Soap Bark	2 g.
Karaya, Gum	2 g.

Porous Cast Refractory
U. S. Patent 2,224,498

Magnesium Oxide	40
Alumina	40
Silica	20

Refractory Cement
British Patent 532,463

Fired Clay	7
Aluminous Cement	10
Mineral Wool	2
Kieselguhr	2

This is applied on top of fibrous heat insulation materials.

Artificial Marble
Swiss Patent 213,857
Formula No. 1

Colored Cement	12
Granulated Marble	12

They are mixed dry and then made into a thick paste with water.

No. 2

White Cement	12
Granulated Marble	12

They are mixed similarly to Formula No. 1.

No. 3

Cement	50
Stone (washed and powdered the size of sand)	75

Ingredients are dry mixed.

No. 4

Cement	50
Stone (crushed as in Formula No. 3)	75
Sand	75

To make the slabs, Formula No. 1 is poured into a mold and No. 2 is poured on top of it. The two are caused to mix so as to produce an effect of "veins." This is followed by Nos. 3 and 4. The whole is compressed at 200–250 atmospheres. The pressed blocks or slabs are kept for 20 days and then polished with powdered (No. 100) silicon carbide. After another 10 days they are polished again with No. 300 silicon carbide.

Chemically Resistant Machinable Sandstone
U. S. Patent 2,227,312

Porous sandstone of apparent density of 2.1–2.4 is heated to 220–240° F. and immersed in the following:

Asphalt	20–40%
Gilsonite	80–60%

at 220–240° F. under 20–100 lb. per sq. in. pressure. After impregnation, remove adhering bitumen.

Rapid Drying Plaster
U. S. Patent 2,245,458

Calcium Oxide	65
Antigorite	35
Sulphur	2

Building Plaster
British Patent 530,001

Asbestos Fibers	8 lb.
Pipe Clay	32 lb.
River Sand	160 lb.
Cement	160 lb.
Water	28–33 gal.

For fine plaster, replace the river sand by marble dust.

Fluorescent Plaster

By impregnation of plaster of Paris, while still in the plastic state, with zinc beryllium silicate phosphor or calcium sulphide phosphor a highly fluorescent solid may be obtained. This can be put to a variety of spectacular uses.

Slow Setting Portland Cement
U. S. Patent 2,290,956

Add 0.2–0.4% casein on weight of dry cement.

Light Weight Cement Mix
Canadian Patent 402,600

Grog	30 g.
Kaolin	30 g.
Water	40 g.

To the above mixture add 148 cc. of a beaten mixture of

Soap Bark	2 g.
Starch	4 g.
Water	194 g.

Cement or Plaster
U. S. Patent 2,227,790

Portland Cement	40
Silica, Ground	35
Calcium Hydroxide, Hydraulic	20
Magnesium Silicate (Colloidal)	5

Cement Setting Retarder
U. S. Patent 2,211,368

Boric Acid	52.5
Borax	26.2
Gum Arabic	5.2
Calcium Tartrate	12.6
Tartaric Acid	3.5

About 0.5% of above mixture is added to the cement.

Quick Setting Magnesia Concrete
U. S. Patent 2,242,785

Magnesium Chloride	1
Magnesic, Caustic Non-sintered	8
Water	15

Porous Light Concrete
U. S. Patent 2,240,191

About 4.5 bags of cement, each bag corresponding to about 1 cu. ft. or 94 lb. are mixed with about 17.5 cu. ft. of sand. About 13 lb. of about

35% bleaching powder with about 1/7 lb. of ground licorice root are added to the cement and sand mixture. These materials are now mixed with the necessary amount of water in a conventional cement mixer, about 2.5 lb. of granular urea carrying approximately 42% nitrogen is added and, after mixing, the mixture is poured into the mold or formed in the conventional manner. About 1 cu. yd. of porous concrete is obtained, having a density of about 0.6 of normal.

Concrete Coating
Swiss Patent 204,185

Portland Cement, White	45
Gravel, Fine	46
Water	9
Color, To suit	

Fluorescent Concrete

By adding several per cent of zinc orthosilicate phosphor to ordinary cement before adding the water and sand, a brilliant green fluorescing concrete is obtained. The phosphor may be varied to get other colors and effects. Many small articles, such as signs and models, may be built with this concrete.

Crystalline Floor Hardener
Formula No. 1

Magnesium Fluosilicate ($6H_2O$)	56.75
Zinc Fluosilicate	25.75
Calcium Fluosilicate	17.50

No. 2

Magnesium Fluosilicate	67
Zinc Fluosilicate	33

Make at least two applications on floors, allowing 24 hours for drying of the first coat. In the first dissolve ½ lb. magnesium fluosilicate crystals in 1 gal. of water and in the second 2 lb. of magnesium fluosilicate crystals in 1 gal. of water. Apply with a broom or whitewash brush. On new concrete, make the application 48 hours or more after the concrete has hardened. On old concrete, apply when the cement is thoroughly clean, dry and free from paint or oil. In general, the 2½ lb. of crystals in the two above applications should treat 100 sq. ft. of surface.

Roofing Composition
U. S. Patent 2,243,494

Lime	165 –195
Tallow	10 – 14
Salt	10 – 14
Glue	¼– ¾

Water to slake the lime.

Roofing and Building Material
U. S. Patent 2,210,348

A woven cotton base is imbedded in the following heated composition:

Rock Asphalt	200
Asbestos Fiber	50
Rubber	5
Rosin	5
Cottonseed Oil	30

Patching Interiors of Furnaces
U. S. Patent 2,124,865

Furnace interiors, while in operation, are sprayed with a water slurry of

Sandstone (200 Mesh)	97
Fireclay	3

Repairing Rough and Textured Walls

If a test of the surface reveals the fact that the plaster surrounding the crack is sound—not loose—it is a pretty good plan to ignore the advice so freely given in trade papers in the past—to cut out the cracks ack to the plaster, for unless you an be certain that you can imitate erfectly the surface on which you re working so that no trace of the epair remains it is better to proceed s follows, bearing in mind that hat is said should not be construed o refer to smooth plaster walls. racks in them should be handled n the usual way.

Repairing Texture (Plastic) Paint Surfaces

With a sharp-pointed jack knife, lean out any loose plaster that may e adhering to the edge of the crack. Take out a little of the plaster in he rear of the surface to a depth of about one-eighth of an inch in a wedge shape so that the crack is widened beneath the surface. Use every precaution not to widen the surface of the crack any more than is absolutely necessary.

When the crack is cleaned thoroughly, work a thin mixture of flat paint into the crack and allow it to dry. This produces a foundation for the filling of the crack which will follow and will prevent the oil from the filler from being sucked into the plaster. This step is vital and upon its proper performance depend the success of and the life of the repair.

Mix together 12½ lb. of heavy paste white lead and ¼ pt. of flatting oil. Stir until a smooth, creamy mixture results. The consistency will be very heavy.

Mix together 5 to 6 lb. of dry whiting and ½ pt. of flatting oil. Add the whiting to the oil a little at a time and stir until a smooth, even batter results.

Mix together the lead mixture and the whiting mixture and stir until one is thoroughly incorporated with the other, then add a small amount of pale drier and work this in thoroughly.

There should now be a thick, almost putty-like mass of material, which, if kept in a covered can with a thin coating of water on top, should keep indefinitely.

Before using pour off the water, take what is required from the can, level up that which remains in the can, and recover it with water.

Place some dry whiting in a shallow pan or in the cover of a lead keg and place the mixture removed from the can on the powdered whiting. Knead the whiting into the mixture, adding whiting if necessary until a stiff putty is formed. This is the material to be used for filling cracks in rough, sand finished or textured walls.

Take a gob of the special putty in the left hand and with a reasonably flexible putty knife or spatula in the right hand work this putty down into the crack which has previously been prepared to receive it. Fill the crack just a trifle more than full and then go over the surface of the crack with a small sash tool and a very little flat paint or flatting oil. This softens the surface of the filler and will allow you to manipulate it to look like the surrounding surface.

To manipulate the surface of the putty to match a sand finished surface have on hand a little fine sand, a whisk broom and the before mentioned sash tool.

Throw a little of the sand onto the wet surface of the crack, then drag the whisk broom over the surface until a sufficient amount of sand has been worked into the surface of the crack to match the surrounding surface.

When certain that it matches, run a thin coat of flat paint over the sand to bind it. Be sure it is a thin coat for if it is thick enough to fill up the sand finish in the immediate vicinity of the crack, it will show through the final coats.

It is a good idea to confine the putty mixture to within an eighth of an inch of the sides of the crack and to brush the thin mixture of flat paint out to a feather edge to a point a foot or so away from the crack on either side.

This permits painting the surface after the thin, feather edged, touch-up coat is dry without a trace of the repair appearing.

Textured Walls

The procedure is similar but instead of the whisk broom use the same tool for manipulating the wet surface of the crack as was used to produce the original texture. Manipulating must be done in such a manner as to blend into the surrounding surface without leaving any trace of the repair.

It may or may not be necessary to put on the final thin touch-up coat and feather edge it out from the crack. Here again use judgment based on the appearance of the repair.

Rough Plaster Walls

Follow the same procedure as for textured walls. A spatula, putty or wide knife can frequently be used to imitate rough plaster surfaces on small areas such as the surface of cracks.

Repairing Holes in Rough or Textured Walls

Naturally, large holes in rough sand finished or textured walls must be repaired in the usual way by cutting away the plaster clear back to the lath in accepted wedge shape fashion.

The holes must then be wet down and filled with patching plaster not quite level with the surface, leaving it just enough below the surface so that the sand finish, rough coat or texture will be level with the surrounding surface when it is later applied on top.

When the patching plaster is thoroughly dry the decorator must texture the surface of the hole to match the surrounding surface. This will require rare care and good judgment, to decide what tools and what type of material were used to produce the original texture.

When the texturing is finished, the new texturing should be touched up with flat paint and allowed to dry, after which the entire wall should be given a coat of white, ivory or pale cream flat paint.

When that is dry, the surface may be reglazed in the usual way.

Holes in Sand Finished Walls

The patching is done in the same way up to the point where the patching plaster has been put in and allowed to dry.

Next a coat of varnish is applied to the patch and when it is dry a reasonably heavy coat of paint is applied and sand is thrown on it and manipulated until it matches the surrounding surface.

Chapter Twelve

METALS, ALLOYS AND THEIR TREATMENT

Magnesium Die Casting Surface Treatments

A chrome-pickle treatment is normally given all magnesium die castings to protect the metal during shipment, storage and machining. The treatment will remove 0.006–0.002 in. of metal from the surface and is not recommended on machined surfaces where close dimensions must be held. The following solutions may be used:

(a) Sodium Dichromate ($2H_2O$) 1.5 lb.
Conc. Nitric Acid (sp. gr. 1.42) 1.5 pt.
Water, To make one gallon.

(b) Chromium Trioxide 1.0 lb.
Conc. Nitric Acid (sp. gr. 1.42) 0.9 pt.
Water, To make one gal.

Temperature of the solution should be 125–135 deg. F. and time of treatment 10 sec. After removal from the treating solution, die castings should be exposed to the air for about 5 sec. before rinsing thoroughly in cold water and then in hot water to facilitate drying.

Sealed Chrome-Pickle

A modification of the chrome-pickle which will give increased corrosion resistance, consists in boiling in a bichromate solution. The die casting, which has been previously chrome-pickled, is immersed for 30 min. in a bath containing 1–2 lb. of sodium, potassium or ammonium bichromate per gal. The bath is maintained at boiling temperature and the solution is controlled to a pH of 4.0–4.4.

Hydrofluoric Acid-Dichromate

This treatment gives practically no change in dimensions. Parts are immersed for 5 min. in a solution containing 15–20% by weight of hydrofluoric acid, and then are rinsed thoroughly in cold water, boiled for 45 min. in a solution containing 1–1.5 lb. of sodium dichromate per gal. and then rinsed thoroughly in cold water followed by a dip in hot water. The concentration of hydrofluoric acid should not be allowed to fall below 10%.

Hydrofluoric Alkaline-Dichromate

Parts are immersed for 5 min. in a 15–20% hydrofluoric acid solution, washed thoroughly in cold running water, and boiled for 45 min. in the following solution:

Ammonium Sulphate 4 oz.
Sodium Dichromate ($2H_2O$) 4 oz.
Ammonia (sp. gr. 0.880) ⅓ fl. oz.
Water, To make one gal.

Depletion of the bath is indicated by non-uniform or pale coatings, slowness of coating formation, and by an increase in the pH to about 6.2. To replenish add equal parts of chromic acid and concentrated sulphuric acid (sp. gr. 1.84) until the pH is decreased to 5.6. The best operating range of the bath is a pH of 5.6–6.0. Wash thoroughly in cold water and boil for at least 5 min. in a solution containing one oz. of arsenious acid per gal. Because of the low cost of this bath no control is advocated.

Chrome-Alum Treatment

Parts are immersed in a boiling solution of the following composition:

Potassium Chrome Alum ($24\ H_2O$)	4 oz.
Sodium Bichromate ($2H_2O$)	13.3 oz.
Water, To make one gal.	

Time of treatment may range from 2–15 min. Upon removal from the bath, the parts should be rinsed thoroughly in cold water followed by a hot water rinse to facilitate drying. The solution may be revivified by the addition of sulphuric acid not exceeding 0.33 fluid oz. per gal. or sufficient to just redissolve a brown precipitate which settles out. The solution should then be boiled. It is best controlled by additions of sulphuric acid to maintain the pH between 2.5–3.5. The pH of a depleted solution is 5.5.

Pickling Sterling Silver

Fire scale on sterling silver can be removed in a solution of 3 parts nitric acid, 1 part of water, by volume. The work will turn gray in color. A few seconds is all the time required. For best results, operate solution hot, by having the crock surrounded by a hot water bath. Silver can be recovered from a spent dip by adding a chloride.

An alternative procedure to the use of the nitric acid dip is to heat the sterling article, after fabrication is complete, to a dull red. Then clean, and follow with a pickle in 8% sulphuric acid. The fire scale will then be uniform all over the article and it can be satin finished and burnished, or color buffed. The silver will not have as white a color or take as high a luster, but will have a good color.

Work made the anode in a solution of 8 oz. sodium cyanide, 4 oz. sodium ferro-cyanide to a gallon of water, will be given a fair luster. The work must be cleaned first. Also keep it in agitation during the anodic treatment. Do not agitate solution by air or stirring.

Pickling of Stainless Steel

Hot-rolled sheet is pickled for 10–20 minutes in 8% ferrisul, 2% hydrofluoric acid; cold-rolled sheet for 5–15 minutes in 6% ferrisul, 1.5% hydrofluoric acid; bars and wire for 10–40 minutes in 6% ferrisul, 2% hydrofluoric acid; and cold-rolled strip for 1–4 minutes in 12% ferrisul, 2% hydrofluoric acid —all at 160–180° Farenheit. These treatments produce an unetched, white, mat, chemically passivated surface, smoother than could formerly be obtained.

NICKEL ALLOYS

Pickling

The types of oxide produced on Monel, nickel and Inconel vary with he heating conditions, and the proper pickling solution for each case must e chosen accordingly.

Cleaning

Before any attempts at pickling are made, all grease, drawing comounds, lubricants, and other adhering foreign matter acquired during nechanical operations must be removed completely. Soluble oils can be emoved with soap and hot water, followed by thorough rinsing in hot r warm water. Tallow, fats and fatty acids can be removed by exposing he work to a hot solution containing 10 to 20% of soda ash (sodium caronate) or caustic soda (sodium hydroxide). Mineral oils and greases are emoved best by organic solvents such as gasoline, kerosene, carbon tetrahloride or other chlorinated solvents, followed by a final dip in a 10 to 0% solution of either caustic soda, trisodium phosphate, or a mixture of oth.

Removal of Tarnish by Flash Pickling (Bright Dips)

This treatment is applied to drawn and spun shapes, cold-headed rivets, old-drawn wire and other cold-worked products that have been bright nnealed in a strongly reducing, sulphur-free atmosphere, and cooled ither out of contact with air or by quenching in a 1 or 2% (by volume) lcohol solution.

Monel and "K" Monel

Best results on Monel and "K" Monel are obtained by the use of two olutions, as follows:

First Dip—Formula M–1

Water	1 gal.	or	1000 cc.
Nitric Acid (38° Bé)	1 gal.		1000 cc.
Common Salt	½ to ¾ lb.		60 to 90 g.
Temperature	70° to 100° F.		21° to 38° C.

Time, Not over 5 sec.
Container—earthenware crocks, glass, or ceramic vessels.

The parts should be cleaned thoroughly in the first dip, using only a short time for each exposure, then rinsed in hot water (180° F.), followed by a rapid dip in the second solution consisting of:

Second Dip—Formula M–2

Water	1 gal.	or	1000 cc.
Nitric Acid (38° Bé)	1 gal.		1000 cc.
Temperature	70° to 100° F.		21° to 38° C.

Time, Not over 5 sec.
Container—18–8 stainless steel, glass, or ceramic vessels.

The second dip should be followed by rapid rinsing and neutralizing in 1 to 2% (by volume) ammonia solution (4 to 8 fl. oz. commercial aqua ammonia in 1 gal. of water). The parts should then be dried by dippin in boiling water, followed by rubbing in dry sawdust, or with a dry clotl

Nickel and "Z" Nickel

Only one dip for nickel is required. Use Formula M-3.

Formula M-3

Water	1 gal.	or	1000 cc.
Sulphuric Acid (66° Bé)	1½ gal.		1500 cc.
Nitric Acid (38° Bé)	2¼ gal.		2250 cc.

Allow to cool and add:

Common Salt	¼ lb.	30 g.
Temperature	70° to 100° F.	21° to 38° C.

Time, 5 to 20 sec.

Container—earthenware crocks, glass, or ceramic vessels.

The parts are warmed first by dipping in hot water, after which they ar immersed in the acid bath.

The fumes from these flash pickles for Monel and nickel are discomfort ing and a hood or ventilating system should be provided.

Inconel

It is impractical for the average shop to bright anneal Inconel an therefore no consideration need be given to bright or flash pickling suc as can be accomplished with Monel and nickel. Flash pickling is used o Inconel only after the oxide or scale has been removed.

Removal of Reduced Oxide

This treatment is applied to forgings, hot-rolled shapes, hot-rolled wir rod in coil, and other hot-worked products that have been oxidized bu where the oxide has been reduced by heating in a strongly reducing sulphur-free atmosphere, followed by cooling out of contact with air or b quenching in 2% (by volume) alcohol-water solution.

Monel, "K" Monel, Nickel and "Z" Nickel

The solution of Formula M-4, following, works equally well for all fou materials.

Formula M-4

Water	1 gal.	or	1000 cc.
Sulphuric Acid (66° Bé)	¾ pt.		95 cc.
Sodium Nitrate (crude)	½ lb.		65 g.
Common Salt	1 lb.		110 g.
Temperature	180° to 190° F.		82° to 88° C.

Time, 30 to 90 min.

Containers—earthenware crocks, glass or ceramic vessels, or, for short time usage, wooden barrels.

After pickling, the parts should be rinsed in hot water and neutralized in a 1 to 2% (by volume) ammonia solution. This pickling operation is often facilitated by occasional intermediate scrubbing with pumice on a fibre brush.

It is advisable to maintain separate solutions for Monel and nickel. hey work better after having been used a while; therefore, in making p new solutions, about 2% (by volume) of spent solution should be added the fresh solution.

Inconel

Since it is not practical to bright anneal Inconel under average shop onditions, no procedure is offered for the pickling of that alloy to remove educed oxide.

Removal of Oxide Film or Scale

This treatment is applied to forgings, hot-headed bolts, all hot-rolled nd hot-formed products, and pieces that have been annealed in oxidizing tmospheres, or allowed to cool in air. Two solutions are required for Ionel and "K" Monel.

Monel and "K" Monel

First Dip—Formula M–5

Water	1 gal.	or	1000 cc.
Hydrochloric Acid (20° Bé)	½ gal.		500 cc.
Cupric Chloride	¼ lb.		30 g.
Temperature	180° F.		82° C.
Time	20 to 40 min.		

Containers—earthenware crocks, glass, ceramic or acid-proof, brick-lined vessels.

The cupric chloride may be omitted if not readily available, although t is recommended that it be used since the pickling process is slow in its bsence.

Rinse in hot water and immerse in the second dip for brightening.

Second Dip—Formula M–6

Water	1 gal.	or	1000 cc.
Sulphuric Acid (66° Bé)	1/10 gal.		100 cc.
Sodium Dichromate	1 1/10 lb.		132 g.
Temperature	70° to 100° F.		21° to 38° C.
Time	5 to 10 min.		

Containers—earthenware crocks, glass or ceramic vessels, or rubber-lined tanks.

Follow with a rinse in cold water, and then neutralize in 1 to 2% (by volume) ammonia solution.

Nickel and "Z" Nickel

The hydrochloric acid-cupric chloride first dip solution (Formula M–5) recommended for Monel can be used for nickel also, but a longer time, 1 to 2 hours, is required.

Rinse with hot water and if brightening is required, dip for a few seconds in Formula M–3 and follow by rinsing in cold water and neutralizing in a 1 to 2% (by volume) ammonia solution.

Inconel

During hot-working or annealing in oxidizing atmospheres Inconel ac-quires a tenacious oxide film that is more difficult to remove than thos formed on Monel and nickel. The following solutions have been foun adequate for the removal of this oxide.

For all types of oxide, both thick and thin, the following solution is usec

Acid Formula—Formula M-7

Water	1 gal.	or	1000 cc.
Nitric Acid (38° Bé)	1 gal.		1000 cc.
Hydrofluoric Acid (40%)	1¼ pt.		150 cc.
Temperature	70° to 100° F.		21° to 38° C.
Time	15 to 90 min.		

Container—carbon brick is best.

If the oxide is thin (greenish chromic oxide), the following procedu should be followed, particularly because of the greater ease of handlin the solutions. Two solutions are required.

First Dip—Formula M-8

Water	1 gal.	or	1000 cc.
Sodium Hydroxide	18 oz.		134 g.
Sodium Carbonate	18 oz.		134 g.
Potassium Permanganate	7 to 11 oz.		52 to 82 g.
Temperature	180° to 190° F.		82° to 88° C.
Time	2 hr.		

Container—steel tank.

The work should not be rinsed; it should be carried directly to th acid solution (second dip).

Second Dip—Formula M-9

Water	1 gal.	or	1000 cc.
Sulphuric Acid (66° Bé)	½ pt.		63 cc.
Copper Sulfate or Copper Nitrate	1½ oz.		12 g.
Temperature	180° to 190° F.		82° to 88° C.
Time	1 hr.		

Containers—wood, earthenware, glass, ceramic, or acid-proof, brick lined vessels.

Follow by rinsing in cold water and neutralizing in 1 to 2% (by vol ume) ammonia solution.

It is difficult to say which method of pickling Inconel is better. Th nitric-hydrofluoric acid pickle works well for shop practice and may b preferred to the alkaline-acid procedure.

Paste-Pickling

It is desirable sometimes to use paste pickles for large pieces that ar difficult to immerse in pickling solutions.

Monel and "K" Monel

The treatment is applied to hot-rolled plate, forgings, tanks, drawn hells and other large pieces. Use Formula M–10.

Formula M–10

Lampblack	1 lb.	or	100 g.
Fuller's Earth	10 lb.		1000 g.
Hydrochloric Acid (20° Bé)	3 gal.		2500 cc.
Nitric Acid (38° Bé)	½ pt.		52 cc.
Cupric Chloride	1 to 2 lb.		100 to 200 g.
Temperature	70° to 100° F.		21° to 38° C.
Time	20 to 60 min.		

Mix in a crock, butter tub, or similar container and apply with a long-andled brush for large surfaces, or an ordinary paint brush for smaller vork. Wash off with a hose and follow with a scrub with sand or pumice. n cold weather, cold wash-water will cause Monel to tarnish. This is revented by spreading a creamy lime paste directly on the pickle and horoughly mixing to neutralize the acid. A cold-water wash may then e used. The lime treatment is unnecessary if hot water is used for washing.

Nickel and "Z" Nickel

This treatment is applied to hot-rolled plate, forgings, tanks, drawn hells, nickel-clad steel plate, tanks and other large pieces.

Use Formula M–10 and follow the same procedure as for Monel. A onger time is required to descale nickel—2 to 4 hours or longer—depend-ng upon the thickness and the nature of the scale. Nickel does not tar-ish with cold-water washing; hence, the lime treatment may be omitted.

Inconel

The treatment is applied to hot-worked and annealed parts carrying xide, Inconel-clad plate, tanks and other large pieces.

Most oxidized Inconel work requires two treatments.

First Treatment—Formula M–11

Lampblack	1 lb.	or	100 g.
Fuller's Earth	10 lb.		1000 g.
Hydrochloric Acid (20° Bé)	3 gal.		2500 cc.
Cupric Chloride	2 lb.		200 g.
Temperature	70° to 100° F.		21° to 38° C.
Time	1 to 3 hr.		

Second Treatment—Formula M–12

Lampblack	1 lb.	or	100 g.
Fuller's Earth	10 lb.		1000 g.
Nitric Acid (38° Bé)	2¾ gal.		2330 cc.
Hydrochloric Acid (20° Bé)	¼ gal.		175 cc.
Temperature	70° to 100° F.		21° to 38° C.
Time	15 to 60 min.		

Wash thoroughly and scrub with pumice and water, either hot or col

Removing Discoloration from Automatic Screw-Machine Products

Monel acquires a brown discoloration at high production machinin speeds due to the formation of sulphur compounds by reaction with sulphu base coolants. This discoloration may be removed easily to give a whi surface.

Degrease to remove all traces of coolant and immerse in:

Formula M–13

Water	1 gal.	or	100 cc.
Sodium Cyanide	½ to 1 lb.		60 to 120 g.
Temperature	70° to 100° F.		21° to 38° C.
Time	5 to 30 min.		

Containers—earthenware crocks, steel, glass, or ceramic vessels.

Handle the parts in a perforated Monel or woven wire dipping baske Rinse thoroughly in hot water, if available, to hasten drying, or in col water followed by shaking or tumbling in sawdust to dry.

CAUTION: *Sodium cyanide is a deadly poison and shops not regularl using cyanide solutions should keep the solution and stock of cyanide sa under lock and key. Under no conditions should acid from pickling oper ations or other sources be carried into the solution. The acid will liberat hydrocyanic acid gas—probably the most lethal industrial substance.*

Removing Copper Flash

Sometimes in the pickling of Monel, nickel and Inconel, especially i the pickling bath should contain copper salts and has been used for som time, a thin copper flash may form on the part. For this reason it is desir able to use separate pickling solutions for nickel and Monel, although th same formula may be called for. If the copper flash should occur, it ca be removed readily by immersing the part in an aerated, 4 to 5% ammo nia solution (approximately 1 pt. of commercial aqua ammonia to 1 ga of water) at room temperature. The time required is short, usually onl a minute or so. This dip is followed by rinsing in water.

DECORATIVE ETCHING AND COLORING

The etching procedures herein described were developed through prac tical experience in commercial operations. They differ distinctly from th etching processes used in the preparation of specimens for microscopic o macroscopic examination.

Applying the Resist

The resists used normally in commercial etching operations are appli cable to Monel, "K" Monel, nickel and "Z" Nickel. Wax and gelati resists are applied by swabbing or brushing, and varnish and lacquer re sists by brushing or spraying. When coating on both sides is desired, thi may be accomplished by dipping.

Rotogravure or carbon tissue transfers should be applied according t the manufacturers' directions. The double transfer method is preferre with Monel. Neither of these resists has been used with nickel or Incone

Etching Solutions

The etching solutions described for Monel, "K" Monel, nickel and "Z" Nickel are used at room temperature. For most consistent results they should be made up freshly for each use.

A. Monel and "K" Monel

1. For etching through beeswax, varnish, or other chemically inert resists.
   a. Iron perchloride (ferric chloride) (38°–42° Bé) or
   b. Iron perchloride (38°–42° Bé) containing 10% (by volume) concentrated hydrochloric acid; or
   c. Iron perchloride (38°–42° Bé) containing 2% (by volume) concentrated nitric acid.
2. For etching through carbon tissue, gelatin, or other light-sensitive resists.
   a. Iron perchloride (38°–42° Bé) to which 10 per cent (by volume) of concentrated nitric acid has been added.
3. For "deep-etch" lithograph plates:—*
   a. Iron perchloride, lumps—25 g.
   Calcium chloride solution (40°–41° Bé) 1000 cc.
   Hydrochloric acid (c.p.) (1.19 sp. gr.)—20 cc.

B. Nickel and "Z" Nickel

Both solutions 1b and 1c, as described for Monel and "K" Monel, work satisfactorily for nickel and "Z" Nickel. Preference is given generally to solution 1b.

C. Inconel

There are no recommended etching solutions for Inconel.

Chemical Coloring

Chemical coloring of Monel and nickel sheets requires a preliminary light etching, graining, or sand blasting. Suitable surfaces for this purpose may be developed readily by brief immersion in a 38°–42° Bé solution of iron perchloride at room temperature.

A. Monel

Monel may be colored to shades from grey to black by immersion in the following solution.

Potassium sulphide—12 g. per l.
Ammonium chloride—200 g. per l.

The bath should be used at about 160° F. Coloring may be effected either by complete immersion, to assure most uniform results, or by surface swabbing. The solution has poor keeping qualities and must be made up freshly for each set of applications.

After coloring has progressed to the desired shade, the Monel specimens should be dried quickly, preferably in a hot-air blast at about 200° F. If the colored articles are to be subjected to handling and wear, their surfaces should be protected further by an application of clear lacquer or varnish.

* Formula developed by Lithographic Technical Foundation.

B. Nickel

Nickel may be coated with an adherent black film by an immersic in the following solution.

Ammonium persulphate—200 g. per l.
Sodium sulphate—100 g. per l.
Ferric sulphate—10 g. per l.
Ammonium thiocyanate—5 g. per l.

The solution is used at room temperature, with immersion for 5 mir utes being adequate generally to develop a hard black film. In use it i necessary to maintain the red color of the solution by periodic small addi tions of ammonium thiocyanate, although otherwise the solution retain its useful properties for long periods.

For most requirements the black film developed by this means will nee no added surface protection. If such is desired, an application of clea lacquer or varnish can be made directly over the colored surface.

Aluminum Dyeing

The aluminum is first thoroughly cleansed and degreased in a hot bath of potassium hydroxide, rinsed hot, then cold.

Next a coating of basic aluminum oxide is formed by electrolysis. The metal is suspended in lead tanks of sulphuric acid and a current of high amperage and low voltage is applied for 20–30 minutes; meanwhile the bath is kept cool with ice.

After a thorough rinse the aluminum is dyed at temperatures between 75° F. and 125° F. for 1–5 minutes depending on depth of shade. This is followed by a sealing treatment produced by dipping the dyed metal in a boiling bath of dilute acetic acid.

A selection of the most suitable colors, mixtures of which will produce practically any desired shade, are known as:

Metal Canary Yellow
Metal Old Gold
Metal Green
Metal Lilac
Metal Delft Blue
Metal Sunflower
Metal Fuchsine
Metal Copper Red
Metal Red B
Metal Dark Wine

Many of the ordinary basic an acid colors will dye deep shades bu are not fast to the sealing bath.

Anodizing Aluminum

U. S. Patent 2,231,086

Formula No. 1

Tungstic Acid	20 g.
Water	30 cc.

Add this mixture to a solution o 300 grams of sodium sulphate an 420 grams of citric acid in 1 liter o water. Operate at 20–30° C., alter nating current, 15–25 volts. Tim about 30 minutes.

No. 2

Oxalic Acid	69 g./l.
Molybdo-Oxalic Acid	40 g./l.

No. 3

Sulphuric Acid	50 g./l.
Crystal Cobalt Sulphate	20 g./l.

Room temperature. 21 volts alter nating current.

No. 4
U. S. Patent 2,231,373

Titanyl Potassium Oxalate	50
Citric Acid	15
Glucose	20
Phosphoric Acid	6
Water	1000

110 volts alternating current. 75° C. 5–6 amp./sq. dm.

Time until 2.5 ampere hours per square decimeter have passed.

No. 5
British Patent 522,571

The aluminum is treated anodically at 110 volts in a hot (95° C.) solution of

Sodium Aluminate	30 g.
Sodium Silicate	10 g.
Sodium Hydroxide	8 g.
Water	1 l.

Then washed with dilute ammonia and dried.

Coloring Metals

Violet to yellow colors (depending on film thickness) are produced on metals from a bath of

Copper Sulphate	60 g.
Sugar, Refined	90 g.
Sodium Hydroxide	45 g.
Water	1 l.

The metal to be colored is made the cathode and anodes of pure copper are used at temperature of 25–40° C. with 0.01 amp./cm^2.

Coloring Aluminum Powder
British Patent 489,574

Aluminum Powder	100 g.
Alcohol, Denatured	10 g.
Sodium Phosphotungstomolybdate	5 g.
Water	1500 cc.

Boil for 45 min., filter and wash with water. Add washed paste to

Acridine Orange R	3
Water	500

Stir and heat at 60° C. for 5–10 min. Filter wash and dry to get a golden yellow powder. Rhodamme B. or other dyes may be used to get other colors.

Coloring Aluminum

Commercial aluminum usually contains small proportions of silicon and iron; these impurities, plus the character of the metal itself, introduce certain difficulties in coloring aluminum, particularly by chemical means.

The most popular color is black, which can be obtained on aluminum by the simple expedient of first freeing the metal from all grease by rubbing with a rag saturated with a strong washing soda (sodium carbonate) solution, rinsing well, then flowing the metal surface with well beaten fresh egg albumen, previously strained through several layers of clean muslin. The albumen coating is allowed to dry, after which the plate is gently heated and the temperature gradually raised until a deep black color is attained.

The color (deposit) so produced will withstand the action of acids and can only be removed by repolishing, features of some interest in the case of etched aluminum name plates bearing black backgrounds, since the resist (image) can be removed (after blackening) without damage to the black deposit.

Olive or linseed oil can be applied instead of albumen, as can also strong solutions of tannin or gallic acid; the applications are permitted to dry and the plate then heated.

Rather expensive substitutes for the above agents are solutions of platinum chloride, either aqueous

(5%) or alcoholic (1%). The plate is immersed in these solutions and the coating left to dry in a temperature of about 305° F.

A mixture of aniline and bichromate solution imparts an adhering black color to aluminum, and the following procedure promotes a matt black surface to the metal. The first step is to mordant the plate by very briefly immersing it in:

Formula No. 1

Sulphuric Acid	3
Water	1

after which the plate is immersed in the following bath, maintained at a temperature between 86–95° F.:

Alcohol (95%)	1000
Antimony Chloride	150
Manganese Nitrate	100
Washed Graphite	20
Hydrochloric Acid	250

The alcoholic solution present on the plate is burned off and the gray metal surface immediately varnished with:

Alcohol	1000
Sandarac	50
Shellac	100
Nigrosine	100

The varnished plate is heated (baked) in an oven and finally rubbed with linseed oil varnish, which brings forth the desired effect.

No. 2

Water	128
Potassium Permanganate	1½
Nitric Acid (20° Bé)	⅓
Copper Nitrate	4

The temperature of the bath should be 175° F. and the plate immersed therein from 20 to 30 minutes, after which the blackened surface is protected with a clear lacquer.

No. 3

Water	128
Caustic Soda	16
Common Salt	4

The solution is kept at 200° F. and the plate immersed for about 15 minutes. It is then rinsed thoroughly and dipped in:

Water	128
Ferrous Sulphate	16
White Arsenic	16
Hydrochloric Acid	128

The plate is dipped for only a few seconds in the cold bath, then rinsed well in hot water and dried by applying fine sawdust, after which the surface is lacquered.

Those possessing electroplating installations may find the following procedure of interest, particularly for etched aluminum name plates. It is based on the deposition of "black" nickel on the etched areas of aluminum, the coating being sufficiently durable for interior service. A typical solution for black nickel plating is:

No. 4

Water	128
Nickel Ammonium Sulphate	8
Zinc Sulphate	1
Sodium Sulphocyanate	2

Nickel anodes are used, with a voltage of 1 volt and a current density of 1 to 2 amperes per square foot. The bath (electrolyte) is kept nearly neutral by the addition of zinc carbonate.

If anodized aluminum is placed in a solution of an organic dye, the dye unites with the anodic coating on the aluminum and forms a colored lake. These colors will not wash out. Thus, by dipping anodized aluminum in a green dye solution, a green coating is obtained; in this way, any desired color may be possible.

Silvering Aluminum

Formula No. 1

A white or silvery finish can be imparted to aluminum by immersing the metal from 15 to 20 seconds in a hot 10% solution of caustic soda saturated with common salt. The plate is rinsed, scrubbed briskly, then reimmersed in the solution for about one-half minute, during which an energetic liberation of gas takes place. The plate is again washed and then dried with sawdust.

No. 2

Water	1000
Silver Nitrate	10
Potassium Chromate	2½
Potassium Carbonate	100
Sodium Bicarbonate	80

The solution must be used in a boiling state, the time of immersal not to exceed 10 to 15 minutes.

Coppering Aluminum

Copper-colored finishes can be produced on aluminum by cooking the metal in a weakly acid solution of cupric tartrate, which should possess a specific gravity of 36° Bé. Another solution for the purpose is one containing 1 part of cupric chloride in 8–12 parts of water, with the addition of 2 parts of potassium chloride, the plate is immersed in the solution (122° F.) until the required deposit of copper is obtained.

Coloring Brass

Black is the most popular color with brass (particularly for name plates). This color is obtainable by four different methods: (1) Those employing electrolysis; (2) those requiring the use of hot solutions; (3) those requiring heating of the metal after application of the coloring bath; and (4) those employing solutions or baths at ordinary (room) temperatures.

When blackening etched brass name plates, one might bear in mind that because of changes in the surface structure of the etched metal, the color may be different from that obtained on unetched brass. Different etching solutions cause the etched surfaces to react differently during blackening, which condition often is attributed to the character of efficiency of the blackening bath, whereas it really is due to the nature or state of the etched metal surface itself. Should any real difficulty be experienced in this direction, it may be necessary to preliminarily copperplate the etched areas before they are subjected to blackening.

The color produced on the metal, not infrequently, is influenced by the metallic constituency of the brass (ratio of zinc and presence of other metals) and its method of manufacture; no one bath is equally efficient for every grade of brass, and different results (as to color) can be expected with different alloys.

Formula No. 1

Copper Carbonate	3½
Ammonia (0.960)	26½
Distilled Water	5½

The copper carbonate is added to the ammonia and dissolved therein by frequent shaking; at the end of 24 hours standing, there must still be some undissolved carbonate present, assuring a saturated solution—

if not, more carbonate must be added. At this stage, the water is added (some craftsmen also prefer the addition of from 25 to 40 grains of graphite), the solution assuming a blue color, which deepens to a darker blue within a few days. The mixture should be contained in a well-stoppered bottle and stored in a cool place.

In use, the clean and grease-free brass is immersed in the bath until the desired depth of color is obtained, the black deposit (color) consisting essentially of cupric oxid. The plate is then withdrawn, rinsed in cold water, briefly immersed in hot water, then dried and the resist removed (in the case of name plates) after which the plate is varnished with clear lacquer. If the color is not sufficiently deep, the immersal may be repeated, obviously before varnishing.

The efficiency of the above bath is enhanced somewhat by using the solution hot (140–170° F.), but here is faced the difficulty of the ammonia volatilizing and escaping into the work chambers in the form of fumes. The solution soon weakens even when cold; heating accelerates this with the ensuing separation (elimination) of cupric oxid from the bath.

A blue-black color can be deposited on brass plates by boiling them in a solution of potassium sulphide (liver of sulphur). This operation must not be carried out in the vicinity of photographic galleries: the fumes from the sulphide bath would irretrievably fog any photographic plate, film or paper with which they came into contact.

A blackish-green color is obtained on brass plates by immersing them in a boiling solution of:

No. 2

Water	1000
Copper Sulphate	100
Ferrous Sulphate	80
Ammonia	10
Glacial Acetic Acid	5

This bath is particularly efficient with brass alloys of a high copper content.

No. 3

Water	1000
Copper Chloride	35
Copper Nitrate	120

After treatment, the plates are rinsed in cold water, then dried either in an oven or by the application of sawdust.

No. 4

Brass can be blackened in:

A

Copper Nitrate	8¼
Water	20

B

Silver Nitrate	8¼
Water	20

The solutions are prepared separately with the aid of heat, then mixed together. After immersal in the bath, the plate must be heated until a black color is obtained.

No. 5

Hydrochloric Acid	16
White Arsenic	4
Ferric Chloride	16
or	
Hydrochloric Acid	100
Ferrous Sulphate	10
Arsenic Acid	10
or	
Water	128
White Arsenic	12
Yellow Antimony Sulphide	¼

The above baths should be used

hot; the blackened plates, after drying, should be lacquered to prevent tarnishing.

No. 6

Water	128
Lead Acetate	8
Sodium Thiosulphate	8

On immersal, the surface of the brass progressively turns yellow, blue and finally black, the latter color due to a deposit of lead sulphide. With a hot bath, the action is quite rapid. After blackening, the plate is rinsed in cold water, then in hot water, and finally dried. If scratch-brushed dry, the black deposit will have a high luster, but it is well to lacquer the surface to prevent oxidation and fading.

Oxidizing Brass

Immerse the piece in a solution at 180° F., made up in the following proportions:

Sodium Thiosulphate	136	g.
Lead Nitrate	24	g.
Ferric Nitrate	17	g.
Water	2.4	qt.

Treat until the surface is completely oxidized, remove from solution, rinse in water and dry thoroughly.

Coloring Bronze

Bronze can be blackened by immersing in:

Formula No. 1

Hydrochloric Acid	1000
Arsenic Acid, Powdered	60
Antimony Chloride	30
Iron Scale	150

All the ingredients are added to the hydrochloric acid, which is then heated to 158–176° F. for one hour and the mixture frequently stirred to promote solution of the arsenic acid. Avoid inhaling fumes. If the bath is not required immediately, solution of the ingredients in the hydrochloric acid can take place in the cold. The mixture is then allowed to stand from 24 to 36 hours, but intermittently shaken during this time.

The bath is used cold and two immersions (not exceeding 15 seconds each) are, in most instances, sufficient. Before immersing the second time, the bronze must be rinsed with water and dried with soft linen, taking care to remove all drops of water, which otherwise would cause spots. After coloring, the metal is dipped in fairly strong soda solution (to neutralize acid traces), then liberally rinsed in water, and dried with sawdust.

No. 2

For blackening phosphor bronze:

Cupric Nitrate	500
Alcohol (90%)	150

To assist solution of the nitrate, the salt is first melted over a low heat; then, *away from flame*, the alcohol is added to the warm mass. The bath is used cold and the plate immersed until a good black is obtained. Lacquering of the blackened bronze is suggested.

No. 3

Blackening of an occasional piece of bronze can be performed by brushing on the metal a 5% aqueous solution of platinum chloride, a much cheaper medium is a potassium sulphide (1:20) solution. With the latter agent, it is necessary to heat the plate bearing the sulphide solution.

Coloring Chromium

Blackening Chromium

U. S. Patent 1,937,629

Formula No. 1

Sodium Cyanide	45
Soda Ash	35
Salt	20

The period of immersion is from 20 to 30 minutes in a bath maintained at a temperature of 1292–1652° F.

No. 2

Black chromium plating can be performed with an electrolyte containing between 250 and 400 grams per liter of chromic acid and 5 cc. per liter of glacial acetic acid. The bath temperature should be kept below 75° F. and a current density between 750 and 1000 amperes per sq. ft. is necessary.

Coloring Copper

A wide array of formulae exist for coloring this metal, but success depends on the character (purity) of the copper, the reaction of the coloring baths, and last but definitely not least—on the skill and patience of the worker. There is no royal (easy) road to success in coloring metals: it is not always possible to exactly duplicate a given color, regardless of the experience of the operator and the care used in preparing baths and treating the metal.

A black finish on copper can be obtained by applying an even coating of

Formula No. 1

Copper Nitrate	2
Water	8

to the surface to be blackened. The plate is then heated over a gas flame to produce the black color; the operation can be repeated if the deposit is not uniform or the color not sufficiently black.

No. 2

Water	20
Copper Carbonate	8
Ammonia (.880)	80
Sodium Carbonate	4

The water is heated to about 120° F., after which the carbonates are added, stirring well for five minutes to aid solution; this is followed by the gradual addition of the ammonia, the mixture being continuously stirred and the bath brought to a final temperature of 120° F., when it is ready for use.

The well-cleaned plate is immersed in the bath for about three minutes, next washed in water, and then placed in a solution of sodium sulphide (3 ounces; water, 60 ounces) until the desired color is obtained. To further harden and deepen the color, the well rinsed plate can finally be immersed in a 10% solution of potassium bichromate.

The copper-silver nitrate bath given under *brass* can also be used for blackening copper, as can a liver of sulphur bath:

No. 3

Water	1000
Potassium Sulphide	20
Salt	20

No. 4

Water	800
Arsenic Acid	20
Hydrochloric Acid	40

Use at a temperature of 122–140° F.

With the following bath colors on copper ranging from yellow and rose to violet and blue, the color depend-ng on the length of immersion, can be obtained.

Water	1000
Selenious Acid	7
Copper Sulphate	13
Nitric Acid	2

The bath should react slightly acid, otherwise the color may be spotty and less permanent.

Coloring Lead

Lead can be blackened in a warm solution of ammonium sulphide, or in the hydrochloric acid-arsenic bath used for blackening bronze.

Coloring Nickel

Because of its inherent nice color and durability, nickel is seldom colored. The operation can easily be performed by electrolysis, particularly the so-called "black nickel" coatings. Black or gray tints can be chemically produced on nickel by treatment with platinum chloride (1%) or ammonium sulphide solutions, the latter containing a small proportion of ammonium chloride.

Other baths for the purpose are the arsenic solution for blackening bronze, and the alcoholic cupric nitrate solution (*q.v.*) for the same metal.

Bronze Powder Effects

Many bronze-powder finish effects can be produced directly on the metal itself, without the use of bronze powder. Light bronze finish effects can be produced on brass, either sheet metal or castings. Darker bronze finishes can be applied to bronze metal or copper plated brass or steel. Copper plated metal gives a darker finish than bronze metal.

Castings are polished with No. 70 emery. One reason for using this grade of emery is that it will take the skin off the casting at a low labor cost, but the most important reason is that the sandblast will remove all of the emery marks.

For light finishes, use brass. For darker browns, use bronze. For brown to black, use copper-plated metal.

The following solution is for light finishes. Add 4 oz. of liver of sulphur to 30 gallons of water heated to 120° F.

After the articles to be finished are sandblasted they are ready to be stained. To stain, immerse the articles in hot water, then in the solution just described. From there transfer them directly into a toning solution made up by adding a gallon of sulphuric acid to 30 gallons of water. (Note: Make sure acid is added to water.) Then give them a cold-water rinse followed by a hot-water rinse. Dry them in hardwood sawdust. Scratch-brush the surfaces with a brass-wire brush.

The coloring will depend upon the number of times you repeat the operations just specified. The more you dip the articles, the darker the finish will be.

After brushing the treated surfaces, give them a coat of matt lacquer for a dull finish or gloss lacquer for a shiny finish.

If a silver color is desired, it can be obtained by: Sandblast first, then bright dip, silver plate, dry out and scratch brush.

If you silver plate a bronze article for about one minute and brush the surface with a steel wire brush, so as to cut it through (that is remove some of the silver so the base metal shows through) you will then have a silver bronze finish.

For a light brown to dark brown color on bronze metal, use the following solution: Mix into one quart of warm water:

Barium Sulphide	2 oz.
Liver of Sulphur	4 oz.
Aqua Ammonia (26°)	8 oz.

When dissolved, add to 30 gallons of water heated to 120° F. Use this the same as the other solution. It is advisable to make a new solution every day for best results.

These finishes can be brushed wet, which gives a soft finish. If the wet method is used, it must be remembered that after completing the brushing—that is, until the color you wanted is gotten the article is still wet—that is the way it will look when it is lacquered.

In order to keep the same color on the whole job, keep the first article brushed as a sample. Keep it wet and match each piece to this one sample.

Articles finished in this manner will keep their color much longer than those coated with regular bronze powder finishes.

Bright Dip for Brass

Sulphuric Acid	2 gal.
Nitric Acid	1 gal.
Water	1 qt.
Hydrochloric Acid (Muriatic)	½ oz.

(Use this solution cold.)

If, for any reason, too much muriatic acid has been added to the solution at any time, the work will come out "sooty." In this event, add a small amount of sodium chloride (salt) slowly. Add enough salt to return the solution to proper balance. Use only the exact amount of water specified in the formula; do not get any more in the dip.

Matte Dip for Brass

Sulphuric Acid	1 gal.
Nitric Acid	1 gal.
Zinc Oxide	2 lb.

(Use hot and keep free of water.)

If the action is too coarse, add sulphuric acid. If too fine, add nitric acid.

Semi-Lustrous Matte Dip for Brass

Sodium Dichromate	3 oz.
Sulphuric Acid	½ pt.
Water	1 gal.

Both solutions are poisonous and corrosive. Handle with extreme care.

Steel and Iron Burnishing Baths

German Patent 704,400

	Formula No. 1	No. 2	No. 3
Sodium Hydroxide	75	60	60
Disodium Phosphate	16½	—	10
Sodium Nitrite	5	40	3
Sodium Iodate	½	—	½
Potassium Iodide	—	½	—
Trisodium Phosphate	—	—	14
Sodium Carbonate	—	—	4
Temperature of Use	128° C.	133° C.	150° C

40% water and 60% of any of above formulae is used.

Bluing of Iron and Steel

The uniformity and general appearance of the color depends considerably on the character of surface finish of the steel previous to the bluing operations. Bright shining colors are produced only on well-polished surfaces. An abraded or mat surface always appears a little darker than a polished surface colored by the same method. Slight variations, however, are less noticeable on an abraded or mat surface. The brightness or luster can, therefore, be controlled to a great extent by a suitable choice of the finishing operation preceding the coloring.

Uniformity of coloring is greatly influenced by the initial cleanliness of the surface. Grease, rust, or any adherent foreign particles usually shield portions of the surface against the chemical action and lead to uneven coloring. Sometimes a very unsatisfactory spotted appearance is produced. Care must be taken to avoid any contact of the hands with the surface during the cleaning or bluing operations. The articles should be supported by wires, or handled in wire baskets or with suitable tongs. Any surface rust should be removed early in the cleaning operations.

For cleaning metals contaminated with grease or polishing compounds employ organic solvents, such as trichloroethylene, ethylene chloride, orthodichlorobenzene or carbon tetrachloride; emulsifiable solvent cleaners, such as sulphonated corn or castor oil, or triethanolamine oleate added to an organic solvent such as high-flash naphtha or kerosene; or aqueous alkaline solutions, applied either by immersion or electrolytically. These methods may be used individually or in conjunction with one another.

The bluing methods may be divided into three general classes: temper-coloring or heat-tinting, coloring in chemical solutions, and coloring by electrolytic methods. Most of these films or coatings provide only slight protective value against corrosion, unless oil, wax or lacquer is applied to the surface subsequent to the coloring.

Temper-Coloring or Heat-Tinting in Air

A good blue may be obtained on iron and steel by heat-tinting (temper-coloring) the work in air or other oxidizing atmospheres. The color is produced by the interference of light in a thin surface film of iron oxide. As the thickness of the oxide film increases, the following sequence of colors is obtained: light straw, straw, dark straw or a light golden brown, brown or bronze, purple, dark blue, and light blue. Further heating (with oxidation) produces the "second order" of interference colors in the same sequence, but they are usually less intense and consequently less pleasing. Further oxidation beyond the second order gives very indistinct colors.

In general, the thoroughly cleaned article is placed in an oven or furnace or on a hot metal plate, and the heating is continued until the desired color is obtained. It is then removed, cooled in water and dried. The dry surface may then be coated with a suitable light transparent lacquer or with a film of oil by immersion in a heated light oil (such as boiled linseed oil). Greater protection against corrosion is secured in this manner.

The period of heating required to obtain any specific color desired depends chiefly upon three factors: the temperature of the oven or furnace, the composition of the steel, and the size and shape of the article.

The rate at which a desired color can be obtained on steel varies somewhat with the composition of the steel. This is not an important factor, however, unless the differences in composition are very pronounced, for example, open-hearth iron vs. stainless steel. The blue may readily be obtained on these steels by heating in the temperature range of 500 to 700° F. (260 to 370° C.), the period of heating varying from about 3 to 4 hours to one minute respectively for the extremes of this temperature range. Cast iron blues at a slightly lower rate than the open-hearth iron or the steels and, consequently, it requires a little longer period of heating, or a slightly higher temperature.

The temper-coloring of steels, as well as the bluing in molten salt baths, influences the physical properties, because tempering of the steel takes place at these temperatures. Hardened steels are softened somewhat by these processes. Consequently bluing in aqueous solutions at lower temperatures, or electrolytic coloring, may be preferable when it is desirable to maintain the full hardness of the steel.

The size and shape of the articles are important practical factors that influence the heat-coloring. A large object requires a longer period of heating than does a smaller one, because of its greater heat capacity and the consequently longer period of heating required to bring the temperature of the object up to that of the oven or furnace. In irregularly shaped articles, the thinner sections reach the temperature of the furnace more rapidly than do the thicker sections and thus color faster. This may result in non-uniform bluing of the article. This effect can be minimized by placing the article in the cool furnace and allowing it to heat up with the furnace.

The selection of the best operating temperature of the furnace should be determined by a few preliminary tests. If the heating period is so short that the bluing occurs in only a minute or less, the color may not be very uniform. In this case heating at a lower temperature with a correspondingly longer heating period generally gives a more uniform blue, especially with irregularly shaped objects.

Steel strip or wires may be blued by passing them through a bath of molten lead or a low-melting alloy. The color develops only after the steel leaves the bath and comes in contact with the air. The amount of the oxidation, and hence the color, are controlled by adjusting the temperature of the bath and by passing the steel strip or wire through a quenching medium, such as a water bath, when it has developed the desired color.

Very good blue can be obtained on iron and steel by immersion in molten salt baths. The surface of the steel is oxidized by the oxygen liberated in the salt bath, and temper colors, the same as those formed by heating in air, are obtained. The molten salt also serves as an excellent means of heating the article uniformly, and hence is an aid in coloring it evenly. The time-temperature relationships for the formation of these colors are very similar to those obtained by heating in air.

For bluing with the molten salts, either sodium or potassium nitrate or nitrite may be used. A mixture of the sodium and potassium salts is preferable to either alone, because of the lower melting point of the mixture. This permits the coloring to be carried on at lower temperatures, and thereby furnishes better control, because the period of heating is increased and the oxidation can be stopped more easily at the desired color. Some of the salt usually solidifies on the surface of the object immediately upon its immersion in the bath, but melts again in a short time as the temperature of the article rises. This causes non-uniform bluing if the solidified salt is not melted from all of the surface at about the same time, because coloring occurs only when the surface of the article is in contact with the molten salt. In a bath having a low melting point, the amount of salt that solidifies on the object is much less, and it melts more readily, facilitating uniform coloring.

A. Potassium Nitrate 100
Sodium Nitrate 100
Use at about 625° F. (330° C.).

B. Potassium Nitrate 100
Sodium Nitrate 100
Manganese Dioxide 5–10
Use at about 625° F. (330° C.).

C. Potassium Nitrite 110
Sodium Nitrite 90
Use at about 660° F. (350° C.).

The salt bath may be melted in a cast iron or steel pot that is free from rust and has a shape and size to suit the articles to be blued. For small articles a short section of a steel pipe closed at one end with a pipe cap is suitable.

The same procedure is followed with all of the different salt baths. The articles should be *thoroughly* clean, free from rust and free from moisture. Moisture on them at the time of immersion may cause severe spattering of the molten salt. The articles are so suspended as to be completely covered by the molten salt until the desired color has been attained. (The article may be temporarily raised out of the bath to facilitate observation of the coloring.) After the desired blue has been obtained, the article is withdrawn and quenched in clean cold water to stop the oxidation, then immersed in boiling water to remove any salt remaining on the work, and finally dried. To obtain greater protection against corrosion or wear, the blued articles may be dipped in hot oil or finished with a light transparent lacquer or varnish. Finishing with a lacquer that has been lightly tinted blue usually improves the appearance.

Salt bath B is the same as A except for the manganese dioxide. This should be added after the temperature of the molten salt bath has been raised to about 900° F. (495° C.), after which the bath is allowed to cool to about 625° F. (330° C.) before using. All suspended matter should be allowed to settle to the bottom of the pot before the articles are immersed, otherwise discolored spots will appear on the blued surface.

In general, the blue obtained with aqueous solutions differs considerably in appearance from that obtained by heat-tinting, although in both cases it is mainly due to interference colors. The specific color of the surface film itself has a great influence on the "over-all" appearance, this being especially noticeable in the case of the "luster washes," in which the precipitated surface film of lead sulphide has a characteristic "soft" appearance. Moreover, the blue obtained with aqueous solutions is always less brilliant and less intense than that formed by heat-tinting. A good light blue can be easily obtained but it is difficult to obtain a one corresponding to the dark blue or violet temper color. The aqueous solutions are used at a much lower temperature than that of the molten salt baths and consequently the correct operating conditions can be maintained much more easily. Hardened steels can be blued by this method with very little, if any shortening.

As in the previous methods, the articles must be thoroughly clean. The rust-free article is first cleaned with an organic solvent (acetone, carbon tetrachloride or trichloroethylene). This is followed by electrolytic cathode cleaning in an alkaline solution, such as is generally used in electroplating. Electrolytic cleaning is faster and preferable to immersion alkaline cleaning. The following solution, at a temperature of about 195° F. (90° C.), is suitable:

	g./l.	oz./gal.
Trisodium Phosphate	30	4.0
Sodium Hydroxide	10	1.5

A steel anode is used and a voltage of 4 to 5 volts is applied with the article serving as the cathode. The articles are then rinsed in water, dipped in a weak sulphuric acid solution (approximately 5% solution) for a few seconds, rinsed thoroughly in clean water and immediately immersed in the coloring solution.

Unless otherwise noted, all the following formulae refer to aqueous solutions, in which the required weights of the materials are dissolved in water and sufficient water is added to make a final volume of 1 liter or 1 gallon.

Luster Wash

A very pleasing blue may be obtained on iron and steel by using "luster washes." An adherent lead sulphide film is precipitated on the surface of the immersed article and this produces interference colors which are modified somewhat by the specific color of the lead sulphide. Hence the appearance is different from the interference colors of the oxide films formed by heat-tinting.

Good "luster colors" are obtained with the following four solutions, the blue produced by solution D being slightly better than that of the other solutions.

	g./l.	oz./gal.
D. Sodium Thiosulphate	200	27.0
Lead Acetate	20	2.5
Potassium Acid Tartrate	25	3.0

Use at approximately 105° F. (40° C.).

	g./l.	oz./gal.
E. Sodium Thiosulphate	70	10
Lead Acetate	20	3

Use at 160 to 175° F. (70 to 80° C.).

	g./l.	oz./gal.
F. Sodium Thiosulphate	10	1.5
Lead Acetate	10	1.5

Use at approximately 195° F. (90° C.).

	g./l.	oz./gal.
G. Sodium Thiosulphate	30	4.0
Lead Nitrate	8	1.0
Ferric Nitrate	3	0.4

Use at 160 to 175° F. (70 to 80° C.).

The salts should be dissolved separately in portions of the water and then mixed just before using. The cleaned article is immersed in the heated solution and moved to and fro until the desired color appears, whereupon it is rinsed, first in cold water, next in boiling water, and then dried and wiped over with a clean soft cloth. The colored surface may be coated with wax, or a transparent lacquer or varnish as a final treatment.

It is advantageous to use the solution at the lowest operating temperature since the bath is then most stable and the lead sulphide film formed at a low temperature adheres more tightly to the basic metal. Although solution D may be used without the addition of the potassium acid tartrate, the coloring rate is decreased and either a longer immersion period or a higher bath temperature is required. Solution F, which has the lowest concentration of sodium thiosulphate, required the highest temperature. The concentration of these solutions or of their constituents may be varied considerably and still furnish good coloring, although this may require a change in the temperature of operation.

A light blue on iron and steel may be obtained with either of the two solutions given below:

H.		
Arsenious Oxide	85 g./l.	11 oz./gal.
Hydrochloric Acid (sp. gr. 1.18)	630 ml./l.	80 fl. oz./gal.

Use at approximately 195° F. (90° C.).

	g./l.	oz./gal.
I. Sodium Nitrate, Potassium Nitrate	400	55
Sodium Hydroxide	400	55

Use at approximately 255° F. (125° C.).

The article, cleaned as given above, is immersed in the heated solution until the desired color is attained, whereupon it is rinsed thoroughly in cold water, then in boiling water and dried. If desired, it is given a final finishing treatment as described previously.

A blue-black or gray-black on iron or steel can be obtained by treatment with the following solution:

	g./l.	oz./gal.
J. Ferric Chloride	90	12
Mercuric Nitrate	90	12
Hydrochloric Acid (sp. gr. 1.18)	90 (77 ml.)	12 (10 fl. oz.)
Alcohol	360 (460 ml.)	48 (40 fl. oz.)
Water, To make	1 l. or	1 gal.

The article (preferably cleaned with an alkaline cleaner) is immersed for 20 minutes, removed and allowed to dry for about 12 hours, after which the immersion and drying operations are repeated. The article is then placed in boiling water for 1 hour and dried again. The colored surface is finished by lightly scratch-brushing and oiling or waxing.

A gun-metal blue or gray can be obtained on iron or steel by immersion in the following solution:

	g./l.	oz./gal.
K. Ferric Chloride	200	25.0
Antimony Chloride	4	0.5
Gallic Acid	4	0.5

This solution is used hot, at a temperature of 160° F. (70° C.) or higher. The cleaned article is immersed in the bath until the desired color is attained, whereupon it is rinsed thoroughly in cold water, then in hot water, and dried. The finishing treatment is similar to those previously described. More precise control is required with electrolytic methods than with heat-tinting or bluing in aqueous solutions. The coloring is influenced greatly by the initial surface condition (cleanliness and type of finish) and also by the chemical composition of the steel. A slight difference in the chemical composition may cause an appreciable variation in the coloring rate and necessitate a change in the voltage and the current density.

The blue obtained on iron and steel with these electrolytic methods is not so brilliant and intense as that obtained by heat-tinting in air or in salt baths. It is obtained, however, at lower temperatures, which is often very advantageous, especially in the case of fully hardened steels.

A light blue can be obtained by cathodic treatment in the following solution:

	g./l.	oz./gal.
L. Sodium Hydroxide	35	5
Arsenious Oxide	35	5
Sodium Cyanide	7	1

The solution is prepared by dissolving the arsenious oxide and sodium hydroxide in hot water and then adding the sodium cyanide to the *cool* solution. A plating tank or other suitable vessel equipped with carbon anodes is used. The carefully cleaned article (electrolytic alkaline cleaning preferred) is made the cathode and a potential of about 2 volts is applied. A current density of 2 to 7 amperes per square foot is required. The current density should be adjusted to suit the size and the composition of the steel and is best determined experimentally. A treatment of 1 to 3 minutes is usually sufficient to blue the article, whereupon it is removed, washed in hot water, dried, and if desired, finished as previously described.

A blue or blue-black can be obtained on iron or steel with an electrolytic method, the article to be blued being used alternately as anode and cathode, U. S. Patent 1,342,910. The following solution is used.

	g./l.	oz./gal.
M. Sodium Nitrate	150	20
Sodium Hydroxide	375	50

The temperature of the bath is held at 250 to 255° F. (120 to 124° C.), that is, just below the boiling point of the solution. The cleaned article is used as anode against a carbon cathode for 5 minutes with a current density of approximately

50 amperes per square foot (4.5 amperes per square decimeter), after which the current is reversed for the same period, the article serving as cathode and the carbon as anode. This cycle is preferably repeated three times. After removal from the bath, the article is washed, dried, and finished in accordance with previous directions. A lighter blue with less gray or gray-black may be obtained by decreasing the coloring periods to 2 or 3 minutes. Stirring of the bath is advisable, as it promotes uniformity of coloring.

Blackening Iron and Steel
British Patent 538,210

The surface is blackened by a two-stage treatment in boiling aqueous 2:1 sodium hydroxide-sodium nitrate mixture, the first bath boiling at 11–14° below the second. E.g., the first bath contains 7¾ pounds of the mixture per gallon and boils at 140°, and the second 9¼ pounds of the mixture per gallon and boils at 154°.

Coloring Iron, Blue-Black
German Patent 706,313

Sodium Hydroxide	1200 g.
Sodium Nitrate	50 g.
Manganese Acetate	40 g.
Water	1 l.

Treat for 5 min. at 140° C.

Coloring Iron

Perhaps the simplest way to blacken iron is to apply to the surface a liberal coating of linseed oil, then heat the plate to the burning point of the oil. The coating so produced can finally be rubbed with benzine or dilute soda solution.

Formula No. 1

Water	20 oz.
Copper Sulphate	240 gr.
Sodium Thiosulphate	360 gr.

The iron surface must be thoroughly cleaned with dilute sulphuric acid, then boiled in the above bath.

No. 2

Water	5000
Mercuric Chloride	250
Ammonium Chloride	250

No. 3

Water	5000
Ferric Chloride (30° Bé)	375
Copper Sulphate	25
Nitric Acid (26° Bé)	100
Alcohol	150

The baths should be used near the boiling point; immersion can be repeated until the desired depth of black is obtained.

Besides a glossy black, various other colors can be obtained on iron in the following bath; and, as with copper, the exact color depends on the length of immersion:

Water	1000
Selenious Acid	100
Copper Sulphate	100
Nitric Acid	40 to 60

The bath soon exhausts itself because of precipitation, therein, of insoluble yellow iron selenite.

Coloring Steel

Closely related to iron is the coloration of steel. In its simplest form, steel is an alloy of iron and carbon, but modern metallurgy has introduced a wide variety of steels, some of which are difficult (if not impossible) to color by simple chemical means. The only recourse with such alloys is the electrolytical deposition of other metals on the surface of the steel.

Steel can be blackened by dipping the thoroughly cleaned metal in a physical solution of sulphur in oil of turpentine (produced by heating the sulphur in the oil on a water-bath). Evaporation of the oil leaves, on the steel, a thin film of sulphur, which is converted to a black color by heating the plate, until dry, in a non-oxidizing flame.

Formula No. 1

Water	1000
Caustic Soda	400
Potassium Nitrate	10
Sodium Nitrate	10

The bath should be used at a temperature of 120° F.

No. 2

Water	50
Hydrochloric Acid	6
Mercuric Chloride	2
Cupric Chloride	1
Alcohol	5

The plate is immersed in the bath for a few minutes, allowed to dry, then placed in boiling water for one-half hour. If the surface is not black enough, the operation can be repeated.

The following process of blackening steel depends on initial coppering of the steel surface by immersion for 10 seconds in a bath of:

No. 3

Water	250
Copper Sulphate	10

to which is added:

Water	100
Hydrochloric Acid	20
Stannous Chloride	15

with water added to make a final 1000.

The adhering film of copper is converted (blackened) to copper sulphide by immersing the plate for 2–3 minutes in:

Water	1000
Sodium Thiosulphate	1500
Hydrochloric Acid	75

The thiosulphate is dissolved in the water by the aid of heat; on cooling, the hydrochloric acid is added to the solution. Should the bath refuse to blacken after use, add a small volume of hydrochloric acid.

No. 4

Coloring Steel Blue

U. S. Patent 2,102,925

Dip steel sheets into solution of:

Mercuric Chloride	2 lb.
Potassium Chlorate	2 lb.
Nitric Acid (6N)	150 cc.
Water	3 gal.

Color and Temperature in Heating Steel

The average repair shop is not equipped with pyrometers with which to accurately determine the heat temperature of iron or steel. They have to depend upon the human eye, and in vision the idea of color frequently varies, but for all practical purposes color can be used as a fairly accurate gauge in heating and tempering steel.

Solid bodies which are heated to a point so that they glow with the intensity of their own heat will emit certain colors, depending on the temperature and this independently of the nature of the heated material. In other words, a piece of iron or steel heated to 1500 degrees Fahrenheit, will have practically the same brightness and color as a piece of firebrick which has been heated to the same temperature.

Consequently, providing that the eye of the individual and his keenness of vision are true, the following table will enable the mechanic who has not a pyrometer to determine approximately the temperatures of iron or steel from the following colors:

Color	*Approximate Temperature F.°*
Lowest red visible in the dark	800
Lowest red visible in twilight	900
Lowest red visible in daylight	950
Faint red	1000
Blood, medium	1050
Dull red	1100
Dark cherry	1250
Full cherry red	1300
Bright cherry	1450
Salmon	1550
Light red	1600
Dark orange	1625
Orange	1700
Full Yellow	1800
Lemon	1900
Light yellow	2250
White	2400
Brilliant white	2600
Dazzling (blueish) white	2800

When the temperature is too low for the material to glow—as in tempering—it is sometimes possible to estimate it by placing a piece of polished steel at a point where it can come approximately to the temperature of the article to be determined upon. After taking the steel away it will be found to have a "temper" color corresponding to the highest temperature reached by it. These colors for the various temperatures are given below and are only for the range between 430° and 600°.

Color	*Approximate Temperature F.°*
Very pale yellow	430
Light yellow	440
Pale straw yellow	450
Straw yellow	460
Deep straw yellow	470
Dark yellow	480
Yellow brown	490
Brown yellow	500
Spotted red yellow	510
Brown purple	520
Light purple	530
Full purple	540
Dark purple	550
Full blue	560
Dark blue	570
Very dark blue	600

From the foregoing table on high heats it is possible to consider the heating and cooling cycles of .20% carbon steel, for example. This is about the average carbon content of the steel most widely encountered in the trade.

Refer to the heat colors in the above chart, and you can then determine the temperatures at which the changes take place in the composition of the steel due to the heat or cooling applied. By this means you can make heat treatment closely accurate in the blacksmith shop.

Stress Relieving, 1150° F.—Around this temperature, there is no change in grain size or composition of the steel. But there is a change in grain structure resulting in plastic flow of the material. In this manner, any internal stresses induced by welding are to all intents and purposes nullified. But if the material has received a permanent set after welding, stress relieving will not return it to normal.

The material should be held at this temperature for an hour per inch thickness. Cooling in still air is recommended.

Critical Temperature, 1330° F.—No change in grain size of steel but change in composition starts. Iron carbides begin to dissolve and austenitic iron makes its appearance. Below this temperature, quenching has no effect on hardness. Above this temperature, quenching increases the hardness but lowers the ductility of the steel.

Critical Temperature, 1520° F.—At this high-critical temperature, full austenitic iron is obtained. Quick cooling just above this temperature results in a fine grained, hardened steel. As the temperature increases, however, grain growth begins and quenching results in large coarse grains with a corresponding loss in ductility. For welds in this state subsequent annealing is strongly recommended.

Hardened steels may be tempered by reheating to a temperature just below the critical range and followed by a convenient rate of cooling.

Top Annealing Temperature, 1625° F.—For full annealing, the steel should be held around this temperature for one hour per inch of thickness and cooled very slowly in the furnace or in some medium such as lime that will prolong the rate of cooling as compared to that of air.

Top Quenching in Water, 1640° F.—Above this temperature (a dark orange), the steel should be quenched in oil or lead.

Top Normalizing Temperature, 1770° F.—The steel is heated at thi temperature and cooled in still ai at ordinary temperatures. The stee is thus given a grain of known struc ture, size and composition. Normal izing is always followed by some de sired heat treatment.

Top Forging Temperature, 2380° F—Above this temperature, the stee should not be forged.

At 2710° F., liquid iron appear in the steel, and at 2770° F., iron be comes completely liquid.

Coloring Silver

In view of the tendency of metal lic silver to discolor in the presenc of gaseous sulphur compounds, i the atmosphere, it is not surprisin that sulphur solutions are used fo blackening the metal.

Formula No. 1

Water	1000
Ammonium Sulphide	4
Ammonium Chloride	8

No. 2

Water	1000
Potassium Sulphide	5
Ammonium Carbonate	10

The baths are used in a hot stat (158–176° F.), and before immersing in the solution, the silver plat should be passed through boiling water.

Silver can be mechanically blackened by rubbing the surface with the following:

No. 3

Lampblack	30
Graphite	5
Silver Nitrate	15
Lead Acetate	10

The ingredients must be finely powdered and used in the form of a thin paste, achieved by mixing together and grinding the mass to the equired state with the addition of ufficient alcohol.

Coloring Tin

Tin-plated articles or sheets can e blackened by rubbing with car-on black mixed with the least pos-ible volume of French polish. Or he plate can be immersed in:

Formula No. 1

Water	10 oz.
Borax	150 gr.
Shellac	300 gr.
Glycerin	150 min.
Nigrosin	600 gr.

With the exception of the nigrosin, he various ingredients are boiled ogether in the water until dissolved, after which the nigrosin is added. This solution really is a form of var-nish, the plate, after dipping, being allowed to dry spontaneously.

No. 2

A black finish on tin can be had by first ridding the plate of all grease by treatment with a boiling caustic potash solution, rinsing, and immediately immersing in:

Water (Hot)	128
Antimony Chloride	6
Copper Chloride	12

The plate is kept in the bath until the desired color is obtained, then rinsed in hot water.

No. 3

A dilute (2–5%) palladium chlo-ride solution produces a nice black color on tin, but a less expensive solution is:

No. 4

Bismuth Subnitrate	5
Nitric Acid	50
Tartaric Acid	80

The bismuth salt is dissolved in the nitric acid, followed by the ad-dition of the tartaric acid, after which sufficient water is added to make a final 1000 parts of bath.

Coloring Zinc

Photoengraving zinc etchers have long known that a simple solution of copper sulphate imparts to the sur-face of zinc a dark gray color, but the following solution gives a more pleasing effect:

Formula No. 1

Water	64
Hydrochloric Acid	8
Copper Chloride	3
Copper Nitrate	2

The solution is slightly blue in color and imparts an intense black to a thoroughly clean zinc surface. After blackening, the metal should be liberally rinsed with water and allowed to dry.

No. 2

Alcohol	800
Antimony Trichloride	90
Hydrochloric Acid	60

The antimony salt is dissolved in the alcohol, followed by addition of the acid. The solution so obtained can be used for dipping, or it can be locally applied to the zinc with a brush. Drying of the treated areas should be hastened to prevent bleaching of the antimony deposit.

Cleaning Lead Castings Prior to Plating

Because of the shortage of zinc for domestic use, the zinc base die cast-ing has been replaced by a 12 to 14% antimonial lead casting; and platers have had to change their cleaning cycle to meet the change in casting metal.

There are several methods in use and all of them use different clean-ing solutions—by that is meant different proprietary cleaners. However, the cleaners have this in common—they are mild in nature, similar to cleaning solutions used for zinc base die castings.

Cycle No. 1
- Degrease.
- Clean cathodically in a mild cleaner at 160° F. (change cleaner often).
- Rinse.
- Soak for 5 minutes in a 10% sodium cyanide solution at room temperature.
- Transfer directly to high speed copper for 5 minutes at 35 amp. per sq. ft.
- Rinse.
- Acid dip.
- Nickel plate.

Cycle No. 2
- Degrease.
- Clean cathodically in a mild cleaner for 1 minute.
- Rinse.
- Dip in 5% by volume acetic acid.
- Rinse.
- Copper strike in a Rochelle salt copper bath at 2 volts or less for 10 minutes.
- Rinse.
- Acid dip.
- Copper or nickel plate.

Cycle No. 3
- Wash in washing machine.
- Clean anodically in a mild cleaner for 5 to 20 seconds.
- Rinse.
- Dip in 3% sulphuric acid for 2 seconds.
- Rinse.
- Copper strike in a Rochelle salt bath at 10 amp./sq. ft. for 5 minutes.
- Rinse.
- Copper or nickel plate.

Cycle No. 4
- Degrease.
- Clean in a mild cleaner cathodically.
- Rinse.
- Reverse current in a 55° Beaumé sulphuric bath (a nickel strip solution will work) for 15–30 seconds.
- Rinse.
- Copper strike for 15–30 seconds at 5 volts.

Cycle No. 5
- Clean cathodically.
- Rinse.
- Soak in hot water for 5 minutes.
- Transfer directly to nickel solution.

Platers' Stripping Solution

This stripping solution is used so that an object may be plated upon the plated portion and then separated from the object forming a mold.

The metallic object, preferably of copper or copper plated is dipped in an aqueous solution of selenious acid of such strength that 1 liter contains .8 g metallic selenium, the object is kept immersed until the copper turns to a deep blue tinge, then washed with water, then plated thereon either with copper, nickel or cobalt, until the desired thickness has been attained, then backed up by pouring on solder. When cool, the original and the plated part can be separated by inserting the point of a knife at the juncture of the original piece and the plating, thereby forming a mold.

Stripping Solution for the Removal of Cadmium Plating

Hydrochloric Acid (37%)	73 cc.

Antimony Trioxide	2 g.
Water	27 cc.

Stripping Tin From Brass

For removing a hot-dipped tin coating from 70/30 brass, use a solution of 10 grams "Ferrisul"* and 10 grams sulphuric acid per 100 cc. This solution removes a 0.0005 inch coating from brass in 5 minutes at a temperature of 75° C. (167° F.) and in 15 minutes at 28° C. (82.4° F.), leaving a matte finish. The relative rates of attack of this solution on tin and brass are:

At 75° C. Tin coating:
8.0 mg. per sq. in. per min.
70/30 brass:
2.7 mg. per sq. in. per min.

At 28° C. Tin coating:
3.4 mg. per sq. in. per min.
70/30 brass:
1.0 mg. per sq. in. per min.

It is known that the rate of attack on the brass is considerably greater for a high brass than for the low brasses or for copper.

Etching Nickel Plated Printing Plates
U. S. Patent 2,233,546

Nickel is removed, where desired, by 40% aqueous copper chloride; the etch is then deepened, without undercutting, by treating with a 40% solution of a 2:1 mixture of iron chloride and nickel chloride.

Stainless Steel Etch
British Patent 541,630

Iron Chloride (Saturated Solution)	4
Nitric Acid, Concentrated	1
Hydrofluoric Acid	1

* Ferric Sulphate.

Color Etching of Stainless Steel
U. S. Patent 2,243,787

Gold Tint Formula No. 1

Ammonium Dichromate	46
Sulphuric Acid	7–11
Water at 85–93°	19–23

Black No. 2

Ammonium Dichromate	10–14
Sulphuric Acid	36–50
Water at 121-127°	30–50

Zinc Etching Fluid
U. S. Patent 2,245,219

Copper Sulphate	10–25%
Sodium Bisulphate	2%
Wetting Agent	0.005%
Water, To make	100%

Etching Agent for Zinc Engravings
U. S. Patent 2,245,219

Copper Sulphate (20% Solution)	95%
Niter Cake (35% Solution)	5%
Wetting Agent	0.005%

Aluminum, Protective Coating for
U. S. Patent 2,234,206

Formula No. 1

Manganese Dihydrogen Phosphate	8
Manganese Silico Fluoride	50
Potassium Fluoride	4
Water	100

No. 2
Italian Patent 368,331

By dipping aluminum or its alloys for 5–10 min. in a solution of a metal fluoride or silicofluoride or borofluoride in the presence of an organic colloid it acquires a protective coating. A suitable solution may be made by mixing a nearly saturated solution of cadmium silico-fluoride with 4–5 times its weight of water and adding 2–5 parts per thousand of albumen or dextrin.

Brass Plating Bath

Cuprous Cyanide	26.2 g.
Zinc Cyanide	11.3 g.
Sodium Cyanide	45.0 g.
Water, To make	1 l.

Use at 30° C.; 1 amp./sq. dm.; pH 10.5 with brass anode of about 25% zinc.

Molybdenum Bronze Plating Solution

Crystal Copper Sulphate	20	oz.
Molybdenum Trioxide	8	oz.
Sulphuric Acid C. P.	7	oz.
Nitric Acid C. P.	½	oz.
Hydrofluoric Acid C. P.	¼	oz.
Sodium Chloride	¼	oz.
Water	1	gal.

Temperature 75°–80° F. Molybdenum-Iron Anodes (60%–70% Mo.)

The Anodes must be tubular and Hydrogen Gas must be slowly sent through these into the plating bath. Proportion of metals must be reasonably maintained. Anodes should be bagged in Blue African Asbestos Cloth.

Immersion Plating of Cadmium
U. S. Patent 2,272,777

Adherent coatings of cadmium are formed on metals of the group consisting of copper, brass, iron, copper plated iron and brass plated iron without the application of an outside source of electric current by means of immersion in the following:

Example for Brass:

Cadmium Oxide or Salt	4
Alkali Cyanide	70–150
Water	1000

Operate at 40°–70° C.

Example for Iron:

Cadmium Oxide or Salt	5– 8
Caustic Alkali	1200–1400
Water	1000

Operate at 125° C. Some alkal[i] cyanide may be added to this bath.

Copper Plating Bath

	g./l.	oz./gal.
Copper Cyanide	120	16
Sodium Cyanide	135	18
Free Sodium Cyanide corresponding to	3.75	0.5
Caustic Soda	30	4
or		
Caustic Potash	42	5.6
Brightener	15	2
Anti-pit Agent	1.5	0.2

Operating Conditions

Cathode current density
10 to 100 amp./sq. ft. (1.2 to 11 amp./dm.^{2})

Anode current density
5 to 30 amp./sq. ft. (0.54 to 3.2 amp./dm.^{2})

E. M. F. on still tanks
1 to 2.5 volts at bus bars

E. M. F. on barrel units
3 to 4 volts

Rate of copper deposition
8.8 amp./sq. ft. (0.95 amp./dm.^{2}) deposits 0.001 in./hr. (0.025 mm./hr.)

Cathode efficiency
100% approx.

Anode efficiency
100% approx.

Temperature
75 to 85° C. (168 to 185° F.)

Cathode bar agitation
2 to 15 ft./min. (60 to 450 cm./min.)
Anode to cathode ratio
approx. 2:1

Copper Plating on Iron, Aluminum and Duraluminum

Copper Sulphate	65 g.
Ammonium Sulphate	20 g.
Ammonia	500 cc.
Water, To make	1 l.

Use following current densities:

For Iron	2–3 amp./sq. dm.
For Aluminum	2.7–3 amp./sq. dm.
For Duraluminum	4.7–5 amp./sq. dm.

Copper Plating Bath Without Cyanide

Copper Sulphate Crystals	26 g.
Ammonium Oxalate	58 g.
Oxalic Acid	33 g.
Ammonia (25%)	150 cc.
Water, To make	1 l.

Use at 50° C. at 3.7 amp./dm.^{2}.

Ammonia Copper Plating Bath

Copper Sulphate	65 g.
Ammonium Sulphate	20 g.
Ammonia (25%)	500 cc.
Water, To make	1 l.

Before plating clean iron, aluminum or duraluminum with 15% caustic soda; the latter should then be treated with 18% hydrochloric acid.

Use 2.5–3 amp./sq. dm. for iron and aluminum and 4.7–5 amp./sq. dm. for duraluminum.

Black Mirrors

Solution A—Dissolve 10 grams of thiourea in a liter of water.

Solution B—Dissolve 40 grams of lead acetate trihydrate in a liter of water.

Solution C—Dissolve 20.2 grams of sodium hydroxide in a liter of water.

First clean the glass surface upon which the mirror is to be produced with sodium hydroxide, with soap and water and finally water. Protect one surface with Scotch tape or the like. Then immerse the glass plate in a mixture consisting of 100 cc. of Solution A, 25 cc. of Solution B and 50 cc. of Solution C (larger quantities, if desirable). Heat the mixture to 40–41° C. and maintain at this temperature for a period of fifteen minutes, when a black mirror will have been produced.

Silver Plating Optical Glass

The surface should be cleaned with hot chromate-sulphuric acid solution and thoroughly rinsed with distilled water. It is unnecessary to use the application of caustic soda so often recommended. If a drop of distilled water will spread evenly over the entire surface, it may be considered free from grease or other organic matter.

About a half liter of (1) a 10% silver nitrate solution containing one or two drops of concentrated nitric acid and of (2) a 10% solution of technical triethanolamine are conveniently made up. Both solutions will keep indefinitely.

The surface to be mirrored is placed face upward in a clean Petri dish of sufficient diameter to accommodate it. To 25 cc. of solution 1 in a large test tube are added 10 cc.

of solution 2; then with constant agitation further additions are made of 2 or 3 cc. at a time, just to the point where the precipitate which forms on the first addition clears completely. The mixed solution is poured immediately over the object to be plated so as to cover it by a layer of at least 0.25 inch. The deposition of silver begins within a few seconds.

For the half-reflecting surface required for interferometers, a layer of silver which transmits about as much light as it reflects is ideal. Such a layer has a distinct violet tinge and appears within 10 minutes at room temperature. It is advisable to have several surfaces in preparation at the same time and being plated with the same mixture of reagent but removed from the bath at graduated intervals, and to select the one which has the proper depth of coating rather than attempt to plate a surface additionally which has been found to have been immersed for an insufficient time.

For completely reflecting surfaces, the immersion may last for 24 hours; this particular bath is unique in that the deposition seems to be continuous for that length of time.

When the desired thickness is deposited, the plated object is taken from the bath, only the edges being handled. The silver is removed from the under surface with dilute nitric acid. The mirrored surface is washed with absolute alcohol and then with xylene, and is covered with isobutyl methacrylate which, when allowed to dry, forms an optically inactive film which protects the silvering permanently.

Silvering Mirrors
British Patent 537,987

Silvering of e.g., glass is carrie out by spraying with a mixture o e.g., silver nitrate 3, aqueous an monia 3, and water 128 ounces an a reducing solution comprising a sa (sulphate) of hydroxylamine 2⅔ aqueous ammonia 1, and water 1C ounces, and 2 ounces of glyoxal in gallon of water.

Plating Silver on Plexiglas (Plastic

Solution A. For pretreatment c Plexiglas
Sodium Hydroxide (C. P.), 35 grams per liter of water

Solution B. Reducing solution
Cane Sugar (Granulated), 90 grams
Nitric Acid (C. P.), 4 cc.
Ethyl Alcohol, 95%, 175 cc.
Water, Distilled, To make, 100 cc.

Dissolve the sugar in 500 cc. c water, add the nitric acid slowl with stirring, then add the alcoh and make up to 1000 cc. with th water. Let stand for one week a room temperature before use.

Solution C. Silver solution
Part I. Silver Nitrate (CP), 10 grams
Water, Distilled, 1000 cc.
Part II. Sodium Hydroxide (CP 100 grams
Water, Distilled, 1000 cc.
Part III. Ammonium Hydroxid (CP, 28%), 400 cc.
Water, Distilled, 600 cc.

Solution D. For removing silver de posits
Nitric Acid, 500 cc.
Water, Distilled, 500 cc.

olution E

Dreft (Wetting agent), 50 grams
Water, Distilled, 1000 cc.

A Plexiglas, glass, hard rubber, r enamel-lined tray is required for ie silvering operation. Leave pieces f Plexiglas to be plated for 24 hours i Solution A. Add 50 parts of Part I (Solution C) and 40 parts of Part II (Solution C) to 1250 parts of istilled water. To this mixture lowly add, with constant stirring bout 150 parts of Part I (Solution), that is, until a clear solution is btained containing a very small mount of light brown precipitate. olution C must be mixed fresh from ay to day.

For each 20 cc. of Solution C, 1 cc. f Solution B is required.

Rinse the Plexiglas with distilled vater. Always keep an unbroken ilm of water or silvering solution n the surface of the piece to be ilvered.

Immediately mix solutions B and and pour over the work in the ilvering tray. Rock the tray until he solution becomes colorless and he flocculent gray silver precipitate as coagulated. Pour off the solu-ion. Rinse the Plexiglas with dis-illed water, using also running wa-er. Immediately pour over mirror second mixture of B and C. Pro-eed as before. Repeat this a third ime. Put mirror on rack to dry. Then paint on a protective coating n back.

Plating Bath Without Cyanide
Silver

Silver Sulphate	3 g.
Potassium Iodide	60 g.
Tetrasodium Pyro-phosphate	6 g.
Ammonia (25%)	7 cc.
Water	100 cc.

Use at 17–60° C. at 0.36–2.15 amp./dm.^{2}

Silver-Plating Without Electricity

Silver Nitrate	25
Ammonium Chloride	11
Sodium Thiosulphate	25
Whiting	225
Tartaric Acid	13
Water	480

Shake well before using. Rub on articles to be silver plated. This will silverplate copper and brass.

Plating Solutions
Green Gold

Potassium Gold Cyanide	¼ oz.
Sodium Cyanide	1 oz.
Silver Cyanide	Tiny pinch only
Cadmium Cyanide	Tiny pinch only
Water	1 gal.

Temperature 75–80°
1/10 Sterling Silver
9/10 Fine Gold Anodes

White Gold

Potassium Gold Cyanide	¼ oz.
Sodium Cyanide	1¼ oz.
Nickel Cyanide	1¼ oz.
Cadmium Cyanide	¼ oz.
Nickel Chloride	¼ oz.
Water	1 gal.

Temperature 120°–130° F.

Anodes
¼ Gold
½ Nickel
⅛ Steel
1/24 Cadmium
5/24 Silver

Do not replace Cadmium and Silver Anodes until practically all the Gold Anode is used up. Use Iron Wire to hold Anodes.

Gold Electroplating Solution
Stock Solution

Gold	1 g.
Sodium Cyanide	2 g.
Water, Distilled	100 cc.

Dissolve gold in aqua regia using 150 cc. per ounce of gold, diluting with equal parts distilled water. Heat to speed reaction to completion. Boil down until thermometer reaches 115° C. Dilute 25 cc. water and 25 cc. concentrated hydrochloric acid per ounce of gold. Boil down to 115° C. to remove nitric acid.

Transfer to 3 liter jar to make 250 cc. per ounce of gold. Add ammonium hydroxide with stirring until alkaline and add small excess. Allow precipitate to settle; filter with suction.

Caution: Do not allow to dry or crack. Gold fulminate is explosive!

Wash with about 3 gallons of water until free of chlorine. Dissolve precipitate in solution of 2 g. sodium cyanide per g. of gold. Filter and wash paper, adding washings to filtrate. Analyze for actual gold content and make up with water to 1 g. per 100 cc. solution.

Plating Solution

Stock Solution A	100 cc.
Disodium Phosphate	2 g.
Sodium Cyanide	2 g.
Water, Distilled	1 l.

Plating Instructions

Volts	1–2
Temperature	50–70° C.
Anode	Platinum
Color	22 Carat

Add cadmium cyanide to get pale tone. Add nickel cyanide to get lower carat value and color. Add copper cyanide to get redder or pinker color.

Plating Solution
Iron (For Antique Effects)

Ferrous Chloride	41 oz.
Calcium Chloride	19 oz.
Sodium Chloride	1 oz.
Ortho-Phosphoric Acid	5 drp.
Water	1 gal.

Cut up about one ounce of Iron Binding Wire and allow it to dissolve in the above solution (for several days) before putting it into use.

Temperature 190° F.

Rolled Armco
Iron Anodes

Anodes should be bagged in Blue African Asbestos Cloth.

Coating Iron with Aluminum or Zinc
British Patent 541,240

A mixture is made of

Zinc (or Aluminum) granular	48.6
Vehicle *	51.4

The metal is coated with this mixture, dried and heated to 600–700 C.

Coating Iron with Lead
U. S. Patent 2,230,602

Immerse iron or steel in a hot solution of:

Lead Nitrate	6
Sodium Cyanide	100
Water	1000

until a sufficient thickness of lead is gotten.

* Vehicle consists of

Zinc (or Aluminum) Stearate	21.4
Toluol	78.6

Lead Plating

Like other plating baths the com-osition of the fluoborate lead bath aries, depending largely on the cur-nt density required. For deposits p to about 0.001 in. (0.025 mm.) ick produced at fairly low current ensities, a bath of the following omposition is recommended:

Formula No. 1

	g./l.	oz./gal.
Basic Lead Carbonate	150	20
Hydrofluoric Acid (50%)	240	32
Boric Acid	106	14
Glue	0.2	0.025

For barrel plating operation or r still plating at higher current ensities in which heavy deposits p to 0.05 in. (1.27 mm.) thick are equired, best results can be secured rom a bath of the following com-osition:

No. 2

	g./l.	oz./gal.
Basic Lead Carbonate	300	40
Hydrofluoric Acid (50%)	480	64
Boric Acid	212	28
Glue	0.2	0.025

The fluoboric acid solution is best repared by placing the necessary mount of commercial hydrofluoric cid in a lead-lined or rubber-lined, r wooden tank and slowly adding rystalline boric acid. Considerable eat is developed during the ensuing eaction. The solution is allowed to ool, after which the lead is intro-luced as basic lead carbonate, in he form of a thick paste with water, dding it slowly with stirring. Con-iderable effervescence results from he dissolving of lead carbonate in the acid and the resultant evolution of carbon dioxide. The solution is then diluted to the required volume and decanted or filtered into the plating tank. Since the commercial materials used in the preparation of the solution invariably contain sulphate, some precipitate of lead sulphate will be left behind in the original tank. The preparation of the fluoborate lead bath thus entails considerable care and caution, and differs from the preparation of most plating solutions in the respect that the chemicals used must be added in a specific order.

The acidity of the fluoborate bath is high enough so that the pH need not be controlled. However, the concentration of free fluoboric acid in the solution, although not critical, should be maintained within fairly definite limits.

The commercial operating characteristics of the fluoborate bath are given below:

Temperature

25° to 40° C. (77° to 105° F.) Above 40° C. (105° F.) use cooling.

Cathode Current Density

Dilute Type Solution for Thin Deposits

5 to 50 amp./sq. ft. (0.54 to 5.4 amp./dm.^{2}) — extreme limits. Average value — 20 amp./sq. ft. (2.2 amp./dm.^{2}) Heavy deposits—10 amp./sq. ft. (1.1 amp./dm.^{2})

Concentrated Type Solution

5 to 70 amp./sq. ft. (0.54 to 7.6 amp./dm.^{2}) — extreme limits. Average value — 30 amp./sq. ft. (3.2 amp./dm.^{2}) Heavy deposits — 20 amp./sq. ft. (2.2 amp./dm.^{2})

Cathode Current Efficiency
100%
Throwing Power
Haring cell, 5:1 ratio 8 to 9%.
Anode Current Density
10 to 30 amp./sq. ft. (1.1 to 3.2 amp./dm.^{2})

For Iron or Steel
U. S. Patent 2,230,602
Formula No. 1

Lead Nitrate	6	g./l.
Sodium Cyanide	100	g./l.

Temp. 180°–200° F.

No. 2

Lead Monoxide	5	g./l.
Caustic Soda	100	g./l.
Sodium Cyanide	10	g./l.
Nekal BX (Wetting Agent)	2.5	g./l.

Temp. 180°–200° F.

No. 3

Lead Monoxide	1.5	g./l.
Caustic Soda	100	g./l.
Sodium Cyanide	50	g./l.
Nekal BX	0.05	g./l.

Room Temperature.

For Copper or Zinc

Lead Monoxide	3	g./l.
Caustic Soda	100	g./l.
Sodium Cyanide	25	g./l.

Room Temperature.

No. 4

Lead Carbonate	2	oz.
Sodium Hydroxide	6	oz.
Water	to 1	gal.
Volts	3–4	
Temperature	175° F.	
Anode	Lead	

Lead Carbonate	20	oz.
Hydrofluoric Acid	32	oz.
Boric Acid	14	oz.
Glue	0.025	oz.
Water	to 10	gal.

Mix hydrofluoric and boric aci[illegible] together. Add the mixture of lea[illegible] carbonate and glue. To the whol[illegible] add water, at room temperature, 1[illegible] to 20 amps. per square foot, lea[illegible] anode.

Nickel Plating Bath
U. S. Patent 2,125,229

Nickel sulphate hexahydrate 3[illegible] nickel chloride hexahydrate 4, bori[illegible] acid 5, selenium dioxide 0.01, an[illegible] naphthalene disulphonic acid 0.[illegible] gram/liter.

Nickel Plating Salt
German Patent 699,021

Nickel Sulphate	90
Sodium Chloride	34
Boric Acid	18
Sodium Naphthalene-Sulphonate	2

Nickel Plating Aluminum
U. S. Patent 2,233,410
Degreasing Solution
Formula No. 1

Manganese Sulphate	3–5%
Sodium Cyanide	2–6%
Ammonia	2–8%
Gelatin	0.5–1%

Cathode at less than 2 volts, pref[illegible] erably 1–1.5 volts.

No. 2

Zinc Sulphate	4–6%
Sodium Cyanide	4–8%
Potassium Hydroxide	2–8%
Dextrin	0.5–1%

Cathodic at a minimum of 0.5–volt.

Plating Solution

Hexaamino-Nickelo-Sulphate	15–45%
Ammonium Acetate	1– 5%
Magnesium Sulphate	10–15%

C. D. 0.25—6 amps./sq. dm.; 2–4 olts; room temp. or hot.

Nickel Dip for Enameling
U. S. Patent 2,265,467

Single Nickel Salts	2–4 oz./gal.
Boric Acid	¼ oz./gal.
Alkali, To pH	5.2–6.4

Immerse the iron for about 5 min. t about 160 or 170° F. Rinse for ½ min. in a solution of 0.1% by eight of sulphuric acid and then eutralize in the usual neutralizing ath.

Bright Ductile Nickel Plating Bath
British Patent 513,634

Nickel Sulphate	280
Nickel Chloride	60
Boric Acid	30
Nickel Benzene Disulphonate	11
Zinc Sulphate	3

Water, 1 liter.

Nickel Brightener
U. S. Patent 2,228,991

Brightener:

Cobalt Chloride	14
Cadmium Chloride	1
Ammonium Chloride	3
Nickel Chloride	7
Sodium Sulphate, Anhydrous	3
Sulphuric Acid	2
Water	70

Plating Solution:

Double Nickel Salts	12 oz./gal.
Single Nickel Salts	4 oz./gal.
Boric Acid	3 oz./gal.
Ammonium Chloride	4.5 oz./gal.
Sodium Chloride	0.8 oz./gal.
Brightener (Above)	0.4 oz./gal.

pH = 5.5
Temperature = 70° F.
C. D. = up to 48 amp./sq. ft.
Deposit has a bluish cast.

Semi-Bright Nickel Plating
U. S. Patent 2,274,112

Nickel Chloride	300 g./l.
Boric Acid	30 g./l.
pH	1.0

Temp. 110° F.
C. D. 75 amp./sq. ft.

Platinum Plating Solution

Ammonium Nitrate	3½ oz.
Platinum Diamino Nitrite	⅓ oz.
Sodium Nitrite	⅓ oz.
Ammonia (26°)	2 oz.
Water	1 qt.

Use at 200° F. with platinum anodes.

Protective Filming of Tinplate
British Patent 282,156

The plain cans are immersed in the filming bath for about five minutes at a temperature of 85–90° C. (185–195° F.). They are then removed and very thoroughly rinsed with water.

The composition of the bath is:

Trisodium Phosphate (Crystalline)	40 g./l.
Sodium Hexametaphosphate	20 g./l.
Sodium Dichromate	12.5 g./l.
Sodium Hydroxide	14 g./l.
Wetting Agent	5 cc./l.

The solution should preferably be

contained in an enameled tank, but a plain, mild-steel tank may be used provided that the cans do not come into contact with the steel. If possible, softened water should be used in making up the solution. The wetting agent will not dissolve properly in the cold, but dissolves on heating.

The pH of the bath must be maintained at 12.5, and although the process consumes a negligible amount of the chemicals, small adjustments have to be made from time to time. A valuable feature of the process is that the bath is essentially an active degreasing agent and no preliminary degreasing is called for.

The film is quite invisible and the cans have a bright and attractive luster; and as the solution has some passivating action on any iron exposed at pores, the cans are less liable to rust during storage or transport.

Tin Plating

Tin Sulphate	54	g.
Sulphuric Acid	100	g.
Cresol or Phenol	20–30	g.
Gum	2½	g.
Water, To make	1	l.

Use at 20–30° C.; at 2–5 amp./sq. dm.

Tin Coating Aluminum Pistons

The first step in the immersion method of the tin coating process, after the piston is machined, is cleaning. This is accomplished by dipping the pistons in a 3.0 oz. per gallon solution of tri-sodium phosphate which is operated at a temperature of between 175° F. and 185° F. The pistons are immersed in the cleaning bath only until gas bubbles form on the aluminum, since t surface of pistons remaining for t long in the solution would be etche An etched surface has been o cause of blistered plate.

Following immersion in the clea ing solution, the pistons are rins in cold water. The pistons are th dipped in a 3% nitric acid soluti for about 30 seconds. This dip used only to prevent any alkali the cleaning solution from contam nating the tin bath. A thorough rin in cold water follows the acid dip.

The pistons are then placed in t sodium stannate bath which is ma up by dissolving 6.0 oz. of sodiu stannate in each gallon of wat used. The tin bath is operated at temperature between 170° F. a 175° F. The metal content must kept between 0.9 and 1.5 oz. per ga lon of tin. There is a tendency f free alkali to build up in the bat however, the free alkali conte must be kept below 0.3 oz. per ga lon, otherwise blistering will devel in the coating. The blisters are sem microscopic and can only be see with the aid of a powerful ha lens or a low power microscope.

The tin content of the bath kept up to standard by the additi of sodium stannate. The stannate dissolved in hot water and added the bath as required. The free alka content is controlled by adding ac tic acid, diluted one part of acid 10 parts of water, to the bath required.

The length of time the work mu remain in the bath depends on ho thick a coating is desired. A min mum thickness of .00018 in. of t is obtained by immersing the wo in the bath for four minutes. Aft ie plating operation, the work is nsed successively in cold and hot ·ater.

After continued operation of the ath, there is a tendency for sludge form in the bath. This sludge, omposed of aluminum, carbonates nd stannic tin, must be removed om the solution periodically; therwise, too great a concentration ·ill cause blistering in the tin coat.

A schedule of sodium stannate nd acetic acid additions can be orked out according to production. 'or example, add 10 lbs. of sodium tannate to the bath after every ,000 pistons are coated and one allon of acetic acid after every ,000 pistons are coated. These dditions keep the metal content ·ithin the specified range and keep he alkali content below 0.3 oz. per allon. Therefore, chemical analyis is not necessary each time tin or cid should be added to the bath, ut an analysis for tin and free alali should be performed at least nce a day in heavy production.

Immersion Tinning of Copper and Brass

U. S. Patent 2,159,510

Sodium Cyanide	50	g. per l.
Stannous Chloride	5	g. per l.
Caustic Soda	5.6	g. per l.

The solution is easily prepared by dding the caustic soda slowly and ·ith agitation to an aqueous soluion of stannous chloride; after adding the sodium cyanide and diluting the required volume the bath is eady for use.

The article to be tinned must be horoughly cleaned to remove grease r dirt. Pickling is preferable, hough not essential. The better the surface before immersion the better will be the coating. A brass or copper vessel can be tinned on the inside by filling it with the solution and allowing it to stand until the required coating has been deposited. The insides of pipes and tubes can be coated either by circulating the solution through them or by filling them with the solution for the required time and then draining and washing.

The thickness of the coating depends upon the time of immersion.

For many purposes it is sufficient to judge the coating by its appearance, but it has been found that an immersion of 2½ minutes in the solution is just sufficient to give a coating of good appearance, but not to protect the copper against attack by moist hydrogen sulphide. Specimens immersed for 6½ minutes or longer acquire coatings of good appearance which are unattacked by hydrogen sulphide.

Zinc Plating

Zinc Cyanide	8.0	oz./gal.
Sodium Cyanide	3.0	oz./gal.
Sodium Hydroxide	7.0	oz./gal.

Zinc Oxide	6.0	oz./gal.
Sodium Cyanide	10.0	oz./gal.
Sodium Hydroxide	2.0	oz./gal.

The temperature of a zinc solution is best kept at between 40 and 50° C. The current density necessary is 10 to 20 amp./sq. ft. The anodes used may be pure zinc or mercury-zinc or aluminum-mercury-zinc. The anodes containing other metals than zinc are far more expensive than the pure zinc. Although they give a better looking deposit than the pure zinc, the pure zinc anodes used with the addition of mercuric oxide in the plating bath will give as good a result and this is probably cheaper.

The addition of mercuric oxide to the bath up to 2% of the zinc content increases the cathode efficiency and gives a whiter deposit. While the ordinary zinc solution will not plate cast or malleable iron, the addition of mercury will make possible the plating of these metals. With the addition of mercuric oxide the current density may be raised to 30 or 40 amp./sq. ft.

Incidentally, about zinc anodes. The aluminum-mercury-zinc anodes are sludge-free while the mercury-zinc and the pure zinc anodes produce sludge. The polarization of the aluminum-mercury-zinc anodes is practically constant at 2.9 volts while polarization of the mercury-zinc anodes and the pure zinc anodes at continuous plating will be 3.0 to 3.1 volts.

Another zinc solution used successfully is made up as follows:

Zinc Cyanide	10.0 oz./gal.
Sodium Cyanide	4.0 oz./gal.
Sodium Hydroxide	10.0 oz./gal.

This solution is analyzed and kept at these concentrations:

Zinc	4– 6 oz./gal.
Total Cyanide	10–12 oz./gal.
Sodium Hydroxide	10–12 oz./gal.

Bright Zinc Coating of Steel and Brass

Zinc Oxide	45
Sodium Cyanide	78
Sodium Hydroxide	75
Glycerin	3–5
Water, To make	1 l.

Plate with 2–4 amp./sq.dm. a 20–25° C. After plating wash for a few seconds with 3–6% nitric acid

Electropolishing of Copper

Use copper anode in not more than 600 g./ liter of aqueous solution of ortho-phosphoric acid, 20-25° C., current density 75–300 A/square foot.

Polishing Powder for Gold and Platinum Final Finishing

Bauxite Ore Concentrate	6 lb.
Red Rouge Powder (Dixon)	35 g.
Putty Powder	1 lb.
Aluminum Chloride	½ lb.

Mix putty powder and aluminum chloride thoroughly and add bauxite ore concentrate and red rouge powder. Mix again.

To use: Make paste with alcohol-water 50% solution and apply to work. Use high color buff.

Plating Touch-Up Solution

Where nickel or chrome plated surfaces have been scratched or cut through in spots by handling or mechanical operations, they may be temporarily obscured by painting with a 10% solution of mercuric chloride.

Plating Non-Conductors

Four hundred cc. of 9% caustic soda, 200 cc. of reducing solution (100 g. of sucrose in 250 cc. of water are heated with 0.5 cc. of nitric acid to development of an amber color, and water is added to 1250 cc.), and 80 cc. of formalin are added to 1 l. of solution containing copper sulphate, 30 g., glyc rol 80 cc., and concentrated aqueous ammonia 20 cc., and the article (galalith, glass, etc.) is immersed n the solution. After a coating of copper has formed, the desired design is applied with cellulose lacquer, 15% nitric acid is applied to remove uncovered copper, the article washed, the lacquer removed by means of butylacetate, and the required metal coating deposited electrolytically.

Corrosion Resistant Coating
British Patent 476,042

Colloidal Graphite	4.5 %
Gelatin	0.75%
Potassium Dichromate	.04%
Water, To make	100 %

This gives a hard, adherent, water resistant coating after subjection to actinic light.

Improving Corrosion Resistance of Metals
British Patent 530,006

Prior to painting, iron or steel is coated with:

Phosphoric Acid (75%)	30 g.
Zinc Chromate	14 g.
Water	8 l.

Metal Degreasing and Rustproofing Solution
U. S. Patent 2,123,856

Trichlorethylene	97.5%
Rosin	2.0%
Linseed Oil, Boiled	0.5%

Rust Inhibitor for Metal to Be Painted
U. S. Patent 2,127,202

Salt	66
Chromium Oxide	33
Cresylic Acid	1

Wash the metal with a solution of 1–2 oz. of the above per gallon of water.

Rust Preventing Compounds
Formula No. 1

Petroleum Asphalt	50
Kerosene or Solvent Naphtha	40

No. 2

Asphalt	52
Benzene	48
Sublimed Red Lead	30
Sodium Chromate	½

Corrosion Prevention
U. S. Patent 2,244,526

A process of coating articles of iron, copper and aluminum and their alloys to form a rust and corrosion resistant film, comprises subjection to a molten mixture of an alkali metal nitrate, an alkali metal hydroxide and manganese dioxide followed by neutralization and finishing or fixing in an aqueous solution of a coloring and sealing medium.

First Bath, for Iron and Alloys

Sodium or Potassium Nitrate, Phosphate or Carbonate	66% or more
Sodium or Potassium Hydroxide	34% or less

First Bath, for Copper Alloys and Aluminum

Sodium or Potassium Nitrate	50%
Sodium or Potassium Hydroxide	50%

The addition of 1–2% manganese dioxide to either bath improves the fluidity and acts as a flux. Crystal copper sulphate in the same amount also has a stabilizing effect. Temperature is from 500–900° F., preferably 700–900° F.

Neutralizer Bath

1–2% hydrochloric acid or up to 5% of either sulphuric acid or oxalic acid at 160° F. A preferable solution is 4 oz./gal. of iron sulphate which not only neutralizes but darkens the color.

Finishing Solution

4 oz. to one Imperial gallon of iron sulphate, crystalline haematoxylin, water soluble nigrosine, tannic acid or logwood chips.

First Bath

	Formula No. 1	2	3	4	5	6
Sodium Nitrate	5	5	—	—	—	—
Potassium Nitrate	1	5	1	—	65	50
Sodium Hydroxide	—	1	1	2	34	50
Trisodium Phosphate	—	—	4	—	—	—
Potassium Carbonate	—	—	—	4	—	—

0.5–10% of manganese dioxide or copper sulphate may be added.

Rust Preventor and Remover
U. S. Patent 2,235,944

Linseed Oil, Boiled (Including Drier)	10
Pine Oil	2
Oleic Acid	2
Petroleum Distillate	86

Rust Remover
U. S. Patent 2,209,291
Formula No. 1

Phosphoric Acid	36
Zinc Phosphate	3
Gum Arabic	2
Manganese Chloride	1
Water	28
Butyl Propionate	30

Australian Patent 109,096

No. 2

Phosphoric Acid	32
Sulphuric Acid (66° Bé)	15
Potassium Dichromate	4
Zinc Chloride	6
Sodium Potassium Tartrate	2
Water	37
Alcohol	4

Rust and Mill Scale Remover

To remove scale and rust from materials which are to be cleaned around a mill, dip the articles in the following, then rinse in plain water and apply a coat of thin oil to prevent further rusting.

Methyl Alcohol	5
Ortho Phosphoric Acid	40
Denatured Ethyl Alcohol	2
Water	53

Cleaning Rusty Tools

The life of tools which will be progressively more difficult to replace may be lengthened by keeping them free from rust. Even heavy deposits of rust can be removed easily and economically by the use of a paste made from the following ingredients:

Glycerin	1
Oxalic Acid	2
Phosphoric Acid	2
Ground Silica	5

The tools should be coated with the paste and allowed to stand in a warm place for about 20 minutes, after which the paste and the rust with it can be washed off, and rust preventive applied.

Removing Rust from Steel Instruments

Lay the instruments, over night, in a saturated solution of chloride of tin. The rust spots will disappear through reduction. Upon withdrawal from the solution, the instruments are rinsed with water, placed in a hot soda-soap solution, and dried. Cleaning with absolute alcohol and polishing chalk may also follow.

Another method is: Make a solution of 1 part of kerosene in 200 parts of benzene or carbon tetrachloride, and dip the instruments, which have been dried by leaving them in heated air, in this, moving their parts, if movable, as in forceps and scissors, about under the liquid, so that it may enter all the crevices. Next lay the instruments on a plate in a dry room, so that the benzene can evaporate. Needles are similarly thrown in a paraffin solution, and taken out with tongs or tweezers, after which they are allowed to dry on a plate.

Corrosion Inhibitors for Automobile Radiators

With ethylene glycol or glycerol anti-freezes use

Morpholine
or Dimethyl Morpholine
or Ethanol Morpholine
about 0.2%

Preventing Corrosion of Brass

U. S. Patent 2,272,216

Rinse in any of the following:

Formula No. 1

Zinc Dihydrogen Phosphate	42 lb.
Ammonium Persulphate	8 lb.
Water	100 gal.

5 minutes immersion at 200° F.

No. 2

Manganese Acid Phosphate Solution	40 lb.
Zinc Carbonate	1 lb.
Sodium Nitrite	4 lb.
Water	100 gal.

15 min. at 210° F.

No. 3

Zinc Dihydrogen Phosphate	40 lb.
Sodium Nitrite	4 lb.
Water	100 gal.

5 min. at 200° F.

Add nitrite continuously for better results as it decomposes rapidly.

Increasing the zinc content in the above solutions such as by adding zinc sulphate or nitrate improves the coating a great deal.

The final rinse may contain 7–21 oz./100 gal. of chromic acid, phosphoric acid, oxalic acid or salt of aluminum, chromium or iron and is operated at 150°–180° F.

Protecting Copper and Brass Against Corrosion

U. S. Patent 2,233,442

For brass: To 100 gal. of water add 26 lb. of magnesium dihydrogen phosphate and 4.5 lb. of sodium bromate. Immerse article at 210° F. for approximately 5 minutes or spray the solution on. The final rinse may contain 7–21 oz. of chromic acid, phosphoric acid, oxalic acid in 100 gal., or various metal salts, at a temperature of 150–180° F. for 1 minute.

For copper: To 100 gal. of water add 25.2 lb. of zinc dihydrogen phosphate and 9 lb. of sodium chlorate. Boil. The increase of the zinc content by adding 9 lb. of a soluble zinc salt improves the coating. The solution should test 2.2 free acid and 15.5 total acid. The operating conditions are the same as for brass.

Corrosion Proofing Iron and Copper
U. S. Patent 2,271,374-5

Three baths are used: first, the molten cleaning, stripping and coating bath; second, a neutralizing bath; and, third, a sealing and/or coloring finishing bath. Crystalline copper sulphate or manganese dioxide in amounts of 0.5%–10% can be added to the formulae below, preferably in amounts of 1%–2% as fluxing and stabilizing agents. Operation is at 600°–900° F.

After coating in the fused bath, the articles are rinsed and neutralized in a weak solution of a suitable acid or aqueous solution of iron sulphate, preferably at a temperature of about 160° F. The preferred bath contains about 4 oz./gal. of iron sulphate. The articles are then rinsed and may be darkened in a third bath containing a dye.

	Formula											
	No. 1	No. 2	No. 3	No. 4	No. 5	No. 6	No. 7	No. 8	No. 9	No. 10	No. 11	No. 12
Trisodium Phosphate	3	3	4	4	2	2	2	2	2	2	2	50
Potassium Nitrate	2	—	1	—	—	—	3	—	—	—	1	—
Sodium Hydroxide	—	2	1	2	4	2	—	3	1	3	3	50
Sodium Carbonate	—	—	—	1	—	—	—	—	—	1	—	—
Potassium Phosphate	—	—	—	—	2	2	—	—	—	—	—	—
Sodium Nitrate	—	—	—	—	—	2	—	—	3	—	—	—

Iron Pipe, Corrosion Protected
U. S. Patent 2,216,376

Immediately after drawing, quench in:

Linseed Oil	82.5
Oleic Acid	9.4
Trisodium Phosphate	3.0
Lauryl Alcohol	1.0
Triethanolamine	1.8

and dry.

Rustproofing Iron and Steel
French Patent 850,642

Trisodium Phosphate	1
Sodium Bicarbonate	2
Gelatin	2
Water	100

Immerse metal in above; withdraw and dry.

Inhibiting Corrosion of Magnesium Alloys

Electrochemical oxidation in baths containing the following protect magnesium alloys:

Potassium Dichromate	10%
Sodium Hydrogen Phosphate	5%

or

Sodium Hydroxide	5%
Sodium Carbonate	5%

The bath is run at 2 to 3 amp. per sq. dm., at 50° or 45–50°, for 60 or 30 minutes. The chromate bath is rapidly spent. Unless it is renewed, scaly, irregular, poorly adhering layers are formed.

Corrosion Prevention of Magnesium
U. S. Patent 2,268,331

A method of producing a non-smudging, adherent coating on a magnesium surface, comprises treating the surface in a hot aqueous solution containing an alkali metal borate and an alkali metal bicarbonate capable of providing available carbon dioxide in the coating solution.

Sodium Bicarbonate	4–12%
Borax	1– 8%

Time: 5–45 min.
Temp.: 200–212° F.

Lubricating, Rust Preventing and Bright Annealing in Manufacture of Nickel Alloy Stampings

Nickel alloys, which are die stamped (the dies being cooled by a soluble oil mix) and then annealed require a protective coating against water, as well as additional lubrication in the forming. Flexo Wax C when applied to the nickel alloy from a solvent solution, either by brushing, spraying or dipping, furnishes lubrication as well as a water resistant finish. It has been found that, when the nickel alloy is annealed, the Flexo Wax C has a reducing action on the oxides and gives a very bright anneal, e.g., a 2% solution of Flexo Wax C in carbon tetrachloride is sprayed on a 78% nickel alloy. The dies do not heat up as much as when only lubricated with a cutting oil and have a longer operating life. After the pieces are formed they are then subjected to a temperature of 1980° F. for the bright anneal.

Anti-Rust Solution

The addition of 0.025% sodium chrome glucosate to water, in contact with ferrous metals during mechanical operations, will prevent rusting.

Rust Inhibitor for Water Systems

Sodium Silicate (40° Bé)	30
Caustic Soda	3½
Trisodium Phosphate	½
Aluminum Oxide	¼

(Use ½ pt. to 4 gal.)

Tarnish and Corrosion Prevention

Ozokerite	20
Paraffin Wax	40
Diglycol Stearate	10

The waxes are dispersed in hot water in proportions of 7 parts of wax to 40 parts of water, giving an emulsion which is applied preferably by spraying, leaving a thin, transparent, adhesive and flexible film. Protection is obtained from tarnish or corrosion of ferrous as well as non-ferrous metal surfaces. For instance, the flat ground steel surfaces and the nickeled intaglio portion of candy molds are both protected effectively. The coating tolerates extremes of high room temperature and very low freezing temperature. A recent report indicates that waxy coatings on steel machinery and parts are much better than grease slushing because waxy coatings apparently seek out the spots particularly susceptible to corrosion or which have already begun to corrode. Greases do not always inhibit the electrochemical action resulting in corrosion, whereas waxes do. Where the coating must be removed, boiling water, hot water and soap, or solvents may be used.

Silver Tarnish Inhibitor

U. S. Patent 2,117,657

Silver articles are dipped in:

Stannous Chloride	1 g.
Hydrochloric Acid	7 cc.
Water, To make	1 l.

Precious Metal Casting Mold

U. S. Patent 2,229,946

Formula No. 1

Silica	40–80%
Plaster of Paris	15–60%
Ammonium Bromide	¼– 3%

U. S. Patent 2,121,969

No. 2

Magnesium Oxide, Fused	75
Calcium Hydroxide	25

Press into shape and heat to 820° C.

Core Binders

Emulsified oleoresinous compositions make good core binders. Linseed, cottonseed and castor oils are effective, usually prepared as a melt in rosin before being emulsified in water. For the usual requirements as to mechanical stability a core may be made up with 1.1% of emulsion binder, calculated on the weight of sand in the core. For stronger cores, to meet unusually severe requirements, it is necessary to use about 1.4% of binder. Since nearly half of the emulsion is oil, the calculated oil consumption is therefore 0.53% for ordinary binders and 0.68% for stronger binders, calculated on the weight of sand. Linseed oil makes stronger cores than do nondrying oils, but for economy in linseed oil it is quite possible to use cottonseed oil instead. In that case 1.6% of emulsion should be used in cores of ordinary strength, instead of 1.1% as in the case of linseed oil. A suitable recipe for a cottonseed oil core binder is: Blend of cottonseed oil and rosin, 100 parts; oleic acid, 2 parts; triethanolamine, 1.2 part. The oleoresin blend in this recipe is made from 71.2% cottonseed oil and 28.8% rosin. Another binder is made by blending 67.24 parts of castor oil with 32.76 parts of rosin and emulsifying the product in 100 parts of water with the aid of Turkey red oil 1, acidol 1 and sodium hydroxide 0.35 part.

Emulsion Core Binder

Castor Oil	67.24
Rosin	32.76
Turkey Red Oil	.50
Acidol	.50
Caustic Soda	.18
Water	100.00

Foundry Core Binder

U. S. Patent 2,215,825

113 pints of ammonium sulphate and 1100 pints of blackstrap molasses are dissolved in 275 pints of water and then 120 pints of a water-soluble glycerol-citric acid resin, an emulsifier comprising 265 pints of tung oil and 15 pints of karaya gum, and 15 pints of beta-naphthol as preservative are added. The core is made by mixing 30 pints of green-sand with 1 pint of binder as made above, molding and baking at 180–300°.

Foundry Core Application for Smooth Metal Castings

Non-ferrous metals can now be cast in smooth, fine cores at a considerable saving of time. This is easily accomplished by the addition of a small percentage of "Sulfatate," a new type of wetting agent. The ordinary core for casting non-ferrous metals is usually a combination of silica, sandclay binder and linseed oil. This type of core is very rough and necessitates a great deal of machining to finish the casting. If, after baking the core, it is dipped into a solution of 1 pt. molasses, 10 lb. graphite, 1 lb. "Sulfatate," and water sufficient to make 5 gal., and then rebaked, the core will present a smooth surface for casting, which will require little, if any, machining on the casting.

The addition of the Sulfatate eliminates the necessity of hanging the core so that the excess graphite-molasses mixture will drip off. The core, when it is withdrawn from the solution, breaks away cleanly. This not only produces a finer and truer core for casting, but also saves a good deal of time in preparation.

Refractory Dental Mold for Casting Alloys
U. S. Patent 2,243,094

Granular Zircon	100.0
Milled Zircon	50.0
Calcined Magnesia	1.5
Zirconium Oxychloride Solution (d. 1.38)	14.0

Foundry Sand Binder
U. S. Patent 2,206,369

Silica Flour	75
Milk, Powdered	15
Soya Bean Meal	10
Sodium Oxalate	1¼
Rosin	5
Borax	1
Mineral Oil	1

Quenching and Descaling Bath
British Patent 532,807

Orange Pectate Pulp	2 kg.
Water	100 l.
Disodium Phosphate	100 g.
Sodium Hydroxide	50 g.

Quenching for Steel Tempering

A mixture of glycerin and water is useful in range between water and oil quenching. The higher the percentage of glycerin the milder the quenching.

Mild Steel Quenching Bath
U. S. Patent 2,216,192

Kerosene	5/6 gal.
Tallow, Molten	1/6 gal.
Salt	2.0 lb.
Disodium Phosphate	0.6 lb.
Sodium Bicarbonate	0.6 lb.
Tartaric Acid	0.3 oz.

Use from 820°.

Tempering Small Tools

For twist drills, taps, dies, small punches, or such articles of tool steel that you wish to keep straight, take as follows: Equal parts of prussiate of potash and common salt, put them together in an iron pot over fire. When it gets to proper temperature it will boil and become a cherry red.

Put the tool in this until it becomes a cherry red. You may leave the tool in all day if you wish, for the longer the more it improves the steel. When you take it out, cool in water or linseed oil, always in a vertical position. Do not draw. But for taps or dies draw to dark straw.

Tempering Cold Chisels

At no time when working steel let it sparkle; just get a nice red heat. Better heat it oftener than get it too hot. Make the chisel the width of the stock and don't let it swell out at the bit. If you do, someone will cut on the outside corner and snap off the corner. It has no back support there.

If in forging, you should get the bit too hot, cut it off as far as it is too hot, and do not fool with it, as it is no good. The higher grade steel you use, the lower you should heat it when tempering. Heat the steel at low red heat and heat at least one-third of the chisel and dip it in cold water.

Try it and if it should not harden very hard, heat it again a little hotter than the first time until it does harden. This way gets the chisel hardened at the lowest heat. The tool should be plunged from 1 to 1½ inches in the water. This will give plenty of time to rub the chisel and the temper will come down slowly and not hurt you. The temper to be right for general purpose service should be purple at the point. There is only one man out of ten who knows how to grind a chisel after he has bought a good one. They should be ground with a very short bevel. The best chisel that can be made won't stand anything that is ground with a long taper. There is no body or support for cutting edge.

Tempering Bath for Files
French Patent 847,979

Barium Chloride	50
Sodium Chloride	20
Potassium Chloride	30
Sodium Cyanide	4–8

Hardening the Thin-Edge Tools

First, the warping and springing of steel is mostly due to the heat, and not as many suppose, to the cooling bath. This can readily be demonstrated by laying a long piece of steel on a fire, where the heat will strike it unevenly, and taking note of the changes by sighting over its top. Thus the first important point is to have the heat strike evenly on all sides of the piece.

In hardening such a piece as an axe (and this applies to all thin-edge tools of the same nature), raise it to the proper heat very slowly, and with as little blast as possible, then plunge it into the quenching bath straight down and head first; in other words, the thin or cutting edge should be immersed last. After cooling, immediately hold it over the flame of the fire until the water rolls up and runs around the surface. Tools treated in this way will never crack.

A very good bath for delicate work is hot water. This, in a great many cases, gives the right degree of hardness without having any draw down afterwards.

To temper knives, or any thin tools, two flat pieces of metal should be used. Lay one plate down flat and spread enough oil on it to envelop the piece to be hardened. After that give the piece its proper heat, quickly lay it on the plate and cover immediately with the second flat piece, or plate. If properly done the work will come out straight.

Metal Cleaning and Annealing Bath
U. S. Patent 2,276,101

Borax	90
Fluorspar	10

Steel Tempering Bath

Molten lead is covered with a 20-30 mm. layer of

Formula No. 1

Sodium Chloride	50
Potassium Chloride	50

No. 2

Sodium Chloride	50
Calcium Chloride	50

Such protective layers protect tempered objects from adherence of lead and oxides.

Steel Hardening Bath
British Patent 527,774

Sodium Cyanide	5
Soda Ash	67
Sodium Chloride	7
Potassium Chloride	7
Barium Chloride	14

Use at 750°–850°.

Calorizing Ferrous Metals
U. S. Patent 2,279,268

Zinc Oleate	21.4
Toluol	78.6

To 48.6 parts of the above solution add 51.4 granular aluminum.

Paint metal with this; dry and heat to 600–700° C.

Carburizing Compound for Case Hardening of Steel
U. S. Patent 2,260,622

Wood Charcoal, Powdered	45
Barium Carbonate	5
Sodium Silicate (d. 1.4)	38
Clay	12

Metal Treating Fused Salt Bath
German Patent 700,904
Formula No. 1

Potassium Dichromate	82
Potassium Nitrate	18

No. 2

Potassium Dichromate	63
Potassium Chromate	37

No. 3

Sodium Dichromate	42
Potassium Dichromate	58

These baths are used at 500–700° C.

Iron Heat Treating Bath
German Patent 700,904

Sodium Dichromate	750
Potassium Dichromate	250
Potassium Chromate	100

Nitrates may be added.
Use at 500°–700° C.

Tempering Gold and Silver Alloys
U. S. Patent 2,127,676

Alloys of gold 51 to 91%, copper 40 to 8%, and silver 9 to 1% are hardened by cooling to not more than 0°, immersing in molten tallow at 193° for not less than 20 seconds and then in boiling soap solution for not less than 20 seconds, drying, allowing to cool to room temperature and repeating the series of processes.

Soldering Fluxes
Formula No. 1

Lactic Acid	15 cc.
Sulphate (Wetting Agent)	0.2 cc.
Water	85 cc.

No. 2
British Patent 368,841

Rosin	20
Aniline Hydrochloride	1
Alcohol, Denatured	79

No. 3

Rosin	20
Lactic Acid	5
Alcohol, Denatured	75

No. 4

Rosin	4
Mannitol	1

No. 5

Rosin	4
Mannitol	1
Lactic Acid	½–1

No. 6

Rosin	48
Mannitol	12

Lactic Acid	10
Alcohol, Denatured	30

No. 7

British Patent 524,572

Colophony 50, soft petrolatum 45, and silicon dioxide (240 mesh) 5, or of aniline hydrochloride 10, ethylene glycol 75, and carborundum (180 mesh) 15%.

Hard Metal Soldering Flux

U. S. Patent 2,267,763

Boric Acid	55–71%
Sodium Fluoride	45–29%

Non-Acid Flux

Petrolatum	65
Ammonium Chloride	3½
Zinc Chloride	25
Water	6½

Soldering Fluids

In making up soldering fluids glycerin is a useful ingredient to give the desired body to the solution and assure better wetting properties. Following are two easily prepared soldering fluids which utilize these glycerin properties. The first, which is satisfactory for use on a variety of metals including copper, brass, steel, terne plate, tinned steel, and Monel metal, consists of:

Formula No. 1

Zinc Chloride	15
Glycerin	25
Water	60

Zinc chloride can be eliminated by using the following soldering fluid:

No. 2

Lactic Acid	1
Glycerin	1
Water	8

No. 3

Belro Rosin	50
Methanol	50

Aluminum Solder

U. S. Patent 2,224,356

Formula No. 1

A fluxed solder for soldering aluminum to aluminum or other metals which will take a coating of tin is made by melting ¼ ounce of zinc, adding ¾ pound of lead, melting (with stirring), and after skimming, adding ¾ pound of tin, which is stirred in; when the mixture is molten, ½ teaspoonful of powdered rosin is added. The rosin ignites and when the flame subsides the metal is skimmed and poured.

No. 2

U. S. Patent 2,243,278

Tin	60
Cadmium	20
Zinc	15
Copper	2½
Silver	2½

Filling Alloy for Aluminum Soft Soldering

French Patent 848,862

Tin	10–60%
Lead	15–45%
Zinc	20–70%
Silver	1–10%

Solder for Zinc Aluminum Alloys

German Patent 695,626

Tin	74
Lead	20
Zinc	6

Woven Wire Fabric Solder

U. S. Patent 2,224,952

Mix together the following metallic powders:

Copper	15
Silver	55
Cadmium	30

Apply to wires to be joined and heat.

Indium Containing Solders

	Formula No. 1	No. 2
Lead	96	95
Silver	3	3
Indium	1	2

Tin-Free Solder
U. S. Patent 2,293,602

Formula No. 1

Silver	1.5–3%
Antimony	0.5–5%
Lead, To make	100%

No. 2

Lead	90
Cadmium	10

No. 3

Lead	90.5
Cadmium	8.0
Zinc	1.5

Soldering Cast Iron

Clean the cast iron perfectly clean and bright with a coarse file. Use ordinary acid for a flux and tin the cast with pure block tin. It can then be soldered with tinner's solder.

Soldering Metal to Ceramics
British Patent 542,101

The roughened ceramic body is sprayed with an alloy of:

Lead	85
Antimony	10
Tin	5

Bake at 160° C. and dip in solder (50–50) for 5 sec. This surface can be soldered to metals in the usual way.

Brazing and Welding Flux
U. S. Patent 2,229,863

Potassium Carbonate	1.5
Ferrous Sulphate	1.0
Borax	20.0

Welding Flux
U. S. Patent 2,284,619

Borax	36
Sodium Bifluoride	23
Sodium Silicofluoride	19
Antimony Fluoride	1
Phosphoric Acid	13
Trisodium Phosphate	8

Electric Arc Welding Flux
German Patent 681,203

Sodium Hydroxide	4
Bentonite	1
Soda Ash	2
Titanium Dioxide	18
Feldspar	53
Water	22

Aluminum Flux
British Patent 542,490

Formula No. 1

Lithium Chloride	18
Zinc Chloride	10
Stearic Acid	4
Potassium Chloride	30
Aluminum Acetate	8

No. 2

Sodium Chloride	60
Calcium Fluoride	20
Zinc Chloride	10
Lithium Chloride	2
Aluminum Acetate	6
Sulphur	2

Aluminum Welding Flux
U. S. Patent 2,243,424

Formula No. 1

Cadmium Chloride	10.0
Cuprous Chloride	29.1
Cerium Chloride	9.0
Potassium Chloride	34.2
Potassium Bifluoride	6.6
Sodium Chloride	6.6
Lithium Chloride	4.5

No. 2

U. S. Patent 2,239,018

Lithium Chloride	10
Sodium Chloride	35
Potassium Chloride	35
Potassium Fluoride	5
Sodium Fluoride	10
Cryolite	5
Cadmium Chloride	20

No. 3

Aluminum tubing is made from 0.5 m.m. strip and filled with:

Potassium Chloride	45
Sodium Chloride	30
Lithium Chloride	15
Lithium Fluoride	3½
Sodium Fluoride	3½
Potassium Pyrosulphate	3

This filled tube is inserted in another aluminum tube 8 m.m. outside and 4 m.m. inside diameter.

Welding Flux for Cast Iron

Iron	14.5
Iron Sulphide	12.5
Sodium Chloride	10.0
Sodium Nitrate	63.0

Mix together and make into a paste with mineral oil.

Flux for Preventing Oxidation of Molten Magnesium Alloys

British Patent 539,023

Magnesium Chloride	34–40%
Calcium Chloride	12–16
Sodium Chloride	15
Potassium Chloride	15
Magnesium Oxide	11–15
Calcium Fluoride	19–23

Aluminum Soldering and Welding Flux

British Patent 513,695

Potassium Fluoride	30–40%
Potassium Sulphate	2–5 %
Potassium Chloride	20–30%
Sodium Chloride	4–8 %
Ammonium Chloride	10–18%
Stannous Chloride	3–8 %
Lithium Chloride	3–8 %

Soldering Flux for Copper

U. S. Patent 2,228,352

Triethanolamine	35
Glycol	40
Abietic Acid	25

Welding Rods for Red Brass

For gas welding:

Copper	97.5
Silicon	1.5
Zinc	1.0

For carbon arc welding:

Copper	97
Silicon	3

Welding Rod for Copper Alloys

U. S. Patent 2,121,194

Nickel	35.0
Zinc	15.0
Copper	48.6
Silicon Copper (15%)	1.7
Borax	1.0

Sinter in hydrogen at 700° C.

Bronze Alloy for Welding to Iron

U. S. Patent 2,129,197

Copper	80.0
Tin	17.3
Nickel	1.5
Silicon	0.5
Iron	0.5

Soldering Iron Tip Alloy

Dutch Patent 51,352

Zirconium	0.1–1.3%
Copper, To make	100%

Steel Brazing Flux

Powdered borax 1 ounce, boracic acid 3 ounces. Melt the borax and when cool powder it. Mix with acid and apply as a paste with water. Use regular brazing spelter.

Brazing Alloy
U. S. Patent 2,125,228

An alloy for brazing brass, copper, steel, or cast iron contains copper 68–84, cadmium 3–24, silver not greater than 3.9, and phosphorus 2–4%, with or without tin not greater than 10 and nickel not greater than 2%.

Steel Welding Rod Coating
U. S. Patent 2,220,954

Potassium Feldspar	8
Wood Fiber	3
Calcium Carbonate	4
Clay	4
Ferro-Manganese	6
Sodium Silicate	13
Titanium Dioxide	30
Talc	6
Asbestos	2

Stainless Steel Welding Electrode Coating

Marble	37
Dolomite	10
Fluorspar	32
Caustic Soda	4½
Ferro Manganese (75%)	5½
Ferro Silicon (75%)	2
Ferro Titanium (20%)	5½
Starch	3½
Water Glass	24

Apply about 15% by weight to electrode; dry and heat at 180–200° C. to form a hard layer.

Welding Rod Coatings
U. S. Patent 2,240,033

	Formula No. 1 Soft Coating	No. 2 Hard Coating
Plumbago	3000	3000
Sodium Silicate	2600	1800
Bentonite	90	90
Water	950	1290

Dissolving Welding Scale from Stainless Steel

This discoloration is removed by grinding and polishing. However, there are some cases where the scale is dissolved with a mixture of 20% nitric acid with 4% hydrofluoric acid, applying the mixture at a temperature of about 125° F. to the welded seam.

Successive applications at intervals of approximately five minutes, with a small hand brush or equivalent, which would permit some scrubbing action removes the scale satisfactorily for soldering operations.

Non-Tarnishing Jewelers Solder
U. S. Patent 2,270,594

Palladium	15–35%
Gold	10–25%
Copper	7–20%
Cadmium	5–15%
Silver, To make	100%

Solder for Precious Metals
U. S. Patent 2,220,961

Lead	54–48%
Tin	43–39%
* Precious Metal	3–13%

Gold Solder
U. S. Patent 2,274,863

Gold	25 –85%
Palladium	2 –20%
Copper	5 –30%
Cadmium	3 –25%
Silver	½–20%
Zinc	1/10–5 %

Electrical Contact Alloys
U. S. Patent 2,221,285

Formula No. 1

Magnesium	2–4%

* Gold, silver or platinum depending on which metal is being soldered.

Lithium	0.05%
Silver, To make	100%

No. 2

Boron	0.01–10%
Silver, To make	100%

Electrical Contact and Welding Tip Alloy
U. S. Patent

Chromium	3
Iron	1½
Cadmium	1
Copper	94½

Electric Fuse Alloy
U. S. Patent 2,293,762

Tin	40–60%
Cadmium	20–35%
Silver	15–25%
Copper	2–5 %

Sea Water Resistant Aluminum Alloy
U. S. Patent 2,261,210

Magnesium	5.5 –6 %
Zinc	4 %
Manganese	0.25–0.5%
Chromium	0.25–0.3%
Aluminum, To make	100%

Age Hardening Copper Alloy
U. S. Patent 2,261,975

Zinc	12 –30 %
Nickel	0.25– 3 %
Phosphorus	0.02– 0.4%
Copper, To make	100%

Precipitation Hardened Copper Alloy
U. S. Patent 2,225,339

Copper containing beryllium 0.1% and chromium 0.4% is quenched from 900°, reheated at 475–500°, cold swaged and again aged at 100° for 90 minutes.

Hot Working Copper Alloy
U. S. Patent 2,245,327

Arsenic	0.5–5 %
Phosphorus	4.5–8.5%
Copper, To make	100%

Amethyst Colored Gold Alloy
U. S. Patent 2,278,812

Gold	70.6–78.1%
Aluminum	19.4–22.4%
Zinc	0.5–10 %

Non-Tarnish Dip for Gold Alloy Plates
Solution A

Sodium Ortho Phosphate	2 oz.
Sodium Carbonate	2 oz.
Potassium Chromate	1 oz.
Water	to 1 gal.

Solution B

Potassium Chromate	8 oz.
Potassium Nitrate	8 oz.
Water	to 1 gal.

Dip plated ware in cold solution A. Rinse in running water and then soak in solution B for 2–3 minutes. Rinse in cold and hot water and dry in sawdust.

Lead Shot Alloy

Antimony	1 –1.5 %
Sodium	0.02–0.04%
Lead, To make	100%

Cast at 400–420° C.

Porous Metal Bearing Alloy
U. S. Patent 2,214,104

Iron, Powdered	78
Lead, Powdered	20
Graphite, Powdered	2
Stearic Acid	1/100

Press at 20,000–60,000 lb./sq.in. and sinter at 1095° C.

Improved Tin Foil Alloy

Tin	100.00
Zinc	8.50
Nickel	0.15

This gives increased strength and ductility.

Bell Casting Alloy
German Patent 698,659

Zinc	90–95%
Copper	3%
Aluminum	4%

Bearing Alloy
U. S. Patent 2,245,459
Formula No. 1

Zinc	40%
Lead	0.1 – 3%
Phosphorus	0.01– 3%
Copper, To make	100%

No. 2
U. S. Patent 2,246,067

Lead	1– 3 lb.
Copper	3 lb.
Silver	0.05–10%

Carbonized Bearing Composition
U. S. Patent 2,131,021

Petroleum Coke	19
Sulphur	7– 8
Lead Oxide	19
Pitch	47–48

Bake mixture at 1000°, until carbonization is complete and metal is reduced from oxide.

Non-Sticking Coating for Metal Sheets
U. S. Patent 2,132,557

Mineral Oil, Light	99
Magnesium Stearate	1

Warm and mix until dissolved.

Permanent Magnet
German Patent 701,528
Formula No. 1

Nickel 15–25%, iron 10–25% and copper 50–75% is formed in the desired shape. The magnet is then heated to over 1000°, quenched, and rolled out over 40%. The magnet is finished by annealing it at 500–700°. By this treatment the coercivity and remanence are increased without impairing any other desirable properties.

No. 2
British Patent 541,474

Aluminum	8–9.25%
Titanium	2 %
Nickel	25 %
Iron, To make	100%

Cast and then forge at 1100-1360°; heat treat for 10 min. at 1320°; quench in hot oil and age harden at 566° for 16 hr.

Permanent Magnet Casting
U. S. Patent 2,121,799

Ferric Oxide	30–50
Aluminum	10–20
Nickel	5–15

Mix the above powders in a graphite crucible and ignite with a small electric arc. Rock crucible and twirl it so that slag freezes on sides. Finally pour into a section carbon mold.

Tinfoil Substitute in Curing Dental Plates

Two solutions are required. The first is prepared by dissolving 5 or 6 grams of a soluble alginate, such as potassium, sodium, magnesium, or ammonium alginate, in 80 ml. of water. This may require several hours of constant stirring. When solution is complete add 80 ml. of glycerin. This first solution can be applied by dipping, spraying, or painting. After the film has dried, a second solution is applied.

The second solution is a saturated solution of calcium chloride.

The halves of the flask, after the wax has been boiled out, are dipped in the first solution and allowed to drain and dry for a few minutes. A second dipping may be necessary if the solution is too thin or if the plaster does not retain a sufficient amount of the solution to form a continuous film. The halves are next dipped in the chloride solution for 5 or 10 minutes and then dried. The film can be removed from the teeth with a sharp needle-hook instrument similar to an explorer.

The flask is now ready for trial packing and with a moderate amount of care will retain the film until the cure is completed in boiling water.

The substitute has advantages over the foil in that it is easier to apply, takes less time at the bench, gives a smoother surface, and yields easier separations of the halves of the flask.

Dental Alloy
U. S. Patent 2,246,288

Chromium	10 –49%
Cobalt	20 –48%
Molybdenum	3 –15%
Nickel	10 –40%
Tungsten	3 –12%
Zirconium	¼– 3%

Denture Casting Composition
U. S. Patent 2,206,502

A compressed mixture of the following powdered materials is used:

Chromium	31.00
Cobalt	61.20
Iron	7.00
Tungsten	0.50
Silicon	0.45
Aluminum	0.20
Borax	0.50

Cutting Steel When Cold

Cutting steel cold is a satisfactory method, when bar steel is to be cut or broken into certain lengths, as when making cold chisels or other similar tools, but the advantage of this method will cease when cutting steel over a certain size. For example, octagon, round or square cast steel ranging from the smallest size up to 1¼ inches in diameter, can be broken very quickly and satisfactorily when perfectly cold, by nicking the bar equally from all sides, afterwards placing the nicked part of the bar directly over the square hole of the anvil, then striking it with a sledge when it will break.

Care must be exercised when breaking steel after this method, as the pieces are very apt to fly and strike the blacksmith or his helper. To overcome this danger place the handle of the chisel on the piece which is to be broken off before striking it with the sledge, which will prevent the piece from flying. When nicking the steel, hold the chisel so as to cut in a straight line and so enable the steel to break off square on the ends.

To enable the steel to break with greater ease pour a little cold water directly on the nicked part of the steel. By pouring cold water on the steel all the heat is taken out, as steel will break more readily when perfectly cold than when it is slightly warm.

Breaking a bar of steel cold is a very good way of finding out the hardness or the quality of the steel. For example take a bar of ¾ or ⅞-inch steel (after being nicked). If the steel breaks with one or two blows from the sledge it denotes ard steel, but soft steel will re-uire five or six blows before it reaks; also hard steel (by looking t the break) will show a fine and lose fracture. The fracture of soft teel will be more coarse and rough. f the steel is of good quality, the reak or fracture will show a very niform and silvery white appear-nce clear through the bar, but if he steel is of poor quality it will how a dull brown appearance.

Hot Cutting: To test steel bars hat are too large to break cold, for xample a bar two inches in diame-er, heat the bar to a deep cherry ed, then cut in from all sides, say alf an inch deep with a hot chisel, hen lay the bar down to cool and vhen it is perfectly cold it may be roken by striking it with a sledge r dropping the bar over the anvil, nd the quality or hardness can be udged as formerly explained. Bear n mind that the steel must not be eated above a deep cherry red (in rder to cut in the nick) or the frac-ure when broken cannot be judged correct, as a high heat in the steel would materially change the ap-pearance and form of the fracture.

Copper Scale Remover
U. S. Patent 2,211,400

Hydrochloric Acid	10 %
Sodium Chlorate	½%
Ferric Chloride	10 %
Water, To make	100 %

Removing Brittle Surface Austenitic Steel
U. S. Patent 2,238,778

Pickle at 82° in:

Nitric Acid (d 1.42)	20
Stannous Chloride	8
Water	72

Passivation of Stainless Steel
U. S. Patent 2,104,667

Dip into following solution:

Nitric Acid	0.1–4%
Sodium Dichromate	0.1–4%

Recovering Gold from Scrap Solution

Precipitate gold with zinc or lead-zinc couple. Filter precipitate. Boil in aqua regia. Boil off nitrogen peroxide. Add ferrous sulphate to precipitate gold. Filter and wash precipitate. Dissolve gold in aqua regia. Remove nitrogen peroxide. Add oxalic acid or sodium bisul-phite. Precipitate is pure gold. Treat as for preparation of stock solution.

Reclaiming Type Metal from Dross
Dutch Patent 50,737

Dross is fused with 1–30% zinc chloride below 600°.

Briquetting Metal Shavings
Russian Patent 55,173

Metal shavings are mixed with 5–10% powdered chalk, heated to 200° C. and treated with 80–90% coal tar pitch and 10–20% anthra-cene oil and pressed.

Porous Metal
Swiss Patent 216,467

Iron wire (0.1–0.4 m.m. diameter) fragments are mixed with 5% cop-per powder and 5% sodium silicate (25–30° Bé). Compress at 200 kg./sq.cm. and gradually heat to 880° C.

Chapter Thirteen

PAINT, VARNISH, LACQUER AND OTHER COATINGS

TRAFFIC PAINTS

Light Green

Chromium Green Oxide	58 lb.
Magnesium Silicate	400 lb.
Vehicle "A" or Vehicle "B"	521 lb.

Yellow

C. P. Lemon Yellow Oxide	200 lb.
Magnesium Silicate	406 lb.
Vehicle "A" or Vehicle "B"	596 lb.

Red Oxide

Red Oxide of Iron C. P.	200 lb.
Magnesium Silicate	390 lb.
Vehicle "A" or Vehicle "B"	596 lb.

Burnt Sienna

Burnt Sienna	200 lb.
Asbestine	368 lb.
Vehicle "A" or Vehicle "B"	596 lb.

Brown

Burnt Umber	200 lb.
Magnesium Silicate	368 lb.
Vehicle "A" or Vehicle "B"	596 lb.

Black

Carbon Black	49 lb.
Magnesium Silicate	480 lb.
Vehicle "A" or Vehicle "B"	581 lb.

The above traffic paints are all of heavy brushing consistency—they will require reduction for most applications.

Red Oxide

Iron Oxide Red	200 lb.
Dicalite L	100 lb.
Magnesium Silicate (325 Mesh)	200 lb.
50% Solution Rosin in Petroleum Spirits	45 gal.
Petroleum Spirits	2 gal.

Paste ground on stone or roller mill.

Thin with:

Petroleum Spirits	27 gal.
Zinol	6 gal.

White

Titanox B	250 lb.
Dicalite L	100 lb.
Magnesium Silicate (325 Mesh)	150 lb.
50% Solution Rosin in Petroleum Spirits	45 gal.
Petroleum Spirits	3 gal.

Paste ground on stone or roller mill.

Thin with:

Petroleum Spirits	27 gal.
Zinol	6 gal.

Black

Bone Black	125 lb.
Dicalite L	100 lb.
Magnesium Silicate (325 Mesh)	100 lb.
50% Solution Rosin in Petroleum Spirits	45 gal.
Petroleum Spirits	5 gal.

Paste ground on stone or roller mill.

Thin with:

Petroleum Spirits	29 gal.
Zinol	5 gal.

The balance of the colors can be made in the same manner.

Another base for traffic paint is casein. The casein paint is not as durable as the cold cut rosin paint and is not good in damp or wet weather, but is inexpensive and can be used for temporary traffic paint.

Black

Casein	10%
Lime	15%
Iron Black	40%
Limestone	20%
Talc	15%

Mix well in Day spiral dry mixer.

White

Casein	10%
Lime	15%
Calcium Carbonate (Precipitated)	15%
Limestone Whiting	30%
Talc	30%

Mix well in Day spiral dry mixer.

Color Base

Casein	10%
Lime	15%
Talc	15%
Calcium Carbonate (Precipitated)	15%
White Clay	15%
Limestone Whiting	30%

Replace 13% precipitated calcium carbonate in the base with ultramarine blue for the blue traffic paint.

Replace 12½% precipitated calcium carbonate in the base with 11½% red iron oxide and 1% lemon yellow ferrite for the red traffic paint.

Replace 5% precipitated calcium carbonate in the base with 2% lemon yellow ferrite and 3% golden cadmium yellow for yellow traffic paint.

Replace 15% precipitated calcium carbonate in the base with chromic oxide green for the green traffic paint.

To reduce foaming in the above formulae replace 0.5% limestone whiting with 0.5% pine oil.

To prevent mildew replace 1% of limestone whiting with 1% fungicide such as chlorinated phenol.

The above casein powder paints can be mixed with water to spraying or brushing consistency and can be inter-blended either dry or wet to get colors in between the nine basic colors.

Yellow—Burnt Umber—Burnt Sienna—Red Iron Oxide Black

Pigment Composition

Color	40 %
Asbestine	60 %

Paint Composition

Pigment	53.50%
Vehicle	35.00%
Drier and Thinner	11.50%

Formula

Color	261
Asbestine	392
Vehicle	425
V. M. P. Naphtha	142

Chrome Oxide Green

Pigment Composition

Color	53.84%
Asbestine	46.16%

Paint Composition

Pigment	53.41%
Vehicle	34.99%
Drier and Thinner	11.60%

Formula

Chrome Oxide Green	350
Asbestine	300

Vehicle	425
Thinner	142

White

Pigment Composition

Titanated Lithopone	80 %
Asbestine	20 %

Paint Composition

Pigment	65 %
Vehicle	32.50%
Thinner and Drier	25 %

Formula

Titanated Lithopone	520
Asbestine	130
Vehicle	325
Thinner	25

Ramapo Green and Blue

Pigment Composition

Color	4.00%
Lithopone	2.80%
Asbestine	93.20%

Paint Composition

Pigment	49.70%
Vehicle	39.76%
Thinner and Drier	10.63%

Formula

Color	20
Lithopone	14
Asbestine	466
Vehicle	400
Thinner and Drier	106

Vehicle No. 1

F7 Amberol (or equal)	100 lb.
Q Linseed Oil	15 gal.
Mineral Spirits	28 gal.

Vehicle No. 2

Rosin	100 lb.
Lime Flour	6 lb.
Q Linseed Oil	5 gal.
Mineral Spirits	17 gal.

A. Manila Vehicle

Manila Resin	80 lb.
Plasticizer	20 lb.
Solvent	100 lb.

The best plasticizer is raw or blown castor oil. Fish oil or soya bean oil may be used as alternate plasticizers—they do not yield as good results as obtained with castor oil.

The solvent may consist of denatured alcohol or any other alcohol or Ketone of acceptable volatility.

B. Batu or Pale East India Resin

Batu or Pale East India Resin	75 lb.
Plasticizer	25 lb.
Solvent	100 lb.

The best plasticizers for use with these resins are dehydrated castor oil, fish oil, soya bean oil, and linseed oil.

The solvent may consist of high solvency petroleum naphthas, coal tar naphthas or mixtures of these strong solvents with ordinary low solvency petroleum naphthas.

Pigmentation

Best results are obtained with a pigment volume of 40% to 45%. Hiding pigment is incorporated in sufficient quantity to yield the required opacity at a reasonable spreading rate. With this feature taken care of, optimum wearing qualities, appearance, etc., are obtained by the addition of Asbestine to the required pigment volume. Paints made with color pigment only were not equal to color-inert combinations in durability.

White

(1)

Titanium-Barium Pigment	73
Magnesium Silicate	43
Vehicle "A" or "B"	100

(2)

Titanated Lithopone	49.5
Zinc Oxide	29.7
Mica	9.9
Magnesium Silicate	9.9
Vehicle "B"	100.0

Dark Green

Monastral Green	75 lb.
Magnesium Silicate	318 lb.
Vehicle "A" or Vehicle "B"	477 lb.

Blue

Monastral Blue	45 lb.
Magnesium Silicate	357 lb.
Vehicle "A" or Vehicle "B"	519 lb.

Traffic Paints

Spirit Type—Manila Type

Manila Resin	80
Blown Castor Oil	20
Solvent	100

The solvent may consist of any of the following solvents used either alone or in combination with each other:

Acetone
Methyl Alcohol
Denatured Ethyl Alcohol
Isopropyl Alcohol
Butyl Alcohol

White

(A)	Titanium-Barium Pigment	70
	Asbestine	30
(B)	Titanium-Barium Pigment	80
	Asbestine	20
(C)	Titanium-Barium Pigment	70
	Barytes or Blanc Fixe	30
(D)	Titanium Dioxide	16
	Barytes	36
	Micaceous Talc	48
(E)	Titanium Dioxide	16
	Barytes	36
	Asbestine	48

The above pigment-inert compositions are on a weight basis. The pigment volume content recommended is 40% for all pigment combinations listed above.

Yellow

(A)	Chrome Yellow	50
	Asbestine	50
(B)	Chrome Yellow	50
	Asbestine	25
	Barytes	25
(C)	Chrome Yellow	50
	Blanc Fixe	50

The above pigment compositions are on a weight basis. The pigment volume content recommended is 35% for the pigment combinations above.

The Manila resin is dissolved by adding it to the solvent which is agitated by means of a high-speed stirrer. Stirring is continued until solution is complete.

The blown castor oil is added to the resin solution and agitation of the solution continued until the mixture is homogeneous.

It is entirely dependent upon the grade of Manila resin used as to whether or not the resin will have to be ground. The resin should be about pea size before it is added to the solvent. Allow to settle or filter solution.

Sufficient vehicle is used with the pigment to make a grinding paste of a suitable consistency for ball or pebble mill grinding. After the pigment paste has been ground sufficiently, the remainder of the vehicle is added to the mill and the mill rotated for a short period to incorporate the materials. The finished paint is withdrawn from the mill.

Spirit Type—Batu Type

Batu Scraped	30
Kettle Bodied Fish Oil (Heavy)	10
Solvent	60

The solvent may consist of high solvency petroleum naphtha or mixtures of high solvency petroleum naphtha and low solvency petroleum naphtha.

White

(A)	Titanated Lithopone	50
	Zinc Oxide	30
	Mica	10
	Asbestine	10
(B)	Titanium Barium Pigment	65
	Zinc Oxide	10
	Asbestine	25

The above pigment combinations are on a weight basis. The pigment-volume content is 40% in all cases.

Yellow

(A)	Chrome Yellow	50
	Asbestine	50
(B)	Chrome Yellow	50
	Blanc Fixe	50
(C)	Yellow Lead Chromate	55
	Zinc Oxide	25
	Asbestine	15
	Diatomaceous Earth	5

The above pigmentations are on a weight basis and a pigment-volume content of 35% is recommended.

The Batu resin is ground to pea size and is slowly added to the solvent which is being agitated by means of a high speed stirrer. Stirring is continued until solution is complete and then the fish oil added, the stirring continued unt the solution is homogeneous. Allo the solution to stand until any in soluble matter settles out or filt the solution.

Grinding Procedure

Same procedure as for the Manila Spirit Type Traffic Paint. Use ball mill.

Cooked Type Traffic Paints

Batu Scraped or Pale East India Resin (Nubs)	100
Kettle Bodied Linseed Oil or Dehydrated Castor V. M. & P. Naphtha to 55% non-volatile content Driers equivalent to 0.5% Pb, and 0.02% Co as metals on the weight of oil	80

Vehicle Preparation:

Heat the resin and 50 of oil t 620° F. in about 1–1¼ hours. Hol until a cold bead is clear. Add th remainder of the linseed oil.

Reheat to 580° F. only if a col bead, taken after the oil addition is not clear. Cool to 250° F. an thin to 55% solids with V. M. & P Naphtha. Add the driers.

White

(A)	Titanium Barium Pigment	56
	Zinc Oxide	24
	Diatomaceous Earth	20
(B)	Titanium Barium Pigment	56
	Zinc Oxide	24
	Asbestine	20
(C)	Titanium Barium Pigment	65
	Zinc Oxide	10
	Asbestine	25

(D)	Titanated Lithopone	50
	Zinc Oxide	30
	Mica	10
	Asbestine	10

The above pigment compositions re on a weight basis. The pigment-olume content recommended is)%–44%.

Yellow

(A)	Chrome Yellow	40
	Zinc Oxide	20
	Asbestine	20
	Lithopone	20
(B)	Chrome Yellow	45
	Zinc Oxide	20
	Asbestine	15
	Diatomaceous Earth	20

The above pigment compositions re on a weight basis. The recom-iended pigment volume is 40%.

Grind the pigment with sufficient ehicle to give a good grinding aste, in a ball mill. When suffi-iently ground, add the remainder f the vehicle and continue grind-ng to incorporate the paste and ehicle thoroughly. The finished aint is drawn from the mill.

White

Paint:

Pigment	62%
Vehicle	48%

Pigment Composition:

Lithopone or Titanated Lithopone or Titanium-barium pigment	55%
Zinc Oxide, Lead-free, American process	25%
Paint Grade Mica	10%
Siliceous Inerts *	10%

* To improve night visibility, the required amount of No. 1 or No. 1½ pumice may be substituted. This is added by stirring it into the finish paint.

Vehicle Composition:

Zinc resinate-linseed oil cold-cut varnishes *	100%
Nonvolatile in vehicle	40% minimum

White

Pigment (66%)

Albalith–351	500	lb.
† Pigment–725	400	lb.
Magnesium Silicate	100	lb.

Vehicle (34%)

‡ Modified Phenolic Varnish	43.75 gal.
Mineral Spirits	27.75 gal.

Prepare a paste using all the pigment, varnish and one half the mineral spirits. Mill through a roller mill or buhrstone mill set loose. Thin with remaining mineral spirits.

* The varnish shall be made of the following ingredients in the proportions indicated:

Z-Body Linseed Oil	22.0
Zinc Resinate **	22.0
V. M. & P. Naphtha	53.8
Lead Naphthenate (24% Pb)	1.9
Cobalt Naphthenate (6% Co)	0.3

** No preliminary solution (cutting) of the resin in solvent is necessary. Usual procedure in preparing the paint is to load the crushed resin as received into a pebble mill along with the other ingredients.

† A combination of 75%—XX Zinc Oxide (American Process)

A combination of 25%—White, Water Ground Mica (Paint Grade)

‡ A 15-gallon modified Phenolic (Amberol F-7 or equal) varnish with 60% Chinawood oil and 40% linseed oil.

Volatile 30%
Acid No. 4.0
Specific Gravity 0.91
Weight per Gallon 7.58 lbs.
Viscosity K

White

Pigment (66%)

Cryptone BT–301	500	lb.
Pigment–725	400	lb.
Magnesium Silicate	100	lb.

Vehicle (34%)

Modified Phenolic Varnish	43.75	gal.
Mineral Spirits	27.75	gal.

Cheap Fire Resistant Paints

Formula No. 1

Sodium Silicate	112 lb.
Kaolin	150 lb.
Water	100 lb.

No. 2

Slaked Lime	2 lb.
Salt	1 oz.
Water	1 pt.

Indoor Fireproof Paint

Formula No. 1

Pigment:

White Lead	41.0
Borax	32.0
Raw Linseed Oil	22.8
Turpentine	3.6
Japan Drier	0.6

No. 2

Pigment:

Titanium-Calcium Pigment	30.0
Borax	35.0
Raw Linseed Oil	30.8
Turpentine	3.6
Japan Drier	0.6

No. 3

Pigment:

Lithopone	24.0
Borax	39.5
Raw Linseed Oil	32.3
Turpentine	3.6
Japan Drier	0.6

No. 4

Pigment:

Zinc Oxide	21.0
Borax	50.0
Raw Linseed Oil	24.8
Turpentine	3.6
Japan Drier	0.6

No. 5

White Lead, Paste	25	lb.
Borax	17	lb.
Linseed Oil	1¼	gal.
Turpentine	1½	pt.
Drier, Liquid	½	pt.

From the standpoint of hiding power and brushing qualities, an unusually high percentage of borax is undesirable, but fire-retarding effectiveness is lost or decreased if the borax content is dropped too low. It is apparent that a balance must be maintained between paint quality and fire retardance. A minimum of at least 25% by weight of borax appears necessary.

To obtain maximum fire protection, heavy applications of the paint (3 or 4 thick coats or at least 8 gallons per 1000 square feet) were found necessary. This is about twice the amount of paint ordinarily applied to woodwork. Coatings of ordinary thickness undoubtedly would provide protection against comparatively weak fires.

Fine grinding of the borax and subsequent milling in oil are necessary for smoothness, consequently the home preparation of borax paints is practical only when a lumpy, rough coating is not objectionable from the appearance standpoint.

Fireproofing Paint

U. S. Patent 2,247,633

Formula No. 1

Water, Soft	5	gal.
Antimony Chloride	5	oz.
Glue, Powdered	2½	lb.

Mineral Pigment	10 lb.
Diatomaceous Earth	2½ lb.

No. 2

Kaolin	150 lb.
Sodium Silicate (Sp. Gr. 1.41–1.43)	112 lb.
Water	100 lb.

Waterproof and Chemical Resistant Paint

Mix the following all together and apply with a smooth camel's hair brush to the parts of wood or metal which are to be protected.

Coal Tar Pitch	80 g.
Dissolved in Xylol	80 g.
Add Benzene	40 cc.
Pulverized Silica	30 g.
Chlorinated Rubber	7 g.

Water Proof Casein Coatings

(1) for black glaze: black paste dye 100.0, dry colorless casein 8.0, nigrosine 0.75, latex 5.0 and water 300 parts; (2) for brown glaze: brown dye 100.0, aniline dye 0.5–0.7, dry colorless casein 10.0, latex 5.0 and water 400 parts. The casein is digested with 50% of the water of the formula for 1 hour, allowed to stand an extra hour after which 2–6% ammonia (25%) is added, and the mixture is gently boiled. The dyes are added to the remainder of the water at 50–60° while agitating and then the casein and latex are mixed therewith.

Water Resistant Coating Film

British Patent 514,696

Zein	100.0
Paraffin Wax	4.0
Methanol (94%)	218.2
Benzene	116.0

Waterproof Wood Primer

Celluloid	1
Butyl Acetate	9–10

Dam Gate Primers

Formula No. 1

Zinc Oxide	16
Zinc Dust	64
Vehicle No. 2 *	20

No. 2

Zinc Oxide	13
Iron Oxide	19½
Zinc Dust	32½
Vehicle No. 2 *	35

The above are thinned prior to application with one pint to one quart of mineral spirits per gallon of paint.

Waterproof Whitewash

Waterproof whitewash for outdoor buildings is made up in the following proportions: slake 62 pounds of quick lime in 12 gallons of hot water, and add two pounds of salt and one pound of sulphate of zinc in two gallons of water. To this add two gallons of skim-milk. An ounce of alum improves this wash but it is not essential. If the whitewash is required for metal surfaces subject to rust, the salt should be omitted.

* Vehicle No. 2.—The vehicle has the following composition in percentages by weight:

Mineral Spirits	28.12
Varnish Oil	17.15
Para Phenyl Phenol-Formaldehyde Resin No. 2	16.65
China Wood Oil	25.98
Dipentine	4.67
Ethylene Glycol Monoethyl Ether	3.84
Toluol	2.38
High Flash Naphtha	1.21

Disinfectant Whitewash

Dissolve 50 pounds of lime in eight gallons of boiling water. To this add six gallons of hot water in which ten pounds of salt and one pound of alum have been dissolved. A can of lye is added to every 25 gallons of the mixture. A pound of cement to every three gallons is gradually added and stirred thoroughly. The alum prevents the lime from rubbing off, the cement makes a creamy mixture easy to apply, and the lye is added as a disinfectant. A quart of creosol disinfectant to every eight gallons would serve the same purpose as the lye, but if a pure whitewash is desired, the lye is preferable. If a real snowy whiteness to whitewash is desired, it can be obtained by adding a very small quantity of washing blue.

Water Resistant Silicate Coating

U. S. Patent 2,234,646

Formula No. 1

Sodium Silicate	½ gal.
Water	½ gal.
Formaldehyde	1–6 oz.
Sodium Aluminate (30% Solution)	⅓–½ oz.
Sodium Abietate	1–3 oz.

No. 2

U. S. Patent 2,234,672

Sodium Silicate	½ gal.
Water	½ gal.
Formaldehyde	½–3 oz.
Latex (40–60%)	1–2 oz.

Waterproofing Compound

Petroleum Jelly	33
Rosin	11
Beeswax	56

Chemical Resistant Coatings

Furfuryl alcohol resinifies in th presence of mineral acids to for a black resin. When this reactio occurs in situ on wood or oth porous materials, a very satisfa tory finish is produced which is re sistant to the attack of solven as well as to most acids and a kalies. Hence this application ma be advantageously used for pro tecting laboratory table tops an other surfaces subjected to chem cal action.

The surface to be finished shoul be scraped and sand-papered unt uniformly smooth. It is the painted or sprayed with a libera application of furfuryl alcohol an after the surface appears to be dr a 10–15% aqueous solution of hy drochloric acid is applied, usuall by brushing. The surface wi gradually darken and will becom black. As many coats as desire may be applied. After the fina application, the surface may b polished and buffed to give a gloss appearance. The finishing opera tion may consist of rubbing severa applications of either linseed o China-wood oil into the surfac until a high gloss is secured. On gallon of furfuryl alcohol will cove from 200 to 300 square feet of sur face.

Note:

1. Do not add concentrated min eral acids to furfuryl alcohol or it solutions since the reaction is quit energetic. However, when dilut acid is used as indicated above, th reaction proceeds smoothly.

2. In place of furfuryl alcoho a mixture consisting of 50–50 furfu ryl alcohol and furfural may b

ed. The furfural is a good diluent nd reduces cost.

3. Sulphuric or phosphoric acid works as effectively as hydrochloric.

4. Some prefer to use an ethyl alcohol solution of the mineral acid in place of the aqueous solution, since the former is reported to give a more uniform coating.

Oil and Grease Resisting Compounds
German Patent 682,241

	Formula No. 1	No. 2
Rosin	70–75%	72.5%
Rubber	10–15%	14.5%
Silica, To make	100%	13.0%

Wood or Metal Container Coating
U. S. Patent 2,122,543

Ester Gum	65
Japan Wax	11
Petrolatum	24

Melt together and apply hot or from a hydrocarbon solvent.

Anti-Corrosive Paint

Linseed Oil (with Drier)	54
Linseed Fatty Acids	6
Ammonia	4
Water	36

Mix until emulsified and grind in ball mill with

Iron Oxide	60

Acid and Alkali Resistant Protective Coatings for Cement and Concrete

Formula No. 1

Thermally Processed Congo	50 lb.
Chlorinated Rubber (125 Cps.)	100 lb.
Solvesso No. 3	350 lb.
Hercolyn	30 lb.

Mix and agitate all ingredients until a homogeneous solution is obtained.

No. 2

Batavia Dammar	50 lb.
Chlorinated Rubber	100 lb.
Solvesso No. 3	350 lb.
Hercolyn	30 lb.

Mix and agitate all ingredients until a homogeneous solution is obtained.

Both of above formulae dry to hard, flexible films which are acid, alkali and alcohol resistant and are recommended for use over concrete.

No. 3

Thermally Processed Batavia Dammar	12.5 lb.
Chlorinated Rubber	12.5 lb.
Hi-Flash Naphtha	100 lb.

Mix all ingredients together until a homogeneous lacquer is obtained.

No. 4

Thermally Processed Brown Kauri	12.5 lb.
Chlorinated Rubber	12.5 lb.
Hi-Flash Naphtha	100 lb.

Mix all ingredients together until a homogeneous lacquer is obtained.

Acid Resistant Pipe Coating
U. S. Patent 2,214,062

Rosin, Petroleum Insoluble	70
Linseed Oil	30

Heat at 300° C. for 20 min. and stir in:

Asbestos	12½

Protective Coating for Rods and Tubes

Asphalt	33
Rosin (Vinsol)	35
Rubber	4
Calcium Carbonate	29

"Cut" the above with gasoline

and mix well during use. Apply with a spray-gun or brush. It dries very quickly to a film without gloss or "tack."

Protective Coating for Porcelain, China and Plated Metal Parts
U. S. Patent 2,219,583

This is for the manufacture of an adhesive paste for applying to newspaper, to act as a protective coating for temporary protection of glossy or polished surfaces during such time as such surfaces are likely to be subjected to mechanical abuse or to the attack of corrosive chemicals such as cleaning acids, paints, etc. Surfaces such as enameled iron, porcelain, vitreous china, and electroplated parts are covered by newspaper to which the adhesive paste is applied, so that the paper and paste can readily be removed by means of water so that neither the paste itself, nor the ink of the newspaper will affect the covered parts.

The following formula for this type of paste is chemically stable, and requires no preservative to prevent decomposition, nor the use of a neutralizing agent to overcome the effects of decomposition:

Bentonite	50
Water	150
Glycerin	10
Borax	15
Abopon Solution (sp. gr. 1.68)	5

Primers for Metals
Vehicle No. 1
25 Gallon Length (Oiticica-Linseed)

Batu, Black East India or Pale East India	100 lb.
Oiticica Oil	20 gal.
Bodied Linseed Oil (Z Viscosity)	5 gal.

Petroleum thinner to 50% noı volatile content. Driers equivaleı to 0.5% Pb, 0.02% Co and 0.01% Mn, as metals on weight of oil.

Heat the resin and 5 gallons c oiticica oil to 600°–610° F. in 60–6 minutes. Hold until a cold "pill indicates complete compatibilit (about 10 minutes). Add remainde of the oiticica oil. Heat to 580° 590° F. Hold for body (viscosity c finished vehicle should be betwee "B" and "D"). Check with the lin seed oil. Allow to cool to abou 400°–420° F. Add the petroleur thinner and then the driers.

The above vehicle is especiall recommended for preparing primer used for underwater applications.

Vehicle No. 2
25 Gallon Length (Linseed or Dehydrated Castor)

Batu, Black East India or Pale East India	100 lb.
Bodied Linseed Oil (OKO 7½ M.) or Dehydrated Castor Oil	25 gal.

Petroleum thinner to 50% non volatile content by weight. Drier equivalent to 0.4% Pb, 0.02% Co and 0.02% Mn, as metals on the weight of oil.

Heat resin and 5 gallons of the linseed oil to 600°–610° F. in abou 60–65 minutes. Hold until cold "pill" indicates complete compatibility (about 10 minutes). Add remainder of linseed oil. Heat to 560° F. Allow to cool to about 400°–420°

. Add the petroleum thinner and hen the driers.

The above vehicle is recomnended for preparing primers for pplications other than underwater urposes.

Vehicle No. 3

25 Gallon Length (Chinawood-Linseed)

Batu, Black East India or Pale East India	100	lb.
Chinawood Oil	20	gal.
Kettle Bodied Linseed Oil (About Z Viscosity)	5	gal.

Petroleum thinner to 50% nonolatile content. Driers equivalent o 0.5% Lead, 0.02% Manganese, nd 0.01% Cobalt as metals on the veight of oil.

Heat resin and 5 gallons chinavood oil to 600°–610° F., in about 0–70 minutes. Hold until cold pill" denotes complete compatiility (about 10 minutes). Allow to ool to 540° F. Add remainder of hinawood oil. Heat to 560° F. Hold or desired body (viscosity of finshed vehicle should be "C" or ower). Check with the linseed oil. llow to cool to about 400°–420° F. dd the petroleum thinner and then he driers.

This vehicle is particularly recomnended for underwater applications.

Protective Metal Primer Paint

Formula No. 1

Pigment (60%)

Zinc Tetroxy Chromate	855	lb.

Vehicle (40%)

Raw Linseed Oil	55½	gal.
Kettle Bodied "Q" Linseed Oil	7¼	gal.
Mineral Spirits and Drier *	12	gal.

Use all of the kettle bodied oil and as much of the raw linseed oil as required to make paste, grind and reduce with remainder of vehicle.

If the paint happens to be somewhat puffy as first made up, it should be remembered that it will thin slightly on aging. The use of a higher percentage of kettle bodied oil helps to reduce the consistency, but it is preferable not to exceed 15% of bodied oil in the paint.

No. 2

Pigment (60%)

Zinc Tetroxy Chromate	560	lb.
Iron Oxide	300	lb.

Vehicle (40%)

Raw Linseed Oil	55½	gal.
Kettle Bodied "Q" Linseed Oil	7¼	gal.
Mineral Spirits and Drier	12	gal.

Metal Priming Paint

Formula No. 1

(Baking Primer)

Pigment (25%)

XX Zinc Oxide	173	lb.
Standard Zinc Dust—22	81	lb.

Vehicle (75%)

Rezyl No. 420–2 Solution	74¼	gal.
Pine Oil	4¾	gal.
Solvesso No. 2	16	gal.

* 5% of liquid drier should be sufficient. High percentages of drier may tend to make a puffy paint. If this should occur the consistency may be reduced by using ¼% soluble litharge (based on the weight of pigment) in making up the paste and reducing the percentage of liquid drier.

Grind the zinc oxide in the Rezyl solution and the pine oil, stir or mix in the zinc dust and thin with the Solvesso.

No. 2

Pigment (32%)

Standard Zinc Dust—22	125	lb.

Vehicle (68%)

Rezyl No. 408	61	lb.
Rezyl No. 1102	211	lb.
Xylol	47¾	gal.
Pine Oil	2⅞	gal.
Dipentene	16	gal.

Add a small portion of the vehicle to a part of the zinc dust, thoroughly stir or mix them, then add zinc dust and vehicle alternately until all the zinc dust is in; thin with the remaining vehicle.

Metal Priming and Finishing Paint (For Priming Galvanized Iron)

Formula No. 1

Pigment (65%)

Zinc Dust	894	lb.
XX Zinc Oxide	224	lb.

Vehicle (35%)

Syntex 32 Solution *	524	lb.
Mineral Spirits	68	lb.
Lead Naphthenate 16%	8¼	lb.
Cobalt Naphthenate 4%	1¾	lb.

Grind the zinc oxide in a portion of the Syntex solution. The zinc dust should be mixed in the zinc oxide paste with the remainder of the Syntex solution, keeping the paste heavy until all the zinc dust is added, then add the mineral spirits and drier.

* Other comparable alkyd resin solutions may be used, but may require adjustments in drier, thinner, etc.

This paint, as a primer, has ex cellent rust inhibitive properties o iron and steel, and exceptional ad herence to galvanized metal, zin and other non-ferrous surfaces. It i free of the tendency to develop a after tack.

No. 2

Pigment (78%)

Standard Zinc Dust—22	1473	lb.
XX Zinc Oxide	368	lb.

Vehicle (22%)

Raw Linseed Oil	60¼	gal.
Mineral Spirits or Turpentine	4	gal.
Liquid Oil Drier	3¼	gal.

Grind a zinc oxide-linseed oi paste (18% oil), mix in the zin dust with part of the remaining oi and not completing the thinning un til all the zinc dust has been added For priming coat applications, th addition of a pint of turpentine o mineral spirits to a gallon of th above paint is recommended. Pure low acid (below 4), moisture-fre linseed oil, turpentine or minera spirits and a lead-manganese oi drier give uniformly good results.

This paint, as a primer, has ex cellent rust inhibitive properties o iron and steel, and exceptional ad herence to galvanized metal, zin and other non-ferrous surfaces. I is also suitable for wire screens i reduced with sufficient thinner t prevent webbing.

No. 3

Pigment (70%)

Standard Zinc Dust—22	648	lb.
XX Zinc Oxide	259	lb.
Iron Oxide	389	lb.

Vehicle (30%)

Raw Linseed Oil	64½	gal.
Mineral Spirits	4¼	gal.
Liquid Oil Drier	3⅔	gal.

Grind the zinc oxide and iron ox-le into a paste containing 21% of he linseed oil. Mix the zinc dust nd part of the added oil into the aste and do not thin to painting onsistency until all the zinc dust as been added.

For priming coat applications, he addition of a pint of turpentine ubstitute to a gallon of the above aint is recommended.

This paint may be used as a metal rimer or finishing coat paint. It has xcellent rust inhibitive and dura-ility properties but not as good ad-erence to galvanized iron as an 0–20:zinc dust-zinc oxide paint. It hould also serve as a good general hop coat paint, especially so if a par varnish is added as part of the ehicle.

No. 4

Pigment (72%)

Standard Zinc Dust—22	890	lb.
XX Zinc Oxide	205	lb.
Magnesium Silicate	275	lb.

Vehicle (28%)

Raw Linseed Oil	61⅔	gal.
Mineral Spirits	4	gal.
Liquid Oil Drier	3½	gal.

Start with a zinc oxide-magne-sium silicate-linseed oil paste con-taining 25% oil, mix the zinc dust and part of the added oil into the paste, not completing the thinning until after all the zinc dust has been added.

For priming coat application the addition of a pint of mineral spirits or turpentine to a gallon of the above paint is recommended.

No. 5

Pigment (80%)

Standard Zinc Dust—22	1570	lb.
XX Zinc Oxide	392	lb.

Vehicle (20%)

Varnish V-172	55	gal.
Solvesso No. 2	10	gal.

Start with a zinc oxide-varnish V-172 paste (20% vehicle), mix the zinc dust and part of the added ve-hicle into the zinc oxide paste, and should not be completely thinned until all the zinc dust has been added.

No. 6

Pigment (60%)

Standard Zinc Dust—22	453	lb.
XX Zinc Oxide	181	lb.
Iron Oxide	272	lb.

Vehicle (40%)

Varnish V-172	67¾	gal.
Solvesso No. 2	12½	gal.

No. 7

Pigment (65%)

Standard Zinc Dust—22	688	lb.
XX Zinc Oxide	159	lb.
Magnesium Silicate	212	lb.

Vehicle (35%)

Varnish V-172	64	gal.
Solvesso No. 2	11⅔	gal.

Grind the zinc oxide and magne-sium silicate to a paste containing 25% of the varnish V-172. Mix the zinc dust and part of the added ve-hicle into the paste and do not thin to painting consistency until all the zinc dust has been added.

For priming coat application, the addition of a pint of mineral spirits to a gallon of the above paint is recommended.

No. 8

Pigment (80%)

Standard Zinc Dust—22	1562	lb.
XX Zinc Oxide	391	lb.

Vehicle (20%)

Glyptal Solution No. 2455	33	gal.
Raw Linseed Oil	9¼	gal.
Mineral Spirits	21	gal.
Lead Naphthenate 16%	6	lb.
Cobalt Naphthenate 4%	2	lb.
Manganese Naphthenate 4%	1	lb.

Grind the zinc oxide in the linseed oil and Glyptal solution (25% vehicle) to make a paste. Mix the zinc dust and part of the vehicle to be added into the paste, continue until all the zinc dust has been added by alternately adding zinc dust and vehicle. Thin with remaining vehicle. For priming coat applications, the addition of a pint of mineral spirits or turpentine to a gallon of the above paint is recommended.

No. 9

Pigment (60%)

Standard Zinc Dust—22	454	lb.
XX Zinc Oxide	182	lb.
Iron Oxide	272	lb.

Vehicle (40%)

Glyptal No. 2455 Solution	41¼	gal.
Raw Linseed Oil	11¾	gal.
Mineral Spirits	26	gal.
Lead Naphthenate 16%	7¼	lb.
Cobalt Naphthenate 4%	2½	lb.
Manganese Naphthenate 4%	1¼	lb.

Metal Priming and Finishing Paste

Pigment (85%)

Standard Zinc Dust—22	1431	lb.
XX Zinc Oxide	331	lb.
Magnesium Silicate	440	lb.

Vehicle (15%)

Raw Linseed Oil	50	gal.

Grind the zinc oxide and the magnesium silicate into a paste containing 25% of oil. Then the zinc dust can be mixed into this paste with the remainder of the oil.

Metal Finishing Paint

Pigment (49%)

Standard Zinc Dust—22	498	lb.
Aluminum Powder	124½	lb.

Vehicle (51%)

Varnish V-172	72½	gal.
Solvesso No. 2	12½	gal.

No grinding is necessary in preparing this paint; just stir part of the pigment with part of the vehicle, add pigment and vehicle alternately in small portions until all the pigment is thoroughly dispersed, then add the remaining vehicle.

The above paint may be also made by stirring one and two-fifths pounds of aluminum powder and five and three-fifths pounds of zinc dust into a gallon of the vehicle.

This paint may be used as a finishing paint where the bright metallic sheen of an aluminum paint is desired, and where greater rust resistance than is present in a straight aluminum paint is required.

High Heat Resistant Metal Paint

Standard Zinc Dust—22	880	lb.
XX Zinc Oxide	220	lb.
Glyptal—2468 Solution	67½	gal.
Mineral Spirits	11¾	gal.

Grind the zinc oxide in a portion of the Glyptal solution. The zinc dust can then be mixed in the zinc oxide paste with the remainder of the Glyptal solution. Add the mineral spirits.

This paint may be brushed or sprayed, and should give good service on metal which becomes sufficiently hot to burn out the binder, the zinc dust fusing on the metal. If the paint is to be used where the surface is not to be heated, suitable driers should be added to make the paint dry. A suggested combination is 0.6% lead and 0.06% cobalt, as naphthenates, based on the resin non-volatile. The vehicle then becomes 87% Glyptal 2468, 10.8% (9¾ gal.) mineral spirits, 1.6% (1⅛ gal.) lead naphthenate (16%), and 0.6% (3¾ pt.) cobalt naphthenate (4%). This paint will air dry with an after tack which it will retain for about a week, even if drier is increased.

This paint has shown excellent performance on super-heated steam inlets to steam turbines, on high temperature electric drying ovens, on breechings from boilers to stacks, on blast furnace stacks and on the metal sides and roof of buildings where pigs are cast. Some of the above exposures are so severe that galvanized iron peels and black iron had to be used. This paint also shows good adherence to galvanized iron.

Tinted Metal Paints

Zinc dust paints may be tinted to a wide variety of attractive colors, as well as black, by merely replacing varying percentages of the zinc oxide with tinting pigments.

The following table suggests the percentage composition of a number of tinted zinc dust paints in a vehicle composed of 90% raw linseed oil, 5% thinner, 5% liquid drier.

	Gray	*Green*	*Brown*	*Blue*	*Red*	*Black*	*Metal*
Standard Zinc Dust—22	80.0	80.0	80.0	80.0	80.0	80.0	80.0
XX Zinc Oxide	20.0	5.0	14.0	18.0	5.0	5.0	5.0
Prussian Blue		5.0		2.0		5.0	
Chrome Yellow, Medium		10.0					
Burnt Umber			6.0				
Iron Oxide (Persian Gulf)					15.0		
Lamp Black						10.0	
Black Iron Oxide							15.0
Pigment %	80.0	78.0	78.0	80.0	80.0	63.0	80.0

Pipe Thread Compound

Pigment (59.5%)

Standard Zinc Dust—22	888	lb.

Vehicle (40.5%)

Cup Grease *	533	lb.
Varnish V-172	9¾	gal.

Mix the zinc dust in the cup grease and thin with the varnish.

This is a compound for use on threaded pipe joints (oil pipe lines, etc.) to seal the joint and resist corrosion.

Red Lead Base Linseed Oil Primer

Pigment (65%)

Red Lead (95% Grade)	416	lb.
Iron Oxide (70% Fe_2O_3 min.)	208	lb.
XX Zinc Oxide	104	lb.
Magnesium Silicate	312	lb.

Vehicle (35%)

Raw Linseed Oil	67¾	gal.
Mineral Spirits and Drier	5	gal.

Coating Screw Threads
U. S. Patent 2,181,835

Latex (66%)	13
Colloidal Graphite	87

Paint for Compass Bowls and Disks
Formula No. 1

White Lead	50	g.
Litho Middle Varnish	20	g.
Mineral Spirits	20	cc.
Lead Tungate	5	g.

No. 2

Antimony, White	150	g.
Pyroxylin	10	g.
Amyl Acetate	180	cc.

* This is a low viscosity petroleum base lubricant of a suitable consistency for use through a grease gun.

Protective Coating for Aluminum Castings and Zinc and Tin Plated Parts

An emulsion is made with the following formula:

Diglycol Stearate	1
Ozokerite	2
Paraffin Wax	4
Water	40

Melt the three waxes together and keep at a temperature of about 65° to 70° C. Heat the water up to about the same temperature or a little above, and add slowly with high speed agitation to the melted waxes.

The emulsion is applied to the castings, etc., by spraying. This enables ease of control, saving in labor, speeding up of process and elimination of odor over the hot melted wax process because the spraying process is a cold one.

Protective Coating for Beer Cans
British Patent 490,851

The metal is made the anode in following dispersion:

Beeswax, White	47	g.
Ceresin	453	g.
Sodium Aluminate	5.65	g.
Sodium Silicate	9.35	g.
Water	2	l.

Current density 30–50 amp./sq. ft. at start, then reduce to 8 amp./sq. ft. After this treatment heat can to coalesce wax particles.

Hot Melt Metal Pipe Waterproofings
Formula No. 1

Asphalt	10
Rosin	10

No. 2

Asphalt	10
Rosin	10
Paraffin Wax	1

No. 3

Asphalt	10
Rosin	10
Neville R-11 Resin	2 1/5

Painting Farm Implements

Before painting, be sure the wood is clean and dry and the metal clean and free from rust. For the first coat on the wooden parts, use a priming coat which is:

White Lead	100 lb.
Raw Linseed Oil	4 gal.
Turpentine	2 gal.
Liquid Drier	1 pt.

Tint with Venetian Red if woodwork is to be finished red, or tint with lampblack to a medium gray if wood is to be black.

For the metal tools and parts, use a red lead for the priming coat, regardless of the color these items are to be finished.

For a bright red finishing coat, on both the primed wood and metal surfaces, use Bulletin Red and Venetian Red. Mix it as follows:

Bulletin Red	½ gal.
Venetian Red	½ gal.
Raw Linseed Oil	2 qt.
Spar Varnish	1 qt.
Liquid Drier	1½ pt.

For a black finishing coat, on both the primed wood and metal surfaces, use lampblack. Mix it as follows:

Lampblack	1 gal.
Raw Linseed Oil	2 qt.
Spar Varnish	2 qt.
Liquid Drier	1½ pt.

If you intend to paint the handles of small tools, in the case of the red finish coat, use just 1 qt. of oil and 2 qt. of varnish with 1 pt. of drier, to each gallon of the Bulletin Red-Venetian Red combination, in order to get a harder coating. For small tool handles that are to be finished in black, it is advisable to use only 1 qt. of oil, 3 qt. of varnish and 1 pt. of drier to each gallon of lampblack.

Paint for New Galvanized Iron

Many times one wishes to paint new galvanized iron but the paint does not have a good tendency to stick. The following mixture applied to the metal and left to dry and get powdery, will form an excellent base.

Cupric Chloride	1 oz.
Cupric Nitrate	1 oz.
Hydrochloric Acid (Tech.)	1 oz.
Water	¼ oz.

Mix all together and apply with a swab and allow to dry until the surface turns powdery. Usually several hours are sufficient.

Paint for Hot Metal Surfaces

U. S. Patent 2,239,478

Sodium Silicate (35–37%)	60 gal.
Gum Tragacanth	15 oz.
Starch	30 oz.
Sugar	80 lb.
Water	30 gal.
Alum	30–120 oz.
Aluminum Powder	90 lb.

Rust Preventative Black Paint

Melt 300 lb. of pitch very carefully over a fire and remove it from anywhere near the fire. Now add 10 lb. of pulverized Portland cement and then 25 gal. of solvent naphtha. Stir well to get a smooth easy flowing black paint which is well adapted for heavy outside machinery, bridges, steel work and the like.

Anti-Corrosive Paint

Formula No. 1

Alcohol	36¼	gal.
Shellac	40	lb.
Turpentine	3	gal.
Pine Tar Oil	3	gal.
Zinc, Metallic Powdered	48	lb.
Zinc Oxide	142	lb.

No. 2

Australian Patent 113,946

Boiled Linseed Oil	1 1/5	gal.
Zinc Dust	22	lb.
Litharge	1	lb.
Calcium Silicate or Cement	2½	lb.

Ship Bottom Paints

Formula No. 1

Zinc Oxide	200.0
Zinc Yellow	80.0
Silica, Dry	45.0
Magnesium Silicate	48.0
W. W. Rosin	145.0
Coal Tar	47.5
Coal Tar Naphtha	380.0
Manganese Linoleate	75.0

No. 2

Lamp Black	25.0
Zinc Oxide	186.0
Indian Red	93.0
Silica, Dry	45.0
Magnesium Silicate	48.0
W. W. Rosin	145.0
Coal Tar	47.5
Coal Tar Naphtha	380.0
Manganese Linoleate	75.0

No. 3

Zinc Oxide	238.0
Indian Red	74.0
Silica, Dry	38.0
Magnesium Silicate	38.0
Cuprous Oxide	145.0
Mercuric Oxide	45.0
W. W. Rosin	248.0
Coal Tar Naphtha	338.0
Coal Tar	63.0
Pine Oil	39.0

Structural Metal Finish Paint

White Base for Tints

Pigment (59–61%)

Cryptone MS	515	lb.
35% Leaded Zinc Oxide	343	lb.
Soluble Litharge	2	lb.

Vehicle (41–39%)

Raw Linseed Oil	55¾	gal.
Kettle Bodied "Q" Linseed Oil *	8¾	gal.
Mineral Spirits and Liquid Drier †	10	gal.

Pebble or stone mills are recommended since steel rolls may discolor white paints due to abrasion.

NAVAL PAINTS

Boot-Topping, Black

Formula No. 1

Zinc oxide, dry, 164; lampblack, dry, 73; raw linseed oil, 31; spar varnish, 435; paint drier, 222.

No. 2

Manganese linoleate, 67; silica, dry, 84; lampblack, dry, 48; zinc oxide, dry, 138; coal tar, 2.5 gal.; formalin resin solution, 79 gal.; cuprous oxide, 109; mercuric oxide, 36.

* The higher percentage of bodied oil improves leveling.

† In whatever form drier is added, sufficient must be used to make paint dry thoroughly. Naphthenic acid types tend to produce the thinnest paints (without puff). Linoleates tend to produce the heaviest paints (with puff) and resinates are intermediate.

Naval Boot-Topping, Light Gray
Formula No. 1

Zinc oxide, dry, 300; lampblack, in oil, 1; ultramarine blue, in oil, 8; spar varnish, 669; petroleum spirits, 33; paint drier, 36.

No. 2

Zinc oxide, dry, 289; titanium pigment, dry, 396; lampblack, in oil, 1.5; ultramarine blue, in oil, 13; raw linseed oil, 538; petroleum spirits, 15; paint drier, 87.

Naval Outside White

Titanium pigment, dry, 628; zinc oxide, dry, 280; raw linseed oil, 515; petroleum spirits, 29; ultramarine blue, in oil, 0.5; paint drier, 76.

Naval Red Lead

Red lead, dry, 185; zinc oxide, dry, 86; venetian red, dry, 355; spar varnish, 315; petroleum spirits, 210; paint drier, 79; aluminum stearate, 9⅜.

Light Gray, Smokestack Paint

White lead, dry, 481; zinc oxide, dry, 188; litharge, dry, 34; lampblack, in oil, 2–6; ultramarine blue, in oil, 2–4; damar varnish, 194; kerosene, 337; paint drier, 55.

Naval Green Paint

Zinc oxide, dry, 100; chrome green, dry, 300; chrome yellow, dry, 36; yellow ochre, dry, 75; lampblack, in oil, 60; spar varnish, 418; paint drier, 181.

Anticorrosive Ship Bottom Paint

Zinc oxide, dry, 186; venetian red, dry, 93; silica, dry, 93; rosin, grade WW, 145; coal-tar naphtha, 380; coal tar, 47.5; manganese linoleate, 129.

Antifouling Ship Bottom Paint

Zinc oxide, dry, 238; silica, dry, 78; magnesium silicate, dry, 72; cuprous oxide, 145; mercuric oxide, 45; rosin, grade WW, 248; coal-tar naphtha, 338; coal tar, 63; pine oil, 39.

Nontoxic Ship Bottom Paint

Zinc oxide, dry, 212; silica, dry, 82; magnesium silicate, dry, 83; Indian red, dry, 160; rosin, 200; coal-tar naphtha, 296; coal tar, 133; pine oil, 78.

Copper Paint for Bottoms of Wooden Boats

Zinc oxide, dry, 165; Indian red, dry, 165; cuprous oxide, 75; gum shellac, 162; alcohol, 500; pine oil, 90.

White Canvas Preservative

Zinc oxide, in oil, 540; litharge, dry, 17; ultramarine blue, in oil, 1; beeswax, 6; crude rubber, 10; gasoline, 430; spar varnish, 32; raw linseed oil, 7; paint drier, 14.

Light Gray Canvas Preservative

Zinc oxide, in oil, 540; litharge, dry, 17; ultramarine blue, in oil, 8; lampblack, in oil, 3–4; beeswax, 6; crude rubber, 9; gasoline, 415; spar varnish, 32; raw linseed oil, 7; paint drier, 14.

Gray Deck Paint

Zinc oxide, dry, 325; lampblack, dry, 5; lampblack, in oil, 2; ultramarine blue, in oil, 2; spar varnish, 456; paint drier, 254.

Yellow Marking Paint for Boat Decks

Zinc oxide, dry, 257; chrome yellow, dry, 168; magnesium silicate, dry, 168; interior varnish, 304; petroleum spirits, 267; paint drier, 43.

Metallic Brown Paint for Potable Water Tanks

Metallic brown, dry, 397; Indian red, dry, 156; zinc oxide, dry, 79; silica, 78; mixing varnish, 472; paint drier, 109.

Naval Inside White Paint

Titanium pigment, dry, 565; zinc oxide, dry, 389; raw linseed oil, 271; damar varnish, 78; petroleum spirits, 194; ultramarine blue, in oil, .5; paint drier, 36.

Naval Flat White (Inside)

Zinc oxide, dry, 1,287; raw linseed oil, 281; petroleum spirits, 231; ultramarine blue, in oil, .5; paint drier, 44.

Naval White Enamel (Inside)

Titanium pigment, dry, 250; zinc oxide, dry, 250; damar varnish, 683; pine oil, 53; ultramarine blue, in oil, .5.

Naval Semiflat Green (Inside)

Titanium pigment, dry, 650; zinc oxide, dry, 300; chrome green oxide, in oil, 4; damar varnish, 386; petroleum spirits, 214.

Naval Light Green, Gloss (Inside)

Titanium pigment, dry, 250; zinc oxide, dry, 250; chrome green oxide, in oil, 3; damar varnish, 585; petroleum spirits, 64; pine oil, 61.

Varnish, Phenolic Type

Phenolic resin, 110; tung oil, raw, 279; coal-tar naphtha, water white, 269; xylene, 116; kerosene, 94; lead manganese drier, 11.5; cobalt drier, 11.5.

Inside White for Submarines

Titanium pigment, dry, 462; zinc oxide, dry, 146; raw linseed oil, 93; varnish, 451; damar varnish, 25; petroleum spirits, 71; paint drier, 13; ultramarine blue, in oil, 0.5.

Zinc Chromate Primer

Zinc chromate, dry, 565; magnesium-silicate, 65; varnish, 653.

Naval White Paint

Titanium Dioxide	251
Zinc Oxide	91
Magnesium Silicate	114
Dipentine	23
Alkyd Resin	420
Petroleum Spirits	189
Lead Naphthenate	11
Cobalt Naphthenate	1½
Manganese Naphthenate	1½

Asbestos Shingle Paint

Priming Coat

White Lead	100	lb.
Lead Mixing Oil	2½	gal.
Linseed Oil	2½	gal.

Second Coat

White Lead	100	lb.
Lead Mixing Oil	4	gal.

Third Coat—Flat Finish

White Lead	100	lb.
Lead Mixing Oil	4	gal.

Third Coat—Glass Finish

White Lead	100	lb.
Linseed Oil *	3¼	gal.
Liquid Drier	1	pt.

* If boiled oil is used, omit drier.

EXTERIOR WOOD PAINTS

		Priming Coat	Body Coat	First Coat Repainting	Finish Coat	Finish Coat	Alternate Finish Coat	First Coat	Finish Coat
Soft Paste White Lead	lbs.	250	300	250	300	300	300	300	300
Raw Linseed Oil	gal.	10	4½	5	9¾	9⅜	...	4½	8
Boiled Linseed Oil	gal.	...	...	...	...	...	9¾	...	...
Turpentine	gal.	5	4½	5	...	...	...	1½	1⅛
Liquid Drier	pt.	2½	3	2½	3	3	...	3	3
Varnish	gal.	...	...	...	...	...	...	2¼	...
Raw Umber	pt.	...	...	...	...	...	...	¾	...

Boiled linseed oil may be substituted for the raw linseed oil called for in these formulae. If boiled oil is used, omit drier.

Outside House Paint
Whites and Tints

Pigment (58–61%)

Cryptone MS—130	500	lb.
Zinc Oxide (35% Leaded)	250	lb.
Magnesium Silicate	83	lb.
Soluble Litharge	2	lb.

Vehicle (42–39%)

Raw Linseed Oil	58	gal.
Kettle Bodied "Q" Linseed	5½	gal.
Mineral Spirits	6	gal.
Liquid Drier	3¾	gal.

Outside House Paste Paint

Pigment (79–80%)

Cryptone MS—130	700	lb.
Zinc Oxide (50% Leaded)	300	lb.

Vehicle (21–20%)

Raw Linseed Oil	26¾	gal.
Kettle Bodied Linseed "Q"	5¼	gal.

Pebble or stone mills are recommended since steel rolls may discolor white paint due to abrasion.

Can be used for *whites* and *tints.*

Thinning Directions for Finish Coat Paint

Paste 100 lb.	Approx. 5	gal.
Raw Linseed Oil	2¾ to 3½	gal.
Turpentine	3 to 5	pt.
P&L Liquid Drier	Approx. 1	qt.

Outside House Paint
(Fume Proof)

Pigment (57–60%)

Cryptone MS—130	490	lb.
XX Zinc Oxide	250	lb.
Magnesium Silicate	83	lb.
Soluble Litharge	2	lb.

Vehicle (43–40%)

Raw Linseed Oil	57¾	gal.
Kettle Bodied "Q" Linseed Oil	5½	gal.
Mineral Spirits	6	gal.
Liquid Drier	3¾	gal.

(For Metropolitan Areas)
Whites and Tints

Pigment (60–62%)

Cryptone MS—130	540	lb.
Zinc Oxide (35% Leaded)	270	lb.
B. C. W. L.	90	lb.

Vehicle (40–38%)

Raw Linseed Oil	58¼	gal.
Kettle Bodied "Q" Linseed	5½	gal.
Mineral Spirits	6	gal.
Liquid Drier	3¾	gal.

(North Atlantic and Western Seacoast)
Whites and Tints

Pigment (60–62%)

Cryptone MS—130	555	lb.
Zinc Oxide (35% Leaded)	300	lb.
Soluble Litharge	2	lb.

Vehicle (40–38%)

Raw Linseed Oil	56¾	gal.
Kettle Bodied "Q" Linseed	5½	gal.
Mineral Spirits	6	gal.
Liquid Drier	3¾	gal.

Outside House Paint
Formula No. 1

Pigment (63–65%)

Cryptone MS—130	405	lb.
XX Zinc Oxide	200	lb.
B. C. W. L.	405	lb.

Vehicle (37–35%)

Raw Linseed Oil	56¼	gal.
Kettle Bodied "Q" Linseed	5½	gal.
Mineral Spirits	6	gal.
Liquid Drier	3¾	gal.

No. 2

Pigment (62–64%)

Cryptone MS—130	395	lb.
Zinc Oxide (35% Leaded)	250	lb.
B. C. W. L.	295	lb.
Titanium Dioxide	50	lb.

Vehicle (38–36%)

Raw Linseed Oil	56½	gal.
Kettle Bodied "Q" Linseed	5½	gal.
Mineral Spirits	6	gal.
Liquid Drier	3¾	gal

No. 3

Pigment (61.5%)

Pigment—725	400	lb.
Titanium Dioxide	135	lb.
Basic Sulphate, White Lead	165	lb.
Magnesium Silicate	200	lb.

Vehicle (38.5%)

Linseed Oil	58⅞	gal.
Q-Bodied Linseed Oil	5⅞	gal.
Drier	8⅞	gal.
Thinner * and Drier, To make	100	gal.

Thinning Directions for New Wood

First or Priming Coat—To one gallon of paint add one quart of raw or boiled linseed oil and one pint turpentine or mineral spirits.

Second Coat—To one gallon of paint add one pint turpentine or mineral spirits.

* The amount required will vary with the efficiency of the drier. In whatever form drier is added, sufficient must be used to make the paint dry thoroughly. Naphthenic acid types tend to produce the thinnest paints. Linoleates tend to produce heavier paints (puffy) and resinates are intermediate.

Third or Finishing Coat—Apply as made. If necessary one gallon of paint may be reduced with one half pint of linseed oil and as much turpentine or mineral spirits as required to produce the desired brushing properties. This amount of linseed oil reduces the Pigment Fixed Vehicle Ratio to 27.3%.

Thinning Directions for Repaint Work

If surface is in good condition use second and third coat thinning directions for a two-coat repainting job.

Ready Mixed Exterior House Paint

Pigment (64%)

Titanium Dioxide Non-chalking Type	139	lb.
B. C. W. L.	149	lb.
35% Leaded Zinc Oxide	445	lb.
Magnesium Silicate and Tinting Colors	257	lb.

Vehicle (36%)

Raw Linseed Oil	59.5	gal.
Heat Bodied Linseed Oil (Z-2 body)	3.4	gal.
Mineral Spirits and Drier	10.1	gal.

Formula No. 1

Pigment (68%)

Titanium Dioxide	83	lb.
35% Leaded Zinc Oxide	450	lb.
B. C. W. L.	532	lb.
Magnesium Silicate	118	lb.

Vehicle (32%)

Raw Linseed Oil	61	gal.
Mineral Spirits and Drier	12.8	gal.

No. 2

Pigment (58%)

Titanium Dioxide	115	lb.
XX Zinc Oxide	230	lb.
Magnesium Silicate	422	lb.

Vehicle (42%)

Raw Linseed Oil	61	gal.
Mineral Spirits and Drier *	12.8	gal.

Semipaste Paint

Formula No. 1

Pigment (79%)

B. C. W. L. or B. S. W. L.	1081	lb.
35% Leaded Zinc Oxide	542	lb.
Magnesium Silicate (and Tinting Colors)	180	lb.

Vehicle (21%)

Raw Linseed Oil	61.8	gal.

No. 2

Pigment (79%)

B. C. W. L. or B. S. W. L.	786	lb.
35% Leaded Zinc Oxide	823	lb.
Magnesium Silicate (and Tinting Colors)	179	lb.

Vehicle (21%)

Raw Linseed Oil	61.2	gal.

Exterior Ready-Mixed Paint

Formula No. 1

Pigment (63%)

Titanium Dioxide (Semi-Chalking Type)	93	lb.
XX Zinc Oxide	231.5	lb.
B. C. W. L. or B. S. W. L.	231.5	lb.
Whiting	139.0	lb.
Magnesium Silicate	231.5	lb.

* Lead Driers not allowed.

Vehicle (37%)

Raw Linseed Oil	59½ gal.
Mineral Spirits and Drier	12½ gal.

No. 2

Pigment (63%)

Titanium Dioxide (Semi-Chalking Type)	93 lb.
50% Leaded Zinc Oxide	463 lb.
Whiting	139 lb.
Magnesium Silicate	231.5 lb.

Vehicle (37%)

Raw Linseed Oil	59½ gal.
Mineral Spirits and Drier	12½ gal.

Semi-Paste Paint

Pigment (72%)

Iron Oxide	1120 lb.
XX Zinc Oxide	198 lb.

Vehicle (28%)

Raw Linseed Oil	66 gal.

Wood Primer

Formula No. 1

Pigment (59½%)

Cryptone MS—130	685 lb.
(35% Leaded) Zinc Oxide	120 lb.

Vehicle (40½%)

Raw Linseed Oil	52½ gal.
Kettle Bodied "Q" Linseed Oil	4¾ gal.
Liquid Drier	2¼ gal.
Mineral Spirits	12½ gal.

No. 2

Pigment (57%)

Cryptone MS	635 lb.
(35% Leaded) Zinc Oxide	110 lb.

Vehicle (43%)

Bodied Oil	43¼ gal.
Raw Linseed Oil	14½ gal.
Liquid Drier	1½ gal.
Mineral Spirits	15½ gal.

As much of the bodied oil as necessary is used in the base, which is then reduced with the remainder of the bodied oil, the raw oil, and the liquid drier. Mineral spirits should then be added to produce the body desired.

Vehicle for Wood Primers

Raw Chinawood Oil	25 gal.
Raw or Non-Break Linseed Oil	75 gal.

Heat raw Chinawood oil and an equal volume of raw or non-break linseed oil to 565° F. in 55 minutes, check with remainder of raw or non-break linseed oil and cool. This is a light bodied oil.

Flat Wall Paint—Dead Flat

Pigment (59%)

Cryptone ZS—800	405 lb.
Magnesium Silicate	220 lb.
Diatomaceous Silica	110 lb.

Vehicle (41%)

Kettle Bodied Linseed Oil (Z-3 Body)	15.1 gal.
10 lb. Cut Low Acid Ester Gum	9.55 gal.
6 oz. Cut Aluminum Stearate Gel.	19.94 gal.
Mineral Spirits	11.81 gal.
Kerosene	15.2 gal.
Lead Naphthenate 24%	0.56 gal.
Cobalt Naphthenate 6%	0.15 gal.

Self Priming Flat Paint

Formula No. 1

Pigment (57%)

Cryptone ZS—800	415	lb.
Magnesium Silicate *	285	lb.

Vehicle (43%)

37-M OKO Oil	19.2	gal.
Mineral Spirits	40.3	gal.
Kerosene	15.62	gal.
Pb Naphthenate 24%	0.465	gal.
Co Naphthenate 6%	0.086	gal.

No. 2

Pigment (58%)

Cryptone BT—301	600	lb.
Magnesium Silicate	150	lb.

Vehicle (42%)

37-M-OKO Oil	19.6	gal.
Mineral Spirits	41.0	gal.
Kerosene	15.9	gal.
Pb Naphthenate 24%	.47	gal.
Co Naphthenate 6%	.10	gal.

Flat Wall Paint—High Dry Hiding Type

(To reduce 8-2 for brushing)

Formula No. 1

Pigment (72.4%)

Albalith—351	1025	lb.
Whiting	175	lb.
Aluminum Stearate	6	lb.

Vehicle (27.6%)

Z-3 Kettle Bodied Linseed Oil	5.06	gal.
Refined Linseed Oil	8.02	gal.
Blown Linseed Oil	4.97	gal.
N-Glo-5	11.03	gal.
Mineral Spirits	13.15	gal.
Kerosene	18.38	gal.
Combined Naphthenate Drier	1.53	gal.
1% Ivory Soap Solution	0.76	gal.

No. 2

Brushing Consistency

Pigment (64.5%)

Albalith 361 or 362	780	lb.
Whiting	150	lb.

Vehicle (35.5%)

T&W 600 Liquid	45¼	gal.
Refined Linseed Oil	1½	gal.
Mineral Spirits	18¼	gal.
Kerosene	5¾	gal.
No. 8 Pale Japan Drier	1⅜	gal.
Soap Solution	¾	gal.

If higher consistency is desired either the pigmentation may be increased or part of the Albalith-361 or 362 may be replaced with Albalith-424.

No. 3

Reduces One Quart to the Gallon

Pigment (68%)

Albalith—362	670	lb.
Albalith—404	155	lb.
Asbestine	206	lb.

Vehicle (32%)

V-169	20½	gal.
N-Glo-5	6¼	gal.
Mineral Spirits	21¼	gal.
V-201	6	gal.
Kerosene	10	gal.
Liquid Drier *	1	gal.

* A portion of the magnesium silicate may be replaced with diatomaceous silica if a reduction in sheen is desired.

* The liquid drier is composed of 100 parts lead naphthenate drier (16% Pb), one part cobalt naphthenate drier (4% Co) and one part manganese naphthenate drier (4% Mn).

1% Ivory Soap Solution	½	gal.
V-214	2	gal.

Prepare a paste using all the pigment, V-169, V-201, N-Glo-5, kerosene and as much of the mineral spirits as needed to produce the desired paste consistency. Grind on roller mill or buhrstone mill. Thin with the remaining mineral spirits and driers.

No. 4

(Eggshell Flat) Paint

Pigment (70%)

Albalith Black Label—11	850	lb.
Whiting	180	lb.
Calcium Linoleate Pulp (30% solids)	32	lb.
Ultramarine Blue	3½	oz.

Vehicle (30%)

Z-3 Bodied Linseed Oil	20	gal.
Diamond K Linseed Oil	½	gal.
N-Glo-5 Gloss Oil (Newport Industries)	4½	gal.
Kerosene	6¾	gal.
Mineral Spirits	31¾	gal.
Lead Naphthenate Drier (16%)	½	gal.
Cobalt Naphthenate Drier (6%)	¼	gal.

No. 5

Pigment (67.8%)

Cryptone BA-19, BT-367 or BT-803	600	lb.
Whiting	335	lb.
Calcium Linoleate Pulp (30% solids)	30	lb.
Ultramarine Blue	3	oz.

Vehicle (32.2%)

Same ingredients, percentages and gallons as No. 1.

Mix all pigment ingredients with the Z-3 bodied linseed oil, Diamond K linseed oil, gloss oil, and sufficient mineral spirits to give a suitable paste. Mill through a roller or buhrstone mill and thin with the remaining vehicle constituents.

No. 6

Red Label Albalith	980	lb.
Asbestine	137	lb.
Aluminum Stearate	5	lb.
Litharge	2	lb.
Zinol	12	lb.
10–lb. Ester Gum Cut	124	lb.
A Soybean Oil	142	lb.
Kerosene	33½	lb.
Mineral Spirits	231	lb.
16% Lead Naphthenate	7	lb.
6% Cobalt Naphthenate	1	lb.

No. 7

Titanium Calcium Pigment	700	lb.
Asbestine	150	lb.
Aluminum Stearate	4	lb.
Litharge	2	lb.
Zinol	12	lb.
10–lb. Ester Gum Cut	124	lb.
A Soybean Oil	142	lb.
Kerosene	33½	lb.
Mineral Spirits	231	lb.
16% Lead Naphthenate	7	lb.

Reduced Alkalized Polymers for Controlled Penetration
Flat Wall Coats

	No. 1	No. 2
Crude Non-break Perilla Oil	1535 lb.	....
Heat-Poly. Perilla ("X")	743 lb.	....
Linseed Fatty Acid	7½ lb.	....
V.M. Linseed (Punjab)*	....	1675 lb.
Heat-Poly. Linseed ("X") *	....	303 lb.
Heat-Poly. Linseed ("Q") *	....	151 lb.
Calcium Hydrate	31½ lb.	13 lb.
Magnesium Oxide	....	13 lb.
Light Kerosene	67 gal.	97 gal.
Mineral Spirit	607 gal.	575 gal.
Cleaners' Naphtha	57 gal.	53 gal.

* From minimum 190 iodine number V. M. oil.

Metropolitan Type Gloss Paint
Heavy Body—High Gloss—Non-Sagging

Pigment (53.0%)

Cryptone BT-367	700	lb.

Vehicle (47.0%)

Gloss Oil (N-GLO-5)	27¾	gal.
10# Cut Low Acid Ester Gum	14¾	gal.
Heavy Bodied Blown Soya Oil	2	gal.
Alkali Refined Linseed Oil	8	gal.
Extra-Heavy Kettle Bodied Fish Oil	19¾	gal.
Mineral Spirits	6	gal.
* Lead Linoleate Drier 16%	1¼	gal.
* Cobalt Linoleate Drier 6%	¼	gal.

Mix the pigment with gloss oil, ester gum solution, blown soya oil, and sufficient mineral spirits to form a good paste. Grind on roller mill and reduce with fish oil, refined linseed, and remaining mineral spirits. Finally add driers previously cut in some of the mineral spirits. Final consistency characteristics will be attained after aging 48 hours.

This is a heavy consistency product which can be reduced (8–1½) with mineral spirits for application.

* Percent metal based on oil—0.7% Pb and 0.046% Co.

Floor Painting and Finishing

If the floor is made of spruce, pine, fir or other soft wood, the paint for the priming coat should be mixed on the following basis:

Soft Paste White Lead	100 lb.
Raw Linseed Oil	3 gal.
Turpentine	2 gal.
Liquid Drier	1 pt.
Oil Color for Tinting, To suit	

This mix makes 8⅜ gal. of paint that will prime about 600 sq. ft. to 800 sq. ft. of new soft wood flooring per gallon.

The paint in this first coat, as well as in all other coats, should be applied with a well-filled brush and must be spread out carefully and rubbed in closely. A prime cause of sticky floors is too much oil in the paint or flowing on the paint in any or all coats so thick that it does not dry properly, then hurrying too much with succeeding coats. Spread the paint out well, rub it in "close" and allow plenty of time for drying —three main points in producing a good floor job.

If the floor is made of birch, maple, beech or some other hard wood, use a priming coat mixed as directed above, but cut the linseed oil down to 2½ gal. Oak and walnut are never painted unless old and much worn, scarred and bruised.

After the primer is dry and hard, all open joints, cracks, nail holes and other defects should be filled with putty—preferably of the white lead, whiting and linseed oil variety. A little floor varnish in this putty makes it set harder and stick tighter. Fill all the holes and cracks a little more than full, pressing the putty in with a knife. When the putty has dried hard, smooth it off with sandpaper and clean up the dust.

For the next coat, on either soft wood or hard wood floors, use a paint mixed on this basis:

Soft Paste White Lead	100	lb.
Floor Varnish	1	gal.
Turpentine	2¼	gal.

This paint may be tinted to about the color of the finishing coat.

The above mix makes 6½ gal. of paint that will cover at the rate of 700 sq. ft. to 800 sq. ft. per gallon.

When this coat has dried and hardened thoroughly, apply a final coat of paint mixed on this basis:

Soft Paste White Lead	100	lb.
Floor Varnish	2	gal.
Turpentine	1¼	gal.
Oil Color for Tinting, To suit		

Painting Old Wood Floors

The first thing to do before painting an old wood floor is to get it in shape to receive paint. If the wood has old paint on it that is blistered, rough and scaly, or is thick and gummy it should be stripped off, using an electric floor sanding machine, a torch and scraper, or a paint remover—paste or liquid. This removal of the old paint may be troublesome and expensive but it is necessary if a good repaint job is to result.

Paint or varnish removed with liquid or paste removers commonly sold for that purpose is most practically handled in less time and with less work this way: Brush the liquid or paste on to three or four boards from wall to wall, or coat in a patch about a yard square. Use a broad scraping knife to roughly skin off the sludge. Then have a half pail of hot water in which about one-half package of Gold Dust, or similar washing powder, has been dissolved. Dip a large wad of No. 3 coarse steel wool in the water and scrub off the stripped surface. That will usually take off everything and leave the wood clean and free from wax from the remover. Sometimes two or more applications of the remover are necessary to cut the old paint or varnish enough.

If the paint is taken off with lye or other caustic remover, the surface will have to be washed with strong vinegar water after all the leavings have been cleaned up, in order to neutralize all remaining traces of alkali. Regardless of the method used to remove the paint, the floor should then be thoroughly washed with clean, warm water and allowed plenty of time to dry out. When dry, make sure all floor boards are securely nailed and are firm and solid. Fill the cracks, let dry, sand the cracks level and brush off the dust. Paint as outlined before.

In repainting old wood floors on which the old paint is in good shape, the priming coat is not needed. The floor should first be carefully cleaned, however, and all defects puttied up.

Old wood floors that have never been painted or treated in any way should, of course, be repaired and prepared for painting as suggested.

Painted floors, when hard and dry, may be waxed like varnished floors. Such a finish, if renewed regularly, will protect the paint from scratches and assure a very long life.

Painting Plywood

For concealing the joints in plywood walls and at the same time providing any desired color decorative scheme, the addition of 5% by weight of wood flour to a flat wall paint has shown interesting possibilities. The resultant mixture has a coarse grained corn meal texture, can be applied with a regulation paint brush at a spreading rate of about 180 to 200 sq. ft. per gallon, and may be textured with various tools.

Painting Plaster Statues

If the statues, whether painted or bare, are old and dusty or dirty and have no defective coating of any kind, the only surface preparation needed is a good washing with mild soapy warm water. If the statues do not appear to be very old, and have not been painted, it will be best to insure the neutrality of the plaster by giving the figures a thorough washing with a solution composed of two pounds of zinc sulphate in a gallon of warm water. If there is paint now on the figures and it is cracking and peeling, remove it, using a liquid paint remover. After the paint is removed, rinse the plaster with weak vinegar.

After proper preparation, allow the plaster to dry out thoroughly. For a first coat of paint apply a wall primer. This is a pigmented varnish-type sealer and primer that also has the adherence, spreading capacity and covering power of an actual coat of paint.

Since statuary is usually done in flat finishes, the next coat may be a white lead and mixing oil or white lead and flatting oil paint. If mixing oil is used the paint should consist of equal volumes of white lead and mixing oil. If flatting oil is used, mix the paint on the basis of 25 lb. of white lead to a half-gallon of flatting oil. In either case tint the paint for the proper portions of each statue with colors in oil. These colors have unusually high tinting strength so be sure to add them to your paint, just a small quantity at a time, until the correct tint is reached.

Cement Paint
Australian Patent 113,168

Cement	24
Diatomaceous Earth	1
Linseed Oil, Raw	14
Turpentine	2
Magnesium Carbonate	2

Cement Coating
U. S. Patent 2,244,449

This new cement type coating is a liquid composition of creamy consistency. It is adapted to form an enamel-like coating on an unpolished non-metallic surface. The composition should be applied while the surface is damp. It is allowed to set and cure while kept damp. The composition consists of the product obtained by mixing together the following ingredients:

Portland Cement	100 pt.
Pigment	10 pt.
Calcium Stearate	½–1 pt.
Calcium Chloride	45 pt.

The pigment used is selected from the group consisting of chromium oxide, iron oxide, cobalt oxide, magnesium oxide and aluminum oxide. The aqueous solution of calcium chloride is obtained from dissolving a quantity of calcium chloride in twice the quantity by weight of water.

Plastic Coating for Brick and Cement
U. S. Patent 2,228,061

Magnesia	20–50
Silica	10–50
Calcium Carbonate	5–25
Fluorspar	1–10
Zinc Oxide	1–15
Magnesium Chloride Solution, To suit (d. 20–40° Bé)	

Wall Protective Coating
U. S. Patent 2,260,882

Transparent in dry state and water-soluble.

Bentonite	1–12
Gum Arabic or Tragacanth	½– 1
Water, To make	100

Stipple Paint
Pigment (77.5%)

Albalith 387 or Black Label 11	1122	lb.
Magnesium Silicate	281	lb.

Vehicle (22.5%)

White Refined Linseed Oil	26¼	gal.
Mineral Spirits	15½	gal.
Kerosene	14¼	gal.
Liquid Drier *	¾	gal.

Mix the pigments with the oil and thinners. Grind and reduce with the liquid drier.

Possesses good brushing and sets sufficiently slow to allow laps to be picked up after 15 minutes. Kerosene substituted for some of the mineral spirits will further extend this time and has no effect on the final hardness of the film. To lower the consistency the following suggestions are offered:

1. Whiting for magnesium silicate.

2. Lower consistency pigment as Albalith 386 or Red Label 14.

3. Kettle bodied linseed oil for part of the refined oil and reduce the total pigment to maintain the same brushing.

* Percent metal based on vehicle non-volatile %Pb 0.5, %Mn 0.005, %Co 0.005.

Painting Over Tar, Asphalt or Creosoted Surfaces

First apply a priming coat of shellac or aluminum paint to act as a seal.

Painting Linoleum

Scrub the floor thoroughly to remove all traces of dirt and grease. If the linoleum has been waxed, remove the wax with turpentine, because no paint will get a good anchorage on wax. With the floor clean, apply a priming coat mixed on this basis:

Soft Paste White Lead	12½	lb.
Raw Linseed Oil	1½	pt.
Turpentine	1½	pt.
Liquid Drier	2	oz.

Spread this primer out thin and brush it in well. Let it dry and harden for at least a week. Then apply a body coat mixed as follows:

Soft Paste White Lead	12½	lb.
Raw Linseed Oil	1	pt.
Turpentine	1	qt.
Liquid Drier	2	oz.

This paint should be tinted an ivory shade with raw sienna. Allow the paint at least a week to dry and harden, then apply a coat of good quality prepared floor enamel of the selected shade.

Repainting Golf Balls

To remove old paint from golf balls, dip the balls for a few minutes in a warm 10% sodium hydroxide solution. Follow this by scrubbing, washing and drying. Refinish with two or three coats of exterior enamel.

White Lead Finish Coat Paints

Formula No. 1—White

Soft Paste White Lead	100	lb.
Raw Linseed Oil	3¼	gal.
Liquid Drier	1	pt.

No. 2—Silver Gray

Soft Paste White Lead	100	lb.
Paste Lampblack	⅛	pt.
Raw Linseed Oil	3¼	gal.
Pure Turpentine	1	pt.
Liquid Drier	1	pt.

No. 3—Medium Gray

Soft Paste White Lead	100	lb.
Paste Lampblack	¼	pt.
Raw Linseed Oil	3¼	gal.
Pure Turpentine	1	pt.
Liquid Drier	1	pt.

No. 4—Foliage Green

Soft Paste White Lead	100	lb.
C. P. Chromium Oxide Green	1	gal.
Raw Linseed Oil	2½	gal.
Spar Varnish	1	gal.
Pure Turpentine	1	qt.
Liquid Drier	1	qt.

No. 5—Gray-Green

Soft Paste White Lead	100	lb.
C. P. Chromium Oxide Green	½	gal.
Paste Lampblack	½	pt.
Raw Linseed Oil	2½	gal.
Spar Varnish	½	gal.
Pure Turpentine	1	qt.
Liquid Drier	1	pt.

Free Flowing Flat White Paint

	lb.	gal.
Unilith 88	800	22.3
Suspenso Whiting	200	8.9
VM-411 Flat Liquid	268	34.8

Varnolene	224	34.0

Weight gallon = 14.92 lbs.

Vehicle non-volatile (by weight) = 35%.

100 gallon oil length.

VM-411 is a 4% limed rosin-chinawood-linseed oil varnish, 66% non-volatile, with a viscosity of w-x. The weight per gallon is 7.75 lbs.

Paint consistency (Stormer) = 80.

Thixotropic Flat Paint

	100 gal.	
"Ti-Cal" TC	533	lb.
Diatomaceous Silica	58	lb.
Whiting	132	lb.
* 50-Gallon Ester Gum Varnish	24.3	gal.
Body Z Perilla	10.0	gal.
Kerosene	5.3	gal.
Mineral Thinner	29.0	gal.
Co Naphthenate 6%	1.75	pt.
Pb Naphthenate 24%	1.5	pt.
Soap Solution 1%	1.45	gal.

Gallon weight = 12.2 lbs.

Pigment volume = 57% = 1/0.75-P/B.

Non-volatile in vehicle = 39% by weight.

Oil length of vehicle = 100 gallons.

Consistency (Stormer) = 132 (choppy).

White Enamel

	lb.	gal.
Tidolith 88 N	700	19.8
D-633 Varnish	625	80.2
Cobalt Nuodex 6%	3	.4

Weight/gallon = 13.28 lbs.

Vehicle non-volatile (by weight) = 57.0%.

* This varnish weighs 7.4 lb./gal. and contains 60% N. V. by weight.

Vehicle oil length = 30 gallons.

Vehicle weight/gallon = 7.50 lbs

D-633 is an ester gum-chinawood-linseed oil varnish of viscosity G.

Enamel consistency (Stormer) = 93.

Outside House Paint

"Ti-Sil"	310	lb.
35% Leaded Zinc Oxide	370	lb.
Magnesium Silicate	147	lb.
Q Body Linseed Oil	4.8	gal.
Alkali Refined Linseed Oil	57.0	gal.
Drier P	2.8	gal.
Mineral Thinner	9.4	gal.

Gallon weight = 13.92 lbs.

Pigment volume = 28.5%.

Non-volatile in vehicle = 89% by weight.

Consistency (Stormer) = 87 (short).

Black-Out Paint

Formula No. 1

Carbon Black (Low Color type—Low Oil Absorption)	7½	lb.
China Clay	145	lb.
Short Oil Rosin Varnish	7½	gal.
Short Oil Gilsonite Varnish	9	gal.
Aluminum Stearate	2	lb.
Mineral Spirits	6½	gal.

Obvious adjustments should be made for other colors.

No. 2

Carbon Black (Low Color type—Low Oil Absorption)	5	lb.
Clay	75	lb.
Vehicle A or B	80	lb.
Dipentine	5	lb.
Mineral Spirits	55	lb.

Vehicle A

Batu Resin	150 lb.
Refined Soya Bean Oil	25 lb.
No. 8 Litho Linseed Oil	25 lb.
Hi-Flash Naphtha	175 lb.
Mineral Spirits	25 lb.

Vehicle B

Batu Resin	150 lb.
Paraffin Oil	50 lb.
Hi-Flash Naphtha	150 lb.
Mineral Spirits	50 lb.

No. 3

Belro Resin-Linseed Oil Varnish

Vehicle A

Belro Resin	200 lb.
23 Viscosity Heat-Bodied Linseed Oil	150 lb.
Varnish Makers' Litharge	3 lb.
Electrolytic Manganese Dioxide	½ lb.
Hydrated Lime	3 lb.

Heat the linseed oil and 75 lb. Belro Resin to 350° F. in about 20 minutes. Add the lime slowly while raising temperature to 450° F. (about 10 minutes). Add balance of rosin and reheat to 450° F. in about 20–25 minutes. Then add mixed driers and work in well. Heat to 565° F. and hold at 565° F. about 1¼ hours for a medium hard pill. Cool to 450° F. and reduce to 50% solids with 75 parts mineral spirits and then 25 parts of VM&P Naphtha.

It is important that the liming operation and additions of driers be *carried out slowly*. Otherwise severe foaming will result. Belro Resin foams to a greater degree than ordinary rosin.

Gilsonite—Linseed Oil Varnish

Vehicle B

Gilsonite Selects	500 lb.
Y Viscosity Heat-Bodied Linseed Oil	320 lb.
Varnish Makers Litharge	1 lb.
Electrolytic Manganese Dioxide	1 lb.

Heat Gilsonite and linseed oil to 450° F. in about 40 minutes. Work in mixed driers. Then heat to 565° F. in about 20 minutes and hold at 565° F. for 30 minutes. Cool to 450° F. and reduce with mineral spirits to 45% solids.

Paint Formulation

Grind

Excelsior No. 1 Carbon Black	75 lb.
No. 1290 China Clay	450 lb.
Vehicle A (Belro Varnish)	620 lb.
Vehicle B (Gilsonite Varnish)	620 lb.
Aluminum Stearate	20 lb.

Reduce

VM&P Naphtha	420 lb.
3% Cobalt Naphthenate Solution	3.3 lb.

The paste may be ground through either a buhrstone or roller mill. Two passes give a good grind and one pass should be sufficient. If a pebble mill is employed the VM&P Naphtha used for reducing would probably be included in the charge to secure efficient grinding.

Paint made according to this formula is suitable for brushing or spraying to produce complete hiding with one coat. Coverage on glass as determined by the Gardner Brushout Test is 750 square feet per gallon. For actual coverage under conditions of use expect about 600 square feet per gallon. It dries dust free in less than one hour on glass.

The dried film is a very flat gray-black finish which will pass all the tests that have been suggested for a paint of this kind. Belro yields better results in this type of paint than ordinary rosin and its use is recommended in preference to any other grade of wood or gum rosin. It is available in large quantities and is the lowest cost rosin available at this time.

Camouflage Paint for Glass

A method for treating exterior glass surfaces of buildings so that they are inconspicuous at a distance has been devised. A film of transparent binding liquid is formed on the surface and particles of an aggregate, which have been coated with coloring material, are embedded in it. The particles are of such size that they are dispersed on the surface so that the reflecting property is diminished while light is permitted to be transmitted through the gaps between the particles.

In application, the binding medium is first applied and while this is tacky, the color-coated aggregate is dusted on, so that the particles are dispersed over the surface. An alternative procedure consists in mixing the color-coated aggregate with the binding medium and then applying this mixture to the glass by spraying. A satisfactory binding liquid can be made on the following formulation:

Formula No. 1

Oil-Rosin Mixture (1:3 to 1:1 by Volume of Melted Rosin to Linseed Oil)	1 gal.
Turpentine (More if Rapid Drying Is Required)	2 pt.
Pure Green (Or Other Color if Desired)	3 lb.

The following mixture has been found to be suitable for the aggregate:

Dry Sand	1 ton
Raw Linseed Oil	1½ gal.
Turpentine	10 pt.
Color	16 lb.

A volatile solvent, such as a petroleum thinner, may be added as a diluent and in some cases alum is desirable as a fixative for the color. The liquids are first mixed and then fed with the sand in the mixing machine. After being intimately mixed, the powder is spread out to dry.

The proportion of color-coated aggregate used may be varied but it has been found most effective to treat the glass with binding liquid prepared as above, in quantities of about 35 to 55 superficial yards per gallon and with aggregate in quantities of about 40 to 80 superficial yards per hundredweight.

No. 2

Dilute with carbon tetrachloride, Graphite Suspension "Dag," Type 304, to 1/10th its original concentration and paint on with fine brush.

Camouflage Paints

Light Green No. 1

Pigment	56 %
C.P. Chromium Oxide	12.0 %
Yellow Ochre	17.3 %
Yellow Iron Oxide	2.9 %
Red Iron Oxide	0.35%
Lithopone	16.5 %
Magnesium Silicate	38.45%
Zinc Oxide	12.5 %
Vehicle	44 %

Z Kettle Bodied Linseed Oil	40 %
Mineral Spirits and Drier	60 %

Dark Green No. 2

Pigment	56 %
C.P. Chromium Oxide	11.8 %
Yellow Iron Oxide	8.5 %
Red Iron Oxide	0.6 %
Mineral Black	1.1 %
Lithopone	10.5 %
Zinc Oxide	12.5 %
Magnesium Silicate	55.0 %
Vehicle	44 %
Z Kettle Bodied Linseed Oil	40 %
Mineral Spirits and Drier	60 %

Sand No. 3

Pigment	63 %
Yellow Ochre	25 %
Mineral Black	0.5 %
Lithopone	62 %
Zinc Oxide	12.5 %
Vehicle	37 %
Z Kettle Bodied Linseed Oil	40 %
Mineral Spirits and Drier	60 %

Field Drab No. 4

Pigment	60 %
Yellow Ochre	28 %
Red Iron Oxide	1.3 %
Mineral Black	1.5 %
Lithopone	38 %
Zinc Oxide	12.5 %
Magnesium Silicate	18.7 %
Vehicle	40 %
Z Kettle Bodied Linseed Oil	40 %
Mineral Spirits	60 %

Earth Brown No. 5

Pigment	58 %
Yellow Ochre	55 %
Mineral Black	2 %
Lithopone	11.5 %
Zinc Oxide	12.5 %
Magnesium Silicate	19 %
Vehicle	42 %
Z Kettle Bodied Linseed Oil	40 %
Mineral Spirits	50 %

Earth Yellow No. 6

Pigment	65 %
Yellow Ochre	19.1 %
Yellow Iron Oxide	17.0 %
Burnt Sienna	0.4 %
Lithopone	51.0 %
Zinc Oxide	12.5 %
Vehicle	35 %
Z Kettle Bodied Linseed Oil	40 %
Mineral Spirits	60 %

Loam No. 7

Pigment	55 %
Yellow Iron Oxide	25.4 %
Mineral Black	4.8 %
Lithopone	6.7 %
Zinc Oxide	12.5 %
Magnesium Silicate	50.6 %
Vehicle	45 %
Z Kettle Bodied Linseed Oil	40 %
Mineral Spirits and Drier	60 %

Earth Red No. 8

Pigment	56 %
Burnt Sienna	11.5 %
Yellow Ochre	17.65%
Mineral Black	0.35%
Lithopone	23.0 %
Zinc Oxide	12.5 %
Magnesium Silicate	35.0 %
Vehicle	44 %
Z Kettle Bodied Linseed Oil	40 %
Mineral Spirits and Drier	60 %

Olive Drab No. 9

Pigment	56 %
C.P. Chromium Oxide	1.9 %
Yellow Ochre	31.0 %

Mineral Black	3.3 %
Lithopone	11.3 %
Zinc Oxide	12.5 %
Magnesium Silicate	40.0 %
Vehicle	44 %
Z Kettle Bodied Linseed Oil	40 %
Mineral Spirits and Drier	60 %

White No. 10

Pigment	62 %
Lithopone	40 %
Zinc Oxide	20 %
Magnesium Silicate	40 %
Vehicle	38 %
Z Kettle Bodied Linseed Oil	40 %
Mineral Spirits and Drier	60 %

Camouflage Paint Pigments

Light Green

Chromium Oxide	86
Yellow Iron Oxide or Yellow Ochre	50
Red Iron Oxide	1

Dark Green

Chromium Hydrate	43
Yellow Iron Oxide or Yellow Ochre	11
Red Iron Oxide	2

Sand

White (Lithopone)	108
Chromium Oxide	7
Yellow Iron Oxide	7
Red Iron Oxide	3

Field Drab

Yellow Iron Oxide	8
Lithopone	54
Chromium Oxide	13
Iron Oxide	3

Earth Brown

Chromium Oxide	60
Yellow Iron Oxide	24
Iron Oxide Red	12

Earth Yellow

White (Lithopone)	40
Yellow Iron Oxide	30
Chromium Oxide	5
Iron Oxide Red	3

Loam

Burnt Umber	11
Chromium Oxide	15
Black Iron Oxide	7
Iron Oxide Red	3
Yellow Iron Oxide	1

Earth Red

White (Lithopone)	45
Yellow Iron Oxide	16
Iron Oxide Red	6
Chromium Oxide	10

Olive Drab

Chromium Oxide	47
Yellow Iron Oxide	13
Red Iron Oxide	6

Black

Mineral Black

White

Lithopone

Military Paint Pigments

Metal Primer

(Brown)

Zinc Yellow	7–20
Iron Oxide, Red	35
Zinc Oxide (Not to Exceed That of Zinc Yellow)	0–20
Tinting Pigments	Less than 1
Extenders	Less than 58

Fluorescent Powder

British Patent 530,021

Formula No. 1

Calcium Oxide	20
Molybdenum Oxide	79
Lead Acetate	1
Samarium Oxide	0.1–3

Fire at 1000° C. for 4 hours.

British Patent 535,897
No. 2

Magnesium Tungstate	10%
Zinc Beryllium Silicate	62%
Cadmium Borate	28%

This fluoresces with a white light in electric discharge lamps.

British Patent 532,941
No. 3

A mixture of 20–45% magnesium tungstate, 40–70% zinc-beryllium silicate, and 6 to 18% cadmium borate is used.

Fluorescent Cadmium Silicate
No. 4

450 g. cadmium oxide and 420 g. silica are mixed together with 10 g. manganese chloride. The mixture is dried well above 100° C. for several hours, then heated at 1000° C. for 2 to 3 hours.

Fluorescent Magnesium Silicate
No. 5

300 g. magnesium oxide, 500 g. silica and 16 g. manganese nitrate are mixed together and heated to 1200° C. for several hours.

Fluorescent Zinc Tungstate
No. 6

225 g. zinc oxide and 700 g. tungstic oxide with an admixture of 1.5 g. lead nitrate are formed into a paste with pure water. The mixture is dried, then ground, and heated to 1000° C. for several hours.

Fluorescent Cadmium Tungstate
No. 7

375 g. cadmium oxide and 400 g. tungstic oxide are mixed together, and heated to 1000° C. for one hour. The product shows fluorescence of moderate intensity. Its brightness can be increased by grinding and refiring at 1100° C. for 1½ hours.

Fluorescent Calcium Molybdate
No. 8

580 g. purified and dried molybdic acid and 400 g. calcium carbonate are mixed together, and heated to 850° C. for about two hours.

Zinc Sulphide, Fluorescing Blue
(Laboratory Method)
No. 9

Pure hydrogen sulphide is passed into a neutral solution of zinc chloride. The precipitate of zinc sulphide is filtered, washed and dried. For each gram of zinc sulphide add 0.05 g. sodium chloride + 0.05 g. magnesium fluoride + 0.0003 g. silver nitrate, grind well, and heat in a crucible for 10 minutes at 900° C.

Fluorescent Zinc Borate
No. 10

600 g. purest zinc oxide and 460 g. purest orthoboric acid are mixed together with 3.4 g. manganese nitrate. The dried mixture is fired at 750° C. for one hour. A material showing deep red luminescence is obtained in this way.

Fluorescent Cadmium Phosphate
No. 11

An aqueous solution of 300 g. very pure ammonium phosphate is added to a similar solution of 630 g. cadmium sulphate. The precipitate thus formed is filtered off, washed, and well mixed with 11 g. manganese chloride. The dry mixture is fired at 1000° C. for about two hours. The resultant powder shows a light red luminescence.

British Patent 521,796
No. 12

A fluorescent material may be composed of zinc oxide (60), silica (40), beryllium oxide (1.5–2.5), manganese oxide (3–5), and one-half part lithium chloride. The manganese acts as the activator and the mixture is fired at 1000° C.

Luminescent Pigment for Ceramics
British Patent 487,520
No. 13

Titanium Dioxide	47 %
Magnesium Oxide	48 %
Beryllium Oxide	5 %
Chromium	0.1 %

Wet grind; dry; mold and fire at 1400–1550° C.

British Patent 536,305
No. 14

Zinc Oxide	324 g.
Beryllium Oxide	100 g.
Silica	276 g.
Manganese Chloride	19.6 g.
Arsenic Trioxide	30 mg.

Mill wet; dry; roast at 1220° C.

Green Pigment
No. 15

Dissolve one ounce of analytically pure zinc chloride in 4 ounces of distilled water. Add a trace of manganese sulphate (usually less than 1%). Bring the solution to a boil and add an equal volume of sodium silicate (the flux). The whole is evaporated to dryness.

The above evaporated solid is then placed in a porcelain crucible and heated to a bright red for 2 hours. When cool, this pigment will show a pale green fluorescence and a bright green phosphorescence when exposed to short wave lengths (cold quartz tube) of ultraviolet. Heating the pigment to a bright red for an additional 3 hours will increase both the fluorescence and the phosphorescence. Longer heating will tend to quench the luminescence.

Red Pigment
No. 16

Substituting cadmium chloride for zinc chloride in the above formula, but giving it exactly the same treatment, it was found necessary to heat the compounds to a bright red for 3 hours before phosphorescence became very intense. Subsequent heating appeared to have no marked effect upon luminescence. The fluorescence is pink and the phosphorescence is dark red.

Orange Pigment
No. 17

Some cadmium sulphate is dissolved in distilled water with a small trace of manganese sulphate (less than 1%) and the whole heated to dryness without adding the flux. The resulting white powder fluoresces pale yellow, and shows an orange yellow phosphorescence of much longer duration than the zinc compounds.

Zinc sulphate may be substituted for cadmium sulphate in the above formula and treated in the same manner. The resultant white powder fluoresces pink and phosphoresces red.

Some "chemically pure" calcium sulphide, as sold on the market, may be found to be already an active hosphorescent compound having a rownish-red color.

U. S. Patent 2,270,105
No. 18

Prepare a solution A of pure sodium tungstate dissolved in distilled water in the proportions of 108 grams of sodium tungstate to 1 liter of water, prepare a substantial equivalent amount of solution B of pure cadmium sulphate dissolved in distilled water in the proportions of 154 grams of cadmium sulphate to 1 liter of water, add approximately 1% of solution A to the whole of solution B and filter off the precipitate, add to the resulting filtrate the remainder of solution A and again filter off the precipitate, then wash the latter precipitate in distilled water, dry it, and heat it at approximately 875° C. for one hour in air.

No. 19

(1) 40 g. barium carbonate, 6 g. sulphur, 1 g. lithium carbonate and 0.47 g. rubidium carbonate are heated at dark red heat for 1 hr., powdered and mixed with a suitable pigment. (2) 100 parts calcium carbonate and 30 parts sulphur are heated at dark red heat for 1 hr. This is then ground with alcohol to which an amount of bismuth nitrate equal to 0.0001 the weight of the melt has been added. Finally the mixture is dried in air, heated at dark red heat for 2 hrs. and allowed to cool slowly.

British Patent 525,379
No. 20

Cadmium Phosphate	100
Manganous Chloride	5
Magnesium Chloride	10

Mix dry and heat in a silica tube, plugged with glass wool, for 3 hours at 800° C. Cool and wash with distilled water, six times, and dry at 180° C.

Luminescent Coatings for Discharge Lamps
U. S. Patent 2,257,667
Formula No. 1

For the preparation of a thallium-activated composition, germanium dioxide or silicic acid is mixed with 1–2 times its quantity of a pure oxide of beryllium, magnesium, or aluminum, up to about 30% of a thallium salt is added and the mixture is heated to 800–1200° C.

U. S. Patent 2,255,761
No. 2

Luminescent particles can be held in a porous structure comprising aluminum oxide or magnesium oxide.

British Patent 530,915
No. 3

The luminescent tungstate material contains a proportion of vanadium, less than 1% but so great that the loss of luminous output during life is materially decreased by its presence.

French Patent 845,284
No. 4

Zinc Oxide	60
Silicon Dioxide	40
Beryllium Oxide	2
Manganese Dioxide	4
Lithium Chloride	1

Calcine for 3 hours at 1000–1100° C. and slake with water.

British Patent 538,379
No. 5

Magnesium Tungstate	15
Zinc Beryllium Silicate	70
Cadmium Borate	5
Zinc Ortho Silicate	10

Fluorescent Paint
British Patent 542,006

Chrysene together with, if desired, not more than 1/30 of its weight of naphthalene (not more than 1%) is dispersed in, e.g., an aqueous solution containing gum arabic 10%, phenol 0.2%, glycerin 3%, and Darvan (a dispersing agent) 1.5%.

Luminous Paint

The pigment content per gallon of finished paint should be about 4½ lb. The vehicle must be free from lead, manganese and cobalt driers. Drying qualities are attained by using a fast drier, such as benzine. No grinding of the pigment and vehicle is recommended, as it injures the light properties. The vehicle should be reasonably clear. Some oils, such as linseed, cause a jellying. Soya bean oil has been found satisfactory in a synthetic alkyd varnish.

The batch is machine mixed and allowed to stand over-night, after which it is given a short mix and placed in cans, preferably of sizes not larger than single gallons.

Calcium Sulphide Luminous Pigment is non-poisonous and non-radioactive. The paint will keep in cans over a long period. There is no chemical action to cause deterioration.

Luminous Paint is best applied over a base coat of white paint fre from lead. This is imperative whe it is applied over old paint whic might contain lead.

A strong afterglow of better tha 12 hours can be had with a goo calcium pigment. The calcium sulphide has advantages over zinc an strontium sulphide luminous paint the most outstanding being lon afterglow and freedom from discoloration when exposed to direc sunlight.

Road Markers for Night Use

Fluorescent signposts (made wit fluorescent pigments) and signs ca be used which are either visible o invisible in ordinary light.

Fluorescent liquids, oils, or powders dropped on the trail or road will facilitate traveling at night by automobile, tank, or other vehicles which are provided with an ultraviolet lamp.

Improving Life of Fluorescent Powders
German Patent 702,995

The resistance of luminescent alkaline earth sulphides to air, light, and weathering is enhanced when about 10% boric anhydride is added.

Television Screen
U. S. Patent 2,260,924

In preparing a fluorescent screen for use in television use is made of manganese activated cadmium silicate (55), silver activated zinc sulphide (30), and manganese activated zinc silicate (15), for a screen of substantially white fluorescence having an afterglow of about 0.02 second for eliminating flicker.

Porcelain Enamel Patching

Formula No. 1

Tenex	35
Gelva (2.5)	45
Titanium Dioxide	20

No. 2

Gelva (2.5)	55
Cumar	20
Titanium Dioxide	25

No. 3

White Rosin	40
Carnauba Wax	20
Titanium Dioxide	40

Melt first two ingredients and sift into it the titanium dioxide. Mix well.

20–Gallon Floor Enamel Liquid

Formula No. 1

Modified Alkyd Resin	112	lb.
100% Phenolic Resin	28	lb.
Perilla Oil	216	lb.
Basic Carbonate White Lead	7	lb.
Mineral Spirits	66	gal.

No. 2

Modified Phenol Alkyd Resin No. 227	150	lb.
Tung Oil	24	lb.
Polymerized Linseed Oil "Z-2"	216	lb.
Basic Carbonate White Lead	3¾	lb.
Mineral Spirits	68	gal.

Floor Varnishes and Floor Enamel Vehicles

15–20 Gallon (East India)

Batu, Pale East India or Black East India	100	lb.
Kettle Bodied Linseed or Dehydrated Castor Oil	15–20	gal.

ENAMELS

	Formula No. 1		*Formula No. 2*		*Formula No. 3*	
Grind	*Lb.*	*Gal.*	*Lb.*	*Gal.*	*Lb.*	*Gal.*
Lithopone (Enamel Grade)	500	13.50	...	...	...	...
Titanated Lithopone	...	...	300	8.60	300	8.60
Extender (Surfex)	...	...	200	9.10	200	9.10
Kettle Bodied Linseed Oil	60	7.50	...	...	...	...
Falkovar 203-UD	...	...	60	7.50	60	7.50
Mineral Spirits	40	6.15	40	6.15	40	6.15
Thin with						
Kettle Bodied Linseed Oil	100	12.50	...	...	...	...
Falkovar 203-UD	...	...	100	12.50	...	...
Falkovar SKA	...	...	...	...	188	23.50
66% Ester Gum Cut	90	11.25	90	11.25	...	...
Mineral Spirits	35	5.40	35	5.40	37	5.70
Blown Soya Bean Oil	8	1.00	8	1.00	8	1.00
Cobalt Naphthenate (6%)	1	.12	1	.12	1	.12
Lead Naphthenate (24%)	2.5	.25	2.5	.25	2.5	.25

Mineral Spirits to 50% non-volatile content. Driers equivalent to 0.10% Ca., 0.10% Co., and 0.20% Pb., as metals based on weight of oil present.

Heat resin and 5 gal. of oil to 620° F., in about 1 to 1¼ hours. Hold until cold "pill" is clear. Add remainder of oil. Reheat to 580° F. Allow to cool sufficiently to add mineral spirits and driers.

Indoor Enamel Vehicles

15 Gallon Length (East India)

Batu, Pale East India or Black East India	100 lb.
Bodied Linseed Oil or Dehydrated Castor Oil	15 gal.

Petroleum thinner to 50% non-volatile content by weight. Driers equivalent to 0.10% Ca., 0.10% Co., and 0.20% Pb., as metals on weight of oil.

Heat the resin and 5 gallons of oil to 600°–610° F., in about 60–65 minutes. Hold until a cold "pill" indicates complete compatibility (about 10 minutes). Add remainder of linseed oil. Heat to 580° F. Hold until a cold bead is clear (about 5–10 minutes). Allow to cool to about 400° F. Add the petroleum thinner and then the driers.

---

15 Gallon (Congo)

Thermally Processed Congo	100 lb.
Kettle Bodied Linseed or Dehydrated Castor Oil	15 gal.

Petroleum thinner to 50% non-volatile by weight. Driers equivalent to 0.10% Ca., 0.10% Co., and 0.20% Pb., as metals based on th weight of oil.

Heat the Congo to 650°–670° F in about 1½ hours. Hold until foaming subsides and the hot resin run off the paddle similarly to hot oil Allow to cool.

Heat the processed Congo and al of the linseed oil to 580° F. in abou 1 hour. Allow to cool to 400° F. an thin. Add the driers.

Vehicle for White Enamels

Batavia Dammar A/D or A/E	75 lb.
Thermally Processed Congo	25 lb.
Kettle Bodied Linseed or Dehydrated Castor Oil	80 lb.

V. M. & P. naphtha to 55% non-volatile. Driers equivalent to 1.0% Pb., 0.04% Co., and 0.02% Mn., as metals on weight of oil.

(1) Prepare the Congo by processing in the usual manner (See 15 gallon Congo-General Utility Varnishes).

(2) Heat the dammar, processed Congo and oil to 560° F. Allow to cool sufficiently to thin. When thoroughly cool, add driers.

Non-Yellowing Utility Enamel

Titanium Dioxide, Color Retentive	288 lb.
Zinc Oxide, Fine Particle Size Reactive	48 lb.
Soya Type Modified Alkyd (32% Phthalic Anhydride), 65% Mineral Spirits Solution	226 lb.
Lead Naphthenate 24%	8 lb.

Grind on 3 roller mill and reduce aste with:

Soya Type Modified Alkyd (32% Phthalic Anhydride), 65% Mineral Spirits Solution	340	lb.
Mineral Spirits	114	lb.
Cobalt Naphthenate 6%	1.25	lb.

Non-Yellowing Super White Architectural Enamel

Easy Lapping and Non-Sagging)

Titanium Dioxide, Color Retentive	288	lb.
Zinc Palmitate	4	lb.
Zinc Oxide, Fine Particle Size Reactive	32	lb.
Soya Type Alkyd (29% Phthalic Anhydride), 85% Mineral Spirits Solution	288	lb.
Lead Naphthenate 24%	3	lb.

Mix, grind and thin paste as follows:

Soya Type Alkyd (35% Phthalic Anhydride), 50% Mineral Spirits Solution	147	lb.
Soya Type Alkyd (29% Phthalic Anhydride), 85% Mineral Spirits Solution	75	lb.
Mineral Spirits	159	lb.
Cobalt Naphthenate 6%	3	lb.
Lead Naphthenate 24%	6	lb.

Tint as desired.

ow Cost Synthetic White Enamel

Titanium Dioxide (Calcium Base)	800	lb.
Surfex	200	lb.
Ester Gum Modified Alkyd (15% Phthalic Anhydride), 80% Mineral Spirits Solution	635	lb.
Mineral Spirits	175	lb.
Cobalt Naphthenate 6%	4	lb.
Lead Naphthenate 24%	9	lb.

Synthetic Trim and Trellis Enamel

Formula No. 1

Brown

Aluminum Stearate	1	lb.
Linseed Oil Type Modified (29% Phthalic Anhydride), 70% Mineral Spirits Solution	113	lb.
Iron Oxide, Brown	200	lb.
Mineral Spirits	52	lb.
Lead Naphthenate 24%	5	lb.

Grind fine and thin paste with:

Linseed Oil Type Modified (29% Phthalic Anhydride), 70% Mineral Spirits Solution	402	lb.
Linseed Oil, Raw	42	lb.
Mineral Spirits	149	lb.
Cobalt Naphthenate 6%	2	lb.

No. 2

Green

Green, Medium Chrome	200	lb.
Zinc Oxide	10	lb.
Linseed Oil Type Modified (29% Phthalic Anhydride), 70% Mineral Spirits Solution	404	lb.

Grind and thin paste as follows:

Linseed Oil Type Modified (29% Phthalic

Anhydride), 70% Mineral Spirits Solution	336	lb.
Linseed Oil, Raw	180	lb.
Petroleum Naphtha, Heavy	184	lb.
Cobalt Naphthenate 6%	2.5	lb.
(Anti-skinning agent is optional)		
Guaiacol, Not over	1.2	lb.

Non-Yellowing Synthetic Flat Enamel

Nuad L	3	lb.
Multifex Pulp	50	lb.
Lithopone, Titanated	400	lb.
Titanium Dioxide	200	lb.
Surfex	450	lb.
Celite 165 S	50	lb.
Castor Oil "Z" Body, Dehydrated	17	gal.
Petroleum Naphtha, Heavy	20	gal.
Soya Type Alkyd (32% Phthalic Anhydride), 65% Mineral Spirits Solution	13	gal.
Mineral Spirits	10	gal.
Grind and then add:		
Mineral Spirits	6	gal.
Cobalt Naphthenate 6%	¼	gal.
Lead Naphthenate 24%	½	gal.

Insulating Enamel

U. S. Patent 2,216,234

Rosin	50.00
Alkyd Resin	150.00
Perilla Oil	205.50
Fish Oil	15.00
Chinawood Oil	73.50
Manganese Resinate	0.30
Ferrous Resinate	0.45
Cobalt Acetate	0.25
Zinc Resinate	1.00

MILITARY ENAMEL PIGMENTS

Lusterless Orange Enamel

Medium Chrome Orange	42
Yellow Iron Oxide	8
Extenders	50

Lusterless Red Enamel

Red Oxide, Bright Synthetic	50
Extenders	50

Lusterless Umber Enamel

Yellow Iron Oxide	8
Lamp Black	2
Red Iron Oxide	13
Extenders	77

Lusterless White Enamel

Rutile Titanium Dioxide	30
Extenders	70

Lusterless Yellow Enamel

Chrome Yellow	25
Medium Chrome Orange	5
Yellow Iron Oxide	3
Extenders	67

Gloss Blue Enamel

Prussian Blue	50
Rutile Titanium Dioxide	45
Tinting Pigments	5

Gloss Dark Green Enamel

Chrome Green	100

Gloss Gray Enamel

Rutile Titanium Dioxide	95
Tinting Pigments	5

Gloss Mustard Yellow Enamel

Chrome Yellow	75
Yellow Iron Oxide	25

Gloss Olive Drab Enamel

Yellow Iron Oxide	7
Lamp Black	7
Rutile Titanium Dioxide	18
Chrome Yellow	68

Gloss Red Enamel

Toluidine Toner	100

Gloss White Enamel

Rutile Titanium Dioxide	100

Gloss Yellow Enamel

Chrome Yellow	90
Medium Chrome Orange	10

Semi-Gloss Black Enamel

Carbon Black	5
Extenders	95

Semi-Gloss Olive Drab Enamel

Yellow Iron Oxide	19
Chrome Yellow	11
Lamp Black	3
Extenders	67

Infra-Red Paint Enamel Pigment

Ferrite Yellow-Orange	14.8
Copper Phthalocyanine Blue	1.0
Red Iron Oxide (Medium Shade)	3.5
Chrome Yellow, Medium	0.7
Diatomaceous Silica	8.0
Asbestine	72.0

Lusterless Blue Enamel

Prussian Blue	8
Rutile Titanium Dioxide	5
Extenders	87

Lusterless Blue Drab Enamel

Copper Phthalocyanine Blue	0.5– 3.0
Rutile Titanium Dioxide	10 –21
Tinting Pigments	as required
Extenders	80

Lusterless Brown Enamel

Yellow Iron Oxide	16
Red Iron Oxide	4
Lamp Black	1
Extenders	79

Lusterless Coronado Tan Enamel

Yellow Iron Oxide	17
Rutile Titanium Dioxide	7
Bone Black	2
Red Iron Oxide	2
Extenders	72

Lusterless Desert Drab Enamel

Yellow Iron Oxide	16.0
Rutile Titanium Dioxide	2.5
Chrome Yellow	2.0
Lamp Black	0.5
Extenders	79.0

Lusterless Earth Drab Enamel

Yellow Iron Oxide	15
Carbon Black	5
Red Iron Oxide	5
Extenders	75

Lusterless Gray Enamel

Rutile Titanium Dioxide	20–27
Tinting Pigments	5
Extenders	78

Lusterless Green Enamel

Chrome Green	18
Yellow Iron Oxide	9
Lamp Black	1
Extenders	72

Lusterless Maroon Enamel

Indian Red	50
Extenders	50

Lusterless Olive Drab Enamel

Yellow Iron Oxide	18
Chrome Yellow	7
Lamp Black	3
Extenders	72

Lusterless Olive Drab Enamel, Infra Red Reflectant

Antimony Sulphide Black	20.0
Chrome Yellow, Light	8.6
Ferrite Yellow	11.4
Asbestine	60.0

White Undercoats for Wood and Walls Indoors

Formula No. 1

Pale East India Vehicle

Pale East India Nubs (Macassar-Hiroe)	80 lb.
Coal Tar Thinner, or high solvency Petroleum Naphtha or combination of either with low solvency Petroleum Naphtha	80 lb.
Zinc Oxide	40 lb.
Titanium Barium Pigment	200 lb.
Asbestine	20 lb.

Dissolve the Pale East India in the coal tar thinner, using agitation but no heat. Strain out the undissolved material or allow it to settle. Grind the pigments in a portion of the vehicle thus prepared. Add the remainder of the vehicle.

Plasticizers may be used if desired. Linseed oil is preferred but other oils may be employed.

No. 2

Manila Vehicle

Manila	60 lb.
Denatured Alcohol or mixture of 50 lb. Denatured Alcohol and 10 lb. Petroleum Naphtha	60 lb.
Titanium Dioxide	260 lb.
Diatomaceous Earth	20 lb.

Dissolve the Manila in the denatured alcohol, using agitation but no heat. Strain out undissolved mat ter or allow to settle.

Grind the pigments in a portio of the above vehicle. Add the re mainder of the vehicle.

Add butanol to proper brushin consistency or more denatured alco hol for spraying consistency.

Plasticizers such as castor oi may be used if desired.

Gloss Enamel

Formula No. 1

(Quick Drying)

Pigment (38%)

Cryptone BT–802 or 366	412.75 lb.
Ultramarine Blue	0.32 lb.

Vehicle (62%)

Beckosol No. 1319 Sol.	64.5 gal.
Boiled Linseed Oil	2.5 gal.
Mineral Spirits	10.0 gal.
Amsco. No. 460 Solvent	9.2 gal.
Cobalt Naphthenate 6%	0.5 gal.
Zinc Naphthenate 8%	1.7 gal.

No. 2

(Slow Drying)

Pigment (53.8%)

Albalith Blue Label–10	598 lb.
XX–55 Zinc Oxide	106 lb.
Ultramarine Blue	0.25 lb.

Vehicle (46.2%)

Kettle Bodied Linseed Oil "Z body"	23.60 gal.
Refined Linseed Oil	8.04 gal.
Water	0.71 gal.
Mineral Spirits	26.74 gal.
10–Lb. Cut Low Acid Ester Gum	11.71 gal.
N-Glo-5Y (Gloss Oil, Limed)	9.14 gal.

Naphthenate Drier 24% Pb	0.65 gal.
Naphthenate Drier 6% Co	0.42 gal.

Calcimine Coater

Formula No. 1

Pigment (67%)

Albalith 351, 361, or 362	780
Whiting	195
Vehicle (33%)	
Limed Linseed Vehicle * (40% Non-volatile)	365
Kerosene	96
Water †	24

The entire vehicle is put into the mixer and the pigment is added gradually. The resulting paste is ground on a roller mill using a loose setting, or on a buhrstone mill. The finished product, after aging overnight, is heavy in body and will take a reduction of about three pints of mineral spirits per gallon for application.

No. 2

Calcimine Coater Liquid

Linseed Oil (Z–3 body)	20½ gal.
Lime	4¾ lb.
Mineral Spirits	36 gal.

Directions

Heat linseed oil to 575° F. (302° C.) in about one hour. Hold between 575° F. and 585° F. (307° C.) until a very heavy string on a cold glass is obtained. This requires about ¾ to 1¼ hours. Allow to cool to 300° F. (149° C.) and slowly add the lime which has previously been mixed to a smooth paste in about ½ gal. of the Z–3 linseed oil. Mix in thoroughly and allow the lime to react. During reaction keep the temperature of the batch between 300° F. (149° C.) and 400° F. (204° C.).

When the reaction is completed, a dry pill on the glass is obtained. This requires about three quarters to one hour. Thin, adding the thinner slowly and in small quantities. It is important to incorporate the first few gallons of thinner very thoroughly. This will facilitate the incorporation of subsequent portions of thinner and ensure a vehicle free of lumps.

The temperature of liming is the governing factor in preparing vehicles of this type. At a temperature of 300° F. a gel-like varnish is obtained which has excellent non-penetration properties but is difficult to handle and is liable to produce progressive bodying in the paint. Liming at a temperature of 400° F., decreases the non-penetrating properties but at the same time reduces handling and after-bodying difficulties.

Best results are obtained when the vehicle manufacturer blends several kettles of vehicle which have been limed at different temperatures and this procedure is recommended. The blend should be made so as to obtain the desired degree of non-penetration and yet produce a material which handles easily and will not produce after-bodying.

* The vehicle is typical of the limed liquids on the market or prepared by paint manufacturers. They may contain linseed, perilla, fish, soya or combinations of oils and vary considerably in body producing properties. Some of these vehicles have normal acid number, others are neutral, and some alkaline. Some are free-flowing while others are gel-like. With some vehicles it may be necessary to include drier in the formulation.

† Some vehicles are too sensitive to water and in that case water should be replaced by kerosene or mineral spirits.

Automobile Finishing Enamel

Non-Volatile		Volatile	
Cryptone ZS–30	70	Ethyl Acetate	15
Titanium Dioxide	15	Butyl Acetate	20
Flor. Green Seal–Zinc Oxide	15	Butanol	12
½ Sec. R. S. Cotton (Wet)	24	Amyl Acetate	5
Glyptal No. 2471 (As rosin)	24	Toluol	43
Dibutyl Phthalate	13	Zylol	5

Non-Volatile	39%
Volatile	61%

For application by spraying, thin 10 parts enamel to 6 parts volatile solvent mixture.

Primer

Iron Oxide	34%
Glyptal No. 2462	61%
Solvent Naphtha	5%

Thinned 60–40 by weight with

Mineral Spirits	65%
Solvent Naphtha	25%
Turpentine	10%

Surfacer

40% Paste		60% Base Solution	
XX–50 Zinc Oxide	55.0%	7–10 Sec. Cotton (Wet)	4.5%
Silica	14.0%	½ Sec. Cotton (Wet)	16.5%
Ester Gum (Acid No. 6–8)	8.0%	Ethyl Acetate	20.0%
Lindol X	19.0%	Butyl Acetate	16.0%
Horcosol "5"	3.5%	Denatured Alcohol	8.0%
AA Black (Lampblack)	0.5%	Butanol	4.0%
		Toluol	41.0%

Semigloss Enamel Vehicle (Varnish) (V–255)

Kettle Bodied "Q" Linseed Oil	100	gal.
Dehydrol (Dehydrated Castor Oil)	100	gal.
Aluminum Stearate	3	lb.
Fused Pb Resinate (16%)	21¼	lb.
Fused Co Resinate (3½%)	9	lb.
Mineral Spirits	185	gal.

Heat the Dehydrol to 282° C. (540° F.). Allow to climb to 296° C. (565° F.) and hold at this temperature for a half inch string from the hot paddle. Add the "Q" oil and regain 296° C. (565° F.). Hold for ten minutes to ensure complete incorporation, cool to 288° C. (550° F.) and add the driers; allow to cool to 232° C. (450° F.) and add to aluminum stearate; cool and thin.

Semigloss Enamel

Cryptone BA–19	880	lb.
Varnish V–255	57½	gal.
Aluminum Stearate Gel (6 oz. cut in Mineral Spirits)	11⅛	gal.
Mineral Spirits	5¾	gal.
Lead Naphthenate (24% Pb)	⅜	gal.
Cobalt Naphthenate (6% Co)	⅛	gal.

This vehicle combination produces a semigloss possessing a high sheen. The sheen can be reduced by increasing the quantity of aluminum stearate gel and decreasing the mineral spirits. Example:—18% aluminum stearate gel and 2.5% mineral spirits result in a semigloss with a moderate to low sheen.

Prepare paste using all the pigment, aluminum stearate gel and approximately 37 gallons V–255, grind on roller or buhrstone mill and reduce with remaining vehicle. It is essential to include the aluminum stearate gel in the grind.

Preparations of Aqueous Shellac Solutions

It is possible to prepare solutions of shellac which are substantially water reducible. The method comprises primarily dissolving 100 g. shellac in 100 g. 74° O.P. methylated spirit and then adding to this solution 15–18 cc. of an ammoniacal solution (prepared by adding 1 part 0.880 ammonia to 2 parts water) with constant stirring. The resulting solution can then be thinned with water, and the films obtained from such solutions are moderately clear, glossy and fairly flexible if plasticized, for example with dimethyl-glycol phthalate.

Water Glass Paints, Aging

A very suitable mixture of this type which greatly shortens the second drying or aging period of water glass paint films consists of equal parts of water and sodium water glass (5 parts altogether) and 1 part of calcium hydrate or slaked special lime. Good results are also obtained with a lime mixture consisting of 5 parts of lean (blue) milk and 1 part of calcium hydrate. The underground must be air dry but still fresh since this promotes the aging of the film considerably.

Water Paint for Wood and Plaster
U. S. Patent 2,226,030

Blackstrap Molasses	3
Glue	3
Whiting	18
Hydrated Lime	24
Soapstone	5
Formaldehyde	½

Add water to use.

Water Paint for Stone, Brick or Cement
U. S. Patent 2,246,620

White Cement	58.8
White Finishing Lime	17.8
Beach Sand	9.4
Zinc Sulphate	1.6
Calcium Chloride	10.0
Glue	0.4
Terpineol	0.4
Salt	1.6

Water Paint
Canadian Patent 398,356

Zinc Oxide	4
Lithopone	10
Barium Carbonate	15
Whiting	18
Silica	5
Pyrophyllite	12
Mica, Silver	3

China Clay	8
Litharge	3
Casein	9
Potassium Carbonate	1
Mercuric Chloride	1/8

Interior Emulsion Paints
Resin Oil Emulsion

Solution "A"

Resin	62
Bodied Linseed Oil	50
Pine Oil	21
Mineral Spirits	62
Cobalt Linoleate Drier (6% Co.)	1 1/4
Lead Linoleate Drier (16% Pb.)	4 1/4
Oleic Acid	24

Solution "B"

Powdered Casein	27
Preservative	3/8
2 Amino–2 Methyl–1 Propanol	15
Water	536

Mix the resin, cut in the mineral spirits, with the bodied linseed oil, pine oil, driers, and oleic acid.

Place the casein in mixer, add the water slowly, stirring so that casein is wetted, add the preservative and emulsifier and continue stirring for several minutes.

Add Solution "A" slowly to Solution "B" with constant high speed agitation until emulsification is complete.

The emulsion contains six types of ingredients listed according to their functions:

(1) Organic binders (resin, oil, casein)
(2) Emulsifier (2 Amino–2 Methyl–1 Propanol plus oleic acid)
(3) Wetting and dispersing agent (pine oil)
(4) Metallic driers
(5) Preservative (Moldex)
(6) Water

The functions of 1, 4, 5 and 6 are obvious.

Pine oil acts as a pigment dispersing agent and aids in the emulsification and, incidentally, tends to prevent frothing and improve brushing.

The casein has a dual role, serving both as a binder and to improve the drying properties of the paint. As far as our own work has gone we have found a percentage of casein is a valuable constituent of resin-oil emulsion paint.

The emulsifier is the critical ingredient upon which depends the success of the emulsion and the resulting paint. After the oil phase is broken up into tiny droplets the emulsifier becomes adsorbed on the surface of the oil droplets, forming a protective layer that prevents them from coalescing and breaking the emulsion. It must provide sufficient stability to the emulsion so that it will withstand grinding in of the pigment, and preserve original consistency in the can under unfavorable storage conditions. The protective layer around the oil phase must also resist rupture under the stress of brushing, and must above all allow the release of the water at the proper time so that a coherent paint film is deposited. It must also be present in low concentration so as not to interfere with drying or subsequent washability of the paint.

With the emulsion stabilized, it is simply a matter of mixing and grinding in the proper pigment (Albalith, Cryptone BA or Zinc Sulphide) into the emulsion on a stone or roller mill, using a loose setting. Illustrative semi-paste paint formulae are:

Semi-Paste Paints
Formula No. 1

Albalith–335, 332 or 836 *	790
Resin Oil Emulsion	622

No. 2

Cryptone ZS–830	376
China Clay	230
Resin Oil Emulsion	625

These semi-pastes are designed to be thinned 2–1 with water for brush application. The paints brush easily over porous or well sealed surfaces, hide well, are flat and reasonably washable as are the average flat wall paints on the market. The emulsions, if properly made, are stable with these pigments, but so far as this work has progressed, this cannot be said of calcium sulphate base pigments, since their solubility tends to break the emulsion.

The illustrative formulae of the emulsion and semi-pastes given here can be modified in many ways, but in making substitutions, or if the percentages of the ingredients are changed, the new system must be balanced or the breaking of the emulsion may follow.

Ester Gum-Oil Emulsion
(V-344)
Solution "A"

High Acid Ester Gum (Acid #18–20)	83½
Mineral Spirits	83½
Kettle Bodied Linseed Oil (Z 4–6)	65⅝
Pine Oil	27½
Cobalt Linoleate Drier (6% Co.)	1⅝
Lead Linoleate Drier (16% Pb.)	5⅝
Oleic Acid	31¾

Solution "B"

Casein B–5	36½
Moldex	⅞
Ammonia (28–29%)	20⅛
Water	453

The ester gum, bodied linseed oil, and mineral spirits may be amalgamated by either cooking the gum and oil and thinning with mineral spirits, or by cold cutting the gum in mineral spirits and mixing in the oil. To the varnish thus prepared add driers and oleic acid.

Place the casein in a pony or other type mixer, add the water slowly, stirring so that the casein is wetted, add the preservative and the ammonia solution and continue stirring for several minutes.

Add Solution "A" slowly to Solution "B" with constant agitation until emulsification is complete.

Ester Gum-Oil Emulsion Semi-Paste
Interior

Albalith–836	1150
Ester Gum-Oil Emulsion V-344	540

The pigment and emulsion should be mixed together and milled on a roller mill or a buhrstone mill at a loose setting.

Reduction for brush application —two volumes of paste with one volume of water.

Note: In making tints, the tinting colors should be mixed or ground into the paste with the white

* Cryptone BA–40 can be substituted for the Albalith if greater hiding power is required.

pigments. Less white hiding pigment will be required than in the white formula. Colors that are not affected by alkalies should be used since the paint must be kept on the alkaline side.

Ester Gum-Oil Emulsion
Semi-Paste
Interior

Albalith–332	915
Ester Gum-Oil Emulsion V-344	432
Water	176

The pigment, emulsion and the water should be mixed together and milled on a roller mill or a buhrstone mill at a loose setting.

Reduction for brush application—two volumes of paste with one volume of water.

Ester Gum-Oil Emulsion
Semi-Paste
Interior

Pigment (60%)	
Cryptone ZS-830	639
China Clay	213
Vehicle (40%)	
Ester Gum-Oil Emulsion V-344	477
Water	91

The pigment, emulsion and water should be mixed together and milled on a roller mill or a buhrstone mill at a loose setting.

Reduction for brush application—two volumes of paste with one volume of water.

"Aquaplex" Resin Emulsion
Semi-Paste
Interior
Formula No. 1

Pigment (58.1%)	
Albalith–836, 335 or 332 *	25
Vehicle (41.9%)	
Aquaplex AM-92 †	36
Casein Solution V-242 or V-243	14
Pine Oil	4
Water	21

The Aquaplex, casein solution, pine oil and water are mixed together and used as a combined vehicle. The desired quantity of combined vehicle is placed in a pony or dough mixer, and the pigment is gradually added with the mixer running. Continue mixing until a smooth paste is obtained.

A mixture of two parts of paste to one part of water by volume will be suitable for brush application on most surfaces.

No. 2

Pigment (56%)	
Cryptone ZS–830	16
Whiting	11.3
Vehicle (44%)	
Aquaplex AM–92 ‡	39.4
Casein Solution V–242 or V–243	15.3
Pine Oil	5.8
Water	12.2

* If Albalith–332 is used the resulting paste may be too heavy. If so, 2% less pigment and up to approximately 2% more water may be added without affecting the reduction properties for brush application.

† Aquaplex S–6 may be substituted for Aquaplex AM–92 if desired. The resulting paint will be slightly thinner in body and will become resistant to washing more rapidly.

‡ Aquaplex S–6 may be substituted for Aquaplex AM–92 if desired. The resulting paint will be slightly thinner in body and will become resistant to washing more rapidly.

Mix and use as for No. 1.

Casein Solution (V-242)
For Use in Cold Water Paints
U. S. Patent 1,893,608

Casein B-5	137.2
Borax	15.0
Sodium Fluoride	12.2
Ammonium Chloride	4.7
Pine Oil	14.0
Moldex	0.9
Water	751.0

The casein, sodium fluoride and ammonium chloride are heated at 168° F. with moderate stirring in about 75 gallons of water for about two hours. The borax dissolved in the remaining water (15.12 gallons) is then added and heating continued with rather vigorous stirring for 1½ hours. The pine oil and preservative are added and the solution allowed to cool.

Casein Solution (V-243)
For Use in Cold Water Paints

Powdered Casein B-5	147
Water	660
Borax	22
Pine Oil	14½
Moldex	1

The borax is put in solution by heating with about 10% of the water. The casein is soaked at 160–170° F. with moderate stirring for 1½–2 hours in the remainder of the water. The borax solution is then added to the soaped casein and heating is continued at 160–170° F. with more vigorous stirring for 1½ hours. The pine oil and preservative are then added and the solution cooled. Any water lost should be added to bring the volatile content up to 80%.

"Piccolyte" Resin-Oil Emulsion
(V-338)

Solution "A" (30.5%)

Piccolyte S–100 Resin	67
Lithographic Oil #8	54
Pine Oil	22¾
Mineral Spirits	67
Cobalt Linoleate Drier (6% Co.)	1⅜
Lead Linoleate Drier (16% Pb.)	4½
Oleic Acid	29¼

Solution "B" (69.5%)

Casein B–5	29¼
Moldex	⅜
2 Amino-2 Methyl-1 Propanol	16¼
Water	513½

Dissolve the Piccolyte resin in the mineral spirits and mix with the lithographic oil #8, pine oil, driers, and oleic acid.

Place the casein and preservative in a pony or other type mixer, add the water slowly, stirring so that casein and preservative are wetted, add the 2 Amino–2 Methyl–1 Propanol, and continue the stirring for several minutes.

Add Solution "A" slowly to Solution "B" with constant agitation until emulsification is complete.

"Piccolyte" Oil Emulsion
(Semi-Paste)
Interior
Formula No. 1

Pigment (60.8%)

Albalith–836, 335 or 332	932

Vehicle (39.2%)

Piccolyte Emulsion V–338	559
Water	42

The pigment, emulsion and water should be mixed together and then milled on a roller or a buhrstone mill at a loose setting.

Reduction for brush application —two volumes of paste with one volume of water.

"Piccolyte" Oil Emulsion
(Semi-Paste)
Interior
No. 2

Pigment (54.5%)

Cryptone ZS–830	446
China Clay	274
Vehicle (45.5%)	
Piccolyte Emulsion V–338	559
Water	42

The pigment, emulsion and water should be mixed together and then milled on a roller or buhrstone mill at a loose setting.

Reduction for brush application —two volumes of paste with one volume of water.

"Piccolyte" Oil Emulsion
Semi-Paste
Interior
Buff
No. 3

Pigment (60.7%)

Albalith–836 or 335	895
Yellow Iron Oxide	35
Vehicle (39.3%)	
Piccolyte Emulsion V–338	559
Water	42

"Piccolyte" Oil Emulsion
Semi-Paste
No. 4

Cryptone BA–40	962 lb.
Piccolyte Emulsion V–338	588 lb.

Resin Base Water Paint
U. S. Patent 2,250,346

	lb.	oz.
Triple Washed Calcium Carbonate	10 –25	—
Fibrous Asbestine	8 –15	—
Cold Water Glue	—	3– 5
Domestic China Clay (Extra Fine)	5 –20	—
Powdered Casein (Nitrogen Factor 7.07)	2 – 3½	—
Fine Steamed Mica (1,000 Mesh)	½– 7	—
Zinc Sulphide	20 –40	—
Talc	1 – 7	—
Alkyd Resin Solution	3 – 8	—
Phenol	—	3¾–4½
Borax (Powdered)	—	4–12
Neutral Soap (Flaked)	—	2– 6
Common Laundry Starch	—	1– 5
Tallow	1½–2½	—
Lead Naphthenate Soap	—	1– 4
Glycerin	—	0– 8
Pine Oil	—	0– 8
Sassafras Oil	—	0–16
100 Mesh Casein (Nitrogen factor 7.07)	1¾– 4	—
Water	½– 3	—
Powdered Gelatin	¾– 1½	—

Paint Oil

Limed Rosin Oiticica Oil
255 g. Heat to 550° F. in 45 mins.
Varnish Makers
815 g. Held 1¼ hrs. for body
Add
Linseed Stand-oil
320 g. Heat to 550° F. cool and
Reduce with
Mineral Spirits
1040 g.

Coating Glass with Luminescents
British Patent 530,531

Luminescent Powder	250 g.
Ethyl Acetate	250 cc.

Grind together for 24 hours and dilute with:

Ethyl Acetate	250 cc.
Methyl Methacrylate (40% Sol.)	200 cc.

Shake for 1 hour and add:

Camphor	20 g.

Apply to glass and heat slowly up to 400–450° C. in a current of air.

Frosting Glass

Sandarac Resin	7
Denatured Alcohol	31
Toluol	47
Xylol	15

Dissolve sandarac resin in the denatured alcohol by heating, then add the toluol and xylol gradually while the solution is still warm.

The solution is quite stable and can be stored indefinitely without the resin separating. The following application procedure has given best results:

The glass to be coated is thoroughly cleaned, then washed with denatured alcohol and rubbed dry with a clean cloth. The frosting solution is diluted with approximately 5% by volume of either toluol or xylol and sprayed on the glass. Two spray coats, one with a vertical and the other with a horizontal stroke, should be applied Using a small touch-up suction type spray gun, good results can be obtained by using 25 to 30 pounds pressure and holding the gun approximately 18 inches from the glass. Areas not to be frosted and the window frames may be masked with Scotch masking tape.

Glass Masking Coat for Spray Painting

Kaolin	2.0
Glycerin	1.5
Water	4.5
Butyl Alcohol	0.25

Mix the above ingredients thoroughly. Apply to the glass with a brush and allow to dry. After painting with a spray gun, and allowing the paint to dry, the mask may be removed with a putty knife, a cloth or by washing with water, leaving the glass free of any paint.

This masking mixture can also be used to protect metal and similar surfaces during spray or brush painting.

Coating for Phonograph Records
U. S. Patent 2,203,983

Tritolyl Phosphate	16.0
Soybean Oil	3.0
Chlorinated Biphenyl	1.5
Butyl Stearate	1.5
Ethyl Cellulose	78.0
Color, To suit	

Air Filter Fiber Coating
U. S. Patent 2,112,799

White Oil (Mineral)	23.0
Water	36.5
Calcium Chloride	36.5
Bentonite	4.0

Electric Cable Coating
U. S. Patent 2,216,435

Rosin	25–75%
Rosin Oil	20–50%
Polymerized Isobutylene	5–25%

Electrical Cable Sheathing

Benzyl Cellulose	100
Tritolyl Phosphate	36
Kaolin, To suit	

Decalcomania, "Cellophane"
U. S. Patent 2,228,281

"Cellophane" or cellulose acetate transfers are made by coating with:

Ester Gum	4
Linseed Oil Varnish	8
Carbon Black	4
Carnauba Wax	1

Transfer Coating ("Decal")
U. S. Patent 2,228,280

Ester Gum	4
Linseed Varnish	8
Carbon Black	4
Carnauba Wax	1
Orange Toner (Optional)	2

Transfer Foil
Swiss Patent 211,932

Flour starch 14, gallic acid 24, tannic acid 13, glycerol 39, tragacanth gum 1, carnauba wax 45, white wax 20, and paraffin wax 30 parts, is applied on paper. This paper is used with a paper treated with an aqueous solution of cane sugar 45, ammonium ferrous sulphate 40, ammonium ferric oxalate 10, ammonium meta vanadate 5, titanium dioxide 15, manganous sulphate 20, magnesium sulphate 5, and water 200 parts.

Synthetic Transparent Foils or Coatings
U. S. Patent 2,255,564

Formula No. 1

"Pliolite"	10
Chlorinated Diphenyl	2½
Naphtha	87½

No. 2

"Pliolite" or "Plioform"	10
Paraffin Wax	2
Cumarone	2
Naphtha	43
Benzene	43

No. 3

"Pliolite"	5
Dammar	7.5
Petrolatum	1
Benzene	86.5

Painting Damp Surfaces

The surface is prepared for painting by removing the loose paint and scale by wire brushing and scraping. The dirt and excess moisture are then wiped off the surface and a liberal coating of turpentine is applied.

The turpentine has a two-fold effect: first, to replace the water remaining on the surface, by virtue of its preferential wetting property, and second, to "lift" any loose rust scale remaining on the surface after the mechanical cleaning. A primer coat of standard zinc dust-zinc oxide paint is then applied directly on the turpentine wet surface, thus providing the necessary rust inhibitive coating. In actual practice, a considerable interval may be allowed between the application of the turpentine and the primer, since the turpentine prevents further condensation of water on the steel.

Brush application of the primer is preferred to spray application, since it insures the intimate contact of the zinc dust with the surface. An 80:20 zinc dust-zinc oxide formula with a linseed oil vehicle has been used, but a synthetic vehicle, such as a phenolic, may be preferable, first, because of the more rapid drying of the synthetic resin formulation, and second, because of the improved water resistance of the film.

The usual drying interval is allowed after which the finish coat paint is applied. Just before the application of the second coat, the moist surface is again wiped with turpentine, but care must be taken to avoid an excess of the turpentine, since it might soften the surface of the primer.

Either an 80:20 zinc dust-zinc oxide paint or other suitable metal paint may be used as the finish coat. Colored zinc dust paints are available for finishing coats where a distinctive color scheme is desired.

It has been found that kerosene may be substituted for turpentine to replace moisture on the pipes, but its action is much slower and its assimilation into the paint less positive.

Bodying Varnish (V-214)

Ester Gum (Acid No., 18)	200
China Wood Oil	950
Linseed "Q" Body	320
Lead Acetate	20
Mineral Spirits	1300

Heat gum, chinawood oil, and 7½ gallons of the bodied linseed oil to 565° F. (296° C.). Hold until sign of string * and chill with the balance of linseed oil. Add the powdered lead acetate on the down heat, cool to 400° F. (204° C.) and thin.

This varnish is used in minor quantities in connection with vehicles of low acid number to body lithopone flat wall paints. Paints containing bodying varnish may be increased in body by

(a) Increasing the per cent of the bodying varnish at the expense of the regular binder.

(b) Using a low acid number vehicle.

(c) Including a small amount of Zinc Oxide.

(d) Using a high consistency Albalith like Albalith-362.

(e) Using a "400 Series" pigment (Albalith-404, 424, Cryptone BA-435, Cryptone BT-445, 447).

Varnish (V-172)
(*Metal Primer Vehicle*)

Bakelite XR-1329	100.00
Chinawood Oil	391.00
Lead Acetate	4.00
Mineral Spirits	341.00
Lead Naphthenate	5.75
Manganese Naphthenate	1.00
Cobalt Naphthenate	2.00

Heat all the resin and one-half of the oil to 540–550° F., chill back with the balance of the oil and the lead acetate, and hold at 450° F. for a six inch string from cold glass. These cooking directions may be modified to suit plant conditions. Reduce with mineral spirits and add

* To increase the stability of this varnish and prevent gelling in the can, it may be necessary to reduce the bodying time somewhat or alter the tung oil-linseed oil ratio slightly.

soluble drier: 14½ oz. of lead, 1¼ oz. of cobalt, and ⅝ oz. of manganese as metal.

This vehicle is not exceptionally fast drying, requiring from six to eight hours. It makes an excellent vehicle for all types of metal primers and finishing coats where color is not a factor. If properly cooked, this vehicle is not reactive with basic pigments.

Varnish (V-173)

(*Metal Priming and Finishing Paint Vehicle*)

Glyptal No. 2455 Solution	412½	lb.
Raw Linseed Oil	112½	lb.
Mineral Spirits	32½	gal.
Naphthenate Lead Drier	9	lb.
Naphthenate Cobalt Drier	3	lb.
Naphthenate Manganese Drier	1½	lb.

The above vehicle may be made by thinning the Glyptal solution with the mineral spirits, then adding the linseed oil and driers. If it is desired to prepare a paste with this vehicle, the linseed oil and part of the Glyptal solution may be used as the grinding vehicle and the paste reduced with the remainder of the vehicle.

Four-Hour Floor Varnish

Pentaerythritol Abietate	166	lb.
Over-Poly. Linseed Oil "ZZZ" *	368	lb.
Litharge	4	lb.
V.M. Linseed Oil *	4	lb.
Mineral Spirits	98½	gal.
½% Soln. Guaiacol in Dipentene	3½	gal.
Manganese Naphthenate 6%	4	lb.
Cobalt Drier 4%	1	lb.

* From minimum 190 iodine number V. M. oil.

31-Gallon Varnish and Vehicle

Over-Poly. Linseed Oil "ZZZ" *	368	lb.
Pentaerythritol Abietate	166	lb.
Litharge	4	lb.
V.M. Linseed Oil *	4	lb.
½% Soln. Guaiacol in Dipentene	3½	gal.
Mineral Spirits	98½	gal.
Manganese Naphthenate 6%	4	lb.
Cobalt Drier 4%	1	lb.

25-30-Gallon General Purpose Mixing Varnish and Vehicle

Formula No. 1

Mild Mod. Phen. Resin No. 78A	227	lb.
Hydrated Lime	5	lb.
Tung Oil	303	lb.
Poly. Linseed Oil "Q"	32	lb.
Poly. Soy Oil "J"	117	lb.
Litharge	6½	lb.
Cobalt Resinate, Fused	4½	lb.
Manganese Resinate, Fused	2¼	lb.
Kerosene	11	gal.
½% Soln. Guaiacol in Dipentene	19	gal.
Mineral Spirits, To make	126	gal.

No. 2

Mod. Alkyd Resin No. 225	226	lb.
Hydrated Lime	10	lb.
Over-Poly. Linseed Oil "ZZZ" *	120	lb.

Poly. Linseed Oil "X" *	344 lb.
Zinc Naphthenate 10%	8 lb.
Lead Naphthenate 24%	24 lb.
Manganese Naphthenate 6%	5 lb.
Cobalt Drier 4%	16 lb.
Kerosene	4 gal.
Mineral Spirits, To make	142 gal.

Floor Varnish and Floor Enamel Vehicle: High Adhesion Primer and Primer Vehicle, 20-Gallon Oil Length

A—Refined Lignin Liquor	526	lb.
Over-Poly. Linseed Oil	263	lb.
Maleic Resin No. 235	68	lb.
Glycerin (Dynamite Grade)	18½	lb.
Zinc Oxide, Lead Free	2½	lb.
Mineral Spirits	31¾	gal.
B—Hydrated Lime	25	lb.
Mineral Spirits	7½	gal.
Lead Naphthenate 24%	9	lb.
Cobalt Drier 4%	9	lb.
Mineral Spirits to C-E Viscosity	105	gal.

Charge "A" into an open kettle. (The acid number of the mixture before reaction is 103.) Raise "A" smoothly and evenly in 2½ hours to 575°. Let cool to 545°. Hold 545° until 15 minutes from the time 575° first was reached. Let cool naturally to 410°. (The acid number is now about 50.) At 410° add the first thinner. Then have the batch at 360°, firing up again if necessary.

Work in "B," consisting of a slurry of hydrated lime and thinner, holding batch at 360°. When "B" is all in, stir additional 15 minutes at 360° and then complete by addition of driers and mineral spirit to a viscosity of C to E as desired.

* From minimum 190 iodine number V. M. oil.

General Purpose Grinding Oil, Mixing Varnish and Paint Oil

A—Refined Lignin Liquor	176	lb.
Over-Poly. Linseed Oil *	88	lb.
B—Calcined Magnesia	3⅛	lb.
V.M. Linseed Oil	10½	lb.
Mineral Spirits	12	gal.
C—Hydrated Lime	14¼	lb.
Mineral Spirits	3	gal.
Mineral Spirits	52½	gal.
Lead Naphthenate 24%	2¼	lb.
Manganese Naphthenate 6%	2¼	lb.
Cobalt Drier 4%	2¼	lb.

Raise "A" to 400°. Cream "B" to a slurry and work into the batch, dispersing well. Gain 525°. Hold 525° until clear (1 to 5 minutes). Let cool naturally to 425°. Add 12 gallons of mineral spirits. Stir well. Hold batch at 360° and lime with "C" (stirred to a slurry). Stir 15 minutes at 360°. Fire off. Complete to "E."

* This is heat-polymerized linseed oil of such viscosity that in a Gardner-Holdt tube the bubble moves only the diameter of one bubble while the bubble in the "Z-6" Gardner-Holdt tube traverses the entire length of the tube.

Modified-Phenolic Varnish

Formula No. 1

111-L Albertol (B/S 1) 100 lb.
China Wood Oil 20 gal.

Heat to 585° F. in 30 minutes. Add:

1 hr. Linseed Oil 5 gal.

Cool to 545° F. and hold 13 minutes for viscosity. Add:

3½ hr. Linseed Oil 5 gal.

Cool to 360° F. and reduce with:

Mineral Spirits 50 gal.

No. 2

Oiticica Oil—Viscosity X-Y 20 gal.
3½ hr. Linseed Oil 5 gal.

Heat to 575° F. in 30 minutes. Add:

111-L Albertol 100 lb.

Regain 575° in 15 minutes more and hold 2 minutes for string. Add:

3½ hr. Linseed Oil 5 gal.

Cool to 360° F. and reduce with:

Mineral Spirits 50 gal.

Quick Drying Varnish

Resin (Lewisol 150)	50
Chinawood Oil	60
Sunoco Spirits	164

Heat oil to 100° C., add resin; bring temperature to 273° C. in about one hour.

Hold at 273° for about 15 minutes, then cool to 232° C.

Add Sunoco spirits and finally add drier—about 85 cc. per gallon.

Drier (per 100 lb.)

Lead Drier	1 lb.
Cobalt Drier	¾ lb.
Manganese Drier	½ lb.

Furniture Varnishes (Rubbing)

8 Gallon Length (Congo)

Thermally Processed Congo 100 lb.
Bodied Linseed or Dehydrated Castor Oil 8 gal.
Petroleum thinner to 50% non-volatile by weight.
Driers equivalent to 0.10% Ca., 0.10% Co., and 0.20% Pb., as metals on the weight of oil.

The thermally processed Congo is prepared as follows: Heat the Congo to 650°–670° F. in about 1½ hours. Hold until foaming subsides and hot resin runs off paddle similarly to hot oil. Allow to cool.

This processed Congo is used to prepare the above vehicle as follows: Heat thermally processed Congo and linseed oil to 580° F. in about 1¼ hours. Hold until a cold "pill" has very little tack (about 5 minutes). Allow to cool to about 400° F. Add petroleum thinner and then the driers.

The oil length may be increased to 10 gallons if greater flexibility and slightly softer film are desired.

General Utility Varnishes

15 Gallon Length (East India)

Batu, Pale East India or Black East India 100 lb.
Bodied Linseed Oil or Dehydrated Castor Oil (viscosity of about Z3) 15 gal.
Petroleum thinner to 50% non-volatile content.
Driers equivalent to 0.20% Pb., 0.10% Ca., and 0.10% Co., as metals on the weight of oil.

Heat the resin and 5 gallons of linseed oil to 600°–610° F. in about

60–65 minutes. Hold until cold "pill" indicates complete compatibility (about 10 minutes). Add remainder of linseed oil. Heat to 580° F. and hold until clear (about 5 minutes). Allow to cool to about 400° F. Add the petroleum thinner and then the driers.

The above vehicles may be used to prepare enamels for indoor applications.

15 Gallon Length (Congo)

Thermally Processed Congo	100 lb.
Bodied Linseed Oil or Dehydrated Castor Oil (about Z3 viscosity)	15 gal.

Petroleum thinner to 50% non-volatile content by weight.

Driers equivalent to 0.10% Ca., 0.10% Co., and 0.20% Pb., as metals on the weight of oil.

The Congo is thermally processed as follows: Heat the Congo to 650°–670° F. in about 1½ hours. Hold until foaming subsides and the hot resin runs off a paddle similarly to hot oil. Allow to cool.

The processed resin is used to prepare the above vehicle as follows: Heat the thermally processed Congo and linseed oil to 580° F. in about 50–60 minutes. Allow to cool to about 400° F. Add the petroleum thinner and then the driers.

The above vehicles may be used for preparing enamels for indoor applications.

Tinplate Baking Varnish

Oiticica Oil	75.0 kg.

Gain 120° C. in about 9 to 10 minutes. Add:

Teglac Z-152	79.8 kg.

in portions to avoid accumulation of resin, bringing to 190°–200° C. in 35 to 40 minutes. Continue heat to 295° C. in 60 to 65 minutes. Add:

Standoil	46.2 kg.

Reheat to 260° C. in 90 to 100 minutes. Cool with turpentine to 180° C. and reduce with

Mineral Spirits	180.0 kg.

Lead and cobalt driers are added in varying proportions, according to the baking conditions which are to be used.

Medium Oil Metal Varnish
U. S. Patent 2,248,961

This is a good, rapid drying coating composition for metal. This coating forms a single, transparent, flexible primer and surfacer coating which tenaciously binds itself to clean metal surfaces.

Medium Oil Varnish	17.6%
Beeswax	1.6%
Paraffin Wax	0.3%
Boiled Linseed Oil	9.6%
Mineral Spirits, Balance	

The mineral spirits has an initial boiling point of 300°–310° F. and an end point of 390°–400° F. Enough mineral spirits is used to make a total of 100% by weight of all the ingredients.

Infra Red Paint Enamel Varnish

Alkyd Resin	50
Ester Gum	20
Drying Oil	30

Varnish

Flexalyn (Diethylene Glycol Abietate)	150
China Wood Oil	156
Bakelite 3360	50
Lead Naphthenate (24% lead)	2

Manganese Naphthenate (6% Manganese)	0
Cobalt Naphthenate (6% Cobalt)	1
Varsol	291

Oil Length	10 gal.
% Solids Loss in Cooking	1.7
Time to 450° F.	25 min.
Time Held 450° F.	15 min.
Time to 565° F.	19 min.
Time Held 565° F.	none
Body	40″ string
Time in thinning	8 min.

Viscosity	E
Color Hellige	6
Color Gardner	12

The cooking procedure involves running the Bakelite 3360, 40 parts of China Wood oil, and the Flexalyn to 450° F. and holding until the reaction is complete. The balance of the oil is added and the temperature raised to 565° F. The varnish is held at this temperature for a 40″ string off the cold pill, cooled, reduced, and the driers added.

Naval Spar Varnish

Ester gum, 87.5; rosin, grade WW, 25; lead rosinate, 25; raw tung oil, 109; varnish oil, 218; cobalt acetate, 0.75; petroleum spirits, 194; turpentine, 105; cobalt drier, 37.5; lead (naphthenate) drier, 45.

Mixing Varnish for Potable Fresh Water Tank Paint

Modified phenolic resin (Amberol No. 226), 105; raw tung oil, 348; petroleum spirits, 405; cobalt drier, 8; paint drier, 8.

Increased Length and Absence of "Buttering" in Varnishes Using Pigment Lakes

Ammonium Linoleate Paste S is applied in the acid phase in the manufacture of lakes after coupling, and gives better length and prevents buttering when the lake is incorporated in the vehicle. Suspension in the slurry is maintained by mechanical agitation. It may also be applied in the alkaline phase and precipitated in acid.

Heavy Kettle Bodied Linseed Oil (V-206) (Non-reactive)

Neutral Refined Linseed Oil	814

Heat the oil to 625° F. (330° C.). (Takes 2 hours.) Pull from fire and allow to cool to 575° F. (302° C.). Hold this temperature for one hour. Allow the oil to cool to 535° F. (280° C.) and hold at this temperature for the desired body (about 8 hours). Total time on fire, based on 250 gallon batch, is 13 hours.

Heavy Kettle Bodied Linseed Oil (V-201)

Neutral Refined Linseed Oil	805

Heat to 575° F. (303° C.) and hold until a viscosity of "Z" (Gardner-Holdt Standard) is reached.

Limed Rosin Oiticica Oil

Oiticica Oil	1.9 g.

Heat to 425° F. in 25 minutes Add:

Hydrated Lime	40 g.

Heat to 425° F. and hold 20 minutes. Add:

W. W. Rosin	400 g.

Heat to 425° F. in 20 minutes more. Add:

Varnish Makers' Linseed Oil	200 g.

Cool.

Honey Oil

Neutral Refined Linseed Oil	685
Chinawood Oil	94

Heat the Chinawood oil and 80 gallons of linseed oil to 575° F. (307° C.) and hold for heavy string. Chill with the remainder of the linseed oil, stir until thoroughly incorporated and cool.

Gloss Oil

Refined Lignin Liquor	400 lb.
Freshly Hydrated Lime	38 lb.
Kerosene	8 gal.
Mineral Spirits	50 gal.

Raise the refined lignin liquor to 400° F. Fire off. Add the lime, moistened with the kerosene, stirring. Fire on, regain 400°. Hold 400°, stirring gently, for 10 minutes. Fire off. Thin.

Paint Grinding Liquid

Refined Lignin Liquor	300 lb.
Hydrated Lime	26 lb.
China Wood Oil	300 lb.
Lead Drier	
Cobalt Drier	
Manganese Drier	
Mineral Spirits	75 gal.

Heat lignin liquor to 450° F. Dust in the hydrated lime, hold 450° and stirring. Add China wood oil. Gain 500° quickly. Advance to 565°. Hold between 535° and 550° to a soft, non-tacky pill (about 45 minutes). Let cool naturally to 400°. Add driers and reduce.

Flat Grinding Liquid

Refined Lignin Liquor	400 lb.
Hydrated Lime	34 lb.
Lead Naphthenate (24% Pb.)	2 lb.
Manganese Naphthenate (5% Mn.)	2 lb.
Mineral Spirits	50 gal.

Lignin liquor to 400° F. Hold 400°, introduce the hydrated lime with stirring until all in. Gain 500° rapidly. Gain 550° easily; hold 550° for 5 minutes. Fire off. Let cool naturally to 400°. Add driers and reduce.

Short-Oil Flat Grinding Liquid

Formula No. 1

WW Rosin	108 lb.
Tung Oil	127 lb.
Over-Poly. Linseed Oil "ZZZ"	39 lb.
Poly. Sardine Oil	33 lb.
Hydrated Lime	5 lb.
Zinc Oxide	2¼ lb.
Magnesium Carbonate	2¼ lb.
Sod. Carbonate, Anhydrous	¼ lb.
Litharge	18 oz.
Mineral Spirits	40 gal.
Cleaners' Naphtha, To make	40 gal.

No. 2

Mild Mod. Phenolic Resin No. 78A	108 lb.
Over-Poly. Linseed Oil "ZZZ" *	190 lb.
Hydrated Lime	2½ lb.
Zinc Oxide	1⅛ lb.
Magnesium Carbonate	1⅛ lb.
Sod. Carbonate, Anhydrous	¼ lb.

* From minimum 190 iodine number V. M. oil.

Lead Resinate (20%)	4½	lb.
Mineral Spirits	40	gal.
Cleaners' Naphtha, To make	40	gal.

Varnish (V-169)

Flat Wall Vehicle

W.W. Rosin	200
Slaked Lime	8
Heavy Bodied Linseed Oil (5 hours at 575° F.)	790
Fused Lead Resinate (16% Pb.)	30
Fused Manganese Resinate (3½% Mn.)	12
Fused Cobalt Resinate (5½% Co.)	9
Mineral Spirits	940

Melt the W.W. rosin and sift in the lime in small portions, beginning at about 400° F. (250° C.). After all the lime has been added, raise the temperature to 525° F. (274° C.), and add the bodied linseed oil slowly, taking care the temperature does not drop below 475° F. (246° C.). Add the driers in the order named and after they have been thoroughly incorporated, cool to 400° F. (205° C.), and thin. These directions may have to be changed slightly depending on conditions and equipment in the plant.

30-32-Gallon Interior Spar

Formula No. 1

Ester Gum, Pale (Acid 12)	189	lb.
Tung Oil	376	lb.
V.M. Linseed Oil	12	lb.
Litharge	4½	lb.
Cobalt Acetate	1¼	lb.
Mineral Spirit, To make	116	gal.
Gum Spirit Turpentine	3¾	gal.

No. 2

Mod. Alkyd Resin No. 225	166	lb.
Over-Poly. Linseed Oil "ZZZ" *	364	lb.
Lead Naphthenate 24%	16	lb.
Manganese Naphthenate 6%	4	lb.
Cobalt Drier, 4%	4	lb.
Mineral Spirit, To make	87	gal.
Semi-Aromatic Solvent	15	gal.

45-Gallon Flat and Undercoat Sealing Vehicle

Formula No. 1

Mild Mod. Phenolic Resin No. 78A	107	lb.
Over-Poly. Perilla Oil	474	lb.
Hydrated Lime	13½	lb.
Lead Resinate 20%	18	lb.
Zinc Naphthenate 10%	8½	lb.
Cobalt Drier	8	lb.
Mineral Spirit, To make	160	gal.

No. 2

Mod. Phenol. Alkyd Resin No. 222A	130	lb.
Over-Poly. Linseed Oil "ZZZ" *	300	lb.
Poly. Linseed Oil "X" *	140	lb.
Hydrated Lime	6	lb.
Zinc Naphthenate 10%	8½	lb.
Lead Naphthenate 24%	24	lb.

* From minimum 190 iodine number V. M. oil.

Cobalt Drier 4%	8	lb.
Mineral Spirit, To make	159	gal.

58-60-Gallon Exterior Spar

Formula No. 1

Mod. Phenol. Alkyd Resin No. 227	76	lb.
Poly. Perilla Oil "X"	287	lb.
Tung Oil	72	lb.
Semi-Aromatic Solvent	16	gal.
Lead Naphthenate 24%	15⅜	lb.
Manganese Naphthenate 6%	2⅜	lb.
Cobalt Drier 4%	2⅛	lb.
Mineral Spirit, To make	43	gal.

No. 2

Mod. Phenol. Alkyd Resin No. 227	80	lb.
Poly. Linseed Oil "X" *	232	lb.
Tung Oil	72	lb.
Poly. Linseed Oil "Z-2" *	51	lb.
½% Soln. Guaiacol in Dipentene	2¼	gal.
Semi-Aromatic Solvent	11	gal.
Lead Naphthenate 24%	10	lb.
Manganese Naphthenate 6%	3	lb.
Mineral Spirit, To make	46½	gal.

Paper Varnish

U. S. Patent 2,139,603

(a)	Manila Copal	80	lb.
	Water	32	gal.
	Ammonia	7	gal.
(b)	Tung Oil Fatty Acids	40	lb.
	Water	8½	gal.
	Ammonia	3	gal.
(c)	Cobalt Chloride	15	g.
	Ammonia	200	cc.
(d)	Pine Oil	½	pt.

Emulsify (a) and (b) separately; dissolve (c) and then mix all together until uniform.

* From minimum 190 iodine number V. M. oil.

Pale Liquid Drier

Gradually heat 84 parts (by weight) rosin, 18.5 parts cobalt hydroxide, 20 parts of a liquid such as kerosene or light solvent naphtha to 190–200° C. or until reaction is complete and all water is driven off. Allow to cool to 160° and thin with petroleum naphtha or coal tar naphtha. A basic zinc naphthenate can be made by heating 20 parts zinc oxide with 86 parts naphthenic acid (A.V. ca. 250). A reaction commences at about 70° C. and is completed at between 130 and 140° C., but heating is continued until all water is removed. The final product possesses zinc content of about 16%, but by using suitable proportions of the reactant materials it is possible to prepare a product with a still higher metal content.

"Piano Finish" for Caskets

First the casket must be puttied with a colored putty to match the desired shade, then sanded smooth with No. 6/0 or 7/0 finishing (abrasive) paper. After sanding, it should be dusted off well; then it is ready to be stained.

Stains

Oriental Walnut

Yellow (Conc. Dye)	1

Orange (Conc. Dye)	1½
Denatured Alcohol	32

Dark Oriental Walnut

Orange (Conc. Dye)	½
Black (Conc. Dye)	½
Denatured Alcohol	32

Light Walnut

Red (Conc. Dye)	2
Orange (Conc. Dye)	24
Black (Conc. Dye)	8
Denatured Alcohol	94

Medium Walnut

Red (Conc. Dye)	2
Orange (Conc. Dye)	28
Black (Conc. Dye)	12
Denatured Alcohol	86

Medium Red Mahogany

Red (Conc. Dye)	1½
Yellow (Conc. Dye)	½
Orange (Conc. Dye)	1½
Black (Conc. Dye)	¾
Denatured Alcohol	16

Dark Red Mahogany

Red (Conc. Dye)	3½
Yellow (Conc. Dye)	½
Orange (Conc. Dye)	3
Black (Conc. Dye)	1½
Denatured Alcohol	32

This stain is sprayed on. After letting it dry for an hour, sand the stained surfaces very lightly; then fill with paste filler, using about 12 lb. of filler to the gallon of reducer. This is a very important operation when making a finish of this kind, because if the pores are not filled properly the finish will look "hungry."

After filling, allow the surfaces to dry over night before applying the sealer coat. Use a lacquer sealer and spray it on freely—what you call a "wet coat." It is left to dry from two to three hours before hand sanding.

Use felt sanding blocks and No. 8/0 sandpaper to sand out all traces of orange peel. Then the surfaces are dusted off with an air duster, being careful to remove all dust from the corners and carvings.

Next comes the clear lacquer—about four wet coats. This should be a lacquer containing about 27% solids, reduced two parts lacquer to one part of good grade of reducer. A cheap lacquer evaporates too fast to allow the lacquer to flow out properly. The last coat of lacquer is followed with a mist coat of clear reducer.

The lacquered hardwood casket should then stand at least four days before being rubbed. Rub with No. 320 wet-or-dry sandpaper, using water as a lubricant with Ivory soap for a softener. After it is rubbed down, finish off with a portable rubbing machine. Apply rubbing compound to the surface and use a rotary disc fitted with a sheep's-wool pad to rub or buff the finish. Then clean up the buffed finish with a good cleaner and polish. This process produces a finish that has depth and a very high polish.

A "piano finish" can also be obtained by using two or three coats of rubbing varnish over a varnish sealer. When varnish is used, it should be left to harden for at least nine days before rubbing. The rubbing is done with pumice stone and water; then the surfaces are polished out with rottenstone.

Alcohol Resistant Finish for Wood

A good finish which is alcohol resistant can be procured on wood by mixing linseed oil with 25% white vinegar and rubbing the mixture thoroughly into the wood. If a high gloss is desired it will be found necessary to apply several coats of the mixture.

Wrinkle Finishes

Varnish

Amberol F-7	640
Bodied Oiticica Oil	422

Heat to 260° C. in 35 minutes. Cool to 180° C.

Reduce with:

Solvesso No. 2	100
Solvesso No. 1	500

Paint

Carbon Black, Superba	14
Varnish	700
Xylol	175
Cobalt Naphthenate (1% Solution)	20

Grind in pebble mill.

Application: One coat directly on bare steel with spray gun. Bake at 110–120° C. for ½ hour or more.

To make finer wrinkle add oiticica standoil to paint after manufacture. Longer oil varnish (maximum 15 gal.) gives finer wrinkle in paint of lower bake.

To dilute use xylol or Solvesso No. 1.

Paints of other colors can be made by grinding in suitable quantities of pigments. Pebble mill recommended.

Concrete Floor Finish

This is a two part finish consisting of a "Primer" coat which is in the nature of a colored concrete sealer and which dries in about a half hour. This is applied to the raw floor.

After this is dry, it is coated with a waterproof type water-wax into which is incorporated the same color as in the primer coat.

Floors treated with this combination are maintained by mopping scuffed or worn spots with additional colored wax finish as they occur. Treated and serviced in this way, floors never require refinishing and are always in perfect condition.

Primer

Loba "C" Gum	100	lb.
Alcohol	27	gal.
Toluol	13½	gal.
Ethyl Acetate	4½	gal.

Mix above together in closed tank. Add colors as desired using iron oxides for reds and browns and chrome oxide for green. Approximately one pound of color per gallon of primer is sufficient. This color must be ground into the primer on a colloid mill or on a closed ball or pebble mill as the primer dries so rapidly as to make other types of mills useless.

A standard gray color is made by adding to the above quantity, the following:

Titanium Oxide	25 lb.
Lithopone	25 lb.
Black Iron Oxide	5 lb.

Wax Formula

The wax finish coat is prepared by incorporating about one pound of color per gallon, to any standard type water-wax and running same through the colloid mill.

Limed Oak Finish

The wood used is selected white oak, therefore the use of any bleach is unnecessary. The work is well sanded and must be inspected carefully for any imperfections. Assuming that the work is in A No. 1 shape, it is given a coat of water stain, depending on the particular shade the specifications call for, such as nigrosene and aniline powdered stains. After this stain dries, the surface is sandpapered carefully and then given a coat of clear lacquer sealer.

When the sealer is thoroughly dry, the surface is again smoothed the same as after the application of water stain. A coat of white wood filler is then applied. Of course, this wood filler can be tinted with any oil pigments such as raw sienna, burnt umber, raw umber, black, etc. This tint, of course, would be to match shades on specifications. This filler coat is applied with a brush, making sure to brush across the grain. After this material sets, but is not dry, it is wiped off with wiping rags, rubbing across the grain.

The pores of the wood then contain this filler, leaving a limed oak finish which is sometimes called silver gray oak finish. This is allowed to dry overnight and is then finished with another coat of clear sealer and one to three coats of clear lacquer. When the clear lacquer is dry, the surface is rubbed down with a rubbing compound and polished with a good wood wax.

Fumed Oak Finish

First, use a toner made up in the proportion of—

Tannic Acid	1	oz.
Pyrogallic Acid	1	oz.
Warm Water	1	gal.

Sponge the oak surface with this and, when dry, sand smooth. Then make a stain of the following ingredients and in the proportion indicated—

Potash Carbonate	3	lb.
Potash Bichromate	2	oz.
Walnut Crystals	2	oz.
Ammonia (28°)	3	qt.
Warm Water	4¼	gal.

Note: Keep this material in an earthenware crock, glass jar or wooden keg.

After staining, let the surface dry and apply one coat of white shellac. Sand lightly with fine sandpaper, and wax. Do not fill the surfaces, as this would spoil the effect.

Weathered Wood Effect

A weathered or driftwood effect on new lumber may be obtained by treating the lumber with one or more applications of medium strong solution of ferrous sulphate (copperas). If desired, use a gray-green paint, reduced with turpentine as a stain and as a final finish use a spar varnish, thinned down with turpentine—about one quart to a gallon.

Flock Finishing

Flock finishing is the way of finishing various articles with rayon, cotton and wool to give the surfaces a felt or suede appearance.

It is applied on wood of the very softest kind, such as white pine, California red cedar and various plywoods. First, the wood has to be coated with a good synthetic oil undercoater. After drying overnight, the surfaces are "spackled" filling up any holes or crevices with a Swedish putty. Then the wood is sandpapered to produce a smooth surface. It is given a mist coat of silk-screen process gloss enamel. This paint is not used as in screen process work but is thinned for spraying purposes.

After leaving the mist coat set for 20 to 30 minutes, follow with a wet coat of the same enamel, making sure in this last application of paint that the article is evenly coated, so that when the flock is applied it will adhere to the enamel uniformly.

The article is now ready to receive the flock. Before this operation is started the flock gun should be put in readiness. That is, the container should be filled only three-quarters full with flock and all adjustments should be made so that too much flock does not escape through the nose of the gun. The reason the container on the gun should only be filled three-quarters full is that the air coming through into the container must have a space in which to blow the flock around to scatter the particles. This is very important.

The air pressure for flocking should be between 40 and 60 lbs.—never more nor less. The air lines should be kept filtered to prevent oil and water accumulations. The best method is to coat the edges with flock and then spray the top surface.

It is best to let the flocked surface dry overnight, or, if the facilities of a baking oven are available, it can be baked dry at a temperature of 120° F. It is best, when anything is to be flocked, that the color, of the paint, match closely to the color of the flock. When the article is well dried the excess flock is then dumped off on paper so as to reclaim that which did not adhere to the surfaces being flocked.

Lacquer for Oxidized Brass

	Formula	
	No. 1	No. 2
Titanium Dioxide (Low Oil Absorption)	4.4	9.15
Carbon Black	1.4	0.35
Talc	9.2	—
Nitrocellulose (Dry Basis)	4.5 (125 sec.)	6.05 (5–6 sec.)
† Dyal #7986 Resin Solution (50% total solids)	16.0 *	—
‡ Dewaxed Dammar Solution	9.4	—
Ester Gum (Acid No. 6–8)	—	9.85
Alkyd Resin Solution (50%)	—	3.55
Glycol Sebacate	—	4.15
Blown Castor Oil	1.5 *	0.25
Raw Castor Oil	—	0.20
Ferrite Yellow	—	0.25
Snow Floss	—	0.60
** Solvent	53.6	65.60

* Up to 10% of the Dyal No. 7986, solid basis, may be substituted with blown castor oil, if necessary, to obtain greater flexibility of the film.

† Short oil, phenol modified alkyd resin.

‡ Dewaxed Dammar Solution

Dewaxed Dammar Resin	43
Xylene	25
Anhydrous Denatured Alcohol (S.D. No. 1, 23G or 23H)	32

** See page 324.

** Solvent, Oxidized Brass Lacquer

	Formula	
	No. 1	No. 2
Amyl Acetate	—	6.4
Butyl Acetate (Normal)	25.0	—
Isopropyl Acetate (or Ethyl Acetate)	—	27.8
Anhydrous Denatured Alcohol (SD #1, 23G or 23H)	10.0	5.0
Butyl Alcohol (Normal)	6.0	—
Lacquer Diluent Naphtha	32.0	14.5
Aromatic Petroleum Naphtha	12.0	35.5
Xylene	15.0	—
Coal Tar Solvent Naphtha	—	10.8

Oxidized Bronze Lacquer Thinner

	Formula	
	No. 1	No. 2
Phosphoric Acid (75%)	1.0	—
* Ethyl Acetate	9.0	—
Butyl Acetate (Normal)	30.0	16.2
* Isopropyl Acetate	—	9.6
Denatured Alcohol (S. D. No. 1, 23G or 23H)	12.0	10.0
Butyl Alcohol (Normal)	12.0	4.0
Lacquer Diluent Naphtha	36.0	36.4
Toluene (or Aromatic Petroleum Naphtha)	—	8.6
Xylene	—	15.2

* Ethyl acetate and isopropylacetate are interchangeable.

Metal Lacquer

Nitrocellulose, Wet (5 sec.)	11
Alkyd Resin (50% Solution)	15
Glaurin	4
Butyl Acetate	10
Butyl Alcohol	5
Ethyl Acetate	7
Toluol	48

Wood (Rubbing) Lacquer

Nitrocellulose, Wet (½ sec.)	15
Ester Gum	10
Glaurin (Diethylene Glycol Monolaurate)	3–4
Butyl Acetate	10
Butyl Alcohol	5
Ethyl Acetate	8
Toluol	48

The wet nitrocellulose contains 35% alcohol.

Congo-Ethyl Cellulose
(For Furniture Lacquer)

Congo (Thermally Processed)	140 lb.

Ethyl Cellulose (Low Viscosity)	35 lb.
Castor Oil or other Plasticizer	7 lb.
Toluol	455 lb.
Denatured Ethyl Alcohol	70 lb.

Dammar-Ethyl Cellulose Lacquer

Batavia Dammar A/E	500 lb.
Ethyl Cellulose (Low Viscosity)	125 lb.
Turpentine	1750 lb.
Butanol	750 lb.

Linoleum Cement

Batu Nubs and Chips	37 lb.
High Solvency Petroleum Thinner	38 lb.
Fuller's Earth (200 Mesh)	25 lb.

Dissolve the Batu in petroleum thinner using agitation, but no heat.

Add the fuller's earth to the above vehicle and stir thoroughly.

Other inert materials may be used in place of fuller's earth if it is so desired.

Batu-Ethyl Cellulose Furniture Finishes

Formula No. 1

Batu Scraped	90
Ethyl Cellulose (Low Viscosity)	10

No. 2

Batu Scraped	70
Ethyl Cellulose (L. V.)	20
Mineral Oil	10

No. 3

Batu Scraped	70
Ethyl Cellulose (L. V.)	20
Dow Plasticizer No. 6	10

No. 4

Pale East India Nubs or Chips	90
Ethyl Cellulose (L. V.)	10

No. 5

Pale East India Nubs or Chips	70
Ethyl Cellulose (L. V.)	20
Mineral Oil	10

No. 6

Pale East India Nubs or Chips	70
Ethyl Cellulose (L. V.)	20
Dow Plasticizer No. 6	10

The solvents used may be high solvency petroleum naphthas or mixtures of these with low solvency naphthas.

Pale East India gives smoother films than Batu, but either may be plasticized with practically any oils, the amount of oil used depending upon individual desires.

Fluorescent Christmas Ornament Lacquer

Many paints used on Christmas ornaments fluoresce brightly and may be used, under ultra-violet light, for this purpose.

A lacquer made from ordinary colorless lacquer and having a few per cent of fluorescent dye present can be used on Christmas ornaments.

Adding zinc orthosilicate or anthracene to the lacquer instead of the dyes produces different effects and more permanent paint.

Glass Fibers; Lacquer for
U. S. Patent 2,215,061

Cellulose Nitrate	14.0
Castor Oil	18.6
Pigment	23.4
Ethyl Acetate	17.6
Alcohol	26.4

Lacquer for Glued Surfaces

Nitrocellulose (Dry Basis)	60
Dibutylphthalate	25
Flexoresin D A 4	15
Solvent, To suit.	

Spraying Pigmented Lacquer

Non-Volatiles	Parts by Weight
¼ Sec. Nitrocellulose (Dry Basis)	16
Ester Gum	8
Dibutyl Phthalate	7
Titanium Dioxide	7
Medium Green	3

Volatiles Per Cent by Volume	Solvent Mixture A	B
Butyl Acetate	30	18
Butanol	15	9
* Ethyl Acetate	—	12
Ethyl Alcohol	10	6
Toluol	—	55
Xylol	45	—

Semi-Pigmented Airplane "Dope" Lacquer

Clear dope: wet nitrocellulose 10 parts, butyl acetate 10 parts, butyl alcohol 20 parts, ethyl acetate 23 parts, toluene 27 parts. Various color pastes, e.g., yellow paste: cadmium lithopone 32 parts, castor oil 11 parts, dibutyl phthalate 11 parts, cadmium yellow to color. Cream paste: zinc oxide 22 parts, castor oil 11 parts, dibutyl phthalate 11 parts, chrome yellow to color. Light blue paste: zinc oxide 22 parts, castor oil 11 parts, dibutyl phthalate 11 parts, prussian blue, etc., to color. Dopes are made from mixture of clear dope and pastes.

* Including the ethyl alcohol ordinarily added with the nitrocellulose.

Acetate Dope for Aircraft Cloth

Cellulose Acetate (20 sec.)	9
Triphenyl Phosphate	1
Acetone	31
Methyl Ethyl Ketone	31
Ethyl Acetate	14
Ethyl Lactate	14

Paper Coating Lacquer

Cellulose Acetate (2–5 sec.)	13.3
Santicizer M–17	6.7
Santicizer 8	1.6
Acetone	20.0
Methyl Ethyl Ketone	8.3
1–Nitropropane	40.6
Alcohol	9.5

Artificial Leather Coating on Cloth

Alcohol	58
Ethyl Acetate	22
Naphtha	18
Xylol	2

To 100 gal. solvent, add 176 lb film scrap.

Polyvinyl Acetal Acetate—Nitro Starch Lacquers

These compositions are valuable as quick drying high gloss and transparent lacquers for adhesives, printing ink base, fingernail lacquer, paper, furniture, and metal lacquers.

A. Volatile Solvent

In general a solvent mixture which will be a solvent for the amounts of nitrostarch and polyvinyl acetal acetate used. A good vehicle is as follows by *volume:*

Ethyl Acetate	2– 5%
Butyl Acetate	25–60%
Butyl Ether of Ethylene Glycol	2– 5%
Butyl Alcohol	3–10%

Toluene	10–35%
Xylene	5–21%
Petroleum Naphtha	5–15%

B. Plasticizer

Tricresyl phosphate, dibutyl phthalate, triacetin, etc.

3 to 5% of total weight of lacquer and about 10% of film forming ingredients.

C. Film Formers

1. Polyvinyl acetal acetate: 10 to 25% of total weight of lacquer
2. Nitro starch: 3 to 15% of total weight
3. Soft resin such as alkyds, urea-formaldehydes, sulfonamides: 5 to 30% of combined weight of (1) and (2).

The % of solids is determined by use and method of application; i.e. spraying, brushing, dipping.

Formula No. 1

Polyvinyl Acetal Acetate	18
Nitrostarch	9
Plasticizer	3
Solvent Vehicle	70

No. 2

Polyvinyl Acetal Acetate	5–15
Nitrostarch	5–15
Soft Alkyd Resin	5–15

Solvent Vehicle, To give concentration desired.

No. 3

Finger Nail Lacquer

Polyvinyl Acetal Acetate	15.5
Nitrostarch	6.0
Dibutyl Phthalate	2.0
Iron Oxide (Red)	1.0
Titanium Oxide	1.5
Solvent Vehicle	

No. 4

Furniture Spray Lacquer

Polyvinyl Acetal Acetate	15
Nitrostarch	5
Tricresyl Phosphate	3
Butyl Acetate	38
Toluene	40

Solvent Resisting Ethyl Cellulose Film

This film is especially valuable in transparent pressure adhesive tapes, in the fabrication of laminated products having at least 1 ply of ethyl cellulose, and in transparent containers for oily and greasy substances.

The film is baked from 140° C. for 10 minutes to 50° to 80° C. for 16–64 hours, after being deposited from the solvent.

A. Solid Contents

65 to 90 parts medium ethoxy ethyl cellulose having an ethoxy content of 43.5 to 46.5%.

35 to 10 parts of a heat hardenable aryl sulphonamide – formaldehyde resin, such as p–toluene sulphonamide–formaldehyde, etc.

B. Solvents

The 2 solids are mutually compatible in a wide variety of common solvents or mixtures thereof, such as ethyl and butyl alcohol, ethyl, butyl, or amyl acetate, toluene, xylene, nitroethane, hexone, etc.

Substitutes for Dibutyl Phthalate

In lacquers, dibutyl phthalate can often be replaced by "Glaurin" (diethylene glycol mono-laurate) 1–3% is used on weight of total solids. In synthetic rubbers, glycerin monoricinoleate (water insoluble grade) is an excellent replacement.

Optimum Surface Tension Lacquer Solvent

Butyl Acetate	33 gal.

Butyl Alcohol	12 gal.
Toluol	20 gal.
Troluoil	35 gal.

Low Surface Tension Lacquer Solvent

Butyl Acetate	33 gal.
Butyl Alcohol	12 gal.
Troluoil	55 gal.

Ethyl Cellulose Lacquer Solvents

While ethyl cellulose is soluble in many single solvents, solvent mixtures are usually desirable to effect the proper combination of viscosity, film characteristics, and evaporation rate. A mixture of toluene and ethyl alcohol in the ratio of 70:30 or 80:20 is satisfactory in many cases. Where faster drying is required in lacquers for roller coating or dipping, a suggested mixture consists of: toluene, 40; ethyl acetate, 40; ethyl alcohol, 20.

Another mixture which is advantageous in spraying lacquers has these proportions:

Xylene	35
Troluoil	40
Ethyl Alcohol	15
Butyl Alcohol	10

Duplicator Solvent
U. S. Patent 2,228,108

Ethylene Glycol Monomethyl Ether	10
Water	10
Alcohol	80

Suspension of Pigments

A frequent difficulty in the storage, shipment and utilization of pulp colors or pigments in water is that of maintaining a uniform dispersion. Much time is lost in redispersing colors caked at the bottom of a container or tank. Often the caking is such as to prevent reformation of the original fine particle size, resulting in complete loss of color.

This problem can easily and inexpensively be solved by the use of Diglycol stearate. As little as 1/10% of this added to the weight of the color mix will produce the desired result. In the case of heavy pigments, such as titanium dioxide a higher percentage is usually desirable. A 2% or 3% stock dispersion of Diglycol stearate is made by adding the stearate to boiling or hot water and stirring until room temperature is reached. The proper amount of this is added to the pulp color with agitation, or to the diluted pulp color. If desired, the color may first be stirred into the stock solution, water being added afterwards. The method of addition is quite flexible and easily managed. A typical color dispersion is as follows: 50% dry color, 49.9% water, 1/10% Diglycol stearate.

An interesting property of the foregoing formula, when coated on paper, is the glaze obtainable by stoning or calendering. This coating shows a higher degree of water resistance than might ordinarily be expected. It, therefore, is of interest in replacing casein as a binder and glazing agent in many paper coating operations.

Exterior House Paint Pigments
White

Formula No. 1

35% Leaded Oxide	45	lb.
White Lead	18	lb.
Titanium Dioxide	15	lb.
Inert	22	lb.

No. 2

Pigment—725	41¼	lb.
White Lead	16½	lb.
Titanium Dioxide	13¾	lb.
Inert	20	lb.

White and Tints

No. 3

White Lead	60	lb.
Zinc Oxide	30	lb.
Inert	10	lb.

No. 4

White Lead	40	lb.
Pigment—725	40	lb.
Inert	9	lb.

Tints

No. 5

White Lead	32	lb.
Zinc Oxide	30	lb.
Non-Chalking Titanium Dioxide	8	lb.
Inert	30	lb.

No. 6

White Lead	19	lb.
Pigment—725	45½	lb.
Non-Chalking Titanium Dioxide	7½	lb.
Inert	23	lb.

Colloidal Shellac Dispersion

Shellac	100	g.
Methyl Acetone	200	cc.

Mix until dissolved. Add with vigorous mixing:

Triethanolamine	10–15	cc.

Allow to stand for 24 hours. Add slowly with stirring:

Water	3–4	l.

Thinning Shellac Solutions

To convert shellac	Use alcohol
In 5-lb. cut to 3 -lb. cut	3½ pints to gallon shellac
In 5-lb. cut to 2½-lb. cut	⅔ gallon to gallon shellac
In 5-lb. cut to 2 -lb. cut	1 gallon to gallon shellac
In 5-lb. cut to 1 -lb. cut	2⅔ gallons to gallon shellac
In 4-lb. cut to 3 -lb. cut	1 quart to gallon shellac
In 4-lb. cut to 2½-lb. cut	3½ pints to gallon shellac
In 4-lb. cut to 2 -lb. cut	¾ gallon to gallon shellac
In 4-lb. cut to 1 -lb. cut	2⅛ gallons to gallon shellac

Cutting Liquid Shellac Without Adding Alcohol

By the addition of 4-oz. of ammonia to one gallon of 5-lb. white liquid shellac, the shellac in this alcoholic liquid solution is rendered sufficiently soluble so that 2½ to 3 gallons of water can be safely added to the mixture, thus giving a resultant three to four gallons.

Telephone Transmitter Membrane Coating

U. S. Patent 2,281,940

Polymerized Isobutylene	100
Stearic Acid	1
Ethyl Cellulose	10
Paraffin Wax	20
Carbon Black, To suit	

Wash for Galvanized Iron Before Painting

Alcohol	60
Toluol	30
Carbon Tetrachloride	5
Hydrochloric Acid (Concentrated)	5

The solution is freely applied to the galvanized iron with a brush.

Cleaner for Aluminum Before Painting

Butyl Alcohol	40
Isopropyl Alcohol	30
Phosphoric Acid (85%)	10
Water	20

The cleaner is applied and allowed to remain on the surface for two minutes, followed by slight scrubbing with a bristle brush and thorough rinsing with water.

Paint Remover
Formula No. 1

Tripoli	26	oz.
Mineralite	30	oz.
Neutral Soap Powder	24	oz.
Water	50	oz.
Denatured Alcohol	13½	oz.

(Mix in a paint mill.)

No. 2

Wax-like materials (e.g. paraffin wax) are incorporated into paint removers mainly to retard solvent evaporation. Certain pigments possessing a leafy structure may be satisfactorily substituted for the paraffin wax. A suggested formula is: film scrap 3 parts, methyl acetate 9 parts, ethyl alcohol 8 parts, methylene chloride 32.5 parts, talc or powdered mica 20 parts. Better results are obtainable with the powdered mica if this be used in conjunction with a vinyl acetate resin, thus, polyvinyl acetate resin of high viscosity 3 parts, ethyl alcohol 8 parts, methylene chloride 32.5 parts, ethyl acetate 9 parts, powdered mica 25 parts.

Benzol-Free Paint Remover

By-Product Alkane Chlorides	25
Acetone	50
Alcohol, Denatured	25
Paraffin Wax	2–3

Varnish and Paint Removing Composition
U. S. Patent 1,483,587

A paste is made by mixing furfural with an inert powder ingredient such as corn starch, using such proportions as to obtain as thick a paste as possible and yet have one that may be applied with a brush. The paste is applied to a varnish surface and allowed to remain for ten or twenty minutes. The outer portion of the paste will have dried out slightly, but the inner portion will be still moist with furfural.

By the use of a knife blade, scraper, stiff brush, cloth or other suitable means, the paste residue and loosened varnish film may be entirely removed, leaving the clean surface of the material. This surface is now ready for the application of a fresh coat of varnish or paint, there being no wax or other deposit to interfere. Varnish coats may be moved from wood by this method without harming the original surface in any way. Although the paste will remain moist for a long period of time, it should not be allowed to remain on the surface for so long a time as to be too dry for satisfactory removal. This method works successfully no matter what the position of the surface. It permits the ready removal of varnish from vertical, inclined, or inverted surfaces as well as from horizontal surfaces and is very economical of solvent material.

Although furfural itself may be made into the starch paste and work satisfactorily for removing many ypes of surface coatings, a mixture of furfural and furfuryl alcohol as the liquid ingredient of the paste is advantageous in removing certain coatings, especially those containing large amounts of shellac or cellulose esters.

Varnish Remover

Acetone	1 qt.
Alcohol	1 qt.
Water Saturated with Soda Ash	½ pt.
Benzene	1 qt.

Shake and apply. Keep wet 20 minutes, wipe and scrape off.

Phenol Formaldehyde Varnish Remover

U. S. Patent 2,127,469

Nitric Acid (d. 1.42)	10
Sulphuric Acid (d. 1.84)	90

Apply at 85° C. and wash off thoroughly.

Improved Crack Filler

Because crack filling materials often work out of floors, the filler itself may not necessarily be at fault. As floors made out of oak and other hard woods are usually waxed, the cracks oftentimes hold more or less wax as well, which prevents the filling material from adhering to the sides of the cracks. The same result may also be caused when varnish or shellac trickles into the cracks.

To prevent such trouble, clean the cracks out thoroughly by scraping and remove all dust. For a satisfactory filling, use sifted sawdust made from the same wood as the floor. Enough sawdust for crack filler may be procured quickly by sawing some of the wood. Mix the sawdust to a paste with the fairly strong solution of glue or casein. By adding coloring matter to the mixture it may be darkened to match the color of the wood. As soon as the filler is pressed well into the crack, smooth it down and clean the surplus away before the filler has time to harden.

Wood Sealer

Pontianak Bold Scraped	50 lb.
Cellosolve	100 lb.

Mix the ingredients until a homogeneous solution is obtained.

A film of this material requires about ten to fifteen minutes for an initial set and it may be sanded lightly at the end of one hour.

Preserving Wooden Fence Posts

Formula No. 1

Mercuric Chloride	1 tbsp.
Sodium Chloride	1 tbsp.

Put into ¾ in. hole in post near ground.

No. 2

Creosote	70
Fuel Oil	30

As much as post will absorb is to be used.

No. 3

Creosote	25
Fuel Oil	75

Impregnate under pressure of 120 lb./sq.in., allowing complete absorption.

Adobe Preservative

U. S. Patent 2,137,247

Paint surfaces with solution of:

Vinyl Acetate	10 g.
Acetone	55–60 cc.
Thinned with Toluol	200–1000 cc.

Fireproof Plaster

British Patent 531,039

Slag Wool	30
Asbestos	25
Iron Filings	15
Potassium Silicate (d. 1.33)	30

Plaster Primer

White Lead, Semi-Paste	100	lb.
Interior Varnish	4	gal.
Linseed Oil, Kettle Bodied	2	gal.
Turpentine	¾	gal.

Chapter Fourteen

PAPER

Grease-Proof Coating for Containers

U. S. Patent 2,122,907

Animal Glue	100 oz.
Glycerin	10–500 oz.
Formaldehyde	50–1500 oz.
Water	200–1500 oz.

In the formation of the coating, the time at which the tanning reaction takes place is controlled by carefully controlling the temperature of the coating. In the mixing of the ingredients forming the coating, the temperature of the mixture is carefully kept below the reacting temperature of the tanning agent during the coating process, the exact temperature used depending upon the length of time of delay required in the operation. In the case of formaldehyde, the temperature is maintained below 130° F. during the coating process, to prevent reaction between the tanning agent and the tannable material. When the coating operation is completed, the surface coated is immediately subjected to temperatures above the reaction temperature of the tanning agent, or above 130° F. if formaldehyde is used. This promotes the tanning reaction, and an insoluble material possessing the required properties is produced. This method obviates the low temperature drying to produce a continuous film, which was necessary in previous methods. The materials are accordingly dried much faster in the present method.

Grease Proof Paper Coating

U. S. Patent 2,256,853

Clay	87.7
Polyvinyl Alcohol	1.3
Thin Boiling Cooked Corn Starch	11.0

Greaseproofing Paper

Australian Patent 113,451

To make paper greaseproof it is treated in a mixture of anhydrous lanolin 6, paraffin wax 2, white petroleum jelly 2, white spirit 4 (or less). To this mixture is added 1/16 part of eucalyptus oil or similar substance to mask the odor. The mixture is heated to over 56°, the paper dipped in it and passed over rolls heated at the same temperature followed by rolls at 4°.

Alkali Resistant Paper or Board

U. S. Patent 2,231,562

Dip in a 30% solution of ammonium chloride or acetate and dry.

Waterproofing Writing Paper

Formula No. 1

The paper is spread horizontally over a glass plate and is then given coating on both sides with a 2% solution of a wetting agent or glyceryl monostearate strength. Thereupon the paper is hung up to dry.

The paper is then spread over a glass plate and painted or sprayed with a vulcanizing latex mixture e.g.

Mixture 1

Creamed Latex	100.0 ml.
Kaolin	15.0 g.
Sulphur	1.5 g.
Zinc Oxide	1.5 g.
Ultra-Accelerator (e.g. P extra N)	0.5 g.
Casein Solution (10%)	5.0 ml.
Wetting Agent Solution (5%)	20.0 ml.
Paraffin Emulsion (20%)	5.0 ml.

The creamed latex is added after all of the other ingredients have been thoroughly mixed in a ball mill.

Before application, this entire mixture is diluted with 10 to 20 ml. of water. It is preferable to apply a diluted mixture twice, than a more concentrated mixture only once. Air bubbles should be avoided as much as possible. A thin rubber layer is more than sufficient. Paper properly prepared is tinted only very slightly yellow.

After the paper has been waterproofed on both sides in the manner here described, the following mixture is applied:

Mixture 2

Refined Glass Powder	30 g.
Wetting Agent Solution (5%)	30 ml.
Creamed Latex	100 ml.
Konnyaku Meal Solution (1%)	50 ml.

One single application suffices.

The solution of Konnyaku meal, either through centrifuging or sieving through a cloth, must be freed of the undissolved grains and other coarse particles. In view of the fact that Konnyaku meal is subject to a very considerable degree of enzymic and bacteriological decomposition through which it becomes inactive, the above-described mixture is best prepared shortly before being applied and, where necessary, kept in a refrigerator. The addition of "Santobrite" (sodium pentachlorophenate), as a preserving agent, is to be recommended.

When the paper has become dry, it is covered with talcum powder, and, if necessary, the edges are neatly cut. It must be remembered, however, that the paper ceases to be waterproof where it is cut or where the rubber layer has been otherwise damaged. Such places will have to be treated once more with vulcanizing latex (Mixture 1).

Paper prepared in the manner above described has been found to be very satisfactory by those to whom samples had been submitted. It can readily be written on with pencil, though it is not well adapted to ink. On the other hand it takes print very well and can also be used for typing.

The appearance of the paper depends in a large measure upon the even application of the rubber layers. The best method would be to do this mechanically. Highly sized smooth paper is less adapted to being treated than the cheaper kinds of paper. Very good results are obtained with ordinary typewriting paper.

If paper is soaked in a 6% solution of Pliolite in toluol, a good waterproof paper is obtained. The superfluous solution that has not soaked into the paper must be removed as well as possible to prevent its producing streaks after drying.

The waterproofing is still better if 10% of solidified paraffin is added to the Pliolite.

Paper so treated has almost the same appearance as ordinary paper.

Integral Waterproofing, Paper Pulp
British Patent 528,308

A hot solution of sodium stearate (15 g.) in water (10) is added to a melt of paraffin wax (3 lb.) and mineral oil (1/3 lb.), and rapidly stirred to give an emulsion which is used in proofing a paper pulp-cork mixture.

Moisture Proofing for Paper
U. S. Patent 2,290,563

Ethyl Cellulose (Low Viscosity)	20%
Hydrogenated Rosin	15%
Paraffin Wax (m.p. 136° F.)	50–55%
Ester Gum	0–10%
12-Hydroxystearin	15– 5%

The above is applied molten as a coating or impregnant.

Insectproof Paper or Felt
U. S. Patent 2,129,659

Apply to paper or felt, the following mixture:

Asphalt Paint	73
Lead Arsenate, Powd.	25
Nicotine Sulphate	1
Clay, Powd.	1

Pest-Resistant Paper
Australian Patent 113,158

Sheets of paper or cardboard are coated with a solution consisting of:

Formalin	5
Glycerin	5
Ammonium Hydroxide	2
Casein	5
Water	83

In the process, several sheets of paper are put together, and in between, is placed one sheet coated with the above solution to which some blue coloring matter has been added. The resultant laminated sheet is then coated on both sides with a somewhat similar solution consisting of:

Formalin	10
Glycerin	2
Boric Acid	1
Sulphuric Acid	1
Water	86

The paper products so treated may be manufactured into packing boxes, bags, wrappings and similar packaging products.

Casein Coating for Paper
U. S. Patent 2,229,620

Clay	90
Calcium Carbonate	10
Casein Solution (23%)	11

Paper-Coating Emulsion

Carnauba Wax	50 lb.
Stearic Acid	7 lb.
Borax Soap	10 lb.
Water, To make	50 gal.

This is added to a clay-casein mixture to produce a coating on paper which can be readily friction-calendered to give a high finish.

Paper Sizing
Canadian Patent 398,980
Formula No. 1

Scale Wax (Paraffin)	608
Rosin	1216
Stearic Acid	36
Corn Starch	122
Caustic Soda	130
Water	3168

No. 2

Starch	20%
Sodium Silicate	15%
Glue, Animal	1%
Water, To make	100%

After treating with above, apply:

Aluminum Sulphate	12%
Water	88%

Paper Sizing Emulsion

Petroleum Resin	37
Clay, Colloidal	18
Casein	1 1/5
Caustic Soda	1
Water, To suit	

Millson's Fluorescent Paper Coating

Formula No. 1

Barium Sulphate
Shellac Emulsion *
Amount to make coating of proper consistency.

Add enough fluorescent dye to make fluorescent in the desired shade and intensity. Fluorescent dyes which can be used are: rhodamine, uranine, fluorescent blue G, brilliant flavine, acriflavine, diazo scarlet PRD, yellow OX, red BX, and red 6G.

No. 2

Barium Sulphate	50	g.
Shellac Emulsion *	50	cc.
Fluorescent dye	0.2– 0.5	g.

Fluorescent Documents

Formula No. 1

The paper to be printed, or lithographed, is dipped or rolled with a fluorescent dye and allowed to dry. The solution of dye is made with:

Water	100
Fluorescent Dye	0.2– 0.6

The amount of dye is governed by the characteristics of the dye itself and the purpose for which it is intended. Dyes used are: acriflavine, uranine, fluorescein, rhodamine, eosine, diazo scarlet PRD, auramine, umbelliferone, and others.

No. 2

By printing as usual and then spraying lightly with a colorless and non-fluorescent lacquer. An ordinary atomizer may be used. Before the lacquer dries the surface is dusted lightly with anthracene or other brightly fluorescing substance. Sheets so treated are laid flat and allowed to dry in that position.

No. 3

The sheet is printed as usual and then dipped or sprayed with ordinary lubricating oil, petrolatum, or a benzene solution of the ester of a polyhydric alcohol (Diglycol Stearate). This method is not suited for tissue-thin maps and papers.

No. 4

An ordinary map which cannot be altered is placed under a layer of brightly fluorescent "cellophane." This "cellophane" can be prepared by dyeing ordinary colorless "cellophane" with organic and fluorescent dyes which retain their luminescence after drying. The "cellophane" may be stapled to the document, stuck with colorless tape, or merely laid on top.

* Shellac emulsion is prepared by dissolving 940 grams of shellac in 5200 cc. water containing 250 cc. ammonium hydroxide.

Self Sealing Waxed Paper
U. S. Patent 2,233,186

Paper is coated with a melted mixture of:

Paraffin Wax	20
Crepe or Synthetic Rubber	2– 3
Ester Gum	80

This gives a transparent, flexible, grease- and waterproof product which can be heat sealed above 70° C.

Heat-Sealable Waxed Paper for Foods
U. S. Patent 2,227,516

Paper is coated with:

Paraffin Wax	100
Rubber or Polymerized Isobutylene	1– 3
Titanium Dioxide, Colloidal	5– 15
Sodium Benzoate	1– 3

Waxed Paper Impregnant

Paraffin Wax (150° F. M. P.)	186
Carnauba Wax No. 1 Yellow	14

Paper Milk Container Coating
U. S. Patent 2,253,655

Water	1 gal.
Glyceryl Monostearate	4 oz.
Glycerin	2 oz.

Increased Translucency of Coated Paper

The addition of 5% YUMIDOL (Sorbitol Syrup) and 1/100% of WETANOL (Wetting Agent) based on the total amount of the moisture content of the paper, will give increased transparency to paraffin wax coatings on paper. The paper should be saturated to a moisture content of 11% and dried down to about 6%. It aids prevention of buckling of the paper on overcoating with paraffin wax, and helps give a more translucent coating.

Paper Transparentizing Oil

Peanut Oil	6.00
Chlorinated Rubber	1.00
Toluol	4.75
Tricresyl Phosphate	0.25
Dehydrated Castor Oil	2.25
Glaurin	0.75

Dilute with toluol or xylol as desired.

Paper-Pulp Board
Australian Patent 112,862

Glue 25, water 75, an aqueous solution of sodium silicate 1, beta-naphthol 0.02, and Venice turpentine 0.01 lb. On top of this come more layers of paper, cloth, or like material and the whole is pressed. When the board is dry its upper side is sized with a composition containing water 40, borax 4, shellac 20 lb., and sulfonated castor oil 10 oz. Pigments may be added if desired.

Stereotype Mat
U. S. Patent 2,112,023

Paper of suitable moisture content is treated with:

Glue	5
Urea	5
Inorganic Filler	40
Water	90

It is then dried and calendered and passed through 1% formaldehyde solution for fixing.

Paper Towel and Napkin Softener
U. S. Patent 2,268,674

Urea	20
Glycerin	10

Talc	10
Water	60

Mix well and apply to paper.

Deciphering Charred Documents

The method consists in treating the charred document with chloral hydrate which appears to have a clarifying action on the burned figures or letters. This is applied in the form of a 25% solution of chloral hydrate in alcohol. This is repeated several times, the document being dried at 60° C. between each application, until a mass of chloral hydrate crystals forms on the surface. At this stage, a similar solution, to which 10% of glycerin has been added, is applied and the document dried as before. It may then be photographed; the most suitable type of plate being a contrasty non-color sensitive one.

The method has proven equally satisfactory for typewritten and printed material. With certain modifications it has also been found to restore writing. Moreover, the reading matter is restored equally on both sides of the paper.

Chapter Fifteen

PHOTOGRAPHY

D-7 Developer

For Professional Films and Plates

Stock Solution A

	Avoirdupois	Metric
Water, 125° F. (50° C.)	16 oz.	500 cc
Elon	¼ oz.	7.5 g.
Sodium Bisulphite	¼ oz.	7.5 g.
Pyro	1 oz.	30.0 g.
Potassium Bromide	60 gr.	4.2 g.
Water, To make	32 oz.	1.0 l.
Stock Solution B		
Water	32 oz.	1.0 l.
Sodium Sulphite, Desiccated	5 oz.	150.0 g.
Stock Solution C		
Water	32 oz.	1.0 l.
Sodium Carbonate, Desiccated	2½ oz.	75.0 g.

Dissolve chemicals in the order given.

Tray Development: Take 1 part of A, 1 part of B, 1 part of C and 8 parts of water.

Develop about 7 minutes at 68° F. (20° C.).

Tank Development: Take 1 part of A, 1 part of B, 1 part of C and 13 parts of water.

Develop about 10 minutes at 68° F. (20° C.).

This developer can be used for two or three weeks if the volume is maintained by adding fresh developer in the proportion of 1 part each of A, B, and C to 4 parts of water. It is usually necessary to increase the development time as the developer ages.

D-8 Developer

For High Contrast on Films and Plates

Stock Solution

	Avoirdupois	Metric
Water	24 oz.	750 cc.
Sodium Sulphite, Desiccated	3 oz.	90.0 g.
Hydroquinone	1½ oz.	45.0 g.
Sodium Hydroxide	1¼ oz.	37.5 g.
Potassium Bromide	1 oz.	30.0 g.
Water, To make	32 oz.	1.0 l.

Dissolve chemicals in the order given. Stir the solution thoroughly before use.

For use, take 2 parts of stock solution and 1 part of water. Develop about 2 minutes in a tray at 65° F. (18° C.).

For general use, a developer which is slightly less alkaline and gives almost as much density can be obtained by using 410 grains of sodium hydroxide per 32 ounces of stock solution (28 grams per liter) instead of the quantity given in this formula.

D-11 Developer

For Contrast on Films and Plates

	Avoirdupois	Metric
Water, 125° F. (50° C.)	16 oz.	500 cc.
Elon	15 gr.	1.0 g.
Sodium Sulphite, Desiccated	2½ oz.	75.0 g.
Hydroquinone	130 gr.	9.0 g.
Sodium Carbonate, Desiccated	365 gr.	25.0 g.
Potassium Bromide	73 gr.	5.0 g.
Cold Water, To make	32 oz.	1.0 l.

Dissolve chemicals in the order given.

Develop about 5 minutes in a tank or 4 minutes in a tray at 68° F. (20° C.).

When less contrast is desired, the developer should be diluted with an equal volume of water.

D-19 Developer

For Rapid Development of Films and Plates

	Avoirdupois	Metric
Water, 125° F. (50° C.)	16 oz.	500 cc.
Elon	32 gr.	2.2 g.
Sodium Sulphite, Desiccated	3 oz. 90 gr.	96.0 g.
Hydroquinone	128 gr.	8.8 g.
Sodium Carbonate, Desiccated	1 oz. 265 gr.	48.0 g.
Potassium Bromide	73 gr.	5.0 g.
Cold Water, To make	32 oz.	1.0 l.

Dissolve chemicals in the order given.

Develop about 5 minutes at 68° F. (20° C.) according to the contrast desired.

It is desirable to use the F-10 Fixing Bath with this developer.

D-32 Developer

For Warm Tones on Lantern Slide Plates

Stock Solution A

	Avoirdupois	Metric
Water, 125° F. (50° C.)	16 oz.	500 cc.
Sodium Sulphite, Desiccated	90 gr.	6.3 g.

Hydroquinone	100 gr.	7.0 g.
Potassium Bromide	50 gr.	3.5 g.
Citric Acid	10 gr.	0.7 g.
Cold Water, To make	32 oz.	1.0 l.
Stock Solution B		
Cold Water	32 oz.	1.0 l.
Sodium Carbonate, Desiccated	1 oz.	30.0 g.
Sodium Hydroxide	60 gr.	4.2 g.

Dissolve chemicals in the order given.

For use, take 1 part of A and 1 part of B. For still warmer tones, 1 part of A and 2 parts of B. Stir thoroughly before use. Develop about 5 minutes at 68° F. (20° C.).

D-52 Developer

For Warm Tone Papers

Stock Solution

	Avoirdupois	Metric
Water, 125° F. (50° C.)	16 oz.	500 cc.
Elon	22 gr.	1.5 g.
Sodium Sulphite, Desiccated	¾ oz.	22.5 g.
Hydroquinone	90 gr.	6.3 g.
Sodium Carbonate, Desiccated	½ oz.	15.0 g.
Potassium Bromide	22 gr.	1.5 g.
Cold Water, To make	32 oz.	1.0 l

Dissolve chemicals in the order given.

For use, take 1 part of stock solution to 1 part of water. Develop not less than 1½ minutes at 70° F. (21° C.).

NOTE: More bromide may be added if warmer tones are desired.

D-61a Developer

For Professional Films and Plates

Stock Solution

	Avoirdupois	Metric
Water, 125° F. (50° C.)	16 oz.	500 cc.
Elon	45 gr.	3.1 g.
Sodium Sulphite, Desiccated	3 oz.	90.0 g.
Sodium Bisulphite	30 gr.	2.1 g.
Hydroquinone	85 gr.	5.9 g.
Sodium Carbonate, Desiccated	165 gr.	11.5 g.
Potassium Bromide	24 gr.	1.7 g.
Cold Water, To make	32 oz.	1.0 l.

Dissolve chemicals in the order given.

Tray Development: Take 1 part of stock solution to 1 part of water. Develop about 6 minutes at 68° F. (20° C.).

Tank Development: Take 1 part of stock solution and 3 parts of water. Develop about 12 minutes at 68° F. (20° C.). Add stock solution (diluted 1:3) at intervals to maintain the volume, or the replenisher, D-61R, to maintain the strength of the solution.

D-72 Developer
For Papers, Films, and Plates
Stock Solution

	Avoirdupois	Metric
Water, 125° F. (50° C.)	16 oz.	500 cc.
Elon	45 gr.	3.1 g.
Sodium Sulphite, Desiccated	1½ oz.	45.0 g.
Hydroquinone	175 gr.	12.0 g.
Sodium Carbonate, Desiccated	2¼ oz.	67.5 g.
Potassium Bromide	27 gr.	1.9 g.
Water, To make	32 oz.	1.0 l.

Dissolve chemicals in the order given.

For papers, dilute 1 to 2 and develop about 45 seconds at 70° F. For films and plates dilute 1 to 1 and develop about 4 minutes in a tray or 5 minutes in a tank at 68° F. (20° C.).

D-76 Developer
For Low Contrast on Films and Plates

	Avoirdupois	Metric
Water, 125° F. (50° C.)	24 oz.	750 cc.
Elon	29 gr.	2.0 g.
Sodium Sulphite, Desiccated	3 oz. 145 gr.	100.0 g.
Hydroquinone	73 gr.	5.0 g.
Borax, Granular	29 gr.	2.0 g.
Water, To make	32 oz.	1.0 l.

Dissolve chemicals in the order given.

Average development time about 16 minutes at 68° F. (20° C.). See individual recommendations listed for each material.

The useful life of this developer can be increased 5 to 10 times by use of the D-76R Replenisher.

A faster working developer can be obtained by increasing the quantity of borax. By increasing the borax quantity ten times, from 29 grains to 290 grains per 32 ounces (from 2 grams to 20 grams per liter), the development time will be about one-half that of regular D-76.

D-82 Developer
High Energy: For Underexposures

	Avoirdupois	Metric
Water, 125° F. (50° C.)	24 oz.	750 cc.
Wood Alcohol	1½ fl. oz.	48.0 cc.

Elon	200 gr.	14.0 g.
Sodium Sulphite, Desiccated	1¾ oz.	52.5 g.
Hydroquinone	200 gr.	14.0 g.
Sodium Hydroxide	125 gr.	8.8 g.
Potassium Bromide	125 gr.	8.8 g.
Cold Water, To make	32 oz.	1.0 l.

Dissolve chemicals in the order given.

Develop about 5 minutes in a tray at 68° F. (20° C.).

The prepared developer does not keep more than a few days in a full bottle or about 2 hours in an open tray. If wood alcohol is omitted and the developer is diluted, the solution is not so active as in the concentrated form.

D-85 Developer

For Kodalith Films, Plates, and Papers

	Avoirdupois	Metric
Water, 90° F. (30° C.)	64 oz.	2.0 l.
Sodium Sulphite, Desiccated	4 oz.	120.0 g.
Paraformaldehyde	1 oz.	30.0 g.
Potassium Metabisulphite	150 gr.	10.5 g.
Boric Acid Crystals *	1 oz.	30.0 g.
Hydroquinone	3 oz.	90.0 g.
Potassium Bromide	90 gr.	6.3 g.
Water, To make	1 gal.	4.0 l.

Dissolve the chemicals in water which half fills a narrow mouthed bottle. After adding each chemical, place the stopper in the bottle so that only a small quantity of air is present during agitation. When all the chemicals have been dissolved, add cold water until the solution reaches the position occupied by the stopper. Insert the stopper tightly to exclude the air. Allow the developer to stand about two hours after mixing. Cool to 65° F. (18° C.) before use. If only a portion of the contents of the bottle is used at one time, it is suggested that the balance be saved by filling a bottle of smaller size which should then be stoppered tightly.

Time of Development: For line negatives, 1½ to 2 minutes at 65° F. (18° C.); for half-tone negatives, not over 2½ minutes at 65° F. (18° C.). With a correctly timed exposure, the image should appear in 30 to 45 seconds at the temperature specified.

This developer has the property of cutting off development very sharply in the low densities thus insuring clear dot formation in the half-tone negatives.

* Use crystalline boric acid as specified. Powdered boric acid dissolves with great difficulty and its use should be avoided.

No. 47 M-H Developer

A long life developer.

Water (52° C.)	750	cc.
Metol	1.5	g.
Anhyd. Sodium Sulphite	45	g.
Sodium Bisulphite	1	g.
Hydroquinone	3	g.
Sodium Carbonate, Monohydrate	6	g.
Potassium Bromide	0.8	g.
Water, To make	1	l.

Do not dilute for use.

Tank Development: 6–8 minutes at 20° C. with occasional agitation. For longer developing times dilute 1 part developing solution with 1 part water, and develop 12–16 minutes at 20° C.

Tray Development: 5–7 minutes at 20° C.

No. 45 Pyro Developer

Stock solutions should be kept in stoppered bottles to prevent oxidation by air.

Solution A

Sodium Bisulphite	9.8	g.
Pyro	60	g.
Potassium Bromide	1.1	g.
Water, To make	1	l.

Solution B

Anhyd. Sodium Sulphite	105	g.
Water, To make	1	l.

Solution C

Sodium Carbonate, Monohydrate	85	g.
Water, To make	1	l.

Tank Development: Take 1 part each of solutions A, B, and C, and add 11 parts water. Develop 9–12 minutes at 20° C.

Tray Development: Take 1 part each of solutions A, B, and C, and add 7 parts water. Develop 6–8 minutes at 20° C.

Store solutions separately until just before use.

No. 72 Glycin Developer

For commercial films in reproduction work. Also suitable for roll, pack, and sheet film.

Water (52° C.)	800	cc.
Anhyd. Sodium Sulphite	125	g.
Anhyd. Potassium Carbonate	250	g.
Glycin	50	g.
Potassium Bromide	1	g.
Water, To make	1	l.

Tank Development: Take 1 part stock solution and 15 parts water, and develop 20–25 minutes at 20° C.

Tray Development: Take 1 part stock solution and 4 parts water, and develop 5–10 minutes at 20° C.

High Contrast Developer
U. S. Patent 2,181,861

Water	16	oz
Hydroquinone	105	gr.
Sodium Sulphite	350	gr.
Sodium Carbonate	560	gr.
Potassium Bromide	63	gr.
Sodium Hydroxide	210	gr.
Potassium Alum	280	gr.

Use full strength at 65° F.

No. 42 M-H Tank Developer

A soft-working tank developer for pack, roll, and portrait films.

Water (52° C.)	750	cc.
Metol	0.8	g.
Anhyd. Sodium Sulphite	45	g.
Hydroquinone	1.2	g.
Sodium Carbonate, Monohydrate	8	g.

Potassium Metabisulphite	4	g.
Potassium Bromide	1.5	g.
Water, To make	1	l.

Do not dilute for use.

Develop 15–20 minutes at 20° C.

No. 17 Borax Tank Developer

Fine grain developer. Soft gradations with Agfa direct copy, portrait, and press films.

Water (52° C.)	750	cc.
Metol	1.5	g.
Anhyd. Sodium Sulphite	80	g.
Hydroquinone	3	g.
Borax	3	g.
Potassium Bromide	0.5	g.
Water, To make	1	l.

Do not dilute for use.

Tank Development: 10–15 minutes at 20° C. for fine grain films. 12–20 minutes at 20° C. for direct copy and portrait sheet films.

Tray Development: 8–12 minutes at 20° C. depending on film and density desired.

No. 70 Hydroquinone Caustic Developer

For process film used in reproduction work.

Solution A

Water (52° C.)	750	cc.
Hydroquinone	25	g.
Potassium Metabisulphite	25	g.
Potassium Bromide	25	g.
Cold Water	1	l.

Solution B

Cold Water	1	l.
Sodium Hydroxide	36	g.

Mix equal parts of A and B immediately before use. Develop within 3 minutes at 20° C.

No. 22 M-H Title Developer

For development of cine title film and positive film to obtain high contrast.

Water (52° C.)	750	cc.
Metol	0.8	g.
Anhyd. Sodium Sulphite	40	g.
Hydroquinone	8	g.
Sodium Carbonate, Monohydrate	50	g.
Potassium Bromide	5	g.
Water, To make	1	l.

Do not dilute for use.

Develop 5–8 minutes at 20° C.

No. 64 Rapid M-H Tropical Developer

A clean working developer, useful for rapid development or development at high temperatures.

Water (52° C.)	750	cc.
Metol	2.5	g.
Anhyd. Sodium Sulphite	25	g.
Hydroquinone	6.5	g.
Sodium Carbonate, Monohydrate	16	g.
Potassium Bromide	1	g.
Water, To make	1	l.

Do not dilute for use.

Develop 3–4 minutes at 20° C., or 2–3 minutes at 29° C.

No. 90 High Contrast M-H Tray Developer

Particularly designed for use with commercial and process films to produce negatives of brilliant contrast.

Water (52° C.)	750	cc.
Metol	5	g.
Anhyd. Sodium Sulphite	40	g.
Hydroquinone	6	g.
Sodium Carbonate, Monohydrate	40	g.

Potassium Bromide	3 g.
Water, To make	1 l.

Do not dilute for use.

Develop 4–6 minutes at 20° C.

No. 40 M-H Tray Developer

A brilliant developer for roll, pack, and sheet film.

Water (52° C.)	900 cc.
Metol	4.5 g.
Anhyd. Sodium Sulphite	54 g.
Hydroquinone	7.5 g.
Sodium Carbonate, Monohydrate	54 g.
Potassium Bromide	3 g.
Water, To make	1 l.

For use dilute 1 part stock solution with 2 parts water.

Develop 4–5 minutes at 20° C.

No. 61 M-H Tray Developer

For use with commercial films to give negatives of normal contrast.

Water (52° C.)	750 cc.
Metol	1 g.
Anhyd. Sodium Sulphite	15 g.
Hydroquinone	2 g.
Sodium Carbonate, Monohydrate	15 g.
Potassium Bromide	1 g.
Water, To make	1 l.

Do not dilute for use.

Develop 4–6 minutes at 20° C.

Rapid Developer

U. S. Patent 2,136,968

Monochloroquinol	1.500
Metol	1.500
Potassium Sulphocyanide	0.003
Sodium Sulphite	80.000
Quinol	5.000
Borax	5.000
Water	1000.000

High Speed Developer

Champlin No. 17

Metol	10 gr.
Sodium Sulphite	1466 gr.
Diethylene Glycol	25 min.
Triethanolamine	30 min.
Chlorohydroquinone	49 gr.
Water, To make	32 fl. oz.

Develop 5.5–6.5 minutes at 70° F.

D-61R Replenisher

For D-61a Developer

Stock Solution A

	Avoirdupois		Metric
Water, 125° F. (50° C.)	96	oz.	3.0 l.
Elon	85	gr.	5.9 g.
Sodium Sulphite, Desiccated	6	oz.	180.0 g.
Sodium Bisulphite	55	gr.	3.8 g.
Hydroquinone	170	gr.	11.9 g.
Potassium Bromide	45	gr.	3.1 g.
Cold Water, To make	1½	gal.	6.0 l.

Stock Solution B

Sodium Carbonate, Desiccated	8	oz.	240.0 g.
Water, To make	64	oz.	2.0 l.

Dissolve chemicals in the order given.

For use take 3 parts of A and 1 part of B and add to the tank of developer as needed. Do not mix solutions A and B until ready to use.

D-76R Replenisher
For D-76 Developer

	Avoirdupois	Metric
Water, 125° F. (50° C.)	24 oz.	750.0 cc.
Elon	44 gr.	3.0 g.
Sodium Sulphite, Desiccated	3 oz. 145 gr.	100.0 g.
Hydroquinone	¼ oz.	7.5 g.
Borax, Granular	290 gr.	20.0 g.
Water, To make	32 oz.	1.0 l.

Dissolve chemicals in the order given.

Use the replenisher without dilution and add to the tank to maintain the level of the solution.

No. 17 Replenisher

Add 14 to 21 cc. replenisher to No. 17 for each 36 exposure 35 mm. film developed. Maintain original volume of developer, discarding if necessary some used developer. No increase in original developing time is necessary.

Water (52° C.)	750 cc.
Metol	2.2 g.
Anhyd. Sodium Sulphite	80 g.
Hydroquinone	4.5 g.
Borax	18 g.
Water, To make	1 l.

No. 47 Replenisher

Add 14–21 cc. replenisher to No. 47 for each 36-exposure 35 mm. film developed. Maintain original volume of developer, discarding some used developer if necessary. No increase in original developing time is necessary.

Water (52° C.)	750 cc.
Metol	3 g.
Anhyd. Sodium Sulphite	45 g.
Sodium Bisulphite	2 g.
Hydroquinone	6 g.
Sodium Carbonate, Monohydrate	12 g.
Water, To make	1 l.

Fine Grain Developer

This formula gives satisfactory enlargements up to 5 diameters.

4 Chloro-2-methylamino-phenol	10 g.
Sodium Sulphite	25 g.
Potassium Carbonate	50 g.
Potassium Bromide	1 g.
Water	1 l.

Develop at 18° C. for 10–15 minutes.

Blue Print Developer

A

Hydrogen Peroxide	10
Sulphuric Acid, Diluted	10
Water	100

B

Chrome Alum	1
Potassium Citrate	2
Water	100

To use, mix 100 parts (vol.) of A with 50 of B, for developing blue prints made with ferric ammonium oxalate and potassium ferricyanide.

Direct Reversal Process on Paper

The steps in processing exposed Eastman Direct Positive Paper and the times required for each chemical treatment are as follows:

1. Development (45 sec.—1 min.) D-88
2. Bleach (30 sec.) R-9
3. Clearing (30 sec.) CB-1
4. Re-exposure to light
5. Redevelopment (30 sec.) Black and White, D-88; Sepia, T-19
6. Fixation (if desired, but not necessary) (30 sec.) F-1

It is necessary to wash the prints well in running water for at least 15 seconds between the different solutions. Where the recommended redeveloper for black and white is used, it is necessary to expose the paper to artificial light or daylight before redevelopment. If convenient, the white light can be turned on as soon as the prints are placed in the clearing bath. If brown tones are desired, the sulphide redeveloper may be used when it will be unnecessary to use white light.

Prints of slightly greater brilliancy may be secured by fixing in the regular fixing bath (F-1) after redevelopment in D-88. It is important to wash for at least 10 minutes to insure removal of the fixing bath from the print. It is not necessary to use the fixing bath after the sulphide redeveloper.

D-88 Developer

For Development and Redevelopment of Direct Positive Paper

	Avoirdupois	Metric
Water, 125° F. (50° C.)	96 oz.	750.0 cc.
Sodium Sulphite, Desiccated	6½ oz.	48.8 g.
Hydroquinone	3¼ oz.	24.4 g.
Boric Acid Crystals *	¾ oz.	5.6 g.
Potassium Bromide	150 gr.	2.6 g.
Sodium Hydroxide †	3¼ oz.	24.4 g.
Cold Water, To make	1 gal.	1.0 l.

Dissolve chemicals in the order given.

Use full strength at 70° F. (21° C.). Develop 45 seconds to 1 minute.

T-19 Sulphide Redeveloper

For Direct Positive Paper

	Avoirdupois	Metric
Sodium Sulphide, Desiccated	290 gr.	20.0 g.
Water, To make	32 oz.	1.0 l.

Use full strength at 65° to 70° F. (18° to 21° C.).

* Crystalline boric acid should be used as specified. Powdered boric acid dissolves only with great difficulty, and its use should be avoided.

† It is desirable to dissolve the caustic soda in a small volume of water in a separate container and then add it to the solution of the other constituents. Then dilute the whole to the required volume. If a glass container is employed in dissolving the caustic soda, the solution should be stirred constantly until the soda is dissolved, to prevent cracking the glass container by the heat evolved.

R-9 Bleach Bath

For Direct Positive Paper

	Avoirdupois	Metric
Water	1 gal.	1.0 l.
Potassium Bichromate	1¼ oz.	9.4 g.
Sulphuric Acid, C.P.	1½ fl. oz.	12.0 cc.

Use full strength at 65° to 70° F. (18° to 21° C.). For more rapid bleaching, however, the quantities of acid and bichromate may be increased.

CB-1 Clearing Bath

For Direct Positive Paper

	Avoirdupois	Metric
Sodium Sulphite, Dessicated	12 oz.	90.0 g.
Water, To make	1 gal.	1.0 l.

Use full strength at 65° to 70° F. (18° to 21° C.).

Motion Picture Positive Film Developer

Sodium Thiosulphate	100 g.
Potassium Bromide	35 g.
Sodium Pyrosulphite	165 g.

Intensifiers

If negatives need intensification or reduction, it is best to give them such treatment immediately after they have been washed. Much time is saved and the negatives, when dry, are ready for finishing.

Precautions: Stains are sometimes produced during intensification or reduction unless the following precautions are observed: 1. The negative should be fixed and washed thoroughly before treatment and be free of scum or stain. 2. It should be hardened in the formalin hardener SH-1 before the intensification or reduction treatment. 3. Only one negative should be handled at a time and it should be agitated thoroughly during the treatment. Following the treatment, the negative should be washed thoroughly and wiped off carefully before drying.

The mercury intensifier is recommended where extreme intensification is desired but where permanence of the resulting image is not essential. If permanence is essential, either the chromium or the silver intensifier should be used.

In-1 Mercury Intensifier

For Films and Plates

Bleach the negative in the following solution until it is white, *then wash thoroughly.*

	Avoirdupois	Metric
Potassium Bromide	¾ oz.	22.5 g.
Mercuric Chloride	¾ oz.	22.5 g.
Water, To make	32 oz.	1.0 l.

The negative can be blackened with 10% sulphite solution, a developing solution, such as D-72 diluted to 1 to 2, or 10% ammonia [1 part concentrated ammonia (28%) to 9 parts water], these giving progressively greater density in the order given. To increase contrast greatly, treat with the following solution:

Solution A

	Avoirdupois	Metric
Water	16 oz.	500.0 cc.
Sodium Cyanide *	½ oz.	15.0 g.

Solution B

	Avoirdupois	Metric
Water	16 oz.	500.0 cc.
Silver Nitrate Crystals	¾ oz.	22.5 g.

To prepare the intensifier, add the silver nitrate, Solution B, to the cyanide, Solution A, until a permanent precipitate is just produced; allow the mixture to stand a short time and filter. This is called Monckhoven's intensifier.

Redevelopment cannot be controlled as with the chromium intensifier, In-4, but must go to completion.

* *Warning: Cyanide is a deadly poison and should be handled with extreme care.* Use rubber gloves and avoid exposure to its fumes. Cyanide reacts with acid to form poisonous hydrogen cyanide gas. When discarding a solution containing cyanide, always run water to flush it out of the sink quickly. Cyanide solutions should never be used in poorly ventilated rooms.

Mercury Intensifier

For increasing the printing density of thin, flat negatives.

Potassium Bromide	10 g.
Mercuric Chloride	10 g.
Water, To make	1 l.

Negatives to be intensified must be thoroughly washed first, or yellow stains may result on the intensified negative. Immerse film in above solution until thoroughly bleached, then wash in water containing a few drops of hydrochloric acid. Redevelop the bleached negatives in 5% sodium sulphite or any standard developer. Surface scum that forms during storage of the bleaching solution does not affect the bleacher, but should be removed before using the solution.

Chromium Intensifier

Convenient to use and gives permanent results.

Potassium Bichromate	9 g.
Hydrochloric Acid	6 cc.
Water, To make	1 l.

Immerse negative until bleached, then wash 5 minutes in water, and redevelop in bright but diffused light in a metol-hydroquinone developer such as No. 47. Wash 15 minutes before drying. Intensification may be repeated for increased effect.

Any blue coloration of the film base after intensification may be removed easily by washing the film 2–3 seconds in water containing a few drops of ammonia, or in a 5% solution of potassium metabisulphite, or in a 5% solution of sodium sulphite. This should be followed by thorough washing in water.

In-4 Chromium Intensifier

For Films and Plates

Stock Solution

	Avoirdupois	Metric
Potassium Bichromate	3 oz.	90.0 g.
Hydrochloric Acid, C.P.	2 fl. oz.	64.0 cc.
Water, To make	32 oz.	1.0 l.

For use, take 1 part of stock solution to 10 parts of water.

Harden the negative first in the formalin hardener SH-1. Bleach thoroughly at 65° to 70° F. (18° to 21° C.), then wash five minutes and redevelop fully in artificial light or daylight (not sunlight) in any quick-acting, non-staining developer which does not contain an excess of sulphite; for example, about 10 minutes at 68° F. (20° C.) in D-72, diluted 1:3. Then rinse, fix for five minutes, and wash thoroughly. Greater intensification can be secured by repeating the process.

Warning: Developers containing a high concentration of sulphite, such as D-76, are not suitable for redevelopment, since the sulphite tends to dissolve the bleached image before the developing agents have time to act on it.

Negatives intensified with chromium are more permanent than those intensified with mercury.

NOTE: See precautions on handling negatives given above.

In-5 Silver Intensifier

For Films and Plates

The following formula is the only intensifier known that will not change the color of the image on positive film on projection. It gives proportional intensification and is easily controlled by varying the time of treatment. The formula is equally suitable for positive and negative film.

Stock Solution No. 1

(Store in a brown bottle)

	Avoirdupois	Metric
Silver Nitrate Crystals	2 oz.	60.0 g.
Distilled Water, To make	32 oz.	1.0 l.
Stock Solution No. 2		
Sodium Sulphite, Desiccated	2 oz.	60.0 g.
Water, To make	32 oz.	1.0 l.
Stock Solution No. 3		
Sodium Thiosulphate	3½ oz.	105.0 g.
Water, To make	32 oz.	1.0 l.
Stock Solution No. 4		
Sodium Sulphite, Desiccated	½ oz.	15.0 g.
Elon	350 gr.	24.0 g.
Water, To make	96 oz.	3.0 l.

NOTE: See precautions on handling negatives given above.

Prepare the intensifier solution for use as follows: Slowly add 1 part of Solution No. 2 to 1 part of Solution No. 1, shaking or stirring to obtain thorough mixing. The white precipitate which appears is then dissolved by the addition of 1 part of Solution No. 3. Allow the resulting solution to stand a few minutes until clear. Then add, with stirring, 3 parts of Solution No. 4. The intensifier is then ready for use and the film should be treated immediately. The mixed intensifier solution is stable for approximately 30 minutes at 70° F. (21° C.).

The degree of intensification obtained depends upon the time of treatment which should not exceed 25 minutes. After intensification, immerse the film for 2 minutes with agitation in a plain 30% hypo solution. Then wash thoroughly.

The stability of the mixed intensifier solution and the rate of intensification are very sensitive to changes in the thiosulphate concentration. A more active but less stable working solution may be obtained by using a stock solution No. 3 prepared with 3 ounces of hypo per 32 ounces (90 grams per liter) instead of the quantity in the formula. The directions for preparing the working solution are the same as before but the mixed intensifier will not keep over 20 minutes at 70° F. (21° C.).

For best results, the intensifier should be used in artificial light; the solution tends to form a precipitate of silver quite rapidly when exposed directly to sunlight.

Re-development Intensifier

A simple method of intensification for negatives consists of bleaching in the ferricyanide and bromide formula used for the sepia toning of prints (T-7a) and blackening with sodium sulphide exactly as in print toning.

Monckhoven's Intensifier

For reproduction film. Gives great intensification and contrast for line drawings and half-tone reproduction work.

Solution A

Potassium Bromide	23 g.
Mercuric Chloride (Poison)	23 g.
Water, To make	1 l.

Solution B

Cold Water	1 l.
Potassium Cyanide (Poison)	23 g.
Silver Nitrate	23 g.

The silver nitrate and potassium cyanide should be dissolved in separate lots of water, and the former added to the latter until a permanent precipitate is produced. Allow to stand 15 minutes and filter, the filtrate being Solution B.

Place negatives in Solution A until bleached through. Then rinse and place in Solution B. If intensification is carried too far, the negative may be reduced with a weak solution of hypo.

Wet Collodion Intensification

As a wartime measure to conserve potassium bromide and iodine, the developed negative can be bleached in:

Sodium Bromide	4
Copper Sulphate	3
Water	80

The bleached image is then blackened as usual with a silver nitrate solution, and the blackened negative in turn bleached with:

Ammonium Bichromate	1
Hydrochloric Acid	2
Water	80

Reduction ("cutting") of the negative can be carried out with a *very dilute* solution of plain hypo, and the negative finally blackened with sodium sulphide solution.

Uranium Intensifier for Movie Film

Many times when amateur movie film is returned from processing, the images are washed out and dull, due many times to overexposure, or shooting against the light. To correct the images and make them project better, use the following mixture:

Potassium Ferricyanide	3	g.
Uranium Nitrate	3	g.
Sodium Acetate	3	g.
Glacial Acetic Acid	30	cc.
Distilled Water	300	cc.

Mix all the above materials and soak the movie film in this solution for 3 to 5 minutes depending on the depth of color desired. Now remove from the toning bath and wash in water for 5 minutes. Dry between cellulose sponges and hang up to dry. This intensifier works best when prepared freshly and will give fine grain images which project very nicely. The above formula gives brownish-red images, on black and white film, but for most purposes this make no difference, and can be used for titles for color films.

Reducers

Reducer formulae may be classified into three types as follows:

A. Subtractive or cutting reducers which remove equal quantities of silver from the high, intermediate, and low densities respectively. They have the effect of clearing up the shadow areas and therefore appear to increase the image contrast but they do not change gamma. They are useful for treating fogged or overexposed images.

R-2 Acid Permanganate
R-4 Farmer's Reducer (Amateur)
R-4a Farmer's Reducer (Professional)

B. Proportional reducers which remove density in an amount which is proportional to the original density. They therefore lower gamma and also visual contrast and correct for overdevelopment.

R-4b Two-Solution Farmer's Reducer
R-5 Acid Permanganate-Persulphate
R-8 Modified Belitzski (also a cutting reducer)

C. Super-proportional reducers which remove density increments which bear a greater proportion to the original density as the magnitude of the original density increases. They therefore reduce the highlight densities without destroying shadow detail and are useful for treating overdeveloped negatives of contrasty subjects.

R-1 Acid-Persulphate

R-1 Reducer

Super-Proportional: For Great Reduction of Contrast

Stock Solution

	Avoirdupois	Metric
Water	32 oz.	1.0 l.
Ammonium Persulphate	2 oz.	60.0 g.
Sulphuric Acid, C.P.*	¾ dr.	3.0 cc.

For use, take 1 part stock solution and 2 parts water.

Treat the negative in the formalin hardener (SH-1) and wash thoroughly before reduction. When reduction is complete, immerse in an acid fixing bath for a few minutes and wash thoroughly before drying. If reduction is too rapid, dilute the solution with a further volume of water.

* *Always add the sulphuric acid to the water slowly with stirring and never the water to the acid,* otherwise the solution may boil and spatter the acid on the hands or face causing serious burns.

NOTE: See precautions on handling negatives given above.

R-2 Reducer

Cutting: For Clearing Shadow Areas

Stock Solution A

	Avoirdupois	Metric
Water	32 oz.	1.0 l.
Potassium Permanganate	1¾ oz.	52.5 g.

Stock Solution B

	Avoirdupois	Metric
Water	32 oz.	1.0 l.
Sulphuric Acid, C.P.	1 fl. oz.	32.0 cc.

The best method of dissolving the permanganate crystals in Solution A is to use a small volume of hot water (about 180° F.) (82° C.) and shake or stir the solution vigorously until completely dissolved; then dilute to volume with cold water. When preparing Stock Solution B, *always add the sulphuric acid to the water slowly with stirring and never the water to the acid,* otherwise the solution may boil and spatter the acid on the hands or face causing serious burns.

The negative must be thoroughly washed to remove all traces of hypo before it is reduced. For use, take 1 part A, 2 parts B and 64 parts of water. When the negative has been reduced sufficiently place it in a fresh acid fixing bath (F-5) for a few minutes, to remove yellow stains, then wash thoroughly.

If reduction is too rapid, use a larger volume of water when diluting the solution for use. This solution should *not be used* as a stain remover as it

NOTE: See precautions on handling negatives given above.

has a tendency to attack the image before it removes the stain. Use S-6 for removing developer stains.

NOTE: If a scum forms on the top of the permanganate solution or a reddish curd appears in the solution, it is because the negative has not been sufficiently washed to remove all hypo, or because the permanganate solution has been contaminated by hypo. The separate solutions will keep and work perfectly for a considerable time if proper precautions against contamination are observed. The two solutions should not be combined until immediately before use. They will not keep long in combination.

A close observance of the foregoing instructions is important. Otherwise an iridescent scum will sometimes appear on the reduced negatives after they are dry; and this is difficult, if not impossible, to remove.

R-4 Farmer's Reducer

Cutting: For Amateur Use for Clearing Shadow Areas

	Avoirdupois	Metric
Solution A		
Water	1 oz.	32.0 cc.
Potassium Ferricyanide	15 gr.	1.0 g.
Solution B		
Water	32 oz.	1.0 l.
Sodium Thiosulphate	1 oz.	30.0 g.

Add A to B and immediately pour over the negative to be reduced. The reducer solution decomposes rapidly after mixing together the A and B solutions and therefore should be used at once. When the negative has been reduced sufficiently, wash thoroughly before drying. Local areas may be reduced by applying the solution with a cotton pad.

This formula is less active than R-4a.

NOTE: See precautions on handling negatives given above.

R-4a Farmer's Reducer

Cutting· For Clearing Shadow Areas

	Avoirdupois	Metric
Stock Solution A		
Potassium Ferricyanide	1¼ oz.	37.5 g.
Water, To make	16 oz.	500 cc.
Stock Solution B		
Sodium Thiosulphate	16 oz.	480.0 g.
Water, To make	64 oz.	2.0 l.

For use take: Stock Solution A, 1 ounce (30 cc.), stock Solution B, 4 ounces (120 cc.), and water to make 32 ounces (1 liter). Add A to B, then add the water and pour the mixed solution at once over the negative to be reduced, which preferably should be contained in a white tray. Watch closely. When the negative has been reduced sufficiently, wash thoroughly before drying.

Solutions A and B should not be combined until they are to be used. They will not keep long in combination.

NOTE: See precautions on handling negatives given above.

R-4b Farmer's Reducer

Proportional: For Lowering Contrast

Farmer's Reducer also may be used as a two-bath formula by treating the negative in the ferricyanide solution first and subsequently in the hypo solution. This method gives almost proportional reduction and corrects for over-development. The single solution Farmer's Reducer gives only cutting reduction and corrects for overexposure.

Stock Solution A

	Avoirdupois	Metric
Potassium Ferricyanide	¼ oz.	7.5 g.
Water, To make	32 oz.	1.0 l.

Stock Solution B

Sodium Thiosulphate	6¾ oz.	200.0 g.
Water, To make	32 oz.	1.0 l.

Treat the negatives in Solution A with uniform agitation for 1 to 4 minutes at 65–70° F. (18–21° C.) depending on the degree of reduction desired. Then immerse them in Solution B for 5 minutes and wash thoroughly. The process may be repeated if more reduction is desired. For the reduction of general fog, one part of Solution A should be diluted with one part of water.

NOTE: See precautions on handling negatives given above.

R-5 Reducer

Proportional: For Lowering Contrast

Stock Solution A

	Avoirdupois	Metric
Water	32 oz.	1.0 l.
Potassium Permanganate	4 gr.	0.3 g.
Sulphuric Acid (10% Solution) *	½ fl. oz.	16.0 cc.

Stock Solution B

Water	96 oz.	3.0 l.
Ammonium Persulphate	3 oz.	90.0 g.

For use, take one part of Solution A to three parts of Solution B. When sufficient reduction is secured the negative should be cleared in a 1% solution of sodium bisulphite. Wash the negative thoroughly before drying.

* To make a 10% solution of sulphuric acid, take 1 part of concentrated acid and add it to 9 parts of water, slowly with stirring. *Never add the water to the acid,* because the solution may boil and spatter the acid on the hands or face, causing serious burns.

NOTE: See precautions on handling negatives given above.

R-8 Reducer

Semi-Proportional: For Lowering Contrast and Clearing Shadow Areas

This is the only single solution reducer which keeps well in a tank.

	Avoirdupois	Metric
Water, 125° F. (50° C.)	24 oz.	750 cc.
Ferric Chloride	365 gr.	25 g.

Potassium Citrate *	2½ oz.	75 g.
Sodium Sulphite, Desiccated	1 oz.	30 g.
Citric Acid	290 gr.	20 g.
Sodium Thiosulphate	6¾ oz.	200 g.
Water, To make	32 oz.	1 l.

Dissolve chemicals in the order given.

Use full strength for maximum rate of reduction. Treat negatives 1 to 10 minutes at 65° to 70° F. (18° to 21° C.). Then wash thoroughly. If a slower action is desired, dilute one part of solution with one part of water. The reducer is especially recommended for the treatment of dense, contrasty negatives.

* Sodium citrate should not be used in place of potassium citrate because the rate of reduction is slowed up considerably.

NOTE: See precautions on handling negatives given above.

Flattening Reducer

Used for lessening density and contrast of heavy negatives.

Potassium Ferricyanide	35 g.
Potassium Bromide	10 g.
Water, To make	1 l.

Bleach in the solution and after thorough washing redevelop to the desired density and contrast in Agfa No. 47 or some other negative developer. Conduct operation in subdued light.

TONING FORMULAE

Three distinct methods of toning are possible:

(1) Toning by direct development.

(2) Toning by replacement of the silver image with inorganic salts (metal tones).

(3) Toning with dyes (dye tones).

1. Toning by Direct Development

The color of the silver image produced by development is determined by the size and condition of the silver particles composing the image. It is possible to control the size of these particles and therefore the color of the image by modifying the nature of the developer.

The range of colors obtainable, however, is not very great and it is usually easier and more certain to produce such slight modifications of color either by delicate dye tinting or by giving a short immersion in one of the diluted toning baths.

2. Toning by Replacement of the Silver Image with Inorganic Salts

Since most toning processes intensify the original silver image, it is best to commence with a slide or positive print which is somewhat on the thin side. Experience will dictate the most suitable image quality to yield the best results with various toning processes.

3. Dye Toning

It is not possible to obtain more than a limited number of tones by the use of colored inorganic compounds owing to the limited number of such compounds. Certain inorganic compounds, however, such as silver ferrocyanide, can be used as mordants for basic dyes such as Victoria Green, Safranine, etc. If, therefore, a silver image is converted more or less to a silver ferrocyanide image and then immersed in a solution of a basic dye, a mordanted dye image is produced.

Immersion of a silver image in an acid solution of potassium ferricyanide will produce a satisfactory mordant image of silver ferrocyanide but, if the image is left too long in the acid ferricyanide bath, the mordanting action of the silver ferrocyanide image is destroyed. By incorporating uranium (uranyl) nitrate in the bath, brown uranyl ferrocyanide is deposited along with the silver ferrocyanide which serves as a signal to indicate when the film should be removed from the mordanting bath. When the black silver image just commences to turn brown, sufficient silver ferrocyanide has been formed to mordant basic dyes strongly but, if the time of immersion is prolonged so that the image is appreciably colored, it will not mordant as well.

Stability of Solutions

All toning baths containing potassium ferricyanide are sensitive to light, the ferricyanide being reduced to ferrocyanide, with the resulting formation of a sludge of the metallic ferrocyanide. When not in use, tanks should be covered to prevent exposure to daylight, and small volumes of solution should be placed in dark brown bottles.

It is important that no metallic surface, however small, come in contact with the solution. Glass trays are best, but enameled trays can be used if they are in good condition, with no cracks or chipped spots. Wooden or stoneware tanks with hard rubber faucets should be used for large scale work. Motion picture film should be wound on wooden racks, free of metal pegs.

Toning Baths for Papers

While most of the toning processes are quite simple, the final tone obtained is affected by a number of factors. Successful production of the desired tone depends upon the proper control of every step from the exposure of the sensitive paper to the final drying of the tone print. Therefore, it should be emphasized that uniform quality of black-and-white prints is a prerequisite to consistent success in toning.

Sepia Tones on Prints

A shade of brown is suitable for many subjects. There are three principal methods of producing pleasing sepia tones. These are: (1) production of brown silver sulphide by direct combination of silver and sulphur with the T-1a Hypo-Alum Sepia Toner; (2) production of silver sulphide indirectly by oxidation of the silver to silver bromide which is then converted to silver sulphide with the T-7a Sulphide Sepia Toner; and (3) deposition of gold on the silver image, with the T-21 Nelson Gold Toner. The different chemical actions of these three processes provide for the treatment of all brands of photographic papers.

T-1a Hypo Alum Sepia Toner

For Warm Tone Papers

	Avoirdupois	Metric
Cold Water	90 oz.	2800 cc.
Sodium Thiosulphate	16 oz.	480 g.
Dissolve thoroughly, and add the following solution:		
Hot Water 160° F. (70° C.)	20 oz.	640 cc.
Potassium Alum	4 oz.	120 g.

Then add the following solution (including precipitate) *slowly to the hypo-alum solution while stirring the latter rapidly.*

Cold Water	2 oz.	64.0 cc.
Silver Nitrate Crystals	60 gr.	4.2 g.
Sodium Chloride	60 gr.	4.2 g.
After combining above solutions		
Add Water, To make	1 gal.	4.0 l.

For use, pour into a tray supported in a water bath and heat to 120° F. (49° C.). At this temperature prints will tone in 12 to 15 minutes depending on the type of paper. Never use the solution at a temperature above 120° F. (49° C.). Blisters and stains may result. Toning should not be continued longer than 20 minutes at 120° F. (49° C.).

In order to produce good sepia tones, the prints should be exposed so that the print is slightly darker than normal when developed normally (1½ to 2 minutes).

The prints to be toned should be fixed thoroughly and washed for a few minutes before being placed in the toning bath. Dry prints should be soaked thoroughly in water. To insure even toning, the prints should be immersed completely, and separated occasionally, especially during the first few minutes.

After prints are toned, they should be wiped with a soft sponge and warm water to remove any sediment, and washed for one hour in running water.

The bath is particularly suitable for use with papers having slightly warm tones, such as Vitava Athena and Azo, in which the images are composed of comparatively fine grains. Vitava Opal and Vitava Projection, also tone well in T-1a.

NOTE: The Silver nitrate should be dissolved completely before adding the sodium chloride and immediately afterward, the solution containing the milky white precipitate should be added to the hypo-alum solution as directed above. The formation of a black precipitate in no way impairs the toning action of the bath if proper manipulation technique is used.

Hypo Alum Toner

Solution A

Water	2350	cc.
Hypo	450	g.

Solution B

Water	30	cc.
Silver Nitrate	1.25	g.

Solution C

Water	30	cc.
Potassium Iodide	2.5	g.

Add B to A and then add C. Then add 105 g. of potassium alum and heat to boiling, or until solution becomes milky.

Tone prints 20–60 min. at 43–52° C., agitating slightly until toning is complete. Be sure blacks are entirely converted before removing prints from toning bath, otherwise double tones will result.

T-7a Sulphide Sepia Toner

For Cold Tone Papers

Stock Bleaching Solution A

	Avoirdupois	Metric
Potassium Ferricyanide	2½ oz.	75 g.
Potassium Bromide	2½ oz.	75 g.
Potassium Oxalate	6½ oz.	195 g.
Acetic Acid (28% Pure) *	1¼ fl. oz.	40 cc.
Water	64 oz.	2 l.
Stock Toning Solution B		
Sodium Sulphide, Desiccated	1½ oz.	45 g.
Water	16 oz.	500 cc.
Prepare Bleaching Bath as follows:		
Stock Solution A	16 oz.	500 cc.
Water	16 oz.	500 cc.
Prepare Toner as follows:		
Stock Solution B	4 oz.	125 cc.
Water, To make	32 oz.	1 l.

The print to be toned should first be washed thoroughly. Place it in the Bleaching Bath, and allow it to remain until only faint traces of the half-tones are left and the black of the shadows has disappeared. This operation will take about one minute.

Rinse *thoroughly* in clean cold water.

Place in Toner Solution until original detail returns. This will require about 30 seconds. Give the print an immediate and thorough water rinse; then immerse it for five minutes in a hardening bath composed of 1 part

* To make 28% acetic acid from glacial acetic acid, dilute 3 parts of glacial acetic acid with 8 parts of water.

NOTE: Particular care should be taken *not* to use trays with any *iron* exposed, otherwise blue spots may result.

of the stock hardener F-1a and 16 parts of water. The color and gradation of the finished print will not be affected by the use of this hardening bath. Remove the print from the hardener bath and wash for one-half hour in running water.

This toning bath tends to give warm tones, an advantage with papers such as Velox. With the inherently warm-tone papers, such as Azo, Vitava Athena, Vitava Opal, Vitava Projection, and Kodalure, it tends to produce rather disagreeable yellow tones. Kodabromide tones well in this bath.

T-21 Nelson Gold Toner

For Warm Tone Papers

The Nelson Gold Toner has the advantage that a variety of excellent brown tones may be obtained by varying the time of toning, that is, the prints may be removed at any time from the bath when the desired color is reached.

Stock Solution A

	Avoirdupois	Metric
Warm Water 125° F. (50° C.)	1 gal.	4.0 l.
Sodium Thiosulphate (Hypo)	2 lb.	960.0 g.
Ammonium Persulphate	4 oz.	120.0 g.

Dissolve the hypo completely before adding the ammonium persulphate. Stir the bath vigorously while adding the ammonium persulphate. If the bath does not turn milky, increase the temperature until it does.

Prepare the following solution and add it (including precipitate) slowly to the hypo-persulphate solution while stirring the latter rapidly. *The bath must be cool when these solutions are added together.*

Cold Water	2 oz.	64.0 cc.
Silver Nitrate Crystals	75 gr.	5.2 g.
Sodium Chloride	75 gr.	5.2 g.

NOTE: The silver nitrate should be dissolved completely before adding the sodium chloride.

Stock Solution B

Water	8 oz.	250.0 cc.
Gold Chloride	15 gr.	1.0 g.

For use, add 4 ounces (125 cc.) of Solution B slowly to the entire quantity of Solution A while stirring the latter rapidly.

The bath should not be used until after it has become cold and has formed a sediment. Then pour off the clear liquid for use.

Pour the clear solution into a tray supported in a water bath and heat to 110° F. (43° C.). During toning the temperature should be between 100° and 110° F. (38° and 43° C.).

Prints to be toned should be washed for a few minutes after fixing before they are placed in the toning solution. Dry prints should be soaked thoroughly in water before toning.

Keep at hand an untoned black-and-white print for comparison during toning. Prints should be separated at all times to insure even toning.

When the desired tone is obtained (5 to 20 minutes), remove and rinse the prints in cold water. After all prints have been toned, return them to the fixing bath for five minutes, then wash for one hour in running water.

The bath should be revived at intervals by the addition of Gold Solution B. The quantity to be added will depend upon the number of prints toned and the time of toning. For example, when toning to a warm brown, add 1 gram (4 cc.) of gold solution after each fifty, 8 x 10-inch prints or their equivalent have been toned. Fresh solution may be added from time to time to keep the bath up to the proper volume.

Uranium, Iron, and Dye Tones on Prints

Tones on paper may be obtained with uranium (T-17) ranging from chocolate to brick red. This formula may also be used as a mordant bath for dye tones. The paper stock usually becomes tinted unless it is protected by squeegeeing temporarily to another support coated with rubber cement. Blue tones may be obtained with an iron toning bath (T-12).

Toned images obtained with these formulae are not absolutely permanent since they consist of a mixture of silver with one or more of the following compounds: silver ferrocyanide, dye, ferric ferrocyanide, and uranyl ferrocyanide. On exposure to the atmosphere, which usually contains traces of hydrogen sulphide, the silver and uranyl ferrocyanides are converted to silver or uranyl sulphide which is usually apparent as a metallic sheen on the surface of the toned print. This sulphiding of the image can be prevented almost completely by treating the prints with Print Lacquer.

In all these processes, the final tone depends not only on the time of toning but also on the density of the original print.

Important: Prints to be toned by any of the following methods should be washed thoroughly, and treated in the HE-1 Hypo Eliminator, to insure freedom from hypo. If hypo is present, inferior tones will result.

T-12 Iron Toner

For Blue Tones on Kodabromide, Velox, and Azo Papers

	Avoirdupois	Metric
Ferric Ammonium Citrate	58 gr.	4.0 g.
Oxalic Acid Crystals	58 gr.	4.0 g.
Potassium Ferricyanide	58 gr.	4.0 g.
Water, To make	32 oz.	1.0 l.

Dissolve each chemical separately in a small volume of water, about 8 ounces (250 cc.) and filter before mixing together. This solution does not keep well except in brown bottles.

Immerse the well-washed print in the toning bath for 10 to 15 minutes until the desired tone is obtained. Then wash until the highlights are clear.

T-17 Mordant Bath

For Dye Toning

Stock Solution

	Avoirdupois	Metric
Uranium (Uranyl) Nitrate	116 gr.	8.0 g.
Oxalic Acid	58 gr.	4.0 g.
Potassium Ferricyanide	58 gr.	4.0 g.
Water, To make	32 oz.	1.0 l.

Directions for Mixing: First dissolve each chemical separately in a small volume of water. Then add the oxalic acid solution to the uranyl nitrate solution and finally add the ferricyanide solution. If the potassium ferricyanide solution is not clear, filter it before adding to the uranium nitrate and oxalic acid solution. If the uranyl nitrate is added directly to the potassium ferricyanide, a brown precipitate will be obtained which will not dissolve readily in the oxalic acid. After mixing, the bath should be light yellow and perfectly clear. The solution should not be exposed to light any more than necessary.

For Use as a Toning Bath (Chocolate to Brick Red): Dilute 1 part of the stock solution with 2 parts of water. As the toning time is increased, the tone changes from chocolate to brown and finally to brick red. The print may be removed at any stage.

Wash until the highlights are clean; this usually requires from 10 to 15 minutes. Prolonged washing should be avoided.

For Use as a Mordant for Dye Toning: Dilute 1 part stock solution with 4 parts of water.

Treat the well-washed print about 2 minutes until the image turns a light chocolate color. Rinse for about 1 minute or less in running water to remove the yellow stain from the highlights. Then immerse for 10 to 15 minutes in the T-17b Dye Bath:

T-17b Dye Bath for Papers

For Use with Papers After the T-17 Mordant Bath

	Avoirdupois	Metric
Dye (1:1000 Solution)	x fl. oz.	x cc.
Acetic Acid, 1% Solution	6¾ fl. dr.	25.0 cc.
Water, To make	32 oz.	1.0 l.

x = Volumes of 1:1000 * Dye Solution for Various Colors as Follows:

		Avoirdupois	Metric
Tone No. 1 Red	Safranine A	3¼ fl. oz.	100 cc.
Tone No. 2 Yellow	Auramine	3¼ fl. oz.	100 cc.
Tone No. 3 Orange	(Equal parts Nos. 1 and 2)	3¼ fl. oz.	100 cc.
Tone No. 4 Blue-Green	Victoria Green	3¼ fl. oz.	100 cc.

* The 1 to 1000 stock solution of the dye is prepared by dissolving one part of dye in 1000 parts water (1 gram in 1 liter, or 15 grains in 32 ounces).

Tone No. 5 Brilliant Green	(Equal Parts Nos. 2 and 4)	3¼ fl. oz.	100 cc.
Tone No. 6 Blue	Methylene Blue BB	3¼ fl. oz.	100 cc.
Tone No. 7 Violet	Methyl Violet	5½ fl. dr.	20 cc.

Mixtures of the following may also be used.

Victoria Green plus Methyl Violet
Victoria Green plus Methylene Blue
Methyl Violet plus Auramine
Methyl Violet plus Victoria Green

The dye toned print should be washed in running water until all extraneous color is removed from the highlights, but prolonged washing should be avoided.

Toning Baths for Slides and Transparencies

T-9 Uranium Toner

For Brown to Red Tones on Slides or Films

	Avoirdupois	Metric
Uranium (Uranyl) Nitrate	35 gr.	2.5 g.
Potassium Oxalate	35 gr.	2.5 g.
Potassium Ferricyanide	15 gr.	1.0 g.
Ammonium Alum	85 gr.	6.0 g.
Hydrochloric Acid, 10% Solution	1¼ dr.	5.0 cc.
Water, To make	32 oz.	1.0 l.

Dissolve chemicals in the order given.

The solution should be perfectly clear and pale yellow in color. *It is light sensitive, however, and should be stored in the dark.*

The maximum toning effect is produced in about 10 minutes, the tone passing from brown to red during this time.

After toning, wash for about 10 minutes; the washing should not be prolonged, especially if the water is slightly alkaline, since the toned image is soluble in alkali.

T-10 Sulphide Toner

For Warm Sepia Tones on Lantern Slides

Solution A

	Avoirdupois	Metric
Potassium Ferricyanide	1 oz.	30.0 g.
Potassium Bromide	½ oz.	15.0 g.
Water, To make	32 oz.	1.0 l.

Solution B

Sodium Sulphide, Desiccated	13 gr.	0.9 g.
Water, To make	32 oz.	1.0 l.

The well washed slide, or film, is thoroughly bleached in Solution A, washed for 5 minutes, and immersed in Solution B for about 2 minutes until thoroughly toned. The slide should then be washed thoroughly for 10 to 15 minutes before drying. The transparency of the tone is much improved by the addition of a little hypo to the B solution, say, 66 grains per 32 ounces or 4.5 grams per liter.

T-11 Iron Toner

For Blue Tones on Slides or Films

	Avoirdupois	Metric
Ammonium Persulphate	7 gr.	0.5 g.
Iron and Ammonium Sulphate	20 gr.	1.4 g.
Oxalic Acid	45 gr.	3.0 g.
Potassium Ferricyanide	15 gr.	1.0 g.
Ammonium Alum	73 gr.	5.0 g.
Hydrochloric Acid, 10% Solution	1/4 dr.	1.0 cc.
Water, To make	32 oz.	1.0 l.

Dissolve chemicals in the order given.

The method of compounding this bath is important. Each of the solid chemicals should be dissolved separately in a small volume of water; the solutions then should be mixed strictly in the order given, and the whole diluted to the required volume. If these instructions are followed, the bath will be pale yellow in color and perfectly clear.

Immerse the slides or films from 2 to 10 minutes at 68° F. (20° C.) until the desired tone is obtained. Wash for 10 to 15 minutes until the highlights are clear. A slight permanent yellow coloration of the clear gelatin will usually occur, but should be too slight to be detectable on projection. If the highlights are stained blue, then either the slide (film) was fogged during development, or the toning bath was stale or not mixed correctly.

Since the toned image is soluble in alkali, washing should not be carried out for too long a period, especially if the water is slightly alkaline.

Mixed Iron and Uranium Tones

By mixing the uranium (T-9) and iron (T-11) toning solutions in different proportions, tones ranging from reddish-brown to chocolate are produced. Analogous results may be obtained by immersing the film or slide in each solution successively for varying times.

T-17 Mordant Bath

For Dye Toning

Stock Solution

	Avoirdupois	Metric
Uranium Nitrate	116 gr.	8.0 g.
Oxalic Acid Crystals	58 gr.	4.0 g.
Potassium Ferricyanide	58 gr.	4.0 g.
Water, To make	32 oz.	1.0 l.

Follow the directions for mixing given under T-17.

For use, take 1 part of stock solution and 4 parts water.

Time of Mordanting: Immerse the film (slide) at 65° to 70° F. (18° to 21° C.) until a very slight chocolate colored tone is obtained; then remove film at once. If mordanting is prolonged much beyond this point, inferior tones will be produced. With a new bath this will require from 1½ to 2 minutes, but the time will need to be increased as the bath ages. The solution may be revived at intervals by adding a little of the concentrated stock solution.

Time of Washing After Mordanting: Wash until the highlights are free from yellow stain; this usually takes about 10 to 15 minutes. Do not prolong the washing for more than 20 minutes or some of the mordant will be washed out.

T-17a Dye Bath

For Use with Slides or Films After the T-17 Mordant Bath

	Avoirdupois	Metric
Dye	3 gr.	0.2 g.
Acetic Acid, 10%	1¼ dr.	5.0 cc.
Water, To make	32 oz.	1.0 l.

Thoroughly dissolve the dye in hot water, filter, add the acid, and dilute to volume with cold water.

The following dyes are suitable for toning:

Safranine A	Red
Chrysoidine 3R	Orange
Auramine	Yellow
Victoria Green	Green
Methylene Blue BB	Blue
Methyl Violet *	Violet

Time for Dye Toning: Immerse the mordanted and washed film (slide) in the dye bath for 2 to 15 minutes at 65° to 70° F. (18° to 21° C.) according to the color desired. The quantity of dye which mordants to the image increases with time. In case an image is over-dyed, some of the dye may be removed by immersing in a 0.2% solution of ammonia; then rinse before drying.

If, after dyeing 10 minutes, the image does not mordant sufficient dye, remove the film (slide), wash thoroughly, immerse again in the mordanting bath, wash, and re-dye.

Intermediate Dye Tones: Intermediate colors may be obtained either by mixing the dye solutions or by immersing the film (slide) in successive baths. For example, if a reddish-orange tone is desired, first tone for a short time in the Safranine bath and then in the Chrysoidine bath, or the two baths may be mixed in suitable proportions and the tone secured with a single treatment.

* For methyl violet use one-quarter the quantity of dye given in the formula.

T-18 Toner

For Double Tones on Slides or Films

Double Tones: This bath tones the halftones white and the shadows blue. If the resulting image is immersed in any of the basic dye solutions, which are used for dye toning (T-17a), the dye is mordanted to the halftones while the shadows remain more or less blue. By varying the dye solution used, the color of the halftones may be varied at will.

	Avoirdupois	Metric
Ammonium Persulphate	7 gr.	0.5 g.
Iron and Ammonium Sulphate	20 gr.	1.4 g.
Oxalic Acid	45 gr.	3.0 g.
Potassium Ferricyanide	15 gr.	1.0 g.
Hydrochloric Acid, 10%	1/4 dr.	1.0 cc.
Water, To make	32 oz.	1.0 l.

The instructions for preparing the bath are the same as for the Iron toning bath (T-11).

Directions for Use: Tone until the shadows are deep blue. Then wash 10 to 15 minutes. Immerse in the basic dye solution used for dye toning for 5 to 15 minutes until the desired depth of color in the halftones is obtained. Wash 5 to 10 minutes after dyeing until the highlights are clear.

T-20 Single Solution Dye Toner

For Slides or Films

	Avoirdupois	Metric
Dye *	x gr.	x g.
Wood Alcohol (or Acetone)	3 1/4 fl. oz.	100.0 cc.
Potassium Ferricyanide	15 gr.	1.0 g.
Acetic Acid (Glacial)	1 1/4 dr.	5.0 cc.
Water, To make	32 oz.	1.0 l.

The nature of the tone varies with time of toning and eventually a point is reached beyond which it is unsafe to continue as the gradation of the toned image becomes affected. Average toning time at 68° F. (20° C.) is from 3 to 9 minutes.

* The quantity of dye varies according to the dye used as follows:

Safranine Extra Bluish	3 gr.	0.2 g.
Chrysoidine 3R	3 gr.	0.2 g.
Auramine	6 gr.	0.4 g.
Victoria Green	6 gr.	0.4 g.
Rhodamine B	6 gr.	0.4 g.

Tinting Slides or Films

Tinting consists of immersing a film or slide in a solution of an acid dye which colors the gelatin layer, causing the whole picture to have a veil of color over it. Motion picture positive films may be purchased in a wide variety of tinted bases which obviate the necessity for actual coloring.

Sheets of water colors or liquid water colors recommended for coloring photographs may be used for tinting slides or film transparencies. In the case of most colors the absorption of the color is hastened by the addition of 1 volume of glacial acetic acid to 1000 volumes of the dye solution. Bathing for 3 to 4 minutes in the acid dye solution is usually ample. After tinting, the slide should be rinsed in water for a few seconds and wiped off with a moist tuft of absorbent cotton. If the color is too strong, it should be washed in water or a 2% solution of ammonia.

Pleasing effects may also be secured by combined tinting and toning such as a blue tone followed by an orange, red, or yellow tint. The clear portions or highlights thus assume the color of the tinting solution while the half-tones and shadows show a color intermediate between the tint and tone used.

Slides fixed in plain or acid hypo (F-24) take colors better than those fixed in an acid hardening fixing bath.

Non-Hardening Citric Acid Short-Stop Bath

Formula No. 1

Water	1 l.
Citric Acid	15 g.

No. 2

Water	1 l.
Sodium Bisulphite	40 g.

SB-1 Stop Bath

Acid Rinse for Papers

	Avoirdupois	Metric
Water	32 oz.	1.0 l.
Acetic Acid (28%)	1½ oz.	48.0 cc.

Rinse prints for at least 5 seconds. Capacity: about 25 8 x 10-inch prints per quart (liter).

SB-1a Stop Bath

Acid Rinse for Films, Plates, and Papers for Graphic Arts

	Avoirdupois	Metric
Water	32 oz.	1.0 l.
Acetic Acid (28%)	4 fl. oz.	125.0 cc.

The action of this bath instantly checks development and prevents uneven spots and streaks when the prints or negatives are immersed in the fixing bath.

SB-5 Stop Bath

Nonswelling Acid Rinse for Photofinishing

	Avoirdupois	Metric
Water	16 oz.	500 cc.
Acetic Acid (28%)	1 fl. oz.	32.0 cc.
Sodium Sulphate, Desiccated	1½ oz.	45.0 g.
Water, To make	32 oz.	1.0 l.

Agitate the films when first immersed in this bath and allow them to remain about three minutes before transfer to the fixing bath.

This bath should be replaced after approximately 25 rolls have been processed per quart (liter), when about 24 ounces (720 cc.) of developer will have been carried into the rinse bath by the film.

Short-Stop and Hardening Bath

Water	1 l.
Potassium Chrome Alum	20 g.
Sodium Bisulphite	20 g.

NOTE: Leave films in hardening solution for five minutes at 68° F. It is important that the solution be prepared immediately before use and be replaced at least once per week, since it does not keep well.

SB-3 Hardening Bath

For Use at 65° to 75° F. with Films and Plates

	Avoirdupois	Metric
Water	32 oz.	1.0 l.
Potassium Chrome Alum	1 oz.	30.0 g.

This bath is intended for use in hot weather after development and before fixation in conjunction with F-5.

Agitate the negatives for a few seconds when first immersed in hardener. Leave them in the bath for 3 to 5 minutes to secure maximum hardening. This bath should be renewed frequently.

SB-4 Hardening Bath

For Use at 75° to 90° F. with Films and Plates

This solution is recommended for use in conjunction with the High Temperature Developer (DK-15), when working above 75° F. (24° C.).

	Avoirdupois	Metric
Water	32 oz.	1.0 l.
Potassium Chrome Alum	1 oz.	30.0 g.
Sodium Sulphate, Desiccated	2 oz.	60.0 g.

Agitate the negatives for 30 to 45 seconds when they are first immersed in the hardener, or streakiness will result. Leave them in the bath for at least 3 minutes between development and fixation. If the temperature is below 85° F. (29° C.) rinse for 1 to 2 seconds in water before immersing in the hardener bath.

The hardening bath is a violet blue color by tungsten light when freshly mixed, but it ultimately turns a yellow-green with use; it then ceases to harden and should be replaced with a fresh bath. The hardening bath should never be overworked. An unused bath will keep indefinitely, but the hardening power of a partially used bath decreases rapidly on standing for a few days.

SH-1 Special Hardener

For After Treatment of Films and Plates

	Avoirdupois	Metric
Water	16 oz.	500 cc.
Formaldehyde (40%)	2½ dr.	10.0 cc.
Sodium Carbonate, Desiccated	73 gr.	5.0 g.
Water, To make	32 oz.	1.0 l.

This formula is recommended for the treatment of negatives which normally would be softened by a chemical treatment as for the removal of stains or for intensification or reduction.

After hardening for three minutes, negatives should be rinsed and immersed for five minutes in a fresh acid fixing bath and then washed thoroughly before they are given any further chemical treatment.

F-1a Hardener

Stock Solution for Preparing F-1 Fixing Bath

	Avoirdupois	Metric
Water 125° F. (50° C.)	14 oz.	425 cc.
Sodium Sulphite, Desiccated	2 oz.	60.0 g.
Acetic Acid (28%)	6 fl. oz.	190.0 cc.
Potassium Alum	2 oz.	60.0 g.
Cold Water, To make	32 oz.	1.0 l.

Dissolve the chemicals in the order given. The sodium sulphite should be dissolved completely before the acetic acid is added. After the sulphite-acid solution has been mixed thoroughly, add the potassium alum with constant stirring.

For use, add 1 part of cool stock hardener solution slowly to 4 parts of a 25% cool hypo solution (2 pounds of hypo to the gallon of solution). If the hypo is not thoroughly dissolved before adding the hardener a precipitate of sulphur is likely to form.

F-5a Hardener

Stock Solution for Preparing F-5 Fixing Bath

	Avoirdupois	Metric
Water 125° F. (50° C.)	20 oz.	600 cc.
Sodium Sulphite, Desiccated	2½ oz.	75.0 g.
Acetic Acid (28%)	7½ fl. oz.	235.0 cc.
Boric Acid Crystals	1¼ oz.	37.5 g.
Potassium Alum	2½ oz.	75.0 g.
Cold Water, To make	32 oz.	1.0 l.

Dissolve chemicals in the order given.

Add one part of the cool stock hardener solution slowly to 4 parts of cool 30% hypo solution (2½ pounds per gallon of solution), while stirring the hypo rapidly.

No. 201 Acid Hardening Fixer

May be used on paper or film, may be stored indefinitely, and may be used until exhausted. When the fixing bath froths, turns cloudy, or takes longer than 10 minutes to fix completely, it must be replaced.

Solution A

Water (52° C.)	500	cc.
Hypo	240	g.

Solution B

Water (52° C.)	150	cc.
Anhyd. Sodium Sulphite	15	g.
Acetic Acid (28%)	45	cc.
Potassium Alum	15	g.

Add solution B to A. Fix 5–10 min. at 20° C. Do not dilute for use.

Acid Hardener for Film

Hypo	220	g.
Sodium Sulphite	15	g.
Sodium Acetate	18	g.
Citric Acid (Crystals)	11	g.
Alum	22	g.
Water	1000	cc.

Photographic Acid Hardening Fixer
U. S. Patent 2,214,216

Sodium Thiosulphate	120
Sodium Bisulphite	30
Aluminum Gluconate	30

Citric Acid, Hardening Fixing Bath

Solution No. 1

Water (125° F.)	500	cc.
Sodium Thiosulphate	250	g.

Solution No. 2

Water (125° F.)	150	cc.
Sodium Sulphite, Anhydrous	11	g.
Citric Acid	5½	g.
Potassium Alum	5½	g.

Add Solution No. 2 to Solution No. 1 and Then Add Water to Make 1 l.

NOTE: The hardening properties of this bath decrease rapidly with age and at temperature above 70° F.

Fixing Baths
U. S. Patent 2,203,903
Formula No. 1

Sodium Thiosulphate (Hypo)	223	g.
Sodium Sulphite	16	g.
Sodium Acetate	18	g.
Citric Acid	11	g.
Potassium Alum	20	g.

For use, add sufficient water to make 1 liter of solution.

No. 2

A

Water	80	oz.
Hypo	2	lb.

B

Water (125° F.)	16	oz.
Sodium Sulphite	2	oz
Distilled Vinegar	16	oz.
Boric Acid, Crystals	1	oz.
Potassium Alum	2	oz.

Allow B to cool, then add A while stirring well.

No. 3

Hypo	350	g.
Sodium Sulphite	20	g.
Sodium Acetate	15	g.
Sodium Hydrogen Sulphate	20	g.
Potassium Alum	15	g.
Chrome Alum	10	g.
Water, To make	1	l.

No. 202 Chrome Alum Fixer

For use in hot weather. Should be used fresh as it does not retain its hardening action.

Solution A

Water (52° C.)	2.5	l.
Hypo	960	g.
Anhyd. Sodium Sulphite	60	g.

Water, To make	3	l.
Solution B		
Water	1	l.
Potassium Chrome Alum	60	g.
Sulphuric Acid, C.P.	8	cc.

Slowly pour B into A with rapid stirring. Do not dilute for use. Do not dissolve alum at a temperature higher than 66° C. Rinse films thoroughly before fixing. Fix 5–10 min. at 20° C.

F-1 Fixing Bath

For Papers

	Avoirdupois	Metric
Water	64 oz.	2.0 l.
Sodium Thiosulphate (Hypo)	16 oz.	480.0 g.

Then add the following hardener solution slowly to the cool hypo solution while stirring the latter rapidly.

Water, 125° F. (50° C.)	5 oz.	160 cc.
Sodium Sulphite, Desiccated	1 oz.	30.0 g.
Acetic Acid (28%)	3 fl. oz.	96.0 cc.
Potassium Alum	1 oz.	30.0 g.

Dissolve chemicals in the order given. (See directions under F-1a).

If it is desired to mix a stock hardener solution, use F-1a.

F-5 Fixing Bath

For Films, Plates, and Papers

	Avoirdupois	Metric
Water, 125° F. (50° C.)	20 oz.	600 cc.
Sodium Thiosulphate (Hypo)	8 oz.	240.0 g.
Sodium Sulphite, Desiccated	½ oz.	15.0 g.
Acetic Acid (28%)	1½ fl. oz.	48.0 cc.
Boric Acid Crystals	¼ oz.	7.5 g.
Potassium Alum	½ oz.	15.0 g.
Cold Water, To make	32 oz.	1.0 l.

Films or plates should be fixed properly in 10 minutes (cleared in 5 minutes) in a freshly prepared bath. The bath need not be discarded until the fixing time (twice the time to clear) becomes excessive, that is, over 20 minutes.

F-16 Fixing Bath

Chrome Alum: for Films and Plates

Solution A

	Avoirdupois	Metric
Sodium Thiosulphate (Hypo)	2 lb.	960 g.
Sodium Sulphite, Desiccated	2 oz.	60.0 g.
Water, To make	96 oz.	3.0 l.

Solution B

	Avoirdupois	Metric
Water	32 oz.	1.0 l.
Potassium Chrome Alum	2 oz.	60.0 g.
Sulphuric Acid, C.P.	1/4 fl. oz.	8.0 cc.

Pour solution B into solution A slowly while stirring A rapidly. This bath, when freshly mixed, is recommended for use in hot weather, but it rapidly loses its hardening properties with or without use, when it should be replaced by a fresh bath. With an old bath there is a tendency for scum to form on the surface of the film. Any such scum should be removed by swabbing with cotton before the film is dried.

F-24 Fixing Bath

Non-Hardening: for Special Processes

	Avoirdupois	Metric
Water, 125° F. (50° C.)	16 oz.	500 cc.
Sodium Thiosulphate (Hypo)	8 oz.	240.0 g.
Sodium Sulphite, Desiccated	145 gr.	10.0 g.
Sodium Bisulphite	365 gr.	25.0 g.
Cold Water, To make	32 oz.	1.0 l.

Dissolve chemicals in the order given.

This bath may be used for films, plates or paper when no hardening is desired.

For satisfactory use, the temperature of the developer, rinse bath, and wash water should not be higher than **68° F. (20° C.).**

Non-Hardening Bisulphite Fixing Bath

Sodium Thiosulphate	240 g.
Sodium Bisulphite	35 g.
Water, To make	1 l.

NOTE: The above fixing bath does not contain a hardening agent. However, hardening of the films in the fixing bath is unnecessary assuming that the films are treated in a fresh hardening short-stop bath prior to fixation.

No. 203 Non-Hardening Metabisulphite Fixer

Used in color work with Printon Film where high accuracy of registration is required.

Hypo	1900 g.
Potassium Metabisulphite	270 g.
Water, To make	4 l.

The metabisulphite should be added only when the hypo solution is cool. Fix 5–10 min. at 18° C.

Pinakryptol Yellow Desensitizer

For orthochromatic films.

Pinakryptol Yellow	1 g.
Water, To make	1 l.

(50% alcohol may be used in place of water to improve the keeping quality of the desensitizer.) Do not dilute for use. Immerse film in total darkness for 2 min. at 18° C. Orthochromatic film may then be handled in bright red light. (Safelight Filter No. 107 with a 25 watt bulb.) Panchromatic film may then be developed in bright green light. (Safelight Filter No. 103 with a 25 watt bulb.)

Pinakryptol yellow must not be mixed with the developer.

Pinakryptol Green Desensitizer

For high speed panchromatic film.

Pinakryptol Green	1 g.
Water, To make	500 cc.

(50% alcohol may be used in place of water to improve the keeping quality of the desensitizer.) For use dilute 1 part stock solution with 11 parts water. Immerse films in total darkness for 2 min. at 18° C. Development may then be carried out in bright red light. (Safelight Filter No. 107 with a 25 watt bulb.)

The stock solution may be used directly in the developer in the proportion: 1 part desensitizer to 30 parts developer. After 2 minutes development in total darkness, bright red light may be used.

Blue Print Paper Sensitizer
British Patent 539,066

Water	1000
Triethylamine Oxalate	150
Sodium Ferrocyanide	40
Potassium Ferricyanide	10
Ferric Ammonium Oxalate	150

Photogravure Spirit Sensitizer

A solution of ammonium bichromate is prepared, of three times the strength required. This should be neutralized by ammonia if the sensitive material is to be kept. Twice its volume of methylated spirits or denatured alcohol should then be added. *Keep in the dark.*

The sheet of tissue to be sensitized is brushed with the solution, the strokes being made to cross each other to avoid irregularity.

Litho Plate Sensitizers
U. S. Patent 2,025,996
Formula No. 1

Process Glue	100 cc.
Water	100 cc.
Pyridin Bichromate (8% Sol.)	50 cc.

No. 2

For sensitizing stainless steel plates for litho purposes:

A

Distilled Water	32
Egg Albumen (Scales)	9

B

Distilled Water	6
Ammonium Bichromate	2¼

C

Process Glue	1½
Ammonia (28%)	1½

Allow the albumen to dissolve overnight, then add B to A, followed by C.

Photolithographic Deep Etch Sensitizer
Formula No. 1

Separately dissolve ¼ pound each of yellow dextrin and gum arabic in as little water as possible to make a paste. The two solutions are then mixed together, and 7½ ounces of ammonium bichromate solution (ammonium bichromate, 2 ounces; water 10 ounces) then added, together with 1 ounce of stronger ammonia water. Distilled water is finally added to the mixture until the solution tests 15° Baume. The speed of the whirler during coating should not exceed 50 r.p.m. Development of the image takes place with lactic acid and calcium chloride solution (37° Bé).

No. 2

Water	500 cc.
Yellow Dextrin	500 cc.
Potassium Bichromate	80 g.
Ammonium Bichromate	40 g.

The bichromates are dissolved in a small quantity of water, then added to the dextrin mixture. In warm weather or when the work-room temperature rises above 72° F., it is recommended to add 5 grams of albumen to the mixture. The color of the bichromated dextrin mixture is rather dark; any foam created in mixing should be removed by filtration.

Litho Washout Solution

For strengthening litho images on metal plates for offset printing:

Asphaltum	2 lb.
Beeswax	3 oz.
Stearin	2 oz.
Transfer Ink	1 oz.
Lavender Oil	2 oz.
Japan Liquid Drier	4 oz.

Mix thoroughly by heating, then add one gallon of turpentine.

Lithographic Plate Conditioning Solution

U. S. Patent 2,231,045

Chromium Phosphate (10% Solution)	½–2 oz.
Gum Arabic (14° Bé Solution)	1–4 oz.
Water, To make	1 gal.

Lithographic Etch

U. S. Patent 2,230,156

A solution for etching chromium coated plates is made of:

Hydrochloric Acid (37%)	1 pt.
Propylene Glycol	3 pt.

Dot Etching

As a substitute for expensive potassium ferricyanide (Farmer's reducer) used in dot etching (correcting) photolitho halftones on dry plates and films, use the following:

A

Water	25
Copper Sulphate	1
Salt	1

Add sufficient ammonia to clear the solution, which should be a rich ultramarine color:

B

Hypo	5
Water	25

For use, take equal parts of A and B.

PHOTOENGRAVING

Aluminum Etching

For etching purposes, the hard variety of aluminum is preferable, since the soft type of metal does not etch as sharply. As photomechanical sensitizers, bichromated albumen (ink top) and cold enamel can be used; glue enamel is not as satisfactory, although the glue print can be burnt in on aluminum with less injury to the metal than would normally be the case with zinc.

Aluminum can be etched in either an aqueous or alcoholic solution of copper chloride; also with a 10 to 15% hydrochloric acid solution, previously saturated with sodium chloride.

Relief and Intaglio Etching

Formula No. 1

Alcohol	4
Acetic Acid	6
Antimony Chloride	4
Water	40

No. 2

Antimony Chloride	15
Tartaric Acid	45
Acetic Acid	15
Water	100

No. 3

Relief Etchings on Hammered Aluminum

Ferric Chloride, Lump	100
Alcohol (96%)	200
Saccharic Acid	1

Oxalic acid may be substituted for the saccharic variety and in the same quantity.

No. 4

Ferric Chloride, Lump	2
Water	40
Hydrochloric Acid	2

The chloride is dissolved first and the acid then added; the strength of the bath should be at least 35° Bé, as weaker baths are likely to attack the acid resist (image) in the increased time required for etching.

No. 5

Ammonium Persulphate	5
Sulphuric Acid	1
Water	100

Brass Etching

Name plate work, on a large scale usually is performed by printing acid-resisting designs on brass sheets by the tin printing (offset) principle, but the metal can be sensitized with any of the photoresists used in photoengraving. For better adhesion of the image, subject the brass plate to action of the following graining bath (instead of immersing the polished sheets in ferric chloride solution):

Formula No. 1

Alum	1
Water	40
Nitric Acid	1

When using glue enamel as a sensitizer, it will be found that the tarnish or discoloration, sometimes present on copper plates after burning-in of the print, is absent on brass, the print, as a result, standing out with greater contrast (brilliance). However, even with this graining bath, the blue print on brass may display a tendency to lift, to prevent which, harden the developed image by immersing it from one to five minutes in the following bath:

No. 2

Water	16
Denatured Alcohol	1½
Ammonium Bichromate	30
Chromic Acid	5

Brass not infrequently etches somewhat faster than copper: the simplest mordant is a solution of ferric chloride. For machine etching of glue prints, the bath should be as strong as possible (at least 45° Bé), since weak (watery) iron solutions soften the enamel image.

No. 3

Tray Etching

Water	45
Hydrochloric Acid	30
Potassium Chlorate	18

No. 4

Water	200
Potassium Chlorate	5
Nitric Acid	10
Ferric Chloride, Lump	15

No. 5

Water	200
Potassium Chlorate	10
Nitric Acid	20
Hydrochloric Acid	40

Etching solutions containing potassium chlorate should stand overnight before use, in order to permit the initial escape of harmful gases. The solutions should not be used after they assume a brown color.

No. 6

Water	40
Chromic Acid	4
Denatured Alcohol	4

The bath is rather slow in action and the high chromic acid content necessitates guarding against chromium dermatitis (bichromate poisoning). In mixing this solution, be sure to first dissolve the chromic acid in the water, then slowly add the alcohol, stirring during the addition. Chromic acid brought into contact with strong alcohol may cause explosion.

No. 7

A two-solution procedure for brass etching consists of giving the first "bite" (etch) in:

Ferric Chloride Solution, 3:1	500
Nitric Acid (45°)	100
Chrome Alum	100
Water	1000

followed by a "finishing" (final) etch with:

Water	1000
Acetic Acid	30
Nitric Acid	75
Chromic Acid	50

Bronze Etching

Because of its hardness and resistance to acids, bronze (broadly considered as an alloy of copper and tin) is more difficult to etch than either brass or copper. Because of the necessity of powerful etching agents, the best photoresist for bronze probably is bichromated albumen (ink top).

Etching is commenced with a strong solution (40–45° Bé) of ferric chloride, after appreciable depth is obtained, the plate is rolled up with stiff etching ink, dusted with powdered asphalt or rosin (colophony), after which the plate is etched either in a strong nitric acid bath, or with the following solution (aqua regia):

Nitric Acid	18
Hydrochloric Acid	82

The solution may be diluted with water according to the speed of etching desired.

Celluloid Etching

This is sometimes etched either in relief or intaglio, employing as sensitizers either bichromated albumen, glue or gelatin. With glue or gelatin, the image obviously cannot be burnt-in, but it can be hardened sufficiently to withstand etching by immersing the developed print for some time in a warm saturated solution of chrome alum.

Difficulty is sometimes experienced with the image failing to adhere to the celluloid; to remedy this, rub the celluloid with a 5% solution of glacial acetic acid in denatured alcohol, the treatment providing a "tooth" to the surface to which the image will adhere.

For etching celluloid bearing ink images, a strong solution of glacial acetic acid can be used. Bichromated glue or gelatin images can be etched with any of the familiar solvents of celluloid: acetone, amyl acetate, or ether-alcohol mixtures. Acetone probably will prove the most convenient for the purpose.

Copper Etching

Formula No. 1

Ferric Chloride, Lump	50
Distilled Water	50
Sulphuric Acid	30

For very delicate work (though slow in action), the so-called "Dutch mordant," often used by artist-etchers (dry point), is used:

No. 2

Water	70
Hydrochloric Acid	10

to which is added in the form of a boiling solution:

Potassium Chlorate	2
Water	20

The solution can be diluted with

water, if desired. The bath acts more rapidly when heated to a temperature of 85° F.

Lead Etching

The low melting point (620° F.) of lead precludes anything but cold enamel as the photoresist, thus dispensing with the necessity of heating the plate in the creation of an acid resist. The metal can be etched in dilute solutions of either nitric or glacial acetic acid, and in the following special mordant:

Water	40
Tin Chloride	2½
Alcohol	4

In practice, the acid solutions will be found most efficient.

Magnesium Etching

As photoresists, either the glue enamel or cold enamel processes can be used; if the latter is employed, the alcohol developing bath must be used only for the magnesium plates, and not for zinc plates.

Magnesium alloys are most efficiently etched in dilute solutions of nitric acid. For tray etching, the strength of the bath should not exceed 3° Bé; for machine etching, the bath should not be stronger than 8° Bé. Even with such dilute solutions, etching of the alloy proceeds so rapidly that the plate is quickly covered with bubbles, which must be periodically brushed off to determine the progress of etching.

Molybdenum Etching

Etchings on this metal are seldom requested, but it can be performed with concentrated sulphuric acid, nitro-hydrochloric acid (aqua regia), or in the following mixture:

Water	1000
Potassium Ferricyanide	360
Caustic Soda	36

Monel Etching

Photoresists can be either glue enamel or carbon tissue images, the medium for etching being a 50° Bé solution of ferric chloride. A solution of this strength is obtained by dissolving 6¾ pounds of lump ferric chloride in sufficient water to make one gallon of solution.

Nickel Etching

Pure nickel can be etched in strong nitric acid solutions, employing any of the familiar photoresists of photoengraving. The metal and its alloys can also be etched in:

Nitric Acid (70%)	50
Acetic Acid (50%)	50

Water may be added as desired for control of etching.

Plastic Etching

Although certain plastics can be etched with alcohol, acetone, bensol, caustic alkalies and modern hydrocarbon solvents, the most reliable method of platemaking with synthetic resins (plastics) is by direct molding from relief or intaglio etchings.

The wide range of plastics and the different chemical nature of such materials render difficult any specific formulae for the purpose.

Silver Etching

The bichromated albumen and cold enamel processes serve as the best photoresists. Decorative designs on silver can be etched to a matt finish in chlorine-free nitric acid of 24° Bé; for deeper etching, the action is continued with 18° nitric acid.

A strong solution of chromic acid is a good mordant for silver, as is also a mixture of potassium bichromate and sulphuric (or nitric) acid, and a mixture of dilute sulphuric acid with potassium permanganate.

A convenient etching bath for silver is:

Nitric Acid	1
Hydrochloric Acid	1
Water	12

Photoink resists capable of withstanding the action of nitric acid on silver are sometimes difficult to obtain, wherefore the following suggestion is made: the ink image (print) is dusted with powdered asphaltum and a short etch given with ferric chloride solution (40° Bé). The silver chloride formed by this action must be removed (dissolved) from the plate with a strong hypo (sodium thiosulphate) solution, after which the plate is moistened with a dilute gum arabic solution (in the lithographic manner) and another application of ink given, followed by dusting with asphaltum.

Etching with ferric chloride and treatment with hypo is repeated, and the entire operation duplicated until it is possible to apply an ink-powder resist sufficiently strong to withstand etching with nitric acid as in ordinary line (relief) etching.

Stainless Steel Printing Plate Etch
U. S. Patent 2,266,430

The following formula has been patented for etching stainless steel plates in the production of printing surfaces:

Saturated Ferric Chloride Solution	4
Concentrated Nitric Acid	1
Commercial Hydrofluoric Acid	1

Stainless Steel Etching

Heavily reinforced (powdered) ink images probably prove the best resist. Solutions for etching stainless steel comprise:

Formula No. 1

Ferric Chloride, Lump	5
Hydrochloric Acid	50
Water	100

No. 2

Copper Chloride	5
Hydrochloric Acid	100
Denatured Alcohol	100
Water	100

No. 3

Potassium Ferricyanide	10
Potassium Hydroxide	10
Water	100

The latter solution must be used in the form of a hot bath.

Steel Etching

The grade of steel best suited for etching is that known as "cemented steel," which possesses a very fine-grained and uniform metallic structure.

Modern practice has found that ferric chloride (42° Bé solution) gives good results, especially with ink and glue enamel images.

For those not partial to ferric chloride, a host of other formulae have been evolved, among which are acid baths:

Formula No. 1

Nitric (or Hydrochloric) Acid	1
Water	8

No. 2

Glacial Acetic Acid	4

Absolute Alcohol	1
Nitric Acid (1.28)	1

The acetic acid and alcohol are mixed together and allowed to stand for one-half hour, after which the nitric acid is added very gradually.

No. 3

Nitric Acid	30
Water	120
Alcohol	120
Oxalic Acid	1

No. 4

Iodine	2
Potassium Iodide	5
Water	40

No. 5

Silver Nitrate	1
Distilled Water	9
Nitric Acid	17
Alcohol	6

This solution improves with keeping. Before etching, the plate is washed, for a few seconds, in a 4% bath of nitric acid, after which the above mordant is applied for three minutes. The plate is then washed with distilled water containing 6% of alcohol and the etching repeated —the operation can be carried out as often as desired.

No. 6

Water	32
Mercuric Chloride	2
Alum	½
Alcohol	2

The chloride and alum are dissolved in the water by the acid of gentle heat, and the alcohol then added to the mixture.

No. 7

A

Nitric Acid, C. P.	5
Distilled Water	5
Metallic Silver	1

B

Nitric Acid, C. P.	5
Distilled Water	5
Mercury	1

The solutions are prepared separately and when the respective metals have been dissolved in the acid mixtures, the solutions are combined and the mixture stored in a glass-stoppered bottle. When ready for use, the mixture is diluted with an equal quantity of distilled water.

The solution has little effect when simply poured upon the plate: galvanic action must first be generated by bending a strip of zinc so that one end comes into contact with a spot of bare steel and the other end dips into the solution on the steel surface. This starts the action of the acid immediately, when the zinc can then be laid aside, as the acid will continue to corrode the steel until the solution is exhausted.

Zinc Etching

Formula No. 1

Sulphuric Acid	12
Potassium Nitrate	4
Water	40

The potassium nitrate is dissolved in the water and the acid then gradually added until bubbling ceases.

No. 2

Hydrochloric Acid	10
Nitric Acid	2½
Water	40

Glass Etching

Certain military equipment requires graticules, gratings and designs etched on glass. Creation of the designs can be performed by engraving on the glass plate covered with an acid-resisting ground (wax), then etching the glass areas exposed in the engraving. A related process is the litho transfer of acid-resisting images on the glass surface, but for quantity production and certainty of result, the photoengraver will prefer strictly photomechanical methods, based on the etching of glass plates bearing photoresists.

Photomechanical glass etching (photohyalography) is performable in several ways; an efficient sensitizer (photoresist) for the glass plate is asphaltum (bitumen), though the comparatively low sensitivity of this agent and the wartime difficulty of obtaining proper grades of light-sensitive asphaltum rather precludes its use. Bichromated gelatin images (carbon tissue) also can be used as etching resists, but their fragility does not long withstand the powerful and penetrating action of hydrofluoric acid (or its salts), with the result that the etching is limited as to depth.

The bichromated albumen process (ink top) can be used to good effect, though it requires heating of the glass plate. The thoroughly cleaned glass surface is sensitized with the albumen mixture (any formula giving good results on zinc will do), and the plate then exposed under a negative (or positive). The exposed surface is next rolled up with ink composed of two-thirds of good transfer ink and one-third good proving ink, taking care to apply a uniform and fairly heavy film of ink.

After development of the print under the tap, the plate is dried and the print dusted with powdered asphalt, whereupon the plate must be heated to fuse the powder with the ink; this heating is best done in a temperature controlled oven.

Etching can be carried out with a strong hydrofluoric acid solution (40%), the action being complete in from 15 to 30 seconds. A slower acting solution, promoting very uniform action, is obtained by mixing:

Water	1000
Potassium Bifluoride	250
Hydrochloric Acid	250
Potassium Sulphate	140

As previously inferred, hydrofluoric acid is a powerful oxidizing agent, inclined to fuming, and causing very ugly burns on human skin. Because of liberated fumes, etching with hydrofluoric acid should not be performed in the vicinity of lenses, prisms, halftone screens or other optical equipment: the fumes from the acid may etch the glass to a matt surface.

Etching should be carried out in a well ventilated chamber: a good idea would be to place the tray of acid next to an open window and to have a draft of air from an electric fan play on the tray, the air blast quickly and continuously carrying the fumes through the window and off the premises. With all the cited precautions, it seems hardly necessary to warn against inhaling the fumes, or of splattering either the clothes or skin with the etching solution.

Of course, stout rubber gloves should be worn while handling the glass and etching bath: the bath itself should be contained in gutta percha or hard rubber trays; possessing the property of etching glass, hydrofluoric acid will also attack the glaze on porcelain, stoneware or earthenware receptacles.

Actual etching time depends on the chemical nature of the glass and the degree of depth desired. Before etching, both the edges of the glass and its back must be protected with asphalt varnish; a point to be remembered is that if the etched areas are to be smooth and transparent, the glass plate must be immersed in the solution; if, however, a matt effect is desired in the etched parts, it is only necessary to expose the glass surface to the fumes of the acid.

For etching in the bath itself, it will frequently be found that the acid etches cleaner and sharper after using it several times; this has led to the suggestion of dissolving a few small pieces of glass in the bath before using it for etching purposes.

On termination of etching, the plate must be washed very quickly and with a very large volume of water: this is necessary because the resist (top) lifts almost in the same moment that the washwater comes into contact with the acid remaining on the plate, after withdrawal from the bath. The ink resist can then be removed with turpentine and the etched lines filled in with a black pigment, should such be desired.

Objections sometimes voiced against the ink resist are the necessity of heating the plate (to fuse or melt the powder) and the tendency of the top to lift during etching, especially if weak acid baths are employed to afford greater latitude (control) in etching.

Preference might therefore be had for a glue image in conjunction with a wax ground, the process carried out by first soaking the glass plate for several hours in:

Water	16
Potassium Bichromate	1
Sulphuric Acid	1

The glass is then scrubbed, rinsed well and dried in a dust-free place. It is next polished on the face side with gasoline, then coated with:

Best Turpentine	16
Asphalt, Powdered	3½
Beeswax	2½

To dissolve the wax, first heat the turpentine; after the wax has been dissolved in the warm fluid, add the asphaltum and stir until completely dissolved.

The dry wax coating is then washed with a 10% solution of ammonia, rinsed in clean water, and the wax-coated plate sensitized with:

Process Glue	5 oz.
Water	16 oz.
Ammonium Bichromate	88 gr.
Ammonia	10 drops

Sensitization of the plate is performed with the aid of a whirler, but no heat is used, as this would melt the wax coating.

The sensitized plate can then be exposed under a line negative (or positive) and the print developed in warm water, the insoluble (light-affected) glue image forming a stencil on the wax-asphalt resist. After drying the print, the plate is immersed in turpentine, which removes the bare areas of the resist, or those not covered with glue.

Etching of the plate follows along the previously outlined procedure, and after completion of etching, the resist and glue image can be removed by soaking the plate in hot turpentine, which penetrates the glue image and attacks the wax coating, permitting its removal from the glass surface by gentle rubbing with a stiff bristled brush.

Glass Blueprints

Production of blueprints on glass as used in photomechanical processes can be performed by sensitizing the glass plate (on a whirler) with:

Water	28
Process Glue	5
Ammonium Bichromate (20% Sol.)	3
Ammonia (28%)	¾

The exposed print is developed in water and the insoluble image then dyed by immersing it for 5 minutes in a bath containing one ounce of dye to one gallon of water.

Half Tone Zinc Etching Mordant
U. S. Patent 2,245,219

Cupric Sulphate (20% Sol.)	95 cc.
Acid Sodium Sulphate (35% Sol.)	5 cc.
Darvan No. 1 (Dispersing Agent)	0.005 g.

Glue Enamel Process
Formula No. 1

A

Distilled Water	10
Process Glue	6

B

Distilled Water	6
Ammonium Bichromate	1

The solutions are made up separately; B is then poured into A and the mixture stirred with an eggbeater. Allow the solution to stand in a cool dark place for 24 hours before filtration and use.

For halftone color etching, an enamel is preferred which has a tendency to "gray-out" and thus promote soft tones.

No. 2

Water	16
Le Page's Process Glue	6
Albumen (Liquid)	4
Ammonium Bichromate	1

No. 3

A

Gum Arabic, Purified	3 oz.
Water	9 oz.

B

Ammonium Bichromate	184 gr.
Water	5 oz.

C

Albusol	3 oz.

Prepare each solution separately: when dissolved, add B to A, followed by C, then mix thoroughly with an eggbeater.

Gum arabic has been used in conjunction with (rosin) colophony (resino-colloid process), the aim being to either eliminate burning-in of the image or to reduce the degree of heat required for the operation, a not inconsiderable factor with the enamel process on zinc. A representative formula is:

No. 4

A

Colophony (Rosin)	15 gr.
Absolute Alcohol	85 min.
Ammonia	85 min.

B

Water	1 oz.
Gum Arabic	123 gr.
Potassium Bichromate	15 gr.
Ammonia	17 min.

The solutions are prepared separately, then mixed together and the whole passed through filter paper. The image is developed with warm water; after drying, the plate is heated to the melting point of colophony (100°–140° C.), whereupon the image is ready for etching.

Gelatin may be used instead of

fish (process) glue as the colloid for photoengraving enamel:

No. 5

Water (125° F.)	20 oz.
Coignet Gelatin	2 oz.
Potassium Bichromate	120 gr.

The mixture must obviously be employed in a warm condition at the time of coating the plate.

Metagelatin, or gelatin which has lost its normal setting power is also used. The metagelatin is prepared by heating this mixture:

Metagelatin

Gelatin	500 g.
Water	800 cc.
Ammonia	125 cc.

for one hour, or until the gelatin no longer sets. It is employed as follows:

No. 6

Water	100 cc.
Metagelatin	10 g.
Egg Albumen	100 cc.
Ammonium Bichromate	6 g.

The developed image can be burnt-in to a dark color in the creation of an acid resist.

An impure form of gelatin, glue, can also be used for some enamel purposes. The first step is to clarify the glue:

Purifying Glue

Cologne Glue	100 g.
Water	600 cc.

Clarification is performed by initial soaking of the glue in water for 12 hours, after which it is dissolved by heating and 3 grams of dry egg albumen (previously dissolved in a little water) is added to the glue solution. The mixture is then heated on a water bath for 15 minutes at a boiling temperature (212° F.), filtered, and allowed to cool. Of the clarified glue, 60 cc. is mixed with 3.5 grams of dry egg albumen (dissolved in 30 cc. of water), to which is added 30 cc. of a 10% ammonium bichromate solution, followed by filtration of the bichromated mixture.

No. 7

Ammonia	3½ oz.
Potassium Carbonate	15 gr.
Casein	½ oz.

The mixture is allowed to stand for some hours, during which time the casein will uniformly dissolve. When in complete solution, the mixture is rendered light-sensitive by the addition of 3½ ounces of a saturated solution of ammonium bichromate, then filtered. The sensitized solution keeps only a few days.

Dextrin is another substitute for fish glue in the enamel process, the dextrin produced by mixing and heating

Wheat Starch	500 g.
Sodium Bicarbonate	25 g.

in dry form until the starch begins to assume a yellow color. The mixture is then cooked in 1000 cc. of water, thus providing a stock solution, which is used as follows:

No. 8

Stock Dextrin Solution	100 cc.
Ammonium Bichromate	14 g.
Ammonia	20 cc.
Distilled Water	300 cc.

Mix the ingredients in a mortar, grinding the mass until the bichromate has been entirely dissolved. After filtration through flannel, the solution is used in the ordinary manner of the enamel process.

No. 9

Water	10 oz.
Le Page's Glue	4 oz.
Albumen	1 oz.

Glucose	40 gr.
Ammonium Bichromate	120 gr.

It is suggested to harden the developed image for 2 to 3 minutes in a chromic acid bath (chromic acid, 22 grains; water, 16 ounces) before whirling dry and burning-in.

Photoengraving Resist
British Patent 512,914
Formula No. 1

Dicinnamyl Acetone	4 g.
Pale Kauri Resin	6 g.
Ethyl Methyl Ketone	60 g.
Toluene	30 g.
Benzyl Alcohol	1 g.
Crystal Violet Base	0.2 g.

This resist can be used in conjunction with bichromated albumen as the active sensitizer. On application to the metal, the resist is almost colorless, but dries to a blue-green color; this changes to a blue or violet tint during exposure to light.

No. 2

Polyvinyl Alcohol (Powdered)	50 g.
Cold Water	500 cc.

Allow the powder to swell in the water until the mixture becomes thick, then heat at a temperature of 185°–194° F. until a clear solution is obtained. Cover the mixture and allow to cool somewhat, then add:

Ammonium Bichromate	5 g.
Water	250 cc.

If desired, the bichromated solution of alcohol can be further diluted with an additional 250 cc. of water. The image is developed in water, dyed in a 1% solution of safranin and methyl violet (2:1), then hardened in a 1% solution of ammonium molybdate. Following this, the image is burnt-in to a brown color before etching.

Photo-Sensitive Coating for
Photo-Engraving

Whites of Two Eggs		
Ammonium Bichromate	75	gr.
Liquid Glue	3/4	oz.
Water	16	oz.
Iron Ammonium Citrate	25	gr.
Ammonia Solution (26° Bé)	15–20	drops

Mix the ammonium bichromate and iron ammonium citrate with one-half the water. Beat up the white of eggs, add the glue, ammonia and remaining water, stirring well. Mix the two resulting solutions. Let stand in a dark place for 24 hours; then decant. Keep the clear liquid in a dark place.

Photo-Engravers' Glue Ink Top

A

Water	50	oz.
Le Page's Glue	3½	oz.

B

Water	50	oz.
Ammonium Bichromate	1¼	oz.

Mix the solutions separately, then add B to A, and filter as with regular glue enamel. The exposed print is rolled up lightly with etching ink, after which the image is developed with water in the manner of bichromated albumen prints.

Photomechanically Etched Name
Plates

The preferred metal for name

plate work is "etching" brass, though for military name plates, aluminum, stainless steel, terneplate and zinc also are used. The metals range in thickness from .025 to .070 inch and must be free from deep scratches and other surface defects. Specific etching depth is not often required; the depth must, however, be sufficient to provide good relief to the letters and permit convenient blackening (or coloring) of the etched areas. Actual depth of etching ranges from .002 to .010 inch, depending on the metal, its thickness and the nature of the name plate design.

As previously stated, the procedure given is for brass, but the general routine can also be carried out with other metals, substituting cold enamel or bichromated albumen (ink top) as sensitizers for aluminum, zinc and metals of the ferrous type. To facilitate subsequent handling, a quarter-inch hole is perforated in each of the four corners of the sheet, the holes accommodating hooks whereon the sheets are hung during the various phases of name plate making.

The first step in the operation is to provide the "gang" or master negative bearing the required number of duplicate images, which can then be printed down (exposed) on the sensitized metal as a single unit. The position and spacing of the images must be very accurate, so as to permit convenient blanking (guillotining of stamping out) of the individual name plates from the sheet after etching. Since the name plate design must "read right" (appear in normal position) on the metal, lateral reversal of the negative image (as required in photoengraving) obviously is not necessary.

Image-duplication in the master negative can be carried out in several ways, the simplest being to make velox or photographic reproductions to the proper size and required number from the original copy, then accurately arrange and paste the prints to respective positions on a heavy cardboard or thin metal support. While requiring no equipment other than a camera, printing frame and layout table, this method naturally is somewhat laborious and time-consuming, although it is practically the only procedure open to those not possessing photocomposing (step-and-repeat) equipment. This latter equipment may take two forms: a "step-and-repeat back" or controlled-movement plateholder in the rear of a darkroom camera of the process type, or a photocomposing machine, wherewith the photographic surface may be sectionally and progressively exposed under a negative (or positive) until the required number of duplicate images have been recorded.

The step-and-repeat camera is the more convenient arrangement since it permits direct recording of negative images, whereas with a photocomposing machine, either a positive master image would be required (a "negative" drawing or negative print could be used), or the "gang" image would have to be converted to a negative by contact printing on another plate, or by chemical reversal (transformation) of the positive into the required negative image. Chemical reversal is not to be generally recommended: not all brands of plates lend themselves to this treatment, and should failure ensue, not only the plate but also the time of step-and-repeat exposures would be irretrievably wasted.

The number of images in the master (gang) negative will naturally be governed by the size of the name plate design, also by the number of duplicate images to appear on the metal sheet. Although film could be used, gelatin dry plates of the process type are the safest medium, since film negatives are liable to size-alteration and image-distortion because of atmospheric influence, and wet collodion plates are prone to gradual drying of surface during the time required for exposing a number of images.

The master negative should be protected against abrasion and scratches liable to occur when repeated exposures must be made from the same negative. A "hard" negative varnish could be applied to the plate, but there is merit in the idea of protecting the face of the negative with a thin (.001 inch) sheet of Kodapak, a transparent cellophane-like material impervious to humidity and moisture. This treatment has the advantage in that the protecting sheet can be removed after surface damage, whereas a varnished negative would probably require makeover should its surface be badly scratched or otherwise damaged.

With an accurately composed master negative at hand, the next step is cleaning of the metal plates prior to their sensitization. With brass plates first immerse the sheets in:

No. 32B Cleaner (International Chemical Company)	4 oz.
Water	1 gal.

The time of immersal is five minutes with a bath temperature of 200° F. The sheets are then rinsed in cold water, after which they are immersed in:

Sodium Cyanide	4 oz.
Water	1 gal.

remaining in the bath (75–100° F.) for ten seconds. Following a rinse in cold water, the plates are scrubbed with No. 400 pumice, then rinsed thoroughly in cold water, whereupon the metal is ready for sensitization.

As a sensitizer and photoresist for brass plates, use

Photoengraving Glue	32
Water	85
Ammonium Bichromate	3

The solution should be carefully filtered through absorbent cotton and should possess a specific gravity of approximately 1.07. The stock mixture is best contained in brown or non-actinic bottles.

Coating (sensitizing) of the metal plates is performed on a centrifuge or whirler, a specific quantity (depending on the size of the plate) of the bichromated glue mixture being poured on the center of the plate and allowed to distribute. After removal of the plate from the whirler, the coated sheet is dried for 30 minutes at 100–110° F., the operation of coating and drying performed in a clean, dimly lit room, so as to prevent "fogging" of the light-sensitive coating and resultant scummy prints.

(While the glue enamel process gives good results on brass, cold enamel or bichromated albumen may be preferred by some as photoresists for aluminum, steel and zinc, primarily because burning-in of the developed image is dispensed with; further, with albumen-ink images, the photoresist, after etching, is more easily removed than is the case with either glue or cold enamel images.)

The metal plate bearing the light-sensitive glue coating is next locked in contact with the master negative in a vacuum printing frame, and the plate exposed to the light of an arc lamp in the regular manner of photoengraving. After exposure, the print is developed in lukewarm water as is done with glue prints on copper and zinc, the plate being further cleaned and the image sharpened by lightly swabbing the print with wet cotton to aid removal of all soluble and partially soluble glue particles.

The developed print is then dried at room temperature, whereupon the image is burnt in (baked) at a temperature of 400–450° F., with heating continued until the print assumes a light brown color. The photoresist, so created, is then ready for etching.

Prior to etching, the back of the plate is painted with a shellac solution to protect this area against action of the etching solution. The shellac solution is prepared by first mixing a stock solution (orange shellac, 4 pounds; denatured alcohol, 1 gallon), which is used in the following proportion:

Shellac Solution	3 qt.
Denatured Alcohol	1 qt.
Methyl Violet	1/4 oz.

The methyl violet dye has, for its purpose, greater visibility of the applied coating, thus aiding in proper coverage of the back of the plate.

On drying of the protective backing, the glue print is given a scum-removal treatment by wiping its surface with a soft cloth dampened with the following solution:

Hydrochloric Acid	3
Denatured Alcohol	7

At this stage, hooks are inserted in the holes at the corners of the sheets and the plates then placed in a pitch lined wooden tank containing a ferric chloride etching bath at 40° Bé (ferric chloride, 7 pounds; water, 1 gallon). The bath is agitated by air during the etching period and the sheets brushed lightly several times during this period in order to remove possible oxidation and foreign substances from the surface of the metal. Time of etching depends on the etchability of the brass, the depth required and the prevailing room temperature of the bath as well as the strength of solution.

Etching completed, the glue photoresist is removed from the sheets by immersing the plates in:

Sodium Hydroxide	4 oz.
Water	1 gal.

The solution is used at a temperature of 200–212° F. and the plates kept in the bath until the glue image has thoroughly loosened or left the plate, when the sheets are then rinsed in cold water and immersed in:

Sodium Cyanide	4 oz.
Water	1 gal.

This bath is used at a temperature not to exceed 100° F. and the plates immersed until all vestiges of the glue image have been removed. The heets are again rinsed in cold water, dried, and are then ready for acquering and finishing.

Stripping Film Gelatin Solution

A solution to aid adhesion. Fill a 32-ounce bottle half full of water. Add ½ ounce granulated gelatin (stir while adding), and 2 ounces 28% acetic acid. Add enough cold water to fill the bottle.

Stripfilm Cement

As an adhesive and to prevent line and halftone stripfilm negatives from curling when stripped on glass plates, this cement is recommended:

Water (125° F.)	32
Gelatin	½
Process Glue	½
Acetic Acid (28%)	4

Soak the gelatin in cold water, then dissolve in the above quantity of warm water, after which add the glue and acetic acid. In use, apply a thin film of cement to the glass plate before putting down the stripfilm negatives, then squeegee the films into good contact with lintless blotting paper.

Negative Cleaner

For ridding the surface of gelatin negatives from finger marks and dirt of all kinds, the following is used:

Formula No. 1

Ethyl Alcohol	18
Wood Alcohol	2
Strong Ammonia	1

Rub the solution over the surface of the negative with a pledget of absorbent cotton.

No. 2

Petrolatum	1 oz.
Carbon Tetrachloride	20 oz.
Glacial Acetic Acid	¼ oz.

The solution is rubbed lightly on the negative with a soft chamois skin; after drying, polish the negative surface until none of the cleaner is visible.

Glass Cleaner

For cleaning copyboard and printing frame glasses as well as general photographic glassware, a wetting agent is of utility:

Wetanol or Aerosol OT (Wetting Agent)	1 oz.
Denatured Alcohol	10 oz.
Water	5 gal.

TC-1 Tray Cleaner

For General Use

	Avoirdupois	Metric
Water	32 oz.	1.0 l.
Potassium Bichromate	3 oz.	90.0 g.
Sulphuric Acid, C.P.	3 fl. oz.	96.0 cc.

Add the sulphuric acid slowly while stirring the solution rapidly.

For use, pour a small volume of the tray cleaner solution in the vessel to be cleaned. Rinse around so that the solution has access to all parts of the tray; then pour the solution out and wash the tray six or eight times with water until all traces of the cleaning solution disappear. This solution will remove stains caused by oxidation products of developers and some silver and dye stains. It is a very useful cleaning agent.

TC-2 Tray Cleaner

For Removal of Silver Stains

Solution A

	Avoirdupois	Metric
Water	32 oz.	1.0 l.
Potassium Permanganate	73 gr.	5.0 g.
Sulphuric Acid, C.P.*	2½ dr.	10.0 cc.

* Add the sulphuric acid slowly while stirring the permanganate solution rapidly

Solution B

	Avoirdupois	Metric
Water	32 oz.	1.0 l.
Sodium Bisulphite	145 gr.	10.0 g.

For use, pour Solution A into the tray and allow it to remain for a few minutes, then rinse with water. Apply Solution B, and wash thoroughly.

This formula is satisfactory for the removal of most types of general stains but it is especially recommended for the removal of silver stains.

S-5 Hand Stain Remover

Solution A

	Avoirdupois	Metric
Potassium Permanganate	¼ oz.	7.5 g.
Water	32 oz.	1.0 l.

Solution B

	Avoirdupois	Metric
Sodium Bisulphite	16 oz.	480.0 g.
Water	32 oz.	1.0 l.

Rub the hands with a small amount of Solution A, rinse in water; then pour a small quantity of Solution B into the palm of one hand, rub it quickly over the hands, and, when free of stain, wash them thoroughly with water. If the original stain is not entirely removed, repeat the treatment with Solutions A and B.

If it is desired to immerse the hands in Solution B, 1 part of the solution should be diluted with 4 parts of water.

S-6 Stain Remover

For Removal of Developer Stain on Negatives

Developer or oxidation stain may be removed by first hardening the film for 2 or 3 minutes in the formalin hardener (SH-1), then washing for 5 minutes and bleaching in:

Stock Solution A

	Avoirdupois	Metric
Potassium Permanganate	75 gr.	5.2 g.
Water, To make	32 oz.	1.0 l.

Stock Solution B

Cold Water	16 oz.	500.0 cc.
Sodium Chloride	2½ oz.	75.0 g.
Sulphuric Acid, C.P.	½ fl. oz.	16.0 cc.
Water, To make	32 oz.	1.0 l.

Use equal parts of A and B. The solutions should not be mixed until ready for immediate use since they do not keep long after mixing. All particles of permanganate should be dissolved completely when preparing Solution A, since undissolved particles are likely to produce spots on the negative. Bleaching should be complete in 3 or 4 minutes at 68° F. (20° C.). The brown stain of manganese dioxide formed in the bleach bath is best removed by immersing the negative in 1% sodium bisulphite solution. Then rinse well and develop in strong light, preferably sunlight, with any non-staining developer such as D-72 diluted 1 part to 2 parts of water. Then wash thoroughly.

Warning: Developers containing a high concentration of sulphite (such as D-76) should not be used for redevelopment because the sulphite tends to dissolve the bleached image before the developing agents act upon it.

Washing Test and Hypo Elimination

Very small traces of hypo retained in films or prints greatly accelerate the rate of fading of the image. It is extremely difficult to test for small quantities of hypo but the following test (HT-1a) will indicate when the film or prints may be considered reasonably free from hypo. If prints give a negative reaction by this test it is no guarantee that they may not ultimately fade.

HT-1a Hypo Test Solution

For Testing Thoroughness of Washing

	Avoirdupois	Metric
Distilled Water	6 oz.	180.0 cc.
Potassium Permanganate	4 gr.	0.3 g.
Sodium Hydroxide	8 gr.	0.6 g.
Water (Distilled), To make	8 oz.	250.0 cc.

Films or Plates: Take 8 ounces (250 cc.) of pure water in a clear glass and add ¼ dram (1 cc.) of HT-1a. Then take a 6-exposure film, size 3¼x4¼ in. or equivalent from the wash water and allow the water from it to drip for 30 seconds into the glass of test solution. If a small percentage of hypo is present the violet color will turn orange in *about 30 seconds* and with a larger concentration the orange color will change to yellow. In either case the film should be returned to the wash water and washed until further tests produce no change in the violet color.

Papers: Take 4 ounces (125 cc.) of pure water in a clear glass and add ¼ dram (1 cc.) of HT-1a. Pour ½ ounce (15 cc.) of this diluted solution into a clear 1-ounce glass container. Then take six prints, size 4x5 in. or equivalent, from the wash water and allow the water from them to drip for 30 seconds into the ½ ounce of the dilute test solution. If a small quantity of hypo is present the violet color will turn orange in *about 30 seconds* and become colorless in one minute. In either case the prints should be returned to the wash water and washed until further test shows no change in color.

NOTE: Oxidizable organic matter if present in the water reacts with the permanganate solution and changes its color in the same manner as hypo. The water should therefore be tested as follows:

Prepare two samples of the permanganate test solution using distilled water. Then add a volume of the tap water to one test sample equal to that of the wash water drained with the film or prints into the other sample. If the sample to which tap water has been added remains a violet color this indicates the absence of organic matter and it will be unnecessary to make the test in duplicate. If the color is changed slightly by the tap water, however, the presence of hypo in the film or prints will be shown by the relative color change of the two samples. For example if the tap water sample turned pink and the wash water sample became yellow it would indicate the presence of hypo, while if both remained the same shade this would indicate the absence of hypo.

When complete removal of hypo is important, prints should be treated in the HE-1 Hypo Eliminator.

HE-1 Hypo Eliminator

For Professional and Amateur Use

	Avoirdupois	Metric
Water	16 oz.	500.0 cc.
Hydrogen Peroxide (3% Solution)	4 fl. oz.	125.0 cc.
Ammonia (3% Solution)	3¼ fl. oz.	100.0 cc.
Water, To make	32 oz.	1.0 l.

NOTE: Prepare the solution immediately before use and keep in an open container during use. Do not store the mixed solution in a stoppered bottle or the gas evolved may break the bottle.

Directions for Use: Wash the prints for about 30 minutes at 65° to 70° F. in running water which flows rapidly enough to replace the water in the vessel (tray or tank) completely once every 5 minutes. Then immerse each print about 6 minutes at 70° F. in the HE-1 Hypo Eliminator solution and finally wash about 10 minutes before drying. At lower temperatures, increase the washing times.

Life of HE-1 Solution: About fifty 8x10-inch prints or their equivalent per gallon (4 liters).

Test for Hypo: Process with the batch of prints, an unexposed *white* sheet of photographic paper (same weight and size as majority of prints in batch). After the final wash, cut off a strip of this sheet and immerse it in a 1 per cent silver nitrate solution for about 3 minutes; then rinse in water and compare *while wet* with the *wet untreated portion.* If the hypo has been completely removed, no color difference should be observed. A yellow-brown tint indicates the presence of hypo. The depth of the tint increases with increased hypo content. A positive test with silver nitrate may also be obtained in the absence of hypo if hydrogen sulphide or wood extracts are present in the water supply.

Occasional Effects when Using the Peroxide-Ammonia Treatment (HE-1)

1. Slight tendency for prints to stick to belt on belt driers. To prevent this effect bathe the prints about 3 minutes in a 1 per cent solution of formaldehyde prior to drying.

2. An almost imperceptible change in the image tone. To prevent this effect, add 15 grains of potassium bromide to each quart (1 gram per liter) of the peroxide-ammonia bath (HE-1).

3. A very faint yellowing of the whites (undetectable on buff papers). To minimize this effect, bathe the prints in a 1 per cent sodium sulphite solution for about 2 minutes immediately after treatment in HE-1 and prior to the final wash.

NOTE: With *buff papers,* it is possible to use a higher concentration of peroxide (maximum about 500 cc. of 3% solution per liter) and thus extend the exhaustion life to about eighty 8x10-inch prints per gallon. This more concentrated bath is *not* recommended for use with white papers because the yellowing would be objectionable.

GP-1 Gold Protective Solution

For Increasing the Permanency of Silver Images

	Avoirdupois	Metric
Water	24 oz.	750.0 cc.
Gold Chloride (1% Stock Solution)*	2½ fl. dr.	10.0 cc.
Sodium Thiocyanate	145 gr.	10.0 g.
Water, To make	32 oz.	1.0 l.

Add the gold chloride stock solution to the volume of water indicated. Dissolve the sodium thiocyanate *separately* in 4 ounces (125 cc.) of water. Then add the thiocyanate solution slowly to the gold chloride solution while stirring the latter solution rapidly.

For Use: Immerse the well washed print (which preferably has received a hypo elimination treatment) in the Gold Protective Solution for 10 minutes at 70° F. (21° C.) or until a just perceptible change in image tone (very slightly bluish black) takes place. Then wash for 10 minutes in running water and dry as usual.

Approximate Exhaustion Life: Thirty 8x10-inch prints per gallon. For best results the GP-1 solution should be mixed immediately before use.

Films and Plates: The above procedure may also be used with fine-grained images of films and plates when maximum permanency is desired.

* A 1% stock solution of gold chloride may be prepared by dissolving the contents of one tube (15 grains) in 3¼ ounces of water (1 gram in 100 cc. of water).

Diazo Printing

Diazo type printing is very useful for making dry, accurate copies of typewritten, printed, or sketched material in a very short time.

Formula No. 1
French Patent 794,590

The paper on which reproduction is to be made is coated with a small portion of a mixture of 22 parts of stannic chloride double salt of 4 diazo 1 dimethyl aniline, 20 parts of boric acid, 50 parts of tartaric acid, 2 parts of phloroglucinol, and 1000 parts of water. The paper is dried in the dark and exposed to ultra-violet light under a transparent positive. The paper is then treated with gaseous ammonia and the image is developed.

No. 2
British Patent 413,093

The paper is coated with a small portion of a mixture of 5 parts metol, 4 parts potassium nitrate, and 200 parts water. It is then dried in the dark and exposed to ultra-violet light under a transparent positive. A light sepia print is obtained.

No. 3
British Patent 297,363

The paper is coated with a solution of the diazo compound of H acid, dried in the dark, exposed to ultra-violet light under a transparent positive and developed with ammonia.

Making Photographic Prints on Wood

Beat thoroughly the whites of one or two eggs, and let stand for one or two hours. In this interval the albumen will be reduced to a liquid form. Mix it with powdered English oxide of zinc, and place a smooth coating on the surface of the wood, making it as thin as possible. After the surface is thoroughly dry, sensitize it with a solution of Eastman Silver Nitrate (Eastman Hydrometer test about 50). The sensitized surface should be thoroughly dried in a dark place, after which it will be ready for printing, in sunlight or under strong artificial light. After being printed, it should be fixed with a weak solution of plain hypo for about five or ten minutes, and then washed.

Ferrotype Polish
(For Metal Photographic Plates)

Beeswax	40 gr.
Carbon Tetrachloride	4 oz.

Fingerprinting by Photography

Several methods for rendering the hands luminescent prior to photographing them can be employed. The hands may be dusted with a brightly fluorescing phosphor such as zinc-beryllium silicate or magnesium tungstate, illuminated with intense ultra-violet light, and photographed with a camera having two 4A filters. Or, the hands can be dipped in a fluorescent dye such as neutral acriflavine and after drying photographed in the same manner. The whole hand may be thus photographed and reduced on microfilm. On enlargement the fingerprint becomes clear enough for the usual study. This method is applicable for rapid fingerprinting (without soiling) of a great many persons, the record including the palm and its characteristics.

Saving Overexposed Film

Bathe the negative for 5 min. in:

Potassium Bromide	110 gr.
Water	40 oz.

Then develop for 10–15 min. at 65–68° F. in:

Water	16 oz.
Potassium Metabisulphite	4 gr.
Metol	3 gr.
Sodium Sulphite	180 gr.
Hydroquinone	90 gr.
Sodium Carbonate	110 gr.
Potassium Bromide	16 gr.

Motion Picture Screen
U. S. Patent 2,215,061

Glass cloth is coated with:

Cellulose Nitrate	14.0
Castor Oil	18.6
Pigment, White	23.4
Ethyl Acetate	17.6
Alcohol	26.4

Photo Flash Lamp Filler
U. S. Patent 2,252,241

Aluminum	85%
Zinc	6%
Cadmium	9%

Flashlight Powder
German Patent 705,162

Magnesium, Powdered	2.4 g.
Zirconium, Powdered	6.5 g.
Potassium Chlorate	5.0 g.
Nitrocellulose Solution	5.0 cc.

Photographic Hints

1. To eliminate air bells on photographic films during developing, add 1–5 cc. of glycerin to each liter of solution.

2. To prevent curling and cracking, particularly in cold weather, add about one ounce (30 cc.) of glycerin to each quart (liter) of final rinse water. Even better is the procedure advocated in a Government photography manual. Here it is advised that the prints be placed in a glycerin bath after the completion of the final washing. This bath is composed of 1 part of glycerin to from 5 to 10 parts of water, and the print is allowed to remain in this bath for at least five minutes. Then without further washing, the print is placed on the drier.

Do not discard these solutions; they may be used over and over again. From time to time, however, it will be necessary to add a little glycerin to overcome the diluting effects of the water carried over on the prints from the washing and rinsing procedures.

3. To lend brilliance and durability, a print varnish, especially suitable for use on bromide prints, can be made from:

Borax	30 gr.
Pale Shellac	60 gr.
Sodium Carbonate	10 gr.
Glycerin	30 min.
Water	1 fl. oz.

Boil and allow to cool, then add:

Alcohol	1 fl. oz.

Add a small quantity of whiting or powdered pumice to precipitate the gum wax. Shake well at intervals and allow to stand for several days. Decant and filter the clear liquid. Bottle until needed.

Chapter Sixteen

POLISHES AND ABRASIVES

Oil Polish

U. S. Patent 2,275,596

Formula No. 1

Mineral Spirits	169.00
Mineral Oil, Light	187.50
Glyceryl Monoricinoleate	93.60
Ricinoleic Acid	11.03
Water	624.75
Caustic Potash Solution (8.6%)	19.25

No. 2

Sulphonated Petroleum Oil	2
Sodium Oleate	4
Light Mineral Oil	38
Red Oil	3
Water	53

No. 3

Mineral Oil	50
Ammonium Oleate	3
Water	47
Pine Oil, To suit	

No. 4

Light Blown Castor Oil	10
Light Mineral Oil	20
Xylene	9
Potash Soap	1
Water	60

No. 5

Mineral Oil	25
Naphtha	10
Stearic Acid	4
Triethanolamine	2
Water	59

No. 6

Kerosene	5
Light Mineral Oil	5
Glycerin	4
Gum Arabic	1
Gum Tragacanth	1
Water	84

No. 7

Light Mineral Oil	56
Potassium Stearate	1
Paraffin and Carnauba Waxes	3
Water	40

Liquid Wax Polish

Heavy Panoline	165
Mineral Seal Oil	75
Paraffin Wax	13½
Carnauba Wax	1½
Varnolene	24

Emulsion Polish

1. Carnauba Wax	10
2. Petromix No. 5	6
3. Deodorized Kerosene	34
4. Water	50

1 and 2 are melted together. Add 3 and heat to 180° F. Heat water to 200° F. and add to above with stirring.

Pure Carnauba Polish

1. Carnauba Wax	10
2. Petromix No. 5 (Emulsifier)	4
3. Butyl Cellosolve	1
4. Water	85

Melt 1 and 2, then add 3. Add 4 at 200° F. with constant stirring.

Carnauba-Shellac Polish

1. Carnauba Wax	10
2. Petromix No. 5	4

3. Butyl Cellosolve 1
4. Water 65
5. B Solution 20

Directions as for pure Carnauba polish.

B Solution

Dry Powdered Shellac	20
Water	79
Ammonium Hydroxide	1

Heat to boiling. Makes a clear solution.

Naphtha Emulsion Shoe Cleaner

High Viscosity Methyl Cellulose	1.0 g.
Water	100.0 cc.
2-Amino-2-Methyl-1, 3-Propanediol	2.3 g.
Stearic Acid	6.2 g.
Low-Boiling Naphtha	200.0 cc.

The methyl cellulose is dispersed in the water and to this dispersion is added the amino-methyl-propanediol. The stearic acid is dissolved in the naphtha and this solution is poured into the methyl cellulose dispersion and thoroughly agitated.

Shoe Polish Paste
(To be packed in tubes)

Candelilla Wax	50
Diglycol Stearate	30
Stroba Wax	25
Varsol	250
Water	225–300

Melt first three ingredients together and, after removing flame, stir in Varsol. Add the water (at 90° C.) slowly, with vigorous stirring, to the wax-solvent mixture.

Polish for Transparent Plastics

Diatomaceous Silica	100
Emulsone B (Emulsifying Gum)	10
Water	600
Naphtha	300

Battleship Linoleum Polish

Ceresin Wax	27.2
Carnauba Wax	25.0
Rezo Wax B	2.2
Naphtha	72.0

Rubbing Type Paste Wax
Formula No. 1

Carnauba Wax	150
Paraffin Wax	166
Beeswax	76
Durez 420 Resin	34

Heat to approximately 300° to 325° F. to melt.

Add the following:

Mineral Spirits	1280
Turpentine	32

Stir and cool to 160° F. Dump into containers.

No. 2

Carnauba Wax	14.0
Beeswax	14.0
Paraffin Wax	14.0
Stearic Acid	5.0
Amino Methylpropanol	2.5
Water	80.0

The first four ingredients are melted together on a boiling water bath. The amino methylpropanol is stirred into the wax melt, the temperature of which should not exceed 100° C. to 105° C. at this point, and the boiling water is added with vigorous agitation.

Bright Drying Waxes for Fruit Coating
U. S. Patent 2,275,659
Formula No. 1

Carnauba Wax	30.0
Soap	6.5
Water	260.0
Bentonite	0.3–2.6

No. 2

Carnauba Wax	30.0 g.
Oleic Acid	3.3 g.
Caustic Soda	0.5 g.

Triethanolamine	2.7	g.
Borax Shellac Solution	60.0	cc.
Water	260.0	g.
Bentonite	0.3–2.6	g.

No. 3

Carnauba Wax	30.0	g.
Oleic Acid	3.3	g.
Caustic Soda	0.5	g.
Triethanolamine	2.7	g.
Borax Shellac Solution	60.0	cc.
Water	260.0	g.
Gelatin	0.3–1.3	g.

No. 4

Carnauba Wax	30.0	g.
Oleic Acid	3.3	g.
Caustic Soda	0.5	g.
Triethanolamine	2.7	g.
Borax Shellac Solution	60.0	cc.
Water	260.0	g.
Irish Moss, Bleached	0.2–0.65	g.

Automobile Polishes

Formula No. 1

Mineral Oil	42
Castor Oil	5
Ammonia Soap	2
Xylol	5
Water	46

No. 2

Mineral Oil	38
Castor Oil	10
Potash Soap	1
Water	51

No. 3

Coal-Tar Naphtha	9
Mineral Oil	27
Blown Castor Oil	6
Ammonia Soap	3
Water	55

No. 4

Coal-Tar Naphtha	16
Mineral Oil	17
Blown Castor Oil	10
Ammonia Soap	3
Water	54

Automobile Polish and Cleaner Formulae

Formula No. 1

Ortho Dichlorbenzene	1½	gal.
Pine Oil	½	gal.
Kerosene or V.M.P. Naphtha	9½	gal.
Diatomaceous Silica (Fine)	11	lb.
Diatomaceous Silica (Finest)	11½	lb.
Cream Tripoli	40	gal.
Bentonite	3½	gal.
Water	22½	gal.
Glycerin or Glycol	⅜	gal.

Mix thoroughly the pine oil, kerosene, etc., and the chlorbenzene. Mix the abrasives and bentonite together with the glycerin and water to a smooth homogeneous paste. Add the oil mixture to the abrasive mixture slowly with vigorous agitation till fully incorporated. (A centrifugal pump can be used to advantage for this thorough mixing.)

No. 2

Spindle Oil	4⅞	gal.
Light Fuel Oil	2⅜	gal.
Diatomaceous Silica (Fine)	45	lb.
Diatomaceous Silica (Finest)	45	lb.
Bentonite	2	lb.
Water	32	gal.

The abrasives and bentonite are thoroughly mixed into the water. The oils are mixed together and added to the abrasive water mixture with vigorous stirring until emulsification is complete.

No. 3

Carnauba Wax	27	lb.
Spermaceti	50½	lb.
Ozokerite	44	lb.
Kerosene or V.M.P. Naphtha	32	gal.
Diatomaceous Silica (Finest)	65	lb.
Cream Tripoli	131	lb.
Bentonite	16	lb.
Coloring Pigment or Dye, To suit		
Water	13¼	gal.
Borax	3¼	lb.
Trisodium Phosphate	3¼	lb.
Soap Chips (Finest Grade)	16¼	lb.
Water	13¼	gal.

No. 4

Mineral Oil	15
Glycerin	2
Gum Tragacanth	0.1
Diatomaceous Earth	11
Water	71.9

No. 5

Kerosene	9
Mineral Oil	5
Gum Tragacanth	0.25
Glycerin	4
Denatured Alcohol	5
Diatomaceous Earth	13
Water	63.75

No. 6

Petroleum Naphtha	5
Mineral Oil	8
Gum Tragacanth	0.7
Glycerin	3
Denatured Alcohol	6
Diatomaceous Earth	14
Water	63.3

No. 7

Light Mineral Oil	7
Water-Soluble Gum	0.5
Glycerin	4
Ammonium Chloride	0.2
Alcohol	13
Diatomaceous Earth	13
Water	62.3

No. 8

Mineral Oil	10
Blown Castor Oil	1
Gum Tragacanth	0.5
Glycerin	4
Ammonia Water	0.1
Diatomaceous Earth	10
Water	74.4

No. 9

Mineral Oil	6
Petroleum Naphtha	4
Glycerin	5
Gum Tragacanth	0.4
Triethanolamine	0.1
Fatty Alcohol Sulphate	0.1
Diatomaceous Earth	10
Water	74.4

No. 10

Mineral Oil	9
Carnauba Wax	0.6
Glycerin	5
Gum Tragacanth	0.5
Ammonia Water	0.3
Diatomaceous Earth	11
Water	73.6

No. 11

Mineral Oil	17
Carnauba Wax	0.5
Glycerin	3
Gum Tragacanth	0.5
Diatomaceous Earth	11
Water	68

No. 12

Mineral Oil	8
Carnauba Wax	1
Glycerin	5
Gum Tragacanth	0.2
Triethanolamine Oleate	0.2
Diatomaceous Earth	17
Water	68.6

No. 13

Mineral Oil	8
Petroleum Naphtha	8
Carnauba Wax	1.5

Paraffin Wax	0.5
Glycerin	5
Gum Tragacanth	0.1
Ammonia Soap	1.5
Diatomaceous Earth	13
Water	62.4

No. 14

Mineral Oil	7
Kerosene	3
Carnauba Wax	2
Glycerin	7
Triethanolamine	0.1
Gum Tragacanth	0.4
Diatomaceous Earth	12
Water	68.5

No. 15

Kerosene	11
Carnauba Wax	1.7
Glycerin	4
Ammonia Water	0.2
Gum Tragacanth	0.4
Diatomaceous Earth	11
Water	71.7

No. 16

Mineral Oil	9
Castor Oil	3
Stoddard's Solvent	18
Carnauba Wax	0.8
Soda Soap	0.5
Bentonite	2
Diatomaceous Earth	7
Water	59.7

No. 17

Mineral Oil (60–80 sec.)	6
Petroleum Naphtha (95–140° C.)	6
Oleic Acid	2
Carnauba Wax	1
Amino Methylpropanol	1
Diatomaceous Earth	11
Water	73

The first four ingredients are melted together and the diatomaceous earth is stirred into this melt to make a thick paste. The amino methylpropanol and water—mixed together and warmed to approximately 75° C.—are then added to the warm melt with very vigorous agitation.

Automobile Paste Cleaners

Formula No. 1

Mineral Oil	6
Kerosene	6
Pine Oil	2
Sodium Oleate	4
Tripoli	45
Water	37

No. 2

Mineral Oil	9
Petroleum Naphtha	21
Carnauba Wax	4
Paraffin Wax	2
Sodium Silicate	3
Sodium Stearate	3
Tripoli	26
Water	32

No. 3

Kerosene	19
Carnauba Wax	5
Paraffin Wax	3
Sodium Soap	2
Gum Tragacanth	1
Tripoli	33
Water	37

No. 4

Soap	4.05
Water	16.35
Glycerin	6.75
Kerosene	19.30
Dibutyl Phthalate	3.55
Mild Abrasive	50.00

Homogenize in a paint mill.

Burnishing Polishes for Automobile Maintenance

Dissolve 1 ounce by weight of oxalic acid in 1 quart of water.

In another container containing 1 gallon of water, dissolve 5 ounces by weight of Nelgin and 1 ounce by weight of Moldex. To it add 1 pound of ammonium hydroxide and while stirring add 72 fluid ounces of Stetsol. Make sure the Stetsol is added very slowly, while being stirred vigorously for at least 1 hour. Then add 2 pounds of diatomaceous earth slowly to which the dissolved oxalic acid had been previously added. Stirring should continue until a creamy consistent mass has been obtained.

This polish is applied to the body of the car with a semi-wet cloth and rubbed slightly for a few minutes, and followed up with a dry cloth.

Silver Polish

A. Pastes

Formula No. 1

Diatomaceous Earth	15–20
Soap, Hard	3–8
Soda Ash	0.5–1.5
Water	72–80

No. 2

Diatomaceous Earth	16.5
Soap, Hard	2.5
Sodium Silicate	1
Water	80

No. 3

Diatomaceous Earth	10
Soap, Hard	4
Soda Ash	4
Sodium Metasilicate	1
Water	81

No. 4

Diatomaceous Earth	30
Soap, Hard	10
Soda Ash	0.5
Trisodium Phosphate	0.5
Water	59

B. Liquid

Diatomaceous Earth	82.5
Soap, Hard	17
Soda Ash	0.5

C. Powder

Amorphous Silica	25
Soda Soap	5
Pine Oil	5

Aluminum Polish

Formula No. 1

Tripoli	53
Ammonium Stearate	5
Stearic Acid	4
Iron Oxide	1
Petroleum Naphtha	36
Water	1

No. 2

Dolomitic Limestone	85.5
Soda Soap	13
Sodium Silicate	1.5

Jewelry Polish

Mix together, thoroughly, 15 grams of powdered ferric oxide and 5 grams of powdered calcium carbonate.

Add enough water to a portion of the mixture to make a paste and then rub the jewelry with the paste by means of a cloth. Rinse with water.

Metal Aircraft Polish

Formula No. 1

Rottenstone, Powdered	45 g.
Alumina, Levigated	15 g.
Paraffin Wax	1 g.
Pine Oil	20 cc.
Kerosene	30 cc.

No. 2

Rottenstone, Powdered	45 g.
Alumina, Levigated	15 g.
Paraffin Wax	1 g.
Pine Oil	15 cc.
Diglycol Laurate	5 cc.
Kerosene	30 cc.

Metal Roller Polish

Chromium Oxide	86

Silica Gel	2
Stearin	10
Kerosene	2

Metal Polish

Formula No. 1

Silica Abrasive	6.0
Triethanolamine Oleate	3.0
Bentonite	0.5
Orthodichlorobenzene	10.0
Oleic Acid	0.5
Water	80.0

No. 2

Soap Flakes	300 g.
Borax	60 g.
Tri Sodium Phosphate	30 g.
Water	3000 cc.
Ammonia (Strong)	150 cc.
Pine Oil	300 cc.
Silica	900 g.
Denatured Alcohol	60 cc.

Heat about ⅔ of the water and dissolve in it the soap and the salts. Put the silica in a pail and mix with the pine oil. Add ammonia to the soap solution and add this slowly, triturating constantly to ensure a homogeneous mixture. (If ammonia is added too early to the hot solution, it will partly evaporate. If trituration is insufficient or solution added too quickly, the suspension must be passed through a sieve, or an emulsion mill.) Once a good paste is formed, it may be thinned with more solution, and when thin enough, the rest can be added, all at once. Stir until it becomes thick again, usually a few minutes. The rest of the water is best added cold to hasten the cooling of the product. When cold, stir in alcohol. The emulsion is perfectly miscible with water, but every such addition requires good mixing. It is then stable. Cleaning qualities are good.

No. 3

Kerosene	1 pt.
Pulverized Silica	20 g.
Concentrated Ammonia	5 cc.
Liquid Soap	10 cc.

Shake thoroughly before using the above.

Military Leather Dressing

Beeswax, Yellow	65
Acrawax C	2
Turpentine	33

Polish for Patent Leather

Yellow Wax or Ceresine	3 oz.
Spermaceti	1 oz.
Turpentine	11 oz.
Asphaltum Varnish	1 oz.
Borax	80 gr.
Frankfort Black	1 oz.
Prussian Blue	150 gr.

Melt the wax, add the borax, and stir until smooth. In another pan melt the spermaceti; add the varnish, previously mixed with the turpentine; stir well and add to the wax; lastly add the colors.

Furniture Polish

Formula No. 1

Turpentine	1 pt.
Beeswax	1 lb.

Melt the beeswax and work in the turpentine.

No. 2

Dragon's Blood	6 gr.
Linseed Oil	½ fl. oz.
Castor Oil	2 fl. oz.
Mineral Oil	3 fl. oz.
Turpentine	½ fl. oz.
Antimony Trichloride	½ fl. oz.
Hydrochloric Acid	¼ fl. oz.
Acetic Acid	¾ fl. oz.
Alcohol	½ fl. oz.
Water	1 fl. oz.

Mix the first five ingredients. Then mix the next five in the order given. Add the first mixture to the second and emulsify by agitation.

No. 3

Non-Rubbing

Castile Soap	10
Water	160
Carnauba Wax	40
Water	150
Formaldehyde	2

Dissolve the soap in the water and heat to boiling. Add the carnauba wax in small pieces. Stir constantly until a smooth homogeneous emulsion is obtained, then cool to 60° C. by adding the remaining water. Filter through cheesecloth and add the formaldehyde as a preservative.

No. 4

Vinegar	123 lb.
Linseed Oil, Boiled	95 lb.
Linseed Oil, Raw	120 lb.
Diglycol Oleate	6 lb.
Alcohol	17 lb.
Turpentine	135 lb.
Petroleum Spirits	226 lb.

Bright Drying Floor Polish
(Non-Rubbing)

These polishes are composed of mixtures of aqueous Manila resin solutions with carnauba wax emulsions. The following resin solution is recommended:

Formula No. 1

Manila DBB or Loba C, Ground	240
Monoethanolamine	45–50
Water	2090

Dissolve the monoethanolamine in 700 of water, stir, and add the resin. Heating the mixture to 50°–55° C. helps the resin to dissolve more rapidly. Finally, add the remainder of the water and allow to cool. Strain.

This resin solution is mixed with a carnauba wax emulsion to give the final product. Established wax emulsion formulae are satisfactory, but the following may be used if desired:

Carnauba Wax	240
Triethanolamine	28
Linoleic Acid	45
Borax	18
Water	2412

Melt the wax to 90°–95° C.

In a separate container, dissolve the triethanolamine in 200 grams of water, stir at 90° C., and add the linoleic acid. Stir the mixture for five minutes at 85°–90° C. and then add the resulting cloudy viscous solution to the wax. Stir the wax-soap mixture at 90°–95° C. for five minutes.

Dissolve the borax in 50 grams of boiling water, add and stir until a hot drop off a spatula is clear (14–15 minutes). The remainder of the water is then added cautiously at a temperature of 95°–100° C. When the mixture starts to thin out, the addition of the water may be made more rapidly. Allow to cool to room temperature and add water to compensate for evaporation loss.

The polish is made by mixing the resin solution and the wax emulsion in the desired proportions. The emulsion is stirred and the resin solution is added. Stirring is continued in order to obtain a homogeneous mixture.

No. 2

Carnauba Wax, Refined	15.0
Durez 420 Resin	15.0
Oleic Acid	4.5

Heat to 300–325° F. until fluid

Cool to 210° F., add the following at boiling point:

2-Amino-2-Methyl Propane	**1.0**
Sodium Hydroxide	0.4
Water	5.0

Stir in until mixture is clear and jelly-like. Add the following slowly with good agitation:

Water, Hot	210.0

Cool rapidly and add:

Casein Solution 10% *	25.0

No. 3

Carnauba Wax	18 oz.
Pine Oil	1½ oz.
Soap	1½ oz.
Hot Water	3 oz.

Cook till transparent.

Sodium Silicate	2¼ oz.
Hot Water	3 oz.

Cook 5 minutes. Make up to 4 quarts.

No. 4

Carnauba Wax	8½ lb.
Candelilla Wax	16½ lb.
Borax	2 lb.
Duponol W.A. Paste	5½ lb.
Dissolved in Water	2 gal.

Cook till smooth. Make up to 50 gal. with boiling water.

In a separate kettle, cook

Gum Loba C	30 lb.
Water	30 gal.
Ammonia (26°)	150 oz.

until the gum is completely dissolved, and add to the wax emulsion. Make up the volume of the finished mixture to 100 gal.

* Casein Solution

Casein	10.0
Borax	1.5
Pine Oil	0.2
Preservative	0.2
Water at 160° F.	15.0
Mix thoroughly and add	
Water at 160° F.	15.0
Mix thoroughly and add	
Ammonia in water, 7% of 28%	1.0
Water at 160° F.	70.0
Mix until smooth.	

No. 5

Candelilla Wax	64
Oleic Acid	24
2 Amino, 2 Methyl, 1 Propanol	11
Water, To make	500

Very translucent, thin emulsion which dries to a transparent, bright film.

Floor Treatments

Wood Floors

Unfinished wood floors should be mopped or scrubbed with warm water and a mild soap. Scouring with powdered pumice or steel wool may be desirable in some cases. A solution of trisodium phosphate, sodium metasilicate (about ½ oz. per gallon of water), or washing soda (about 2 oz. per gallon of water), may be used for cleaning oily or greasy floors. After using soap or other cleaning agent, the floor should be thoroughly rinsed off with clear water and wiped as dry as possible. It is advisable to scrub a small area at one time and to avoid flooding the floor with cleaning solution or rinse water. Strong solutions of soaps, alkalies, alkaline salts, and the too free use of water may darken wood and may in time soften it and raise the grain. Oak floors are readily darkened by strong alkaline solutions. Where wood is badly stained or discolored, bleaches (such as oxalic acid solution; oxalic acid is a poison if taken internally) may be used, or the floor may be scraped or machine-sanded.

Varnished and shellacked floors should be dusted clean with a soft brush or dry mop, and then rubbed with an oiled mop or a cloth slightly moistened with floor oil, kerosene, or furniture polish. In general, varnished and shellacked surfaces should not be treated with water, but if badly soiled they may be wiped with a mop or cloth wrung out of warm, slightly soapy water, then with a rag or mop moistened with clear water, wiped dry at once and polished with an oiled mop or cloth. The appearance of badly worn varnished wood may be improved by rubbing with a floor wax.

Oiled floors and painted floors should be swept with a soft brush and then rubbed with an oiled mop or cloth. Occasionally, they may be washed with slightly soapy water, rinsed off with a wet cloth or mop, wiped dry and then polished with an oil mop or cloth.

Waxed floors may be cleaned with a soft brush or mop free from oil since oil softens the wax. The film of dirt and wax which darkens the surface may be removed with a cloth wrung out of warm soapy water. The use of a rag moistened with gasoline or turpentine would be a better and more rapid procedure; however, these liquids are very inflammable and care should be taken to avoid having open flames in the rooms. Gasoline and turpentine brighten as well as clean the surface, whereas water dulls and whitens wax. If a water-cleaning method has already whitened a waxed floor, the luster and color may be restored by rubbing with a woolen cloth or a weighted brush; if necessary a little wax may be applied. Many kinds of spots on waxed floors may be removed by rubbing with a little turpentine or gasoline and refinishing with a very thin coat of wax. The entire coating of wax (and dirt) can be removed from wood floors by rubbing first with number 00 steel wool dipped in gasoline or turpentine and then with a soft cloth, after which the floor may be refinished.

Varnish or paint can be removed from a wood floor by scraping and planing, or by applying a paint and varnish remover. The first method, although tedious and laborious, is the better and is necessary if the floor has been stained. After a floor has been scraped, planed, and sandpapered, it can be finished as though it were new.

Removing paint or varnish from floors with paint and varnish removers must be done carefully so as not to damage the finish on baseboards and moldings. The commercial "solvent type" of paint and varnish removers are satisfactory for this purpose and are labeled with instructions for using. A solvent type of paint and varnish remover may be prepared as follows: Dissolve 3 parts of paraffin (in shavings) in 50 parts of benzol, then add 25 parts of denatured alcohol and 25 parts of acetone.

After this mixture has been applied to the surface with a brush and allowed to stand for a few minutes, the paint or varnish will be soft so that it can be scraped off with a putty knife or rubbed off with steel wool or excelsior. When a putty knife is used as a scraper it will prove more effective if the end of the blade is ground to a sharp edge. By holding the putty knife in a vertical position and scraping across the grain of the wood, there is no danger of splintering the floor. This paint and varnish remover and others of this type should be used only where there is good ventilation and no open flame of any kind, as they contain highly inflammable materials.

Caustic soda or household lye solutions are also used for removing paint and varnish, but should not be used on oak floors. These solutions should be handled with care and not allowed to come in contact with the skin, clothing or surfaces other than the one being treated. Rubber gloves should be worn. The caustic soda or lye may be dissolved in plain water and the solution applied while hot, but better results will be obtained if the caustic soda is mixed with a starch solution, such as is used in starching clothes. About 3 or 4 tablespoons of caustic soda is generally added to one quart of the starch solution. This mixture is applied while hot to the floor, using a cotton swab, a fiber (not bristle) brush, or a long-handled scrubbing brush. After a few minutes the softened paint or varnish may be scraped or rubbed off. The floor should then be washed several times with clear water, allowed to dry thoroughly, sandpapered or rubbed smooth, and dusted before it is refinished.

Strong, hot solutions of trisodium phosphate (2 or 3 pounds of the salt dissolved in one gallon of water) are also used for removing paint and varnish coatings. This chemical is safer to handle than caustic soda.

If shellac varnish alone has been used on a floor, it can be removed by flooding a small area at a time with denatured alcohol and, after a few minutes, rubbing with steel wool, or scraping as above.

In refinishing an old wood floor, it is first made as tight, level, and smooth as possible. It may need to be planed, sandpapered or rubbed down with steel wool. Any remnants of tacks must be drawn or driven below the surface. Then scrub the wood with hot water and soap or other detergent and rinse with clear water. If the wood is badly stained, spread over it a bleach solution made by dissolving a teaspoon of oxalic acid (poison) in a cup of hot water, and let stand overnight. Then thoroughly rinse the floor with clear water to remove all of the bleach (and cleaning agents) and let dry. The thoroughly dried floor may be stained, varnished, oiled, painted, or waxed. After applying the first coat of finish, it may be necessary to fill cracks and holes with a commercial "crack filler" colored to match the floors.

After cleaning and drying, wood floors are generally waxed. Floor wax should be applied in very thin coats and thoroughly rubbed with a heavy waxing brush or motor-driven brush, or a heavy block wrapped in burlap or carpet. In preparing a new or refinished floor for waxing, it is common practice to apply a coat of shellac varnish or other quick-drying varnish before waxing. If this is done it is better to have a very thin coating of shellac, as thicker coatings will, in time, crack or peel, which will necessitate complete refinishing. The wax can be applied directly to close-grained woods such as maple or pine, or to such open-grained wood as oak, if a "silicate wood filler" is first applied.

This treatment requires more waxing, and therefore more labor, in the original job, but the finish is likely to be more durable. However, floors finished in this way often darken more readily than if the wax is applied over a thin coat of shellac.

Floor Waxes or Polishes

In making either of the following waxes be very careful to heat only by setting the vessel containing the waxes in hot water and to have no flames in the room, since both gasoline and turpentine are very inflammable.

Formula No. 1

Carnauba Wax	2
Ceresin	2
Turpentine	3
Gasoline (sp. gr. about 0.73)	3

Melt the waxes by heating in a vessel placed in hot water, add the turpentine and gasoline, and cool the mixture as rapidly as possible, while vigorously stirring to produce a smooth creamy wax.

No. 2

Turpentine	1 pt.
Beeswax	4 oz.
Aqua Ammonia (10%)	3 fl. oz.
Water	1 pt.

Mix the beeswax and turpentine and heat them by placing the vessel in hot water until the beeswax dissolves. Remove the mixture from the source of heat, add the ammonia and the water, and stir vigorously until the mass becomes creamy. This wax should be applied lightly on varnished or shellacked floors and any excess wiped off at once, as the ammonia may attack the varnish or shellac. When this wax is used on unfinished oak flooring, the ammonia may cause a slight darkening of the wood.

The newer water-wax emulsions, commonly called non-rubbing liquid waxes, are now widely used on wood, cement, linoleum, rubber, tile, cork, asphalt tile, mastic, and other floorings. Many of these preparations dry rapidly and require little or no polishing. These water-wax emulsions usually consist of carnauba wax (and other waxes) dispersed in a water solution of soap. Sometimes emulsifying agents other than soap are used. A small amount of resins is often used in preparing these emulsions. Synthetic as well as natural products may be used in some of these preparations. A simple carnauba wax-soap-water emulsion may be prepared for *experimental* purposes as follows:

Dissolve 1 part by weight of castile soap in 16 parts of clean, soft water, and heat the solution to boiling. Add to the boiling soap solution with constant stirring 4 parts by weight of a good grade of carnauba wax (cut into small pieces). When a smooth homogeneous emulsion is obtained, cool to a temperature of 135° F. by quickly adding, with constant stirring, the necessary quantity of *cold* water. (This should take about 14 to 16 parts more of water.) Let cool, filter through cheese-cloth, and stir in about 0.5% of formaldehyde as a preservative. The product so obtained should be of the color and consistency of cream. A thicker or thinner product may be made by decreasing or increasing the quantity of water used, taking care to maintain the given ratio between soap and wax. This wax mixture may require polishing or buffing after drying in order to obtain a glossy surface.

Carnauba Wax	72.0	g.
Oleic Acid	9.1	cc.
Triethanolamine	10.6	cc.
Borax	5.4	g.
Water (Boiling)	500.0	cc.
Shellac (Dry Flakes)	10.0	g.
Ammonia (28%)	1.75	cc.
Water	100.0	cc.

1. Melt the wax and add the oleic acid. Temperature should be about 90° C. (194° F.). Placing the container in boiling water keeps the polish at a good temperature.

2. Add the triethanolamine slowly, stirring constantly. This should make a *clear* solution.

3. Dissolve the borax in about 5 cc. of the boiling water and add to (2). Stir for about 5 minutes. This gives a clear, jelly-like mass.

4. Add the rest of the boiling water, *slowly* with constant stirring. An opaque solution should be obtained. Cool.

5. Add the 100 cc. of room-temperature water to the shellac and then the ammonia and heat until the shellac is in solution. Cool.

6. Add the shellac solution to the wax solution and stir well. The resulting solution should give a clear film when applied to linoleum, mastic floors, etc., and one that is not too slippery.

Cement (Concrete) Floors

Unpainted cement floors may be scrubbed with hot water and a scouring powder, or with hot water and washing soda (laundry soda or "modified soda"), sodium metasilicate or trisodium phosphate followed by scouring powder. The floor should first be wetted with clear water and then with the hot solution of washing soda (about 2 to 2½ oz. per gallon of water), sodium metasilicate or trisodium phosphate (about ½ oz. per gallon of water), sprinkled uniformly with the scouring powder, rubbed or mopped, and then rinsed thoroughly with clear water to remove alkaline salts and scouring powder. The use of soap on unpainted or untreated cement floors is not recommended, as a scum of lime soap may be formed on or in the surface of the floor. Painted cement floors should be washed or mopped with plain water. If very dirty a slightly soapy water might be used, followed by thorough rinsing with clear water, but such treatment should not be used as a general or frequent procedure.

Occasionally cement or concrete floors are waxed. The waxes commonly used on wood floors can be used on painted or unpainted cement floors. As these waxes generally vary in color from yellow to brown, they should be used sparingly, as any wax or oil that may penetrate into the floor will tend to darken it. Waxes have been applied in a molten condition to unpainted or untreated cement floors as a special floor treatment, but this treatment is not in general use. Water-wax emulsions (such as a mixture of carnauba wax and resins dispersed in water) are used for polishing cement floors and preventing their "dusting." Wax treatments also make the floors water repellent. Concrete or cement surfaces with deposits of oil or grease, such as driveways and the floors of garages, shops and engine rooms, may be cleaned with sodium metasilicate powder (about 4 oz. per gallon of water), trisodium phosphate (about 4 oz. per gallon of water), or a mixture of 60% trisodium phosphate and 40% soda ash (about ½ to 1 lb. of the mixture per gallon of water). An abrasive powder (scouring powder) may be mixed with the above solutions. These detergents should be used with very hot water and the surface should be rubbed with a wire brush, or an abrasive powder and a mop. If the deposits are thick and of long standing, the powdered detergents may be sprinkled over them and moistened with a little water; after standing about ½ hour the surface should then be scrubbed, using very hot water. After scrubbing, the surfaces should in all cases be *thoroughly* rinsed with plain water. Instead of the foregoing treatment, the oil and grease can be mopped off with kerosene and the mopped areas covered with a layer of sawdust for a few days. After sweeping off the sawdust, the surfaces can be further cleaned with the above detergent solutions if necessary.

The following procedure has been found effective in removing old oil stains: First scrub the surface with a hot solution of trisodium phosphate (about 3 to 4 ounces per gallon of water), using an abrasive powder with the solution if there is a dark-colored film on the surface. Then mix whiting with some of the hot trisodium phosphate solution to form a thick paste; cover the stained area with the paste and leave until dry; scrape off the dried paste and rinse the surface with clear, hot water. Repeat this poultice treatment if necessary. In the case of badly soiled concrete or cement floors the appearance may also be improved by sprinkling over the scrubbed floor a layer (about ¼ inch thick) of *dry hydrated* lime, allowing to stand for several hours, and then removing the covering layer. In some cases it may be desirable to repeat the washing and the treatment with lime several times. Fine, dry coal ashes may be used instead of the hydrated lime. Solvents, such as carbon tetrachloride or a mixture of carbon tetrachloride (⅔) and gasoline (⅓) can be effectively used on the washed and dried floor in conjunction with the hydrated lime or ashes.

Vitreous Tile or Ceramic and Terrazzo Floors

It is good practice to clean these floorings periodically with a vacuum cleaner. The routine washing of these floors is usually carried out by first wetting them with clear water and then mopping with hot water containing a small quantity of an alkaline cleaner, such as washing soda (about 2 oz. per gallon of water), trisodium phosphate, or sodium metasilicate (about ½ oz. per gallon of water). Badly soiled areas on the floor may be cleaned with a scouring powder or a little scouring powder may be sprinkled over the soiled areas before applying the alkaline cleaning solution. Occasionally, the entire floor should be scrubbed with a scouring powder or with an alkaline cleaner and the scouring powder. A motor-driven scrubbing machine is a desirable appliance. After cleaning, the floors should be thoroughly rinsed with plain water and wiped dry. If water is left standing on a tile floor it may loosen the cement that holds the tiles in place. Soaps are not generally used on these floors owing to the tendency to "build up" slippery films, especially if the water is not soft or the rinsing has not been thorough. However, such floors are sometimes wiped up with a cloth wrung out of hot, soapy water, rinsed off, and wiped dry. If *soft* water is used and the surfaces are thoroughly rinsed after cleaning, it is believed that soap would be satisfactory, but more expensive, for the routine cleaning of these floors.

Marble and Travertine Floors

Travertine floors should be first cleaned with a vacuum cleaner and then treated the same as terrazzo floors. It is good practice to clean marble floors periodically with a vacuum cleaner.

1. Various cleaning preparations have been studied with a view of determining the effects on marble of certain ingredients from a long period of use. The laboratory experiments, as well as an examination of actual installations of marble, have indicated that injury may result from injudicious use of harsh grits or from such salts as sodium carbonate, sodium bicarbonate, and trisodium phosphate.

2. The usual type of grit employed in trade cleaning preparations is not appreciably injurious to marble floors or other unpolished marble. Polished marble should rarely be cleaned with preparations containing a scouring agent or abrasive which is harder than the marble.

3. As a rule, the volcanic ash grits are less severe in their abrading action than crushed quartz. This is evidently due to the difference in shape of the particles.

4. While it is seldom if ever necessary to use a cleaning preparation of the scouring type on polished marble, when it is in stock for cleaning the floors of a building it is apt to be wrongly used on the polished marblework. For this reason a preparation of the type is desirable which has a grit that will not injure polished marble. Available minerals which seem to meet this requirement are soapstone and talc.

5. A trial preparation consisting of 90% powdered soapstone and 10% soap powder appears to be as effective in cleaning marble floors as any of the present trade preparations. Such a composition can be used on polished marble without appreciable injury.

6. Injury which may result from the frequent use of such detergents as sodium carbonate, sodium bicarbonate, or trisodium phosphate is mainly a physical effect due to these salts crystallizing in the pores. This action has been demonstrated to be severe enough to cause disintegration of marble when such salts are employed without proper precautions.

7. Experiments have indicated that marble work may be safely cleaned with such detergents if the surface is rinsed with clear water before applying the cleaning solution.

8. Although soap has been found objectionable for use on marble in certain instances, nevertheless if used with soft water it will give entirely satisfactory results and prove to be the safest detergent for general service.

9. Preparations containing a coloring ingredient of different color than the marble may gradually impart their color to the marble. This, however, may be prevented by a preliminary rinsing, as described in **7**.

10. Ammonia water has been used to some extent in cleaning polished marble, but a limited number of tests in this investigation have indicated that it may cause yellow discolorations.

11. Acids dissolve marble, and even the use of such weak acids as oxalic will prove injurious. Although cleaning of interior marble with acids has been practiced to some extent, it is usually done through ignorance of the real effects.

12. Stains which have penetrated the marble usually have to be removed by means of a poultice treatment. Several types of stain demand special treatment, and there is no single cure for all cases. Methods have been found for eradicating practically all of the common stains occurring on interior marble.

Linoleum Floorings

Untreated floorings should be swept daily with a soft floor brush or an oil-treated mop. Anything spilled on the flooring should be wiped up as soon as possible with a damp cloth; and occasionally, as the flooring needs it, it should be washed. Care should be exercised in washing these surfaces. Preparations containing free alkali, alkaline salts, or abrasives should not be used. The safest procedure is to use a lukewarm solution (soft water) of a mild or neutral soap and to rinse all soapy water off with plain soft water after washing. The surfaces should be finally wiped dry with mops or cloths. Care should be taken not to flood the surfaces with water, since any water that seeps through the edges of seams may affect the cementing material and may cause the burlap backing to mildew or rot. Linoleum in kitchens, pantries, or entries will look brighter and wear better if the clean, dry, unwaxed surface is given an occasional coat of a pale, quick-drying lacquer ("Linoleum Lacquer"). Never lacquer over wax. The lacquered surface after drying may be waxed, but this is not generally done. The lacquered surface may be cleaned daily with a dust mop or, whenever required, with a damp cloth. These floors should be relacquered occasionally, depending upon the severity of wear.

In rooms or places where the wear on the floors is not particularly heavy, the clean, dry linoleum should be waxed and polished. Any good floor wax is suitable for linoleum. A wax prepared according to formula (1) is satisfactory. Paste wax can be used but "liquid" wax is easier and more economical to apply. The water-wax emulsions are also used. Care should be taken not to put the wax on too thickly as it is likely to smear and give a greasy appearance to the floor. Too much wax will cause the floor to be slippery. After applying the wax, it should be polished for some time with a weighted floor brush or an electric polishing machine. Daily care of a well-waxed and polished floor should consist in going over it with a dry dust-mop. Washing is seldom necessary—perhaps two or three times a year. Waxing not only adds to the appearance of the linoleum but provides a wearing surface and protects the floor. An occasional polishing may be necessary on the main traveled areas. Rewaxing will be required from time to time depending upon the amount of traffic. Depending upon the service conditions, the floor may require scrubbing at times, after which a coat of new wax should be applied. "Linoleum tile" floors and linoleum floors subjected to much wear should be given one or two coats of a linoleum lacquer and then waxed and polished. After the wax has dried for a short time on the "tile" floors, run a polishing machine equipped with a fine-bristled brush over the floor in both directions in order to work the wax into the surface. Sweep off any dust or wax particles and polish with an electric polishing machine. Lacquered and waxed floors should be cared for as outlined for surfaces that have been waxed directly. Before lacquering an old linoleum or "linoleum tile" floor, it must be *thoroughly* cleaned with a cloth dampened with gasoline and frequently squeezed out in fresh gasoline during the cleaning, in order to remove all grease and wax. Then the entire floor should be scrubbed with lukewarm soapy water, rinsed thoroughly with plain water, and allowed to dry.

The life of a linoleum floor covering can be prolonged by applying a pale, transparent varnish but this treatment is not recommended, as many varnish coatings become slightly yellow. This mars the design effect in patterns and may produce a decided discoloration over plain surfaces. Some varnishes will turn white when water is spilled on them. The same objections may apply to shellac. Floor oils and sweeping compounds containing oil should not be used on linoleum, as these materials may leave a film of oil on the surface to collect dust and dirt. Varnished surfaces can be waxed to produce a "dull polish."

Rubber Floors

To clean unpolished floors, brush off loose dirt with a soft push-broom and wash a small section of the floor with a clean mop wrung out of a solution of washing soda ("modified soda" or laundry soda) or trisodium phosphate (about a quarter of a cupful of the cleaner dissolved in 12 to 16 quarts of clear, cold water). Cleaning may also be done by mopping with clear, cold water containing 2 to 4 ounces of ordinary household ammonia per gallon. Rinse the mop in a second pail of clear, cold water, wring the mop, and wipe the section of floor clean of solution. Continue this process until the entire floor is cleaned. After the floor has dried, buff it thoroughly with a rotary electric buffing machine (for large areas) or a weighted hand buffer to which a piece of rough carpet or similar material is attached as a buffing surface. Daily cleaning can often be satisfactorily done by sweeping with a soft, dry brush, or with a soft push-broom, and an occasional washing with a clean mop wrung out of clear, cold water. When very dirty, the floors should be cleaned with a washing solution as outlined above. Frequent systematic buffing of unpolished floors materially reduces the number of washings required. Owing to the development of the bright-drying wax-water emulsions (polishes) free from oils, fats and organic solvents, it is now general practice to polish rubber floors. The wax or polish enhances the appearance of the floors and prolongs their life. After cleaning, drying, and buffing as outlined above, the floor is ready to polish if all dirt and marks have been removed. Pour the polish into a shallow receptacle, dip the applicator * into the polish and apply a thin coat over a small area with a wiping motion. Do not rub hard. Repeat until the entire floor is covered. Let dry until hard (about 30 minutes) and then buff. Immediately apply a second thin coat, let dry, and again buff thoroughly. The polish should be applied in as thin coats as possible to avoid streaking and at least two coats should be applied. If the polish wears off in certain sections of the floor, clean and repolish only these sections rather than the entire floor. Systematic buffing keeps the polished floor in good condition and reduces the number of washings required. When the polished floor becomes soiled, remove the loose dirt with a soft brush or a soft push-broom and wipe the floor with a clean mop dampened with clear, cold water. This treatment should not remove the polish. If this procedure does not clean the floor, it should be treated as outlined above, using a washing solution. When necessary, the wax or polish can be removed from rubber floors with a solution of trisodium phosphate in *warm* water (about 2 ounces per gallon) with the aid of 00 steel wool. The addition of a small quantity of household ammonia to the solution will hasten the removal of the wax (and dirt).

Stains may be removed from rubber floors by rubbing the stained area with a fine abrasive powder or with number 00 steel wool. If this is ineffective, the spot may be rubbed carefully with a clean cloth dampened with acetone (inflammable), gasoline (flammable), or carbon tetrachloride (non-inflammable). Gasoline, acetone, and carbon tetrachloride have a softening action on rubber if in contact with it long, but the softening is not permanent.

Cleaning materials containing oil (certain sweeping compounds and other detergents) and coarse abrasives, or caustic alkali should not be used on rubber floors. Soap may soften and swell rubber flooring. Although this effect is minimized with careful rinsing, the safest procedure is to avoid the use of any soap. Waxes or polishes containing oils, fats, or organic solvents should not be used. Do not use hot water or excessive amounts of water when cleaning rubber floors. The floor should not be flooded with water. Avoid using more of the cleaning compound than specified and thoroughly mop or rinse the surface in order to remove all of the cleaning solution. Rubber floors should not be varnished. Buffing machines should not be used for scrubbing with water and cleaning agents. The advice or recommendation of the manufacturer of the rubber flooring should be secured before using un-

* Made of lamb's wool, soft absorbent cloth, or felt. The felt applicator, made of 4 or 5 strips of felt ½ inch thick stood on edge and bound together, is very satisfactory.

known cleaning preparations or applying untried methods.

Cork Tile and Cork Carpet Floorings

Many cork floorings are installed without any surface treatment other than sanding to a smooth surface. These floorings are commonly referred to as "natural" cork and often may be cleaned by dry sweeping with a hair floor brush. The entire surface is then buffed or polished with suitable pads, a polishing machine or floor-polishing brush. Care should be taken to overlap the polished sections. If the floor can not be cleaned in this manner, it should be swept with a soft brush and then mopped with a *lukewarm* soapy solution made with a mild or neutral soap and clean, soft water. A separate container of clean, lukewarm, soft water and a separate mop should be used for rinsing. The rinse water should be changed frequently, so that both rinsing water and mop are always clean. Only a small area (say about 50 sq. ft.) of flooring should be washed at one time and it should be thoroughly rinsed with clean water and wiped dry. Water should not be left on the floor.

Where cork flooring is subjected to much heavy and dirty traffic, it should be varnished with a pale, transparent spar varnish, and then waxed and polished. With cork tile it is good practice to apply a filler before varnishing. The varnish usually requires overnight to dry. A dull or mat finish may be secured in place of the glossy varnish by scrubbing the varnished floor with a neutral soap and water and rottenstone or finely powdered pumice. These treatments form protective coatings, but the treated floors require careful and frequent attention. The floors should be washed clean and allowed to dry thoroughly before applying filler, varnish or wax. The waxes used for linoleum may be used on cork floors. Liquid floor wax is generally used and is applied in a thin, even coat and rubbed in. When the wax is nearly dry (tacky), apply a second coat of wax in the same manner. When the second coat becomes tacky, polish thoroughly with a polishing machine, a weighted brush, or a clean soft cloth. The appearance of the surface improves with frequent polishing. The water-wax emulsions that dry rapidly and require little or no polishing are also widely used for waxing cork floors.

Waxed cork floors are cleaned by rubbing with a dry mop or polishing brush, followed by sweeping with a brush or a vacuum cleaner.

A polishing machine is a useful appliance for large areas. Cork floors should be polished or buffed with a brush or machine whenever they appear dull or dingy. The floors should be rewaxed from time to time, depending on traffic conditions and exposure to dirt. The floors should be rewaxed before the old wax is worn off. Areas (doorways, traffic lanes, etc.) subjected to the most wear may be rewaxed when necessary without going over the entire floor. Before rewaxing the floor all old wax must be removed by rubbing the surface with a cloth dampened with gasoline (inflammable), followed by vigorous scrubbing with warm water and soap, rinsing with clean water and drying. If this does not remove the wax, the dried surface should be resanded.

Lacquer suitable for linoleum may be used on cork floors in lieu of waxing. The surface should be thoroughly cleaned and dried before applying the lacquer. Two or three coats of lacquer are used for the first treatment in order to fill the pores of the flooring and give a smooth, dirt-resistant surface. Further coats may be applied as the appearance requires. "Natural" cork floors which have been varnished may be kept clean by sweeping with a dust-mop, soft brush, or a damp cloth or mop.

Stains and spots may be removed from cork floors by rubbing with fine emery paper or number 00 steel wool. In some cases the spot or stain may be rubbed with a cloth dampened with acetone (inflammable) or with carbon tetrachloride. The cleaned areas should then be buffed and waxed, or varnished and then waxed.

Asphalt Tile and Mastic Floors

These "soft composition" floors are sold under various trade names and, in general, have a base of asphalt, bitumen or resin. Cleaners and polishes containing abrasives, oils or organic solvents (gasoline, turpentine, carbon tetrachloride, etc.) should not be used. These floors should be washed by mopping with a neutral soap and lukewarm soft water. Scrubbing machines with soft polishing brushes have been used for large areas. After cleaning and drying, these floorings (especially the asphalt tile) are generally waxed in order to cover the surface with a protective film. The water-emulsion waxes free from oils and volatile organic solvents are the safest waxes to use on these floorings. This type of wax can be applied with a cotton cloth mop or wool applicator. The wax should be spread as thinly as possible on the surface of the floor, using the mop or applicator in one direction only. In a short time the wax should dry to a hard, lustrous finish.

Asphalt tile floors should not be buffed or burnished until the wax or other treatment is completely dry. The treated floors may be maintained by sweeping with a brush, a dry mop, or by buffing with mechanical buffers. Scrubbing with water and a neutral soap may be required at times (probably two or three times a year). Oils, soaps or other detergents containing abrasives, and sweeping compounds containing free oil should not be used on the untreated or the treated floors. The floor treatments should be renewed at intervals, depending upon the severity of wear.

Before treating an asphalt tile floor with an unknown preparation, moisten a white cloth with the preparation and rub over the surface of one tile. If the color of the tile shows on the cloth it indicates that the solvent in the preparation has dissolved part of the surface of the tile and shows conclusively that the preparation would not be safe to use. It would be safer to get the advice of the manufacturer of the flooring before using unknown cleaning preparations.

Plastic Magnesia Cement Floors

These floors contain magnesium oxychloride as the cementing material and are known by various names ("Sorel Cement," "Hard Composition," "Magnesite," "Woodstone," etc.). These floors may be cleaned by first wetting with clear water and then mopping with a neutral soap and warm soft water (a soft soap, such as a linseed oil-potash soap is often used). After cleaning and drying, the floors may be given a protective coating of wax (water-wax emulsion) or of special varnish or lacquer. The treated floor may be cleaned by sweeping with a dry mop or soft brush, and occasional washing, and buffing with a weighted brush or polishing machine. The waxing or other treatment is renewed as conditions require.

Slate Tile Floors

These floors may be cleaned by mopping with a neutral soap (a paste soap is often used) and warm soft water. After cleaning and drying, the floors may be waxed with a water-wax emulsion or given a coat of pale varnish or lacquer. The treated floor can be kept in good condition by sweeping with a dry mop or soft brush, and an occasional buffing with a weighted brush or polishing machine.

Cheap Abrasive Wheel
U. S. Patent 2,124,279

Abrasive	40
Iron Filings	40
Ammonium Chloride	2–10

Add enough water to wet mass thoroughly; press and allow to set until cold.

Bonded Abrasive Wheels

The use of a good grade of animal hide glue is essential. Bone glue, fish glue, and cold glue preparations are relatively inefficient.

Mix by weight only and never heat more than a three hours' supply at any time. For average conditions in the polishing room the following table of mixtures should act as a guide for setting up wheels.

Size of Alundum Grain	% Glue	% Water
24— 36	50	50
46— 54	45	55
60— 70	40	60
80— 90	35	65
100—120	33	67
150—180	30	70
220—240	25	75

Soaking the glue allows it to dissolve more readily on heating. Use cold pure water and soak:

Ground glue—One hour or more
Flake glue—Six hours or more
Cake glue—Twelve hours or more

The glue should be melted in a water-jacketed heater. When the wheels and grain are preheated, apply the glue at a temperature of 140° F. With wheels and grain at room temperature, use glue at 160° F. Keep a thermometer in the glue-pot as a constant check on temperature, even though the pot is regulated by a thermostat.

It is important that all equipment used be kept clean.

Composition for Impregnating Grinding Wheels
U. S. Patent 2,240,302

Hydrogenated Cottonseed Oil	90 --97 %
Salt	0.75– 7.5%
Sulphur	0.75– 7.5%
Turpentine	0.75– 7.5%

Removing Scratches from Glass

Scratches on glass may be eradicated by using a paste made of glycerin, water and rouge (iron oxide) mixed to the desired consistency. A hard felt pad is dipped in this paste and rubbed briskly back and forth over the scratched surface until the markings disappear. The paste can be washed away by simply flushing with water. This paste is particularly suitable for removing shallow scratches. Deeper gougings require more specialized treatment and coarser abrasives to start with. For this latter purpose emery powder will often serve, and glycerin makes a satisfactory medium in which to suspend the powdered abrasive.

Dental Abrasive Paste

Carborundum (100–200 Mesh)	70
Glycerin	30

Lapping Compound for Brass or Bronze

Silica (325 Mesh)	16.5
Soap Flakes, White	2.5
Sodium Silicate	1.0
Water	80.0

Chapter Seventeen

PYROTECHNICS AND EXPLOSIVES

Meal Powder

Mount a 50 gallon wood barrel on two uprights so that it will revolve freely on centers fastened to the heads. On one center attach a crank and cut a hole (closed by a suitable plug) into side of barrel for putting in and removing the necessary ingredients. Place in the barrel 300 to 500 lead balls about one inch in diameter. When it is desired to make meal powder, put into the barrel a thoroughly mixed composition as follows:

Saltpeter, Double Refined	15 lb.
Willow Charcoal	3 lb.
Sulphur, Flour	2 lb.

The barrel is now revolved for about 500 turns. The longer it is turned, the stronger the powder will become. Great care must be exercised to see that no foreign matter such as nails, gravel, etc. find their way into the barrel as this might result in an explosion.

Green Parade Torches

Barium Chlorate	..	..	5
Barium Nitrate	40	30	4
Potassium Chlorate	11	..	..
Potassium Perchlorate	..	6	..
K. D. Gum	6	2	..
Sulphur, Ground	..	3	..
Sal Ammoniac	1	..	..
Shellac	..	..	1
Calomel	..	..	2

Blue Parade Torches

Potassium Perchlorate	5	24	24
Paris Green	2	..	..
Copper Ammonium Sulphate	..	6	..
Copper Ammonium Chloride	..	..	6
Dextrin	1	..	..
Calomel	1	..	..
Sugar of Milk	..	2	..
Sulphur	..	9	..
Stearine	..	..	2
Asphaltum	..	..	1

Purple Parade Torches

Strontium Nitrate	7
Potassium Perchlorate	9
Black Oxide of Copper	6
Calomel	3
Sulphur	5

Amber Parade Torches

Strontium Nitrate	36
Sodium Oxalate	8
Shellac	5
Sulphur	3
Potassium Perchlorate	10

Aluminum Torches

Potassium Perchlorate	13
Fine Aluminum Powder	6
Flake Aluminum	5
Dextrine or Lycopodium	1

A beautiful modification to this is the

Red and Aluminum Torch

This should be ⅞″ diameter, 18″ long and of the following composition:

Formula No. 1

Strontium Nitrate	35
Potassium Perchlorate	7
Shellac	4
Coarse Flake Aluminum	4
Lycopodium	1

No. 2

Strontium Nitrate	13
Sulphur	3
Mixed Aluminum	3

Before ramming, this formula should be moistened with a solution of 1 part shellac in 16 parts of alcohol and 1 part of this solution used to every 36 parts of composition. As this mixture is somewhat difficult to ignite it is necessary to scoop out a little from the top of torch and replace it with starting fire.

Starting Fire

Saltpeter	6
Sulphur Flowers	4
Fine Charcoal	1

Extra Bright Torch

An aluminum torch of heretofore unheard of brilliance and giving an illumination, in the 1 inch diameter size, of what is said to be 100,000 candle power is made as follows:

Barium Nitrate	38
Mixed Aluminum	9
Sulphur	2
Vaseline	1

Rub the Vaseline into the barium nitrate; mix sulphur and aluminum separately; then mix with barium nitrate and vaseline. A starting fire for this also is required, as follows:

Starting Fire

Barium Nitrate	4
Saltpeter	3
Sulphur	1
Shellac	1

Port Fires

These are small torches ⅜″ diameter, 12″ long, used in exhibitions for lighting other pieces of fireworks. They are rammed with rod and funnel and a good mixing is:

	Formula No. 1	No. 2
Meal Powder	1	..
Sulphur	2	4
Saltpeter	5	5
Charcoal	..	1

White Fire

	Formula No. 1	No. 2	No. 3	No. 4
Saltpeter	3	12	8	7
Sulphur	1	2	2	2
Metallic Antimony	1	..	..	..
Antimony Sulphide	1	1	..	..
Realgar	..	..	1	1½

Blue Fire

	Formula No. 1	No. 2	No. 3	No. 4
Potassium Perchlorate	..	..	..	24
Potassium Chlorate	6	16	8	..
Paris Green	4	..	6	..
Shellac	..	..	½	..
Stearine	1	..	1	2
Barium Nitrate	4	..	7	..
Calomel	..	12	1	..
Sal Ammoniac	1	..	..	..
Copper Ammonium Chloride	..	4	..	6
Asphaltum	..	..	..	1
Lactose	..	6	..	..

Red Fire

	Formula No. 1	No. 2	No. 3	No. 4
Strontium Nitrate	80	10	16	14
Potassium Chlorate	20	4	8	4
Shellac	..	..	3	..
Red or Kauri Gum	12	3	..	..
Asphaltum	..	..	..	3
Charcoal	..	1	..	..
Dextrin	1	..	..	..
Fine Saw Dust	12	..	..	..
Rosin	..	1	..	..
Lampblack	1	..	..	..

Pink Fire

	Formula No. 1	No. 2	No. 3
Strontium Nitrate	48	16	18
Saltpeter	12	4	7
Sulphur	5	2	2
Charcoal	4	1	½
Red Gum	..	3	2
Dextrin	..	..	½

Yellow Fire

Barium Nitrate	36
Sodium Oxalate	6
Sulphur	3
Red Gum	5

Green Fire

	Formula No. 1	No. 2	No. 3
Barium Nitrate	8	9	4
Chlorate of Potash	4	3	2
Shellac	..	1	1½
Red Gum	2	..	..
Dextrin	..	1/16	..
Fine Saw Dust	..	½	..
Sal Ammoniac	1	..	..

Smokeless Tableau Fire

For theatrical or indoor use, colored fires are very objectionable on account of the choking smoke they give off. The following mixings give a fire, producing very little smoke, which quickly dissipates after fire is burned.

Red

Strontium Nitrate	8
Picric Acid	5
Charcoal	2
Shellac	1

Green

Barium Nitrate	4
Picric Acid	2
Charcoal	1

Railway Fuses

	Formula No. 1	No. 2	No. 3	No. 4
Potassium Perchlorate	..	..	..	2½
Strontium Nitrate	48	18	16	18
Saltpeter	12	7	4	..
Sulphur	5	2	5	2¼
Fine Charcoal	4	½	1	..
Red Gum	10	2	..	..
Dextrin	..	½	..	..
Sawdust	..	..	..	1

Moisten with kerosene before ramming.

Fuses are provided with a slip cap which is used for igniting them. The end of the torch is capped with paper onto which is painted a mixture of

Potassium Chlorate	6
Antimony Sulphide	2
Glue	1

while the end of the cap is similarly painted with a paste of

Black Oxide of Manganese	8
Amorphous Phosphorus	10
Glue	3

Red

Strontium Nitrate	30
Potassium Chlorate	8
Red Gum	7

Saw-dust may be added as needed. The torches are usually ¾″ diameter and 12″ long and should burn, with the above mixing, 8 to 10 minutes.

Bengola or Blue Light

Saltpeter	12
Sulphur	2
Antimony Sulphide	1

Distress signals are the same except that they burn red. The regulation Life Boat equipment consists of 6 to 12 enclosed in a water tight copper can. The following formula is suitable:

Distress Signal (Red)

Potassium Chlorate	5
Strontium Carbonate	1½
Shellac	1
Dextrin	½

Roman Candle Composition

Powdered Saltpeter	18 lb.
Fine Powdered Charcoal	11 lb.
Flowers of Sulphur	6 lb.
Dextrin	1 lb.
Water	1 gal.

Sky-Rockets

	1 to 3 oz.	4 to 8 oz.	1 to 3 lb.	4 to 8 lb.
Saltpeter	18	16	16	18
Mixed Coal	10	9	12	12
Sulphur	3	4	3	3

Tourbillions

(*Geysers, Whirlwinds, Table Rockets*)

This is a modification of the sky rocket and ascends to a height of about 100 feet, in a spiral manner and without a stick. They are made by ramming a 3 lb. rocket case with one of the following mixtures:

	Formula No. 1	No. 2
Saltpeter	8	5
Meal Powder	7	12
Charcoal	2	3
Sulphur	2	3
Steel Filings	3	..

Pin Wheels

Meal Powder	..	10	8	2
Fine Grain Powder	8	5	8	..
Aluminum	..	..	3	..
Saltpeter	14	4	16	1
Steel Filings	6	6	..	..
Sulphur	4	1	3	1
Charcoal	3	1	8	..

Serpents or Nigger Chasers (*Squibs*)

Meal Powder	3	3
Saltpeter	2	5
Sulphur	1	1
Mixed Coal	1½	¾
FFF Grain Powder	4	3

Starting Fire for Gerbs

Meal Powder	4
Saltpeter	2
Sulphur	1
Charcoal	1

Gerbs

Meal Powder	6	4
Saltpeter	2	..
Sulphur	1	..
Charcoal	1	1
Steel Filings	1	2

Fountains

Meal Powder	5
Granulated Saltpeter	3
Sulphur	1
Coarse Charcoal	1
FF Rifle Powder	¾

Flower Pots

Saltpeter	10
Sulphur	6
Lampblack	3
FFF Rifle Powder	6

Cascades

	1½″ case	2″ case
Granulated Saltpeter	18	16
Mixed Charcoal	4	4
Sulphur	3	3
Iron Borings	6	7

Triangle Composition

Saltpeter	18	12
Sulphur	2	8
Mixed Charcoal	5	5
Rifle Powder FFF	12	12

Wheel Cases (Drivers)

Meal Powder	8	3
Saltpeter	3	2
Sulphur	1	1
Mixed Charcoal	1	1
F Rifle Powder	1	..
Lampblack	½	..
Steel Filings, To suit		

Saxons

Meal Powder	4
Sulphur	2
Saltpeter	2
Mixed Charcoal	1

Japanese Stars

	Formula No. 1	No. 2
Lampblack	12 oz.	6 oz.
Potassium Chlorate	8 oz.	4 oz.
Saltpeter	1 oz.	..
Water	18 oz.	9 oz.
Alcohol	4 oz.	2 oz.
Dextrin	1 oz.	..
Gum Arabic	..	½ oz.

Mix the dextrin and saltpeter (formula 1) well together and add sufficient water to make a gummy liquid. Boil the balance of the water and add the chlorate of potassium to it. Put the lampblack in a large pan and pour the alcohol over it working it in as well as possible. Now add the chlorate of potassium dissolved in the hot water and stir with a stick until cool enough for the hands. Lastly add the dextrin and saltpeter. Remember that you cannot mix it too well and the effect will be in proportion to the evenness with which this has been done.

White Stars

	Formula No. 1	No. 2
Saltpeter	50	54
Sulphur	15	15

Red Arsenic	15	9
Dextrin	3	3
Black Antimony	..	15
Red Lead	..	6
Shellac	..	1

Red Stars

	Formula No. 1	No. 2
Potassium Chlorate	6	24
Shellac or Red Gum	1	3
Fine Charcoal	2	4
Strontium Carbonate	..	4
Strontium Nitrate	6	..
Dextrin	½	1½

Blue Stars

Potassium Chlorate	24
Paris Green	9
Barium Nitrate	8
Shellac	5
Dextrin	1½

Green Stars

Potassium Chlorate	6
Barium Nitrate	6
Fine Charcoal	2
Shellac or KD Gum	1
Dextrin	½
Calomel	½

Yellow Stars

	Formula No. 1	No. 2
Potassium Chlorate	16	16
Shellac or Red Gum	3	3
Fine Charcoal	4	1
Barium Nitrate	6	..
Sodium Oxalate	1	7
Dextrin	1½	1

Exhibition Pumped Stars

Green (Not for Shells)

Barium Chlorate	12
Potassium Chlorate	8
Calomel	6
Shellac	2
Picric Acid	2
Lampblack	1½
Dextrin	½

Red

(For hand pumps, not suitable for Shells.)

Strontium Nitrate	8	..
Potassium Chlorate	4	10
Picric Acid	1½	1½
Shellac	1½	¾
Fine Charcoal	1	1
Dextrin	½	¾
Strontium Carbonate	..	3

Exhibition Blue Stars (Pumped)

Potassium Chlorate	48	18	16
Calomel	18	6	12
Black Copper Oxide	6	..	..
Asphaltum	6	..	..
Dextrin	1½	1	..
Paris Green	..	4	..
Stearine	..	2	..
Copper Ammonium Chloride	..	..	4
Lactose	..	..	6

Box Stars

Red

Strontium Nitrate	3
Potassium Chlorate	3
Shellac	1
Dextrin	¼

Green

Barium Nitrate	3
Potassium Chlorate	4
Shellac	1
Dextrin	¼
Barium Chlorate	9

Yellow Twinklers

Potassium Chlorate	8
Lampblack	12
Stearine	1½
Saltpeter	1

Electric Spreader Stars

Zinc Dust	36
Potassium Chlorate	7½
Granulated Coal	6
Potassium Bichromate	6
Dextrin	1

Granite Stars

Saltpeter	14
Zinc Dust	40
Fine Charcoal	7
Sulphur	2½
Dextrin	1

Gold and Silver Rain (Cut Stars)

	Formula No. 1	No. 2	No. 3
Meal Powder	16	..	4
Saltpeter	10	1	1
Sulphur	10	1	..
Fine Charcoal	4	1	2
Lampblack	2	..	..
Red Arsenic	1	..	..
Shellac	1	..	..
Dextrin	1	..	..
Lead Nitrate	..	3	..

Aluminum Stars (Box Stars Only)

	Formula No. 1	No. 2
Potassium Chlorate	..	8
Potassium Perchlorate	8	..
Aluminum Powder, Medium	4	4
Shellac	..	1
Lycopodium	1	..

Magnesium Stars

Saltpeter	5
Magnesium Powder	2

Comet Star Composition

	Formula No. 1	No. 2
Saltpeter	6	..
Meal Powder	6	3
Sulphur	1	..

Fine Charcoal	3	1
Powdered Antimony	3	1
Lampblack	..	2

Lance Compositions

Red Lances

	Formula No. 1	No. 2
Potassium Chlorate	16	16
Strontium Nitrate	3	..
Strontium Carbonate	..	3
Shellac	3	2
Lampblack	1/8	1/4

Green Lances

	Formula No. 1	No. 2	No. 3
Potassium Chlorate	7	16	..
Barium Nitrate	7	4	4
Barium Chlorate	..	..	5
Shellac	2	4	1
Calomel	..	3	2
Lampblack	..	1/8	..
Picric Acid	..	1	..

White Lances

	Formula No. 1	No. 2	No. 3	No. 4
Saltpeter	9	14	5	8
Sulphur	1	4	2	2
Antimony Sulphide	2	..	..	..
Antimony, Metallic	..	3	1	..
Meal Powder	..	..	1	..
Red Arsenic	..	..	..	1

Blue Lances

	Formula No. 1	No. 2	No. 3	No. 4
Potassium Chlorate	20	16	12	..
Potassium Perchlorate	..	..	..	24
Paris Green	..	5	..	..
Copper Sulphate	6	..	..	..
Copper Ammonium Sulphate	..	..	3	..
Copper Ammonium Chloride	..	..	..	6
Shellac	4	..	1	..
Stearine	..	1½	½	2
Calomel	4	3	3	..
Dextrin	1	..	..	..
Asphaltum	..	..	..	1

Yellow Lances

	Formula No. 1	No. 2	No. 3
Potassium Chlorate	16	4	4
Sodium Oxalate	2	2	2
Shellac	3	1	1
Charcoal	½	..	..
Barium Nitrate	..	..	1

Cannon Cracker Composition

	Formula No. 1	No. 2	No. 3
Potassium Chlorate	60	6	6
Washed Sulphur	23	3	2
Sulphuret of Antimony	5	..	..
Metallic Antimony	..	..	1
Charcoal	..	1	..
Saltpeter	12	..	..

Chinese Fire Crackers

	Formula No. 1	No. 2
Saltpeter	50	45
Sulphur	25	18
Charcoal	25	25
Chlorate, Potassium	..	8
Sand, Fine	..	4

Flash Crackers

	Formula No. 1	No. 2	No. 3
Saltpeter	50	..	..
Sulphur	30	25	30
Aluminum Powder, Fine	20	25	40
Chlorate, Potassium	..	50	30

A very important as well as extremely difficult part of the Chinese cracker is to make the fuse. Very tender and skilled fingers are required to produce this insignificant looking yet most requisite adjunct. A thin strip of the finest Chinese tissue paper, about ¾″ wide and 14″ long is laid on a smooth damp board; a little stream of powder is poured down its center from a hollow bamboo stick and with the tips of soft skinned fingers which seem to have an attraction for the paper and placed against the right hand lower corner, a rolling motion in the general direction of the upper left hand corner causes the paper to roll up into a twine like fuse. The slightest touch of paste secures the end and prevents unrolling. When dry it is cut into the required lengths and is ready for use

Flash Crackers

	Formula No. 1	No. 2	No. 3
Potassium Perchlorate	..	50	30
Potassium Chlorate	6	..	..
Washed Sulphur	3	25	30
Pyro Aluminum	1	25	40
Charcoal	1	..	..

Magic Serpent (Black)

Naphtha Pitch	10
Linseed Oil	2
Fuming Nitric Acid	7
Picric Acid	3½

Reduce pitch to fine powder; add linseed oil and rub in well; add strongest fuming nitric acid, little at a time. Allow to cool for one hour. Wash several times with water, the last time allowing mass to stand in the water for several hours. Thoroughly dry; powder fine and add picric acid, rubbing it in well. Moisten with gum arabic water and form into pellets about the size of a No. 4 star.

Ruby and Emerald Shower Sticks

Strontium Nitrate	6
Coarse Aluminum	6
Potassium Perchlorate	2
Shellac	1

Dissolve shellac in alcohol and add other ingredients, previously well mixed. Stir thoroughly to consistency of thick glue and dip sticks, previously arranged in holder, so they may be placed in drying rack.

Steel Sparkler
Formula No. 1

Fine Steel Filings	12
Fine Aluminum Powder	1
Potassium Perchlorate	6
Dextrin or Gum Arabic	2
Water, To suit	

The steel must be protected from corrosion with paraffin. The gum should be made of the consistency of mucilage. Mix the ingredients thoroughly and add gum solution until a mixture is obtained that will adhere to the wires when they are dipped into it. This varies in different sections and with different runs of ingredients. In practice, bunches of wires are dipped at once and slowly withdrawn in a current of warm, dry air which causes the mixture to adhere evenly.

No. 2

A sparkler of great brilliance and which is very effective may be made as follows: Take 3 lb. dextrin and add to same, little at a time, 12 pt. of water, stirring continually so as to avoid lumps. Mix intimately 10 lb. potassium perchlorate with 7 lb. pyro-aluminum or finely powdered aluminum and add this to the gum water, stirring until a perfectly smooth mixture is obtained. Wood sticks may now be dipped into it to the desired depth while it is contained in a deep vessel, and placed in a suitable rack for drying. It may be necessary to dip the sticks several times dependent on how much composition it is desired to have on them. In this case they should be dried with the composition end up, the first time so that not too much composition accumulates on the end beyond the stick.

Spark Pot

Meal Powder	2
Fine Charcoal	1
Sawdust	1

Smoke

Formula No. 1

Saltpeter	4
Lampblack	1
Charcoal	1
Realgar	1
Rosin	1

No. 2

Saltpeter	12
Pitch	8
Borax	2½
Chalk	1¼
Sand	1
Sulphur	1

White Smoke

Formula No. 1

Potassium Chlorate	3
Lactose	1
Sal Ammoniac, Finely Powdered	3

No. 2

Sulphur Flowers	16
Saltpeter	12
Fine Charcoal	1

Black Smoke

Hexachloroethane	24
Alpha Naphthylamine	6
Anthracene	2
Aluminum Powder	4
Roman Candle Composition	6

Red Smoke (Bright)

Potassium Chlorate	1
Lactose	1
Paranitraniline Red	3

Brown-Yellow

Potassium Nitrate	45
Sulphur	12
Pitch	30
Borax	9
Glue	4

Waterproof Flare Composition

U. S. Patent 2,123,201

Tin Stearate	0.5–10%
Magnesium, Powdered	5 –25%
Barium Peroxide	94.5–65%

Mix cautiously.

Military Smoke Screens

Formula No. 1

Zinc Dust	25
Carbon Tetrachloride	50
Zinc Oxide	20
Kieselguhr	5

No. 2

Zinc Dust	28
Hexachlorethane	50
Zinc Oxide	22

Non-Poisonous Safety Match Composition

U. S. Patent 2,103,698

Reduced Iron	28
Potassium Chlorate	25
Manganese Dioxide	1
Burnt Clay	18
Collodion Solution (4%)	80

Waterproofing Matches

One of the simplest ways to waterproof matches requires only that they be coated with a well-mixed solution of:

Glycerin	1
Collodion	50

(N.B. The coating should be applied only to the heads of the matches. In working with the liquid collodion care should be taken to keep open flames away.)

Black Powder

Potassium Nitrate	74.0
Charcoal	15.6
Dextrin	10.4

Quick Match, Military

Boil a mixture of water and starch in the proportion of 2 quarts of water to 4 ounces of laundry starch until the mixture forms a jelly. Add the black powder for the impregnating mixture in the proportion of 6 pounds to this quantity of starch mixture, and stir until the mass is thoroughly incorporated. The strands of wick should be thoroughly impregnated with the mixture of black powder, starch, and water.

When mealed quickmatch is specified, the wick shall be thoroughly impregnated with the impregnating mixture and coated while wet with the black powder which passes through a screen having 0.0041 inch openings.

Smokeless, Flashless Gunpowder
U. S. Patent 2,228,309

Nitrocellulose (13% N.)	76 –79 %
Dinitrotoluol	21 –24 %
Diphenylamine	0.8– 1.2%

Cellular Smokeless Powder
U. S. Patent 2,230,100

Nitrocellulose	225.00
Sugar	11.25
Butyl Acetate, Secondary	83.00
Toluol	90.00
Diphenyl Amine	1.35

Tumble until completely colloidal and then extrude and cut. Finally dry and age.

Ammunition Primer
U. S. Patent 2,136,801

Mercury Fulminate	30
Potassium Dinitrophenyl Azide	5
Barium Potassium Nitrate	35
Antimony Sulphide	25
Lead Bihypophosphite	5

Primary Detonator
U. S. Patent 2,127,106

Nitrogen Tetrasulphide	65–35
Potassium Chlorate	35–65

Explosive Powder

Sugar (50% Solution)	24
Potassium Nitrate	33
Coal, Powdered	10
Potassium Chlorate	33

Granulate and dry carefully.

Safety Explosive
U. S. Patent 2,215,608

Potassium Chlorate	65.0
Manganese Dioxide	1.3
Sodium Nitrate	5.0
Pyridin	1.4
Sulphur	1.3
Sugar	26.0
Iron Filings	1.0

Coal Mining Explosive
British Patent 530,818

Potassium Chlorate	58–60
Coal Tar Pitch	10–12
Sugar	3
Potassium Dichromate	1
Sodium Chlorate	26

Mining Explosive
U. S. Patent 2,215,608

Potassium Chlorate	65.0
Manganese Dioxide	1.3
Sodium Nitrate	5.0
Pyridine	1.4
Sulphur	1.3
Sugar	26.0
Iron Filings	1.0

Chapter Eighteen

RUBBER, RESINS, PLASTICS AND WAXES

Rubber from Plants
U. S. Patent 2,221,304

Lucerne, Clover, Rag-Weed, Golden-Rod, Sugar Cane or Corn Stalks	20
Crude Petroleum	64

Mix together for 16 days, then add:

Benzol	1½

and mix; then draw off liquid and mix with:

Carbon Black	3
Benzol	1¼
Talc	4
Magnesium Oxide	4
Caustic Soda	½
Water	1½

Mix well and add slowly, with mixing:

Sulphur Chloride	10

Mold with heat and pressure.

Non-Sticking Rubber Powder
Dutch Patent 48,223

Add 1% tryptic enzyme to latex containing ammonia. Heat to 40° C. and leave for 24 hours, then spray dry.

Preserving Rubber
Formula No. 1

Glycerin	8
Alcohol	8
Water, To make	100

No. 2
British Patent 489,103

Lemon Juice	5–15%
Glycerin	85–95%

No. 3

Glyceryl Diricinoleate	10
Alcohol	10

No. 4
U. S. Patent 2,083,176

Glycerin	50
Water	50
Graphite, Powdered	1/5

No. 5

Glycol Bori-borate	10
Water	20

No. 6

Polyglycol Hexaricinoleate	10
Isopropyl Alcohol	10

No. 7
British Patent 489,103

Glycerin	80
Almond Oil	4
Lavender Oil	4
Sugar	4
Petrolatum	3
Lemon Peel	2
Lemon Juice	3

Reconditioning Rubber Articles

Rubber articles such as boots, mats, hose and the like that have become cracked and brittle may often be restored to some degree of usefulness and value by the employment of a so-called "Glycerin process." This process is particularly applicable for reconditioning rubber articles which have grown hard and lost their elasticity because they were stored under adverse conditions such as dry heat or left exposed to extreme cold.

The process is quite simple. First clean the article by thoroughly scrubbing it with a fairly stiff brush dipped in warm water to remove all dirt and grime. This should be continued until the washings are clean and bright. Next, place the item in a solution made of one part of ammonia to two parts of water. Allow it to remain in this solution for an hour or two, then rinse the article with a dilute solution of glycerin and water. Wipe off and dry thoroughly. Such rubber goods should be stored in a cool place, away from the light.

Cheap Eraser Rubber

Rubber	10.00
Factice, White	14.50
Paraffin Oil	11.00
Lithopone	13.00
Lime	1.50
Sulphur	1.40
Diphenylguanidine	0.15
Hexamethylene Tetramine	0.05
Pumice Stone	14.00
Glass, Ground	8.50
Whiting	25.90

Cure 20 minutes in press at 50 pounds steam pressure.

Leatherlike Stiff Rubber

Tire Reclaim, Whole	32.00
Rubber	25.00
Carbon Black	30.00
Zinc Oxide	5.00
Stearic Acid	1.00
Pine Tar	3.00
Sulphur	1.30
Diphenylguanidine	0.40
Shellac	2.30

Cure 24 minutes at 50 pounds steam pressure.

Synthetic Rubber Gear and Pinion Composition

British Patent 513,991

Perbunan	100
Aldol α-Naphthylamine Resin	5
Dibenzyl Ether	3
Pine Tar	2
Zinc Stearate	5
Zinc Oxide	10
Magnesium Carbonate, Light	10
Clay, Active	10
Carbon Black, Active	60
Sulphur	½
Tetramethylthiuram Disulphide	2½
Carbon Black, Semi-Active	25

Rubber-Like Material

(for use as backing, doubling or anchoring coating requiring no strategic materials)

Solution A

Add 5 pounds casein to 19 pounds water, stirring well to break up any lumps. Add 1.1 pounds borax and heat solution to 160° F. with constant agitation. On reaching temperature, shut off heat and continue stirring until casein is completely dissolved.

Solution B

To 43 pounds of water, add 1½ pounds Darvan and stir well. Add 33 gallons of Tysenite (60% dispersion) and stir vigorously.

Add solution A to solution B and stir well. Now add 33 pounds Flexalyn and stir until a homogeneous mix results. This solution can now be spread on fabric or paper by means of a doctor knife and dried to form a rubbery material.

Rubber Substitute
(Plasticized Poly Vinyl Alcohol)

In tubing, linings, coatings that retain flexibility and resistance.

The process consists of making dispersions of mechanical plasticizers with water, with or without soluble plasticizers for PVA (Poly vinyl Alcohol) by means of emulsification. This dispersion is then thoroughly mixed with powdered PVA and sheet molded in a press at elevated temperature. Ten to twenty per cent mechanical plasticizer from group of oxidized linseed oil, turpentine, mineral oil Vistanex, or Neoprene latex is preferred.

Formula No. 1

Water	38.5
PVA	1.5
Oxidized Linseed Oil	10.0
Turpentine	4.0

No. 2

Water	24.0
PVA	1.0
Oxidized Linseed Oil	10.0
Turpentine	4.0
Glycerin	25.0

No. 3

Water	38.5
PVA	1.5
Mineral Oil	10.0
Turpentine	4.0

No. 4

Water	24.0
PVA	1.0
Mineral Oil	10.0
Turpentine	4.0
Glycerin	25.0

Total weight of each of the above is mixed with 100 g. powdered PVA.

Rubber Substitute
U. S. Patent 2,213,549

Gluten	12 oz.
Spirits of Turpentine	100 cc.
Sulphuric Acid	28 drops
Glycerol	100 cc.

Reclaiming Rubber
Australian Patent 110,487

Old rubber is shredded and put in a mixer where it is treated with superheated steam at 400–500° F. and atmospheric pressure for 3–6 hours. At the end of this period the heat is shut off and the mass allowed to cool to 300° F. The whole mass is then further cooled to 200° F. During the entire process the mass is agitated. In case a highly plastic product is desired, 5–10% of a mineral oil is added to the shredded rubber.

Artificial Rubber Latex
British Patent 541,539

Wool Wax	75
Bentonite	40
Mill the above into	
Rubber	400

at 50–70° C. Then emulsify with water. Soap and compounding ingredients may be added.

Latex Stabilizer and Pigment Disperser

A ½–1% dispersion of glyceryl monostearate in water will stabilize latex and permit the suspension of 300 times its weight of whiting.

Golf Ball Cover Composition
U. S. Patent 2,109,948

Deresinated Balata	20
Vulcanized Rubber Latex	15
Glue	5
Pigment	6

Increased Adhesion of Rubber to Cloth

About 4% ammonium stearate paste, when used in raw rubber mixes, increases the adhesion of the rubber to cloth on vulcanization. Replacing stearic acid incorporated into raw rubber mix, by ammonium stearate paste, also has a tendency to prevent the dry material from flying off in mixing.

Tack Elimination of Raw Rubber Sheets

Diglycol stearate is being used very successfully to prevent tack of raw rubber sheets. A fluid dispersion of Diglycol stearate in water, concentration between ½–2%, is made by heating and stirring. This mixture may be applied either by spray or dip, and will not affect future working of the rubber. A similar dispersion is also being used as a lubricant in the cutting of rubber gaskets.

Preventing Adhesion of Rubber to Metal

British Patent 528,959

Graphite	35
Castor Oil	61
Alcohol	4

Synthetic Rubber

German Patent 658,172

Buna N

Butadiene	20 lb.
Acrylonitrile	20 lb.
Water	50 lb.
Sodium Phosphate	175 g.
Citric Acid	100 g.
Aquarex D	280 g.
Potassium Cyanide	20 g.
Carbon Tetrachloride	250 g.
Sodium Perborate	15 g.
Acetaldehyde	60 g.

pH should be about 6.5

Buna S

Butadiene	20 lb.
Styrene	20 lb.
Water	50 lb.
Aquarex D	1,300 g.
Sodium Phosphate	680 g.
Sodium Perborate	135 g.
Carbon Tetrachloride	510 g.
Acetaldehyde	60 g.

The polymerization takes place very readily at room temperature and is completed in approximately three to five hours for making Buna N and in about 40 hours for Buna S.

The reaction vessel should contain a stirrer in the center; and after all materials except butadiene are weighed out and put into the vessel, the lid is tightly closed. The butadiene is added afterward and filled into the reaction vessel through a tube and valve on the side or bottom of the vessel. This arrangement works very satisfactorily as the butadiene, being a gas at normal temperature, is shipped in steel containers under pressure and therefore can be transferred easily into the reaction vessel.

Aquarex D, for instance, is used as emulsifier for the monomers and has proved very satisfactory and better than Nekal, sodiumoleate, or other metal-soaps formerly used as emulsifiers. The amount and the choice of emulsifiers and catalysts have a decided effect on the softness of the final product; therefore this has to be borne in mind, and the amounts in the recipes should not change too widely.

The pH of the system should be 6.5, i.e., not exactly neutral all the time; if this is not the case, separation of the emulsion and coagulation occur sometimes. At the end of the reaction the buna latex is separated from the unfinished part by decantation. The remainder is always used again by adding it to a new batch. To stop the reaction one must add phenyl-beta-naphthylamine to the buna latex. This addition should never take place in the reaction vessel itself as phenyl-beta-naphthylamine would have a pronounced effect as negative catalyst on the new batch. When exposed to sunlight and air, natural rubber oxidizes; while buna-type rubbers, when undergoing the same conditions, tend to polymerize further and also to cyclicize. This action is prevented by adding a stabilizer, the same phenyl-beta-naphthylamine as mentioned above, which also acts as antioxidant and age resister.

Buna S—Basic Formula
Formula No. 1

Buna S	100
Pine Tar	5
Zinc Oxide	5
Micronex	50
Sulphur	2
Captax	1½

No. 2

Buna S	100
Sulphur	2
Stearic Acid	1
Zinc Oxide	5
Thionex	½
Channel Black	65

Buna S Compound—High Heat Resistance

Buna S	75.00
Synthetic 100	25.00
Altax	0.75
Captax	0.75
Monex	0.10
Aminox	1.00
Stearic Acid	2.00
Zinc Oxide	5.00
Micronex	50.00
Sulphur	2.00

For frictioning, add 5 parts of pine tar to above stock.

Butyl Rubber—Cable or Wire Insulation

Butyl Rubber	100
Zinc Oxide	5
Stearic Acid	3
Sulphur	2
Tuads	1
Captax	½
Clay "W" (Moore & Munger)	100
Whiting	50
Paraffin Wax	3
Process Oil	5

Butyl Rubber (Molded Goods)

Butyl Rubber	100
Zinc Oxide	5
Stearic Acid	3
Sulphur	2
Gastex	50–100
Tuads	1
Captax	½

Perbunan Balloon Fabric Compound

Perbunan	100
Zinc Oxide	5
Stearic Acid	¾
Gastex	50
Sulphur	2
Altax	1¼
Diphenylguanidine	½
Pine Tar	25
Rosin, Wood FF	20

Break down Perbunan by passing 30 times through cold tight mill. Then add fillers in regular manner. Dissolve in methyl ethyl ketone.

Thiokol Printing Roller ("25" Shore Durometer)

Thiokol F or Thiokol D	100.0
Neoprene E	20.0
Rapeseed Oil Brown Sub, Hard	20.0
Diphenylguanidine	0.1
Altax	0.2
Cycline Oil	40.0
Gastex	20.0
Zinc Oxide	10.0

Cure: 2½ hours at 278° F. under pressure. Cool under pressure before removing to prevent blowing.

Reclaiming Synthetic Rubber

Mix the ground scrap (2–3 mm.) with pine tar (softener) at 80–90° C. for 1 hr.; heat 4 hr. in steam at 45 lb.; mill cold for 15–20 min.

Opaque Resin

Limed Rosin	25
Stearic Acid	25

Heat and stir until uniform.

Polystyrene Resin
U. S. Patent 2,232,930

Styrene	99.95–90%
Cinnamic Acid	0.05–10%

Polymerize at 100–140° C. for three days.

Polymerization Inhibitor
British Patent 531,973

Stabilization of vinyl acetate is effected by 0.01–0.1% copper abietate.

Synthetic Resin
British Patent 533,288

Dimethylol urea and a monohydroxy carboxylic acid ester, the carboxylic acid residue of which contains less than 8 carbon atoms in the molecule, are heated together. In particular not more than ½ molecule proportions of dimethylol urea is added slowly to 1 molecule proportion of a monohydroxy monocarboxylic acid ester heated to 100° and the heating is continued. Methyl, ethyl and isobutyl glycolates are used. Phthalic acid is a good catalyst.

Water White Synthetic Resin
U. S. Patent 2,230,326

Adipic Acid	1 mol.
Dioctylamine	1 mol.

Heat at 136–270° C., with stirring, for 30 min.

Varnish Resin, Synthetic
U. S. Patent 2,126,242

A mixture of glycerin 1 part and tung oil 4.8 parts is heated at 230° until the product (glyceryl tungate) is soluble in an equal volume of methyl alcohol. The varnish base is then made by heating a mixture of glyceryl tungate 100 parts, ester gum 120–290 parts, and ortho dibutyl phthalate 40 parts at 230° until it sets to a homogeneous resin on cooling.

Casting Phenolic Resins

A heat-setting phenolic liquid resin is poured into a suitable mold of glass, rubber or lead, and then is heated at 172° for two days. This product will be slightly yellow. However, objects of various colors may be obtained by adding pigments before casting. If a clear resin is not desired wood flour may be added as a filler before casting to yield an opaque product at a cheaper cost.

This resin may be set at room temperature by adding hydrochloric acid or this composition may be heated in order to obtain a solid casting in a shorter period of time.

Resin Coating for Food Cans

Cellulose Aceto-Butyrate	57–47
Tricresyl Phosphate	18
Glyceryl Phthalate	35

Warm and mix in dough mixer.

Thermoplastic
U. S. Patent 2,284,432

Plasticize sulphur with about 8% of aluminum stearate, 20% of silica powder, 12% of woodflour, 4% of sugar and up to 9% of rubber.

Adhesive Thermoplastic Coating

Ethyl Cellulose	47
Tricresyl Phosphate	18
Glyceryl Phthalate	35

Mix in heated dough mixer.

Electrical Conducting "Cellophane"
U. S. Patent 2,211,582

Cellophane sheets are first coated with powdered 70:30 brass suspended in a suitable vehicle and then exposed to hydrogen chloride gas to remove oxide films from the particles and render the film conductive. The sheets are then plated in the ordinary way.

Heat Stable "Vinylite" Coating
U. S. Patent 2,130,924

Less than 2% quinoline (based on "vinylite") is used in the coating.

Polyvinyl Butyral Coating (Non-Curing) for Raincoats
Olive Drab

Polyvinyl Butyral	100
"Kronisol"	58
Castor Oil, AA Raw	2
Dibutyl Sebacate	20
Chrome Yellow	4
Thermax	3½
Titanium Dioxide	4
Iron Oxide	½
Stearic Acid	¾
Clay No. 33	100
Acrawax C, Powdered	2

Mix together powders and liquids to form a uniform mix. Mill on hot mill about 230° F.

Calender with rolls at about 230° F. Dust with talc. For spreading work, dissolve in alcohol-butanol 50–50. Dilute with toluol. Coating does not crack at 0° F. or stick at 180° F.

Vinyl Chloride-Acetate Copolymer Raincoat Material Coating
Olive Drab

Vinylite VYNS	12 lb.	8 oz.
Vinylite VYNW	12 lb.	8 oz.
Castor Oil, Baker's P6	3 lb.	2 oz.
Dioctyl Phthalate	15 lb.	12 oz.
Tricresyl Phosphate	6 lb.	4 oz.
Lead Sulphate, Basic	2 lb.	8 oz.
Acrawax C, Powdered		6 oz.
Clay No. 33	20 lb.	
Chrome Yellow	2 lb.	
Thermax	1 lb.	12 oz.
Titanium Dioxide	2 lb.	
Iron Oxide		8 oz.

Mill, after mixing all ingredients together, on a hot mill about 220° F.

Calender with rolls at about 220° F. Dust with talc. For spreading work, dissolve in methyl ethyl ketone-cyclohexane mixture. Film does not crack at 0° F. or stick at 180° F.

Plastic
U. S. Patent 2,284,432

Sulphur	25–45
Aluminum Stearate	2–6
Silica, Powdered	5–9
Wood Flour	3–7
Sugar	1–2
Rubber	1–4

Cheap Plastic
British Patent 542,794

Sawdust	45
Starch	45
Rosin	10

Boil together, with sufficient water, to make a plastic mass. Mold or roll under heat and intermittent pressure to allow escape of surplus water.

Plastic, Self-Hardening
British Patent 522,171

Cement	99
Bentonite	10
Mixture of Rubber Latex (e.g., Revertex) with water	50

Inexpensive Plastic from Wood Pulp or Sawdust

Wood	50
Cresol	60
Hydrochloric Acid	10

A very plastic product is formed by heating the above materials for 6 hours at 70° C., washing with water, drying, grinding, pressing at 125° C. and 5000 pounds per square inch for 15 minutes. This may be baked to further harden it.

Strong and hard products may be formed by adding castor oil or glycerin, about 20 parts, to the above mixture before heating.

Other waste cellulosic material, such as corn-stalks, peanut hulls, may be used in place of wood.

Plastic Dentures

Methyl Methacrylate Powder	10
Methyl Methacrylate Liquid	8

To form a methyl methacrylate molded piece, mix until viscous and stringy, placing in a mold such as a hand screw press, and applying pressure while heating in boiling water for 30 minutes. This may be used for a denture material and for other objects which are to be transparent or colored.

Acrylic Resin for Mounting Specimens

Methyl methacrylate liquid, containing 1% of benzoyl peroxide, is placed in a test tube in warm water and heated at about 80° C. until it becomes syrupy. The object to be mounted is then placed in a suitable container, covered with the methyl methacrylate syrup and heated in an oven at 60–70° C. until the liquid solidifies (about 24 hours). The specimen, for example a butterfly, is thus permanently preserved in a hard, transparent, glass-like mounting. Sunlight accelerates the hardening.

Ethyl Cellulose Plastic
U. S. Patent 2,215,249

Ethyl Cellulose	99–90%
Esparto Wax	1–10%

Vegetable Cellulose Plastic
U. S. Patent 2,130,783

Sawdust, Bagasse or Other Vegetable Fibers

(Ground)	100
Water	100
Aniline	20

Heat with steam at pressure of 160 lb. per sq. in. for 3 hr. Wash with water, dry and grind. Mold at 135–190° C. at 1500–3500 lb. sq. in. with or without

Furfuraldehyde	7

Phonograph Record Composition
U. S. Patent 2,206,636

Vinyl Resin	90.0
Chromium Oxide	7.5
Carbon Black	50.0
Calcium Stearate	1.0
Carnauba Wax	1.5
Chlorinated Naphthalene	4.5

Cast Styrene Plastic

Styrene	200
Benzoyl Peroxide	1

The benzoyl peroxide is dissolved in the liquid styrene, the solution is poured into the desired mold which is preferably sealed and the whole is heated for five days at 45° C.

It is also possible to remove the syrupy product before complete polymerization, and use this to impregnate various materials.

Luminous Plastic
British Patent 534,008

Luminous Calcium Sulphide	1
Methyl Methacrylate, Granular	6

Mix and heat in a mold at 150° C.

Optical Lens Plastic
British Patent 530,834

Ninety per cent of monomeric methyl methacrylate and 10% of methacrylic anhydride give a resin for optics when polymerized at 60° for 24 hours, then subsequently heated to 130° for 8–12 hours.

Shellac Injection Molding Compound

Shellac	300
Jute	200
Kaolin	100
Pigment	15
Calcium Stearate	9

Mix and knead in dilute ammonia, dry, run through steam heated rollers, crush and cure at 85–90° for 1½ hr. before molding.

Plastic Compounds

A coutchouc-like mass can be prepared from a mixture of glue, gelatin, and glycerin. These ingredients should be liquefied under gentle heat. (1 part gelatin, 0.5 part glue and 1.1 to 1.4 parts glycerin.) To the liquefied mixture should be added 0.03 to 0.05 part of sodium chromate dissolved in 0.1 part of water.

A permanently elastic plastic mass may be prepared by dissolving glue in a strong sodium chloride solution, heating the liquefied mass for 10 minutes at 212° F. and finally adding a quantity of formaldehyde to "set" the mass. The more formaldehyde used the harder the mass will be.

French Plastic Mass

Gelatin	43.00 g.
Magnesium Sulphate	2.25 g.
Magnesium Carbonate	5.00 g.
Zinc Oxide	5.00 g.
Hydrogen Peroxide	2.25 cc.
Carbon Dioxide, Saturated Aqueous Solution	4.75 cc.
White Mineral Oil of Medium Density	4.75 cc.
Water	40.00 cc.

Add the water to the gelatin. Let

stand in a warm place until a completely uniform viscous mass has formed (frequent stirring of the mass will facilitate a steady formation of the liquid, also application of gentle heat will be of assistance in this respect). Replace the water which may have been lost through evaporation. When the liquid is perfectly uniform, add the carbon dioxide solution, hydrogen peroxide, and the mineral oil. Stir until the mass is uniform. Then add the powdered magnesium sulphate, magnesium carbonate, and zinc oxide to form a perfectly smooth paste. If desired, either 1 g. of sodium chromate (dissolved in 3 cc. water) or 2 cc. formaldehyde (40%) may be added which will render the mass moisture resisting.

Artificial Wood

Formula No. 1

Magnesium Carbonate	1000
Hydrochloric Acid	100
Calcium Carbonate	10
Casein	100
Wood Flour	800
Turpentine	20
Resin	100
Venetian Turpentine	25
Powdered Cork	75
Lead Acetate	10
Water	1650

No. 2

Powdered Straw	1
Wood Flour	1
Asphalt	2
Magnesite, Calcined	1
Boric Acid	½
Ammonium Chloride	½
Rosin	½

These ingredients are converted into a solid paste by adding sufficient magnesium chloride solution (30° Bé) and compressed under application of heat.

Wall Board and Coating Plastic

U. S. Patent 2,269,509

Cook oat flour in water in the proportion of one pound of oats to twenty pounds of water. The mixture of oats and water is cooked until the mixture assumes a consistency like that of hot jelly. To prevent the liquid binder from souring, and as a mold retardent, an oil ingredient is added to the mixture of oats and water, either during or after the cooking operation, as a preservative. Sassafras oil or oil of cloves has been found to have excellent preservative properties when added to the mixture of oat flour and water in the proportion of one-half teaspoonful of the oil to the mixture of oats and water.

A preferred material for use as a base for the composition is newspapers. These may be subjected to the action of a comminutor to condition them for mixing with the binding solution. Sander dust, i.e., the residual product which comes from a sanding machine in the form of dust, is also an excellent base material for the composition, although because of the desirable properties inherent in newspaper flour, better results are had when the sander dust and newspaper flour are mixed in proportions of about half and half of each. In the place of newspapers, however, any other fibrous materials may be used, such as asbestos, straw, wood or rag pulp, wood shavings, etc.

In the preparation of one formula, the product of which is intended for use for interior purposes, the oats is ground to a fine flour and the flour stirred into cold water. Heat is thereupon applied to the mixture to bring the temperature of the mixture to 212° F. At some time either before, during or immediately after application of heat, a preservative, such as sassafras oil, is added to the mixture. Thereupon, finely divided newspapers or other fibrous material is mixed into the oat flour solution in sufficient quantity to bring the final product to the desired consistency for application. The fibrous base material may be mixed with the oat flour solution while the latter still is hot, although better results will be had if the solution is first cooled to room temperature. For use as plaster to be applied with a trowel, approximately three pounds of fibrous material is used with a solution consisting of one pound of oats and twenty pounds of water. If a stiffer mixture is desired which can be handled with the hands and molded into various desired shapes, fibrous material is added as found necessary. For use as wall board and the like, or when intended to be cast into shapes, approximately four pounds of fibrous material is used with a solution consisting of one pound of oats and thirty pounds of water.

The second formula is designed to produce a material intended primarily for exterior use, such as in place of stucco, shakes, shingles, and the like. In the preparation of this formula the oats is ground to a fine flour as before, and stirred into a cold solution of magnesium chloride having a density of approximately 10° Bé. Heat is thereupon applied to the mixture to bring the temperature of the mixture to 212° F. Thereupon, the solution is cooled to room temperature and calcined magnesite is stirred into the solution, the amount of calcined magnesite used being in the proportion of 2½ pounds of calcined magnesite to each one pound of oat flour used in the solution. Thereafter, the selected fibrous material is mixed into the solution in sufficient quantity to bring the final product to the desired consistency for application. If the compound is to be applied with a trowel, approximately three pounds of fibrous material is used with a solution consisting of one pound of oat flour, twenty pounds of magnesium chloride solution, and two and one-half pounds of calcined magnesite. If the mixture is to be handled with the hands and molded or cast into various shapes, fibrous material may be added as found necessary without increasing the liquid content of the binder solution.

Molded Wood Flour Products

Non-shrinking molded products can be made from wood flour by a binder of Portland cement and "GC" powdered silicate of soda, with a little water.

Wood Flour	10
Portland Cement	10
"GC" Silicate of Soda	10
Water	3½

Mix thoroughly, add water to dampen the mass, and mold under heavy pressure. Such products are durable and water resistant, though they are not refractory. Similar products can be made with part or all mineral filler instead of wood flour, but the proper amount of such heavier material should be much greater than of the flour.

A mixture which sets up to form a hard product capable of being tooled like hard wood is prepared by mixing wood-flour and fine ball clay in proportion of about three to two, adding water, a little at a time, until the mass is of molding consistency and then adding one part of "SS-C-Pwd." silicate of soda. This mixture is lighter and less rock-like than the above formula, but it is not water resistant. The addition of some zinc oxide would improve it in this respect.

Molding Composition
U. S. Patent 2,221,304

Petroleum	64
Alfalfa	20
Sulphur Chloride	10
Talc	4
Magnesium Oxide	4
Caustic Soda	2½
Water	1½
Carbon Black	3
Benzene	1¼

Mold under heat and pressure.

Beeswax Compound, Cable Impregnating

Beeswax (Yellow or Brown)	50
Paraffin Wax	50

Electrical Insulating Wax
British Patent 538,411

Hydrogenated Castor Oil	80
Paraffin Wax	20

Sticky Non-Oxidizing Wax
Formula No. 1

Hydrogenated Rosin	90
Flexo Wax C	5
Partially Polymerized Petroleum	5

No. 2

Hydrogenated Rosin	80
Flexo Wax C	10
Partially Polymerized Petroleum	10

Cobbler's Thread Wax
(Sticky Wax)

Rosin (X Grade)	74
Glaurin (Plasticizer)	1
Petrolatum	20
Water	5

Hard Synthetic Wax

Durocer	95
Glyceryl Tristearate	5

Slow Burning Wax

Durocer	90
Arochlor	10

Hot Melt Wax Coating
(Non-Blocking)
U. S. Patent 2,297,709

Ethyl Cellulose	5–20%
Hydrogenated Castor Oil	25–55%
Rosin or Synthetic Resin	15–40%
Paraffin Wax (m.p. > 40° C.)	15–30%

DENTAL WAXES

Sticky Wax

For temporarily cementing small pieces of plaster or similar materials together in order to make molds from broken plaster impressions. Also to hold small pieces of metal together for investment before soldering.

Formula No. 1

Rosin	16
Yellow Beeswax	8
Vermilion	1

Melt on water bath and stir together, pour on a glass slab and roll with wet fingers into pencils or pour in molds.

No. 2

Yellow Beeswax	4
Rosin	1
Gum Dammar	1

No. 3

Gum Dammar	1
White Beeswax	4
Light Yellow Rosin	7

Pink Base Plate Wax

Formula No. 1

White Beeswax	50
Paraffin Wax	25
Alkanet Root, Whole	1

Melt the waxes and add the alkanet root. Leave on the fire until the desired shade of pink is obtained, strain through cheese cloth into tin molds, about 1/32 or 1/16 of an inch thick. Have the molds coated with a film of glycerin. To polish the sheet wax, pass between the rubber rollers of a wash wringer.

No. 2

White Beeswax	40
Gum Turpentine	10
Cotton-Seed Oil	3
Vermilion	4

No. 3

(To be used in hot weather.)

White Beeswax	20
Crude Turpentine	4
Cotton-Seed Oil	1
Vermilion	2

No. 4

To be used in cold weather.

White Beeswax	20
Crude Turpentine	6
Cotton-Seed Oil	2
Vermilion	2

No. 5

Hard Base Plate Wax

Yellow Beeswax	50
Gum Mastic	6
Prepared Chalk	3
Vermilion	4

No. 6

Rosin	1
Ceresin	3
Paraffin Wax	6

Wax for Partial Impressions

Rosin	1
Yellow Beeswax	16

Gold Inlay Impression Waxes

Formula No. 1

Yellow Beeswax	10
Gum Dammar	10
Yellow Ceresin	20
Hard Paraffin (120° F.)	30
Carnauba Wax	30
Dye, To suit	

Melt the beeswax, ceresin, paraffin and carnauba wax in a porcelain dish on a water-bath, add the gum dammar in small portions and stir constantly until a uniform mass is obtained. Remove from the fire and add the dyestuff.

Gold inlay waxes should be colored deeply with a dye, especially suitable to the needs of the operator. Lamp black or an oil-soluble aniline dye are best suited for this purpose. (The red and blue "Cerasine" or "Sudan Red" aniline dyes are recommended.)

No. 2

Price's

Stearic Acid	110
Paraffin Wax	10
Beeswax	15
Tamarack (Larch Turpentine)	10
Gum Dammar	110

No. 3

National Bureau of Standards

Carnauba Wax	25

Paraffin Wax (55° C.)	60
Ceresin	10
Beeswax	5

Oil soluble aniline dye in suitable colors may be added.

No. 4

Carnauba Wax (Light)	10
Paraffin Wax (70° C.)	35
Beeswax, Pure	55

Melt the carnauba wax on a low flame, avoiding foaming; add the paraffin and after complete melting add the beeswax.

Carving Wax for Tooth Forms

Carnauba Wax	80
Ceresin	128
White Wax	64
Zinc Oxide	48
Saturated Solution Acriflavine in Alcohol	1

Engraver's Wax

Yellow Beeswax	1
Tallow	1
Burgundy Pitch	2

Jewelers Wax

Heat common rosin in a suitable vessel until it flows freely, then add, slowly, and with constant stirring, plaster of Paris until on dropping a small portion on a plate or marble slab and allowing it to cool, then prying it off with the point of a knife it springs off with a metallic ring. With too little plaster it is soft and bends; with too much it is hard and brittle and loses its adhesiveness. This is used by jewelers for a variety of purposes; to cement pieces of thin metal to chucks for turning, to cement stones into fittings, and tools into metal and wooden handles. It is handy in a dental laboratory for cementing felt and brush wheels to the chucks of the polishing lathe.

Modeling Compounds

Formula No. 1

Stearin	25
Gum Dammar	50
Powdered Soapstone	85
Carmine, Enough to color	

Melt the stearin on a water-bath, add the gum dammar, and, when melted, stir in the powdered soapstone, tinted with the carmine.

No. 2

Stearic Acid	20
Oleic Acid	4
Gum Copal	19
Seed-Lac	17
Powdered Soapstone	40

No. 3

Manila Copal	30
Light Colored Rosin	30
Carnauba Wax	10
Stearic Acid	5
Powdered Soapstone	75
Carmine, Enough to color	

No. 4

Best Light-Colored Rosin	50
Gum Copal	2
Yellow Ceresin	8
Gum Turpentine	5
Powdered Soapstone	50
Menthol	½
Fresh Slaked Lime	10

Color with Florentine Lake (for red).

Brake-Block Composition

U. S. Patent 2,268,280

Silica Sand	23
Asbestos Powder	11
Asbestos Cement	33
Coke Dust	22
Linseed Oil	11

Brake Lining Impregnant

Formula No. 1

China Wood Oil	1
Bakelite No. 7112	1
Petropol No. 2138	1
Xylol	3

No. 2

Asphalt	10
Pitch (A-1)	10
Gilsonite	180
China Wood Oil	70
Linseed Oil	20
Petropol No. 2138	50
Petroleum Naphtha	440

After impregnation of lining, bake for 2 hours at 250° F. and 1 hour at 350° F.

Shoe-Filler

British Patent 532,471

Soft wax tailings 24, and B rosin 15 lb. are melted together and cooled to 100° C. One pound of oleic acid is then added, followed by 1/5 lb. of sodium hydroxide in water then a hot paste of ⅔ lb. of bentonite in 8 lb. of water. This mixture is poured into a charge of 11 lb. of granular cork and well mixed therewith. Castor oil, if added up to 2 lb., confers increased flexibility.

Shoe Stiffener

U. S. Patent 2,238,337

Rubber	130
Resin Containing Box Toe Scrap	400
Cumarone Resin	100
Asbestos Fiber	200

Artificial Asphalt (Bitumen)

Addition of 30–40% 200-mesh coal to a heavy hydrocarbon oil (150–300 seconds Furol) and heating at 400° F. for at least one hour produces an asphaltic material quite satisfactory for many purposes where a bituminous material must be employed.

Increasing Softening Point of Asphalt

Melting Acrawax C with asphalt raises its softening point. 10% Acrawax C raises softening point of asphalt from 150–250° F.

Synthetic Brewers' Pitch

U. S. Patent 2,122,543

(Flexible impervious lining for barrels, drums and cans.)

Ester Gum	65 lb.
Japan Wax	11 lb.
Petrolatum	24 lb.

Apply hot.

Elastic Stuffing Box Packing

Australian Patent 112,147

For steam, acids, and ammonia the following composition is used: asbestos fiber 20–30, ozokerite 5–10, graphite 5–10, magnesium oxide, 20–30, and amorphous carbon or graphite 20–30%. For water and compressed air the 20–30% of asbestos is replaced by asbestos 10–15 and rubber chips 10–15%, the other ingredients remain the same. For oil, gasoline and kerosene 10% of granulated cork is added to the first composition.

Neoprene Gasket and Packing

British Patent 516,777

Neoprene	400
Pine Tar	30
Magnesia, Light Calcined	60
Softener	140
Anti-Oxidant	8
Soft Black	200
Channel Black	250
Sulphur	1
Wood Wool	500
Zinc Oxide	40

Compressed Fiber Board
U. S. Patent 2,215,245

Wood fibers, in suspension in water, are compacted then compressed at approximately 200 to 250° and not more than 3000 pounds per sq. in. into a coherent mat, and impregnated (before or after compacting or compressing) with a fatty acid pitch solution or emulsion (e.g., pitch 300, kerosene 700, and some paraffin), and the pitch is hardened by an oxidizing bake at approximately 150°.

Flooring Base Composition
German Patent 698,682 (1940)

Shredded Waste Paper	6.5 l.
Glue	0.5 l.
Cement	2.0 l.
Water	1.0 l.

Mix thoroughly; press into forms and dry. This is elastic and holds nails firmly.

Hydrocolloid Dental Impression Composition
U. S. Patent 2,234,383
Formula No. 1

Agar-agar (No. 1 Kobe, Strip)	30.00
Borax	0.42
Potassium Sulphate	4.20
Water, To make	210.00

The agar-agar is washed in fresh water for about half an hour and then drained. The wet agar-agar is heated and stirred in a steam-jacketed kettle until a smooth sol is obtained. The borax (as a saturated solution in water) is then added slowly, with constant stirring, and the potassium sulphate is then added. The stirring and heating is continued until the product is smooth and homogeneous, and water is added or evaporated until a final weight of 210 lb. results. The material may then be packaged for shipment. The resulting pH value of the above mixture will be in the neighborhood of pH 8.3, depending somewhat upon the character of the agar-agar, the pH of this raw material having been found to vary to some degree, and the pH of the water used.

No. 2

Agar-agar (No. 1 Kobe, Strip)	30.00
Manganous Borate (Powdered)	1.05
Potassium Sulphate	4.20
Water, To make	210.00

The agar-agar is washed in fresh water for about half an hour and then drained. In this washing the agar-agar soaks up almost all the water needed for the batch, so that ordinarily no additional water is necessary. The wet agar-agar is heated in a steam-jacketed kettle and stirred constantly at a moderate rate. After an hour or two of heating and stirring, the agar-agar will have formed a fairly smooth sol. Hot water may be added at this point, if necessary, to bring the total weight of the material to a little over 210 pounds. The manganous borate powder is added at this time, as by sifting through a fine mesh screen (or by ball-milling in a small quantity of water, and adding the wet mix), and added to the agar-agar sol under continuous stirring. The potassium sulphate may then be added in the form of fine crystals or powder or in water solution. The stirring and heating is continued until the manganous borate is uniformly distributed and the potassium sulphate is dissolved (if added as a solid), and longer if necessary to evaporate any excess water which may have been added, so as to end with a final weight of 210 pounds. The hot sol may then be filled into suitable containers and packaged for shipment.

No. 3

Agar-agar (No. 1, Kobe, Strip)	30.00
Borax	.70
Barium Chloride	.93
Potassium Sulphate	4.20
Water, To make	210.00

The agar-agar gel may be made up as described in connection with the first example, and the barium chloride added in crystal form and the mixture stirred until the barium chloride is dissolved, after which the dry borax is added a little at a time. This method requires a strong stirring apparatus to keep the material properly stirred, as the borax will cause the material to become extremely viscous if added too rapidly. A slow and careful addition of the borax is desirable. As a modification of this method, the barium borate may be prepared by separately reacting the barium chloride with the borax, filtering and washing the material, and adding the white precipitate to the gel as in accordance with the first described example.

Chapter Nineteen

SOAPS AND CLEANERS

Transparent Soap

Formula No. 1

An inexpensive soap may be made from 45 stearin, 55 castor oil, 30 sugar solution (sugar 1 and water ½), 50 caustic soda 37° Bé, and 35 alcohol. Fatty acid content of such soap is from 60–70%.

No. 2

50 kg. cocoanut oil fatty acids, 40 kg. tallow fatty acids, 20 kg. distilled vegetable oil fatty acids, about 56 kg. (38° Bé) caustic soda, 52 kg. denatured alcohol, 2.4 kg. potash, 3.2 kg. common salt, 2 kg. sodium nitrate, 1.6 kg. calcined soda ash, 44.8 kg. water. Careful adjustment to neutrality and formulation at a low temperature are very important in producing the desired clarity.

Soft Soap (Dark)

Soybean Oil Fatty Acid	680 lb.
Arachis Oil Fatty Acid	130 lb.
Tallow	80 lb.
Rosin	30 lb.
Caustic Potash (50° Bé)	270 lb.
Sodium Carbonate	50 lb.
Water	350 lb.

Liquid Soap

Cocoanut Oil	225 lb.
Sunflower Oil	75 lb.
Caustic Potash (36° Bé)	220 lb.
Calcium Chloride	18 lb.
Boric Acid	5 lb.
Water	170 lb.

PECTIN SOAPS

Liquid Soap

Soap	50.0
Pectin	22.5
Sodium Silicate (38° Bé)	27.5

Paste Soap

Tallow Fatty Acids	20
Olive Oil Fatty Acids	20
Palm Oil Fatty Acids	20
Caustic Soda (38° Bé)	27
Pectin	45
Water	15
Bentonite or Sodium Silicate	55

Soap from Waste Fat

Although waste fat is always of commercial value, in some instances it might be found an economic advantage to convert the fat into a soap suitable for household cleansing purposes. This can readily be effected by the saponification of the fat by means of caustic soda. The following process will give a satisfactory soap for general cleansing purposes: fat 36 lb., caustic soda 4¾ lb. (may be increased slightly to complete saponification of fat), water 4 gallons. The soap should be prepared in an iron vessel capable of being heated. The caustic soda should be dissolved in the water and heated to boiling. The fat should then be added slowly to avoid frothing and when all has been added, the mixture should be boiled for three hours. The solution can be tested for complete saponification of the fat by treatment of a small quantity with water as no oily globules should separate. When saponification is complete, the soap solution can be poured into a suitable mold. Either individual molds for the bars can be employed or the soap can be poured into a trough about 3 inches deep. When cold the soap mass produced can be cut into suitable pieces. It will usually be found desirable to stack the soap manufactured in this way for a few weeks before use to enable the bars to dry out and harden. The character of the soap naturally will depend to some extent on the nature of the fat used. Soap produced in this way differs chemically from soap as produced commercially, in that it still contains the glycerin produced by the saponification of the fat.

Saddle Soap

Soap Powder	15
Water	72
Neatsfoot Oil	5
Beeswax	8

Dissolve the soap in hot water. Heat the neatsfoot oil and wax until melted and pour hot mixture into soap solution. Stir until it begins to thicken, pour into cans.

Anti-Perspirant Soap
French Patent 841,802

To 100 parts of soap are added 3–5 parts chromic anhydride, 1–2 parts hexamethylene tetramine and 1 part aluminum acetate.

Fluorescent Soap

For catching thieves a small amount of fluorescein can be incorporated in ordinary soap and, when the thief washes his hands with the soap, they will be stained bright greenish which is more visible in ultraviolet light.

Although the above soap is visible, an invisible soap stain may be made by adding beta-methylumbelliferone to ordinary soap. This stain will not be seen in white light but under ultraviolet light a bright blue color will be noted.

Trick soap can be prepared by adding uranine and acriflavine to ordinary soap. The more a person washes his hands the more the staining.

Soap Pad

Soap pads are composed of flannel pads impregnated with jelly cleanser and with ammonium-oleate and other soaps.

Pumice Soap

A pumice soap is prepared by the cold process from 15 parts of cocoanut oil, 10 parts of tallow and 12.5 parts of 38° Bé caustic soda. After saponification, 10 parts of crystalline soda are dissolved in the soap and enough pumice crushed in to give a sufficiently firm product when cold.

Soapless Soap
German Patent 696,126

10 kilograms methyl cellulose are mixed with 100 kilograms of saponin extract. This is brought to the boil and then 150 kilograms of saponin extract added cold. The emulsion formed on standing is stated to be

excellent for washing fine fabrics and is also recommended for toilet use. By the addition of sodium lauryl sulphate, approximately 10%, the detergent properties of the mixture are still further improved.

Acid Cleaning Emulsions

A cleansing emulsion suitable for use on porcelain, metals, or cement is made of the following composition:

Polymerized Glycol Oleate	15
Oleic Acid	5
Mineral Oil	30
Hydrochloric Acid	6
Water, To make	100

Another cleaning emulsion that may be used is of the following composition:

Polymerized Glycol Distearate	10
Stearic Acid	5
Mineral Oil	15
Hydrochloric Acid	3
Water	67

Preventing Gelling of Soap Solutions

Soap Flakes	10
Water	990
Ammonium Sulphamate	1

The above may be used hot and will not gel on cooling.

Cleaner and Polisher
(*For transparent sheet*)

Diatomaceous Silica (Snowfloss)	100
Emulsifying Agent *	10
Water	600
Naphtha	300

* Emulsone B. (Vegetable Gum.)

Non-Inflammable Cleaning Fluid
U. S. Patent 2,245,052

Water	94 fl. oz.
Oxalic Acid	5⅝ oz.
Potassium Chromate	¼ oz.
Glycerin	⅛ fl. oz.

Hand Cleaners
(*Mechanic's*)

Formula No. 1

Soap, 90% (Titre: 30°)	15
Tetrasodium Pyrophosphate	10
Silica Sand, Fine, Round Edged	40
Ash, Finest Volcanic	35

No. 2

Borax	20
Sodium Silicate, "G" Type, Fine Mesh	15
Soap, 90% (Titre: 36°)	20
Sodium Metasilicate, Fine Mesh	5
Silica Sand, Fine, Round Edged	40

No. 3

Sodium Carbonate	3.5
Borax	11.5
Sodium Metasilicate	5.0
Soap, 90% (Titre: 36°)	28.0
Wetting Agent (Sulfatate)	2.0
Silica Sand, Fine, Round Edged	15.0
Ash, Fine Volcanic	35.0

Jelly Hand Cleanser

Agar-Agar	0.2
Psyllium	0.3
Glycerin	5.0
Soda Ash	5.0
Soap	5.0
Ammonia	2.5
Javelle Water	0.5
Water	81.5

Hand Stain Remover

Water	8	oz.
Soap Powder	4	oz.
Silica (150 Mesh)	½	oz.
Pumice Flour	¼	oz.
Chlorox (Chlorine Water)	1	dr.
Glycerin	1	oz.
Rubbing Alcohol	1	oz.

Heat the soap in the water until dissolved, and add the abrasives. When cool add the chlorox (or the like) the glycerin and alcohol.

Bottle and use in the same way as any liquid soap on well-wetted hands.

Abrasive Cleanser

Soap Flakes	12.5
Mineralite	28.5
Water	59.0
Sassafras Oil, Trace	

Waterless Hand Cleanser

Powdered White Soap	1	lb.
Potassium Carbonate	2	oz.
Tri-sodium Phosphate	1	oz.
Finely Powdered Asbestos	1	lb.
Water	10	pt.
Perfume, To suit		

Dissolve the soap and the chemical salts in hot water, allow to cool, stir in the asbestos, mix thoroughly, and cool for packing. This is used without water.

Industrial Hand Cleaners

Formula No. 1

Diethylene Glycol Oleate	97
Wetting Agent (Sulfatate)	2
Perfume	1

No. 2

Propylene Glycol Laurate	44
Water	43
Wetting Agent (Wetanol)	2
Perfume	1

These have excellent cleansing and skin-softening properties and do not abrade or de-fat delicate skin.

Denture Cleaner

British Patent 506,451

A denture cleanser is made from 2 parts of sodium peroxide and 3 parts of sodium chloride. Sodium carbonate, calcined soda, phosphates and other agents may be added.

Laundry Washing

Though no one formula is suited to every washroom, the following proportions for "Stock Soap Solutions" have been found to give excellent results under many conditions. The exact formula to be used with these solutions should be worked out with the aid of pH and titration figures. For a 300 pound dirty load of nets (42 x 84 washer) it is advisable to add one pound sodium metasilicate (dry) to break, following with the stock soap solution as below until the desired suds are obtained.

Stock Soap Solution

Soap	3
or	
Soap	2
Sodium Sesquisilicate	1
Sodium Metasilicate	1

White Work Formula

Soap mixture used: 50 parts high titer soap and 50 parts sodium metasilicate. Extra metasilicate used on the break (pH 11.0). (For net washing use 2 more inches of water and one additional rinse.)

Operation	Water, Inches	Temperature, Degrees F.	Time, Minutes	
1. Break	6″	90°	10 min.	Good suds
2. Suds	4	130–140	10	Good suds
3. Suds	4	160–170	10	Good suds
4. Suds and bleach	5	160	15	Javelle water first; after 5 min. add enough soap to give light suds *
5. Hot rinse	10	160	5	
6. Hot rinse	10	160	5	
7. Hot rinse	10	160	5	
8. Warm rinse	10	140	5	
9. Sour	5	140	5	Dump the water
10. Blue	10	Cold	5	

* Probably preferable to add soap first, then add Javelle water after running 5 minutes.

Fast Colored Work Washing Formula

Soap mixture used: 70 parts soap and 30 parts sodium metasilicate (pH 10.5–10.8 in the break). (For net washing use 2 more inches water.)

Laundry Compounds

Formula No. 1

Sodium Carbonate	50
Sodium Bicarbonate	50

This compound can be used in the wash wheel and does not injure fiber nor does it develop any heat.

No. 2

Formula No. 1 plus any desired amount of an alkali free soap.

No. 3

Soap	65
Sodium Carbonate	15
Sodium Bicarbonate	5
Trisodium Phosphate	10
Sodium Metasilicate	5

No. 4

Sodium Carbonate	50
Sodium Metasilicate	40
Soap	10

No. 5

Sodium Carbonate	40
Sodium Metasilicate	50
Soap	10

No. 6

Sodium Carbonate	50
Sodium Metasilicate	13
Trisodium Phosphate	2
Borax	5
Soap	30

No. 7

Sodium Sesquicarbonate	50
Borax	50

No. 8

Sodium Carbonate	40
Sodium Metasilicate	45
Trisodium Phosphate	15

No. 9

Sodium Carbonate	10
Sodium Bicarbonate	35
Trisodium Phosphate	40
Soap	15

No. 10

Sodium Carbonate	75
Sodium Metasilicate	22
Soap	3

No. 11

Sodium Carbonate	70
Sodium Bicarbonate	10
Borax	5
Soap	15

No. 12

Sodium Carbonate	55
Sodium Metasilicate	45

No. 13

Sodium Carbonate	17
Sodium Metasilicate	32
Pine Oil	1

No. 14

Sodium Carbonate	40
Sodium Metasilicate	44
Disodium Phosphate	16

No. 15

Soap (90%)	55
Sodium Metasilicate	20
Sodium Carbonate	15
Sodium Pyrophosphate	10

No. 16

Sodium Carbonate	20
Sodium Metasilicate	30
Sodium Pyrophosphate	10
Soap	40

No. 17

Sodium Carbonate	50
Silicate, Liquid (40%)	50

No. 18

Sodium Carbonate	5
Sodium Metasilicate	35
Soap	60

No. 19

Sodium Carbonate	45
Sodium Metasilicate	20
Trisodium Phosphate, Anhydrous	30
Caustic Soda	5

No. 20

Sodium Carbonate	55
Sodium Metasilicate	30
Trisodium Phosphate, Anhydrous	15

No. 21

Sodium Carbonate	50
Sodium Metasilicate	15
Trisodium Phosphate	35

No. 22

Sodium Carbonate	65
Sodium Bicarbonate	35

No. 23

Sodium Carbonate	25
Sodium Metasilicate	50
Trisodium Phosphate	25

No. 24

Sodium Carbonate	40
Sodium Metasilicate	24
Trisodium Phosphate, Monohydrated	32
Caustic Soda	5–7
Wetting Agent	1–2

No. 25

Sodium Carbonate	36
Trisodium Phosphate	57
Caustic Soda	6
Wetting Agent	1

No. 26

Soap Powder (95% Soap)	18
Sodium Carbonate	82

Linen Alkali

Sodium Carbonate	75
Caustic Soda	25

Break Soap

Sodium Carbonate	55
Trisodium Phosphate	20
Soap	25

Laundry Detergents
(*Hand or Open Washing Machine*)

Sodium Sesquicarbonate	43
Trisodium Phosphate	17
Tetrasodium Pyrophosphate	15
Soap, Neutral, 90% High Grade, (Titre: Approx. 32°)	25

Soap Builders

Formula No. 1

Sodium Carbonate	90
Tetrasodium Pyrophosphate	10

No. 2

Sodium Carbonate	95
Sodium Metasilicate	4
Sodium Hydroxide	1

Laundry "Breaks"

Formula No. 1

Sodium Carbonate	65
Sodium Metasilicate	22
Sodium Hydroxide	3

No. 2

Sodium Carbonate	22
Sodium Metasilicate	13
Trisodium Phosphate	40
Sodium Bicarbonate	15
Tetrasodium Pyrophosphate	10

Textile Sodas

Formula No. 1

Sodium Carbonate	53
Sodium Bicarbonate	47

No. 2

(*Super Strength*)

Sodium Carbonate	58
Trisodium Phosphate	42

Laundry Soap Builder

Sodium Carbonate	28
Sodium Metasilicate	35
Tetrasodium Pyrophosphate	8
Sodium Hydroxide	29

Laundry Detergents

(*Open or Wheel Washing*)

Formula No. 1

Sodium Carbonate	52
Sodium Metasilicate	38
Soap (Titre: 30°)	10

No. 2

Sodium Carbonate	45
Sodium Metasilicate	47
Sodium Hydroxide	18

No. 3

Trisodium Phosphate	10
Tetrasodium Pyrophosphate	40
Sodium Bicarbonate	5
Wetting Agent (Sulfatate)	30
Sodium Sulphate	15

No. 4

Sodium Bicarbonate	5
Sodium Tetraborate	5
Tetrasodium Pyrophosphate	40
Sodium Sulphate	25
Wetting Agent (Sulfatate)	25

No. 5

Soap (Titre: 30°)	35
Wetting Agent (Sulfatate	15
Tetrasodium Pyrophosphate	40
Trisodium Phosphate	10

No. 6

Sodium Sulphate	50
Trisodium Phosphate	10
Tetrasodium Pyrophosphate	20
Wetting Agent (Sulfatate)	20

No. 7

Soap, 90% (Titre: 30°)	25
Wetting Agent (Sulfatate)	15
Trisodium Phosphate	10
Tetrasodium Pyrophosphate	50

No. 8

Soap, 90% (Titre: 30°)	12
Wetting Agent (Sulfatate)	3
Sodium Bicarbonate	30
Sodium Metasilicate	5

Trisodium Phosphate	10
Tetrasodium Pyrophosphate	40

Laundry Stain Removal

The fabric must stand up under the treatment proposed whether the stain is stripped or not. Chlorine bleach and strong alkalies, while applicable to cotton and linen with adequate control, should be kept away from wool and silk. Strong mineral acids, which can be used carefully and with short exposure time on wool and silk, should not be used on cotton and linen. Stains on wool and silk are in general more resistant to removal than those on cotton and linen. Mixed goods, of course, require that the specific precautions pertaining to each different type of fiber present be taken for successful, safe stain removal treatment.

Colored goods introduce the factor of dye behavior under the influence of the stain removal reagent in addition to that of effect on fiber. In this case, the reagent must be safe to color as well as to fiber. A preliminary test at some out of the way point on the fabric is always imperative to determine just what effect on color the proposed treatment will have. Some staining substances have in themselves the power of altering color and care should be taken not to extend any damage of this nature which has already taken place.

The age of a stain is perhaps the most important factor relating to its ease and completeness of removal. Old stains of any kind are always more difficult to remove than fresh similar ones. The fixative chemical and physical reactions which take place in the fabric after the solvent or vehicle of the staining substance has dried or spread out through capillarity are progressive in practically every case. If simple "First Aid" could always be rendered soon after the staining contact was made, the later removal of the stain would be greatly simplified. Such simple treatments as a little cold water on spots of food products, body fluids and any other substances known to contain water or be soluble in it, or a small amount of kerosene, naphtha, turpentine or similar light oily solvent for paint, grease and oil spots are vastly worthwhile.

Regarding reappearance of stains decolorized by oxidizing or reducing methods, it may be said, in general, that stains removed completely but with difficulty, seldom reappear. This is particularly true of oxidizing bleaches which do their color stripping by forcing oxygen into the chemical makeup of the stain body to change its composition and destroy its chromatic or coloring power. The opposite or reducing conditions necessary to change back a stain once decolorized by oxidation is rarely met with in service.

On the other hand, stains decolorized by reducing agents, which bring about their decolorizing action by taking oxygen away from the stain body, do sometimes reappear, since the oxygen necessary to reproduce the chromatic or coloring matter is found everywhere in the air. To break this chain, it is sometimes efficacious to actively oxidize a stain after stripping with a reducer such as hydrosulphite, by the action of permanganate, and then re-reduce.

The chemical action of forced reversal seems to move the stain body so that the final reduction is more permanent than with a single reducing treatment.

With the above general considerations in mind, the specific procedure in spotting or stain removal is as follows:

1. The first step may well be called "detective work," since the logical first move is to determine by inquiry, visual examination with or without the aid of magnification or by general deduction from the type of article and location of the stain, just what its nature is. A well thought out plan is as valuable as any of the reagents.
2. Try solvents such as naphtha or gasoline, carbon tetrachloride, trichlorethylene, acetone, etc.
3. Try alcohol.
4. To the dry fabric apply Kerosene-Oleic-Benzol cleaner. Work in and wash out with a solution of a safe alkali.
5. Try the oxidizers such as chlorine bleach, hydrogen peroxide and permanganate. (Type of oxidizer contingent on type of fiber.)
6. Try the reducers such as hydrosulphite, bisulphite and titanous chloride. (The last is an acid solution. Rinse well after spotting.)

Acriflavin Stain Removal

The very resistant orange yellow stain from this dye is most frequently found on hospital work since it is in common hospital use. Its fastness makes three separate successive treatments necessary:

1. Soap and water wash.
2. Acetic acid—alcohol.
3. Permanganate—hydrosulphite.

Argyrol (Silver Antiseptic) Stain Removal

Like all other silver stains, argyrol is not removed in the regular washing process but requires separate treatment.

The best remover for silver stains resulting from any of the many silver compounds used in medicine, dentistry and photography is a mixed solution of bichloride of mercury and ammonium chloride. The reagent is made by dissolving 5 grams of bichloride of mercury and 5 grams of ammonium chloride in 100 cc. of water. Heat the solution to just below boiling, immerse the article and follow by thorough rinsing after the stain disappears.

Iodine in the form of ordinary tincture of iodine is recommended for silver stains in some cases. This converts the silver residue to silver iodide which is soluble in ammonia water.

Spot the stain with tincture of iodine, using a glass rod. Allow to stand a few moments for reaction to take place. Apply sodium thiosulphate solution to remove both excess iodine and the silver iodide formed. Follow by the regular wash.

Automobile Grease Stain Removal

This common source of oily stains produces some difficult ones that the ordinary laundering operation will not remove, especially the new long life high pressure greases featured for chassis lubrication. Local treatments with dry solvents such as carbon tetrachloride, trichlorethylene, gasoline, chloroform, etc., will sometimes suffice for removal. Frequently, however, a more efficient treatment is required and spotting with Kerosene-Oleic-Benzol cleaner, followed by a hot 1 to 2% solution of sodium metasilicate. Follow by the regular wash.

Balsam of Peru Stain Removal

The brown stain due to this resinous medicinal material responds to acetone, xylol, chloroform and alcohol.

Benzoin, Tincture of (Stain Removal)

Treatment with a sufficient quantity of alcohol will completely remove the brown stain due to this medicinal compound.

Bleaching Silk or Wool

It is frequently necessary to whiten silk or wool articles, which have become yellowed. The best method for the stain removal operator to use is a treatment with 1 volume hydrogen peroxide made alkaline with ammonia or Escolite.

Make up the 1 volume peroxide bath in a crock or non-metallic container by diluting one part of drugstore hydrogen peroxide (10 volume, 3%) with 9 parts of water or 1 part of the (100 volume, 30%) material with 99 parts of water. Add the ammonia water or Escolite to produce an alkaline bath and immerse the articles to be bleached. The time necessary for bleaching is a function of the depth of off-color to be taken care of.

Iron in any form should be absent and heat should not be applied.

Sulphites will also bleach silk and wool but the control necessary precludes their use in laundry practice for this purpose.

Never use a chlorine bleach on silk or wool.

Carbon Stains

The more finely divided the form in which the carbon in a stain occurs, the more difficult it is to remove. Some of the carbon blacks produced by burning natural gas or oil and used as pigments in polishes, paints, etc., are so fine that their removal from stain residues is difficult. The regular wash is usually not sufficient to take them out without assistance.

Saturate the dry goods containing the carbon stain with Kerosene-Oleic-Benzol cleaner and follow with the regular formula in the washwheel.

Cement Stains

Cement stains appear to be the result of chemical action between the cellulose of the fabric and the lime in portland cement and are formed when goods are allowed to lie on a wet concrete floor. This discoloration is extremely resistant. While part of the lime can be removed by an acetic acid treatment and the residual color lightened by treatment with chlorine bleach, the fabric never completely regains its original condition.

Chocolate and Cocoa Stains

The coloring matter of these stains is bound with fatty matter, sugar and milk solids. Spots remaining after the regular laundering can be stripped with one of the solvents for grease.

Coffee Stains

Ordinarily the brown stains caused by coffee respond completely to the regular washing process. Occasionally, however, the effect of alkali changes them over to a bright yellow color in which form they are no more difficult to remove than before but show up more plainly. The Permanganate-Oxalic treatment will take care of any coffee stains, brown or converted by alkali to yellow, remaining after normal laundering.

Contact or Loose Dye Stains

The first step in attempting to strip bled dye taken up by other goods in the washwell from a loosely dyed article such as a sock or cheap print, accidentally included in a load, is to give the stained articles a thorough treatment with chlorine bleach.

If this treatment does not succeed, the contacted dye is usually of the type requiring the action of a reducing agent. Treatment with sodium hydrosulphite or sodium sulphoxalate formaldehyde is therefore recommended. Either of these strippers may be also used on silk or wool if carried at lower concentrations than that given for cotton or linen goods.

The intensity of the stain determines the amount of reducing stripper required, but a good average can be taken as 4 ounces of sodium hydrosulphite or sodium sulphoxalate formaldehyde for a load of clothes, added to 2 inches of cold water in the wheel and run for 10 minutes. The treatment may be repeated to keep the sulphite bath definitely alkaline since hydrogen sulphide is formed if the solution should become acid and will blacken the metal.

For local spotting of contacts, the same strength solution can be used in a dish. Pyridine-sulphite is also effective on some types of dye contacts.

One of the most powerful reducing stripper treatments for dye stains known is the acid solution of either titanous chloride or titanous sulphate. These are violet colored liquids, always marketed in solution form (20%) and act by changing chemically from the trivalent or titanous salt to the more highly oxidized quadrivalent or titanic equivalent. In this change, the violet color is lost, indicating when the reducing power of the solution has been expended.

For contact dye stains, make up a 1 fluid ounce per gallon solution of titanous chloride (20%) or titanous sulphate (20%). Immerse the stained articles for 15 to 30 minutes, remove, wring out and rinse thoroughly with water.

Heavy contacts require the addition of about 1 tablespoonful of oxalic acid crystals to the above bath at the start.

The treatment with titanous reducers should not be carried out in a metal container. Crockery, glass, wood or vitreous enamel should be used to prevent attack by the acid present.

Blue vat dyes which have become green in the course of bleaching with white goods processed with an indanthrene blue in manufacture, are restored to their original shade by an acid titanous chloride or sulphate treatment.

Vat dyes are normally very fast to alkalies and bleaching since their colors are developed by oxidation in the dyeing process. When, however, they have been transferred by con-act, strong alkali, heat and pressure to another article or to a white adjacent area in the same garment, they are very difficult to remove. Treatment with a reducer causes temporary decoloration but the stain returns on contact with the air, after rinsing, unless the dye is actually destroyed.

The following treatment (*with care*) is effective for vat dye contacts:

Spot the stained area with a 10% solution of potassium dichromate. Allow to stand for 5 minutes, then spot (without rinsing) with a mixed solution of 1% oxalic acid and 2% sulphuric acid. Follow with several thorough rinses. For complete safety from possible acid attack, add a little ammonia or put the article through the regular wash.

Fish Oil Stains

These stains produced by fish oils such as cod liver, haliver and others in present day use as vitamin sources in diet give rise to yellow or brown stains and an unpleasant, fishy odor in goods from which they have not been completely removed.

In instances where fish oil stains persist after washing or as a precautionary measure on badly stained unwashed goods, saturation with Turkey Red Oil (sulphonated castor oil) or Kerosene-Oleic-Benzol cleaner, followed by the regular wash, will bring about removal.

For the final traces of organic coloring matter left after the oily body of some of the dark, high vitamin fish oil concentrates used in infant feeding has been removed, the peroxide-soap soak is very effective.

Fruit Stains

Stains from fresh fruit have greater resistance to removal from fabric than those from cooked fruit since the cooking process coagulates some of the natural fixative substances present. Consequently the summer season brings more trouble from fruit stains than the balance of the year, both because fruit is more plentifully used at that time and because a great deal of it is used uncooked. In contrast to fresh fruit, cooked fruit contains added sugar from the preserving process which aids in keeping the spots soluble and more easily removed in washing.

With cotton and linen goods, no difficulties are usually encountered in the complete removal of fruit stains by normal laundering. Silk, wool and fugitive colors, however, require special treatment:

1. Warm water.
2. Hydrogen peroxide made alkaline with ammonia.
3. Rinse.
4. Permanganate-oxalic.
5. Thorough rinsing.

Sodium hydrosulphite and sodium sulphoxalate formaldehyde are also useful when reducing conditions are required.

Faint, brownish tannin residues from fruit are effectively treated with Pyridine-Sulphite.

Gentian Violet

Violet stains from this medicinal dye are treated as follows:

1. Soak in alcohol-acetic for 5 minutes.
2. Rinse.
3. Light chlorine bleach.

Old stains may leave a yellow color after this treatment. These are removed by drying the fabric and again applying the same reagents.

Glue

Ordinary animal or fish glue is usually removed by normal laundering. Heavy spots may require a period of soaking in hot water followed by bleach, before washing.

Rubber cement requires treatment with a dry solvent such as gasoline, carbon tetrachloride or trichlorethylene.

Cellulose nitrate cements (airplane dope, collodion, etc.) are readily removed by acetone or amyl acetate. (Banana oil.)

Grass Stains

Except for those which are heavily ground in, grass stains usually require only normal laundering for removal from cotton and linen. Increased bleaching for heavy stains is called for since this phase of the formula is the most effective part of removal.

On silk, wool or fugitive colors where chlorine cannot be used, the green coloring matter (chlorophyll) is removed by the use of acidulated alcohol or gasoline in which it is soluble. Sponge the dry spot with alcohol-acetic mixture or with gasoline, using no water since the green natural dye body is not water soluble.

Grease and Oil Stains

Ordinary grease and oil stains are readily removed by treatment with solvents such as gasoline, ether, xylol, chloroform, etc. These reagents are also effective on chewing gum, paraffin, tar and various waxes.

A very common type of resistant stain encountered in the laundry consists of unsaponifiable grease and oil carrying finely divided pigment. These stains originate in so many ways that it is impossible to enumerate all or describe the many greases, oils and pigments composing them. Typical of these are lubricants which have picked up finely divided metal, graphite, carbon, etc., in service, tar oil and stove polishes.

In contact with textiles the greases and oils of such stains are taken into the fiber by absorption and capillary action, carrying with them the finely divided pigments, which become so embedded in the fiber that ordinary washing and solvents are ineffective. In fact, the removal of the grease carrier in this way greatly lessens the chances for complete stain removal.

In recommending a method for removal, let us say at the start that aged stains and those which have previously been set or treated to remove all carrier are not likely to be removed completely by any method safe to the fabric. They are often quite permanent.

For these stains use the Kerosene-Oleic-Benzol treatment in conjunction with one of the colloidal detergents. The Kerosene-Oleic mixture has previously been offered as a treatment for this type of stain, but the outstanding effectiveness of these colloidally active builders as after-treatment in place of ordinary alkali gives this method a decided advantage. The addition of the small amount of benzol increases penetration and greatly extends the scope of stain solvency.

When a dry stained article is

soaked in the Kerosene-Oleic-Benzol cleaner, penetration takes place and the oily matter of the stain is dissolved, diluted and homogeneously mixed with saponifiable oleic acid. This action brings about a loosening of the pigment particles, acts as a lubricant, and envelops the particle with a readily saponifiable film. When, after allowing sufficient time to react, the article is passed without rinsing into a hot solution of one of the colloidal detergents, the stain is subjected to both physical and chemical forces. Peptization, adsorption and colloidal action take place simultaneously with saponification, with the result that the pigment is caught in an active, live emulsion which is readily washed out of the goods.

Ichthyol Stains
(Black or Gray)

Its fresh stains are removed in the regular wash.

For old hardened stains a presoftening by dipping in cresol solution is effective, before laundering.

Indelible Pencil Stains

Such stains as do not come out in the washing process may be loosened with acetic acid and alcohol.

Ink Stains

(a) *Writing Ink*

There are so many different kinds of writing ink that it is impossible to prescribe a remover that will be effective on all ink spots. The following treatments should be tried in the order listed:

1. Dilute bleach solution.
2. Dilute hydrofluoric or oxalic.
3. Permanganate-oxalic.
4. Commercial ink removers.

Stains on wool and silk should be treated with hydrogen peroxide followed by dilute hydrofluoric acid.

(b) *Marking Ink*

Some marking inks contain silver nitrate or other silver compound.

Treat the stains with bichloride of mercury and ammonium chloride solution as described under Silver Stains.

Marking ink of the aniline black type cannot always be successfully removed. Carbolic acid containing a small amount of nitrobenzol and aniline oil is often effective on fresh stains.

(c) *Waterproof India Ink (Drawing)*

Spot or immerse the dry stained fabric in carbon disulphide. Follow with regular soap and water wash.

Iodine Stains

When iodine is dropped on unstarched material, it makes a brown or yellow stain, but on starched material a deep blue or black. For removal, spot with 5% solution of sodium thiosulphate (hypo).

Iron Rust Stains

An iron rust stain is often formed after only a momentary contact of the cloth with iron, particularly if the cloth is white. If, for instance, goods are washed at home and then laid to dry over a radiator, it is no uncommon thing to find a series of rust spots wherever the cloth came in contact with exposed metal on the radiator. Treatment with oxalic acid or hydrofluoric acid with careful rinsing is the most effective means of removal.

Removal of Rust Spots from Fibrous Materials

Twenty-five parts 65% phosphoric acid in 75 parts glycerol of 28° Bé; glycerol added to a 15% aqueous citric acid solution until the specific gravity is 17–18° Bé, phosphoric acid added to specific gravity 30° Bé.

Lacquer and Nail Polish Stains

The use of lacquers in replacing paints and varnish not only in industry but also in the household is reflected in the frequent appearance of stains from this source upon linen and wearing apparel.

The lacquers differ widely from paints and varnish in that they do not react chemically with alkaline washing solutions as do paint and varnish. The base of the lacquers and some of the popular nail polishes is nitrocellulose which is dissolved in such solvents as amyl, butyl and propyl acetates together with a suitable pigment. Upon application, the solvent evaporates leaving the insoluble, water resistant nitrocellulose. The very best solvents or stain removers for this type of composition are, therefore, amyl acetate or acetone.

Acetone is recommended because it has a more agreeable odor and can be applied readily without discomfort to the operator. However, both acetone and amyl acetate are excellent for this work and spotting the dry fabric will prove very satisfactory in the majority of cases. These solvents should not be used on cellulose acetate fibers (celanese, etc.).

Leather Stains

Good washing usually removes leather stains such as are most commonly caused by friction of shoe linings on light colored stockings. However, if any amount of tannin compound is present, it may leave a stain which can sometimes be removed by treatment with potassium permanganate. Treatment with pyridine-sulphite is also highly recommended where tannin is encountered.

Leather Grease Remover

Carbon Tetrachloride	79
Petroleum Ether	15
Amyl Alcohol, Tertiary	4
Diglycol Laurate	1
Butyl "Cellosolve"	1

Lime Soap Stains

Treatment of the stained articles with strong, boiling hot solution of acetic acid followed, after a rinse, with a strong solution of one of the colloidal detergents will prove very effective in the removal of these stains.

For spotting, gasoline or carbon tetrachloride is recommended. Treatment with a strong solution of sodium metaphosphate is also effective. Oleo Calcaire (olive oil and lime water) and other insoluble soap ointments should also be treated as above.

Medicine Stains

In general treat with bleach or with potassium permanganate solution, followed by oxalic acid and thorough rinsing.

There are, of course, many staining substances which are classified as medicines, e.g., Argyrol, Mercurochrome, iodine, etc. The treatment for each is given elsewhere in this discussion. Wherever medicines are in use the precaution of rinsing the goods in warm or cool water after staining is highly advisable.

Mercurochrome Stains

The red dye stain can be removed from cotton through oxidation. When the stain is fresh, bleach will usually bring about removal. When old and set, it is more resistant and the following method is recommended:

Work the stained articles in a 2% solution of potassium permanganate at 100°–130° F. for 5 minutes; rinse and then work in a 2% solution of oxalic acid at 120°–130° F. until all brown coloration is gone; rinse thoroughly.

This treatment will remove all ordinary Mercurochrome stains and can be repeated in special cases where found necessary.

Mercurochrome stain on animal fibers is subject to reactions which make it permanent. Substantial lightening of stain, if not complete removal, can be obtained by the use of acidified alcohol (50:50 alcohol and acetic). For cleaning up stains on hospital equipment a mixture of 90% alcohol and 10% sulphuric acid is generally employed, but this, of course, should never be used on fabrics.

Methylene Blue Stains

Methylene Blue is a basic dyestuff used in commercial and pharmaceutical preparations. Alcohol is usually effective on fresh stains. Pyridine-sulphite treatment is recommended for old stains.

Mildew Stains

Mildew is due to micro-organisms commonly known as fungi, which have the property of attacking textile fibers and producing various colored stains. They grow rapidly when the goods are in a damp condition, and especially when the articles contain soil which serves as food. It has recently been discovered that improperly operated air conditioning in drygoods stores during summer months, resulting in excessive humidity, causes extensive amounts of mildew on new goods. Mildew stains respond to oxidizing agents, and for their practical removal there is no more efficient reagent than bleach. In the treatment of mildew stains in the laundry, a formula can be devised which will remove them in a regular wash, but to treat an entire load with bleach for the purpose of removing stains from a comparatively small percentage of the work would be very poor practice. It is recommended that the mildewed articles be culled out and given a special stain removing treatment, when a quantity sufficient to make a load is obtained. For treatment a short bleach suds or plain bleach, followed by sufficient rinses, will serve to remove the stains.

As mentioned above, there is no washing formula which can be used to remove mildew stains without the use of bleach, and from the standpoint of economy, as well as the life of the linens, we do not recommend that attempts be made to remove resistant deep-seated stains in the regular wash.

Mud Stains

Mud stains constitute a very difficult type of stain since, for the most part, the staining material is chemically inert and must be broken up and suspended by colloidal action. Plenty of mechanical washing action in the presence of soap and one of the colloidal detergents, plus iron-removing sour, is the best method of attack.

Oil Stains

For the removal of the common oil stains see the treatments described under Grease and Oil Stains.

The preparations used in hair treatment which contain an appreciable percentage of crude (mineral) oil give rise to a very conspicuous resistant brown stain. These stains are largely found on bed linen and especially pillow cases. The ordinary oil solvents such as gasoline and carbon tetrachloride are effective before washing, but the residual set stains after laundering are practically impossible to remove completely.

The previously mentioned Kerosene-Oleic-Benzol cleaner is usually successful, but not always positive.

Ointment Stains

Ointment is often water repellent, which accounts for the difficulty in removing stains in the ordinary washing process. To overcome this it is necessary to remove the greasy material as outlined under Grease and Oil Stains and then follow with the recommended treatment for the residual stain.

Paint Stains

Oil paint stains generally consist of a finely divided pigment held in the fiber by vegetable oil, varnish gums and turpentine. The hardening or drying of both paint and varnish is due to oxidation of the oil, which converts it to a solid state, thus holding the pigment of the paint or the gum of the varnish firmly within the fibers, and renders the removal of old stains a difficult job.

Light paint stains may be removed by the ordinary washing formula used for cotton and linen goods. Colloidal detergents are highly effective on oxidized oil. When the stains are on delicate goods, the use of special solvents suggested for automobile grease and road tar stains is necessary. A special treatment may also be required if the paint stains are old and have been thoroughly hardened in the fiber. Castor oil containing a small amount of strong acetic acid is valuable in spotting. Oftentimes aniline followed by removal with alcohol is also excellent. (See Lacquer.)

Perspiration Stains

The composition of perspiration has never been definitely determined, one of the principal reasons being that it varies in different individuals and also under conditions of exercise and repose.

Its reaction from most parts of the body is acid, while as a rule it is alkaline in the armpits. Organic acids, inorganic salts such as sodium and potassium chloride, phosphates, small amounts of iron and albumin are invariably present.

It has a peculiar and destructive action on many dyes. If a perspiration stain is fresh it will largely be removed in the ordinary washing process. On white goods, a yellowish tinge sometimes remains (gilt edges on collars), which is removed by reducing in acid solution as in oxalic acid souring. The older a stain is the more fixed it becomes and consequently the more difficult to remove. It is next to impossible to restore color in a garment in which the dye has been loosened or changed chemically by perspiration. A salt solution acidified with formic acid is often useful in spotting delicate fabrics.

Picric Acid Stains

Picric acid dyes cotton a permanent yellow and produces stains which are exceptionally fast to washing. For removal, immerse the dry stained fabric in a 50:50 mixture of alcohol and ammonia water (regular household 5%).

Printing Ink Stains

These inks commonly consist of finely divided carbon suspended in oils with rosin, turpentine, etc. They virtually constitute a paint or varnish. An extra thorough washing will generally prove effective. For spotting, use a mixture of equal parts of alcohol, amyl acetate, benzol and acetone. This formula should not be used on celanese, etc. (See Carbon Stains.)

Prontosil Stains

This stain is a bright red dye commonly used in medicine. It is generally removed in the regular laundry process, but if a slight residual stain persists it can be entirely cleared up with a light hypochlorite bleach.

Red Clays Stains

Red clays produce a very intensive stain because of their high iron content. Treat as recommended under Mud Stains and follow with hydrofluoric or oxalic for iron removal.

Road Tar Stains

Unfortunately, road tar stains are not usually noticed before they have been wet with water, or in other words, before they have passed through the laundering process. An ordinary washing formula has but little effect upon them. The use of special solutions such as carbon tetrachloride, carbon bisulphide, gasoline, turpentine, or chloroform is required. (See Grease and Oil Stains.)

Tar Remover

For use in removing tar from clothes and autos.

Benzol	50
Carbon Bisulphide	10
Carbon Tetrachloride	40

Toluol can be used in place of benzol. Use methyl salicylate as perfume.

Rouge Stains

Rouge stains can be most readily overcome by removing the carrier (binder) with a suitable solvent before washing with soap and alkali. The carriers sometimes used in rouge compounds are oil and grease. Carbon tetrachloride is an excellent solvent for such carriers. In persistent cases acetone may be more effective. The coloring matter in rouge is usually iron oxide or dye such as carmine. Washing directly with soap and alkali sets these in the fabric. There is one type of cosmetic color known as carmine red, which is very resistant, and for complete removal requires severe rubbing, after moistening, with amyl alcohol. The common lipstick stains do not offer difficulty unless set by age or previous treatment. The iron pigment yields to the common rust removing sours in concentrated form, and the basic dyes offer no difficulty to solution in alcohol after the binder is removed.

Scarlet Red Stains

This stain is a bright red medicinal dye and therefore frequently found on hospital work. The color is removed or considerably reduced in the regular wash. It readily responds to acetone when heavy stains are present.

Scorch Stains

It is easier to remove scorch from cotton and linen than from wool and silk. In a great many cases surface stains can be quickly removed, especially on white goods, by treatment with ammoniacal hydrogen peroxide, which procedure follows: Dampen a small cloth with commercial 3% hydrogen peroxide to which a few drops of ammonia have been added and place it over the scorch stain. Cover with a dry piece of bleached muslin and iron with a medium hot iron. Repeat if necessary. For heavy stains the only helpful treatment is repeated washing with adequate bleaching, as no procedure can be given which would remove heavy scorch stains successfully. To try to remove heavy scorch stains is usually a needless waste of time.

Shoe Polish Stains

Treat stain with gasoline and follow with alcohol-acetic mixture.

Silver Stains

Silver stains which usually show up brown or black do not respond to washing or the common stain removers. They usually develop from colorless silver salts, such as silver nitrate, which are not noticed until the damage is done.

For the removal of silver stains the bichloride of mercury and ammonium chloride treatment, which is described under Argyrol, is highly recommended.

Smoke Stains

Smoke stains are insoluble, consisting chiefly of carbon and generally accompanied by creosote, a substance which produces a yellow stain and has the characteristic odor of smoke. The carbon in most cases can be emulsified and washed away. In order to remove the yellow creosote stain, the material is treated for a reasonable length of time with cresol, the excess being saponified by the alkali detergent in the washing process.

Tannin Stains

Tannin stains are among the most resistant encountered in the laundry.

Leather, tobacco, tea, medicine, Coca-Cola, fruit juices, coffee, liquor, ginger ale and an innumerable quantity of common household preparations contain tannin in greater or lesser amounts. Articles coming in contact with tannin-bearing substances should be rinsed as soon as possible in order to avoid permanent stains.

Tannin reacts with metals producing insoluble compounds. Iron reacts to form black tannin ink, which at one time was a very common source of writing fluid.

Tannin is used extensively as a mordant for cotton in dyeing with basic dyestuffs. Because of its mordant properties, some laundry blues take to tannin stains producing blue spots and streaks which impair quality.

A peculiar characteristic of these stains is their invisibility when first formed, with a gradual development, depending upon atmospheric conditions and age.

Fresh and moderately aged stains are most readily removed by the pyridine-sulphite treatment. Pass the stained article into a hot 5% solution of pyridine; remove and pass directly into a hot 5% solution of sodium hydrosulphite. Alternate from one solution to the other until the stains are completely removed.

Old set tannin stains are very apt to prove permanent.

Tea Stains

The brown coloring matter in tea is not difficult to remove when fresh, but becomes very persistent on standing. Stains from tea to which milk or cream has been added are usually more easily removed than those from clear tea. Potassium permanganate and ordinary chlorine bleach are effective on tea stains that have resisted ordinary washing. Residual tannin stains should be given the pyridine-sulphite treatment.

Verdigris Stains

The blue-green tarnish on copper, brass, and bronze, consisting of basic copper carbonate, is commonly known as verdigris. Stains that are not removed by the ordinary washing process can be dissolved by using a dilute acid such as acetic.

The green deposit which develops on brass and bronze washwheels is not, chemically speaking, verdigris, but a complex conglomeration of metallic soaps and salts. Nevertheless, all green-colored deposits on copper-bearing metals are commonly referred to as verdigris, and the term as used is generally so understood.

Soap, or more accurately soap fatty acids, give green, insoluble copper soaps, which adhere firmly to metal parts.

Regardless of the care exercised, some insoluble lime, magnesium or other metallic soaps are formed in washing. Being insoluble and sticky, they have a tendency to cling to surfaces. When goods are soured, fatty acid is liberated from these soaps, which, in turn, is set free to react with copper on neutralization.

Sours, especially acetic and oxalic, rapidly form verdigris, and most of the alkalies, such as silicates, carbonates and phosphates, yield salts which are green in color.

To prevent staining by contact with verdigris in washing, the machine should be periodically cleaned. An acid bath of sulphuric acid, or less preferably hydrochloric, is highly recommended. In cleaning, run two or three inches of water in the machine; add about one quart of sulphuric acid, depending upon the wheel size; start the machine and allow to run until all deposit is removed. An old broom held against the revolving cylinder greatly facilitates removal. After cleaning, the machine should be rinsed; given an alkali bath and rinsed again.

The usual precautions should be taken in adding concentrated sulphuric acid, i.e., run the acid very slowly into the water.

Water Spots

In sponging some goods, notably silk and wool, an irregular ring or spot usually remains. Under such circumstances it may be necessary to dampen the entire material evenly and press while still damp. Sponging the surface with alcohol is also an effective treatment. A moist chamois is useful in drying the dampened goods.

Yellow Stains

The term "yellow stain" is often used too broadly. The "yellow" may be due to unrinsed alkali or soap; it may be rust, it may be oxycellulose, it may be a contact stain from a loose dye.

In general, such of these stains as do not identify themselves readily as rust, alkali or soap, will usually respond to a strong reducing bleach. Unless the reducing action is quite intensive, however, there is a possibility of the stain reappearing through atmospheric oxidation. Potassium permanganate followed by oxalic acid is usually very effective.

Spot Remover

Carbon Tetrachloride	25
White Gasoline	25
Solvent Soap (Diglycol Laurate)	50

Cleaners' Spotting Mixture

Formula No. 1

Aerosol MA	1	lb.
Toluol	3	qt.
Perchlorethylene or Toluol	1	qt.
Water	½	pt.

No. 2

Aerosol OT (100%)	1	lb.
Stoddard Solvent	2	gal.
Water	3	pt.

Grease Spot Remover

A paste of benzinized magnesia is suitable for this purpose, and is prepared by mixing calcined magnesia with just sufficient pure benzine so as to moisten it without being pasty. It should be just wet enough so that when the mass is pressed between the fingers, a small quantity of liquid benzine is squeezed out. In this state, it forms a crumbly mass which is kept for use in a well-corked, somewhat wide-mouthed, glass bottle. Apply by spreading the paste quite thickly over the stains and rub it thoroughly to and fro with the tip of the finger. Brush off the small lumps of earthy matter thus formed, lay on more of the paste, allowing it to remain until the benzine has entirely evaporated, and then brush off the adhering particles of magnesia.

Laundry Bluing Tablets

Make a thick paste with 6 ounces of ultramarine, 4 ounces of sodium carbonate, and 1 ounce of glucose. Roll into sheets and cut into tablets.

Laundry Bleaching

Cotton, Linen and Other Vegetable Fibers

Bath 1. Permanganate (2 oz. potassium permanganate per gallon of water).

Bath 2. Oxalic Acid (2 oz. oxalic acid crystals per gallon of water).

Immerse the stained articles in Bath 1 at a temperature of 120° F. until thoroughly saturated (1 to 2 minutes). Remove and rinse well.

Transfer to Bath 2 at 140–150° F. and work until all traces of brown coloration are removed. Add more oxalic acid to Bath 2 if the goods do not clear in 2 minutes or so. Rinse thoroughly with hot water to insure removal of all traces of oxalic acid.

Wool, Silk and Other Animal Fibers

Wool, silk and other animal fibers respond better to acid permanganate and in their case, the procedure above is modified only by the addition of a small amount of sulphuric acid to Bath 1. Otherwise the treatment is exactly the same.

Colored goods should not receive the permanganate-oxalic treatment without a preliminary test with it at an inconspicuous point to determine whether or not the dye is fast to the reagents.

Many contact types of dye stains and other organic color residues which respond to oxidizing measures are very effectively stripped by the permanganate-oxalic treatment.

In working with hydrogen peroxide for spotting or small scale bleaching where an alkaline builder is not required, the drugstore material (3% or 10 volume) which may have been made slightly acid in manufacture to hold its strength on the shelf, should be rendered alkaline with ammonia water. This neutralizes the acid, removing a source of later danger to the fabric by acid attack and liberates the nascent oxygen more rapidly for bleaching.

Concentrated hydrogen peroxide (30% or 100 volume) should be diluted with water to 3% strength before use on fabric and should be neutralized as above if acid is present.

Permanganate-hydrosulphite is interchangeable with permanganate-oxalic, the oxidizing stripper. The replacement of the oxalic acid by sodium hydrosulphite results in a quicker discharge of the brown manganese hydrate formed and this feature is of advantage in spotting work. A further advantage of this method over permanganate-oxalic in some instances, is that sodium hydrosulphite itself is an active reducing stripper and the swing from active oxidizing conditions in the presence of permanganate to active sulphurous reducing conditions in the presence of the hydrosulphite, often results in better stripping of a mixed organic stain.

Quantities and procedure are the same as for permanganate-oxalic except that the sodium hydrosulphite bath must be made with cold water.

Kerosene-Oleic Acid-Benzol Cleaner

This mixture is composed of:

Kerosene	1 gal.
Oleic Acid (Red Oil)	4 oz.
Benzol	4 oz.

Its use is in removing deep seated

grease stains and others of similar type as described under the heading Grease and Oil Stains.

Dye Stain Solvent

A mixture of 9 parts alcohol and 1 part glacial acetic acid. Useful for some types of dye stains. Example: methyl violet.

Laundry Starch

Wheat Starch	1
Corn Starch	2

Laundry Gloss

Gloss preparations to be used with starch are widely employed in the laundry, not only to lend a fine smooth finish, but also to prevent scorching and sticking of the iron. The following is a typical formula:

Acacia, Gum	1 oz.
Borax	2 oz.
Glycerin	1 oz.
Water	32 oz.

Soak the acacia in the water for six hours, add the borax, then heat to a boil and add the glycerin, let cool and strain. To use, add two ounces of this preparation to 3 quarts of the usual starch solution.

Textile Soap

(1)	Olive Oil Foots	600
	Potash Lye (20° Bé)	660
(2)	Corn Oil	800
	Rosin	200
	Potash Lye (27° Bé)	790
	Water	340

The oils are heated to 190° F. and then the potash lye is added in the crutcher and the mixing is continued until the soap begins to bunch. When this occurs the soap is run directly into barrels, which should be kept in a warm place for a day or two. The saponification process proceeds in the barrels.

Wool Thrower's Soap

Olive Oil Foots	12
Corn Oil	46
House Grease	20
Soda Lye (36° Bé)	3
Potassium Carbonate (Dry)	5¾
Potassium Hydrate (Solid)	23

Rayon Scourings

Formula No. 1

Box Scouring

Sodium Metasilicate	20 lb.
Soap, Low Titer	10 lb.
Water	1000 gal.

No. 2

Continuous Scouring

The scouring section is charged at the rate of 2 lb. of sodium metasilicate and 1 lb. of low titer soap per 100 gallons of water. Additions are made so that the total alkalinity never drops below 0.04% as Na_2O. The pH (alkali pressure) is maintained above 10.2 at all times; with regenerated cellulose rayons, a pH above 11 is economical and gives rapid scouring.

No. 3

Jig Scouring

Sodium Metasilicate	4 lb.
Soap, Low Titer	2 lb.
Water	50 gal.

No. 4

Sodium Metasilicate	1 lb.
Soap, Low Titer	2 lb.
Sodium Pyrophosphate	1 lb.
Water	100 gal.

Textile Mill Stain Removal

Oil and grease stains are perhaps the most prolific, due in some cases to careless or over-liberal use of the oil can, with consequent dropping upon goods in course of processing; and in others to inefficient machinery. Dye stains are another fruitful source of trouble, as, indeed, are also many of the chemicals employed in modern practice.

The labor element is responsible for introducing certain of the stains with which the technician has to contend; tobacco, gum and foodstuffs can cause a great deal of worry.

Then, again, most mills have their own constructional departments or, at any rate, carry stocks of such raw materials as paint, tar, asphalt and creosote, thus introducing a further stain hazard.

As these stains frequently occur on delicate fabrics, which require very careful treatment, it is disastrous to try various methods haphazardly in the hope that one will prove successful. The best method of treatment is to have a clear jar, basin or other receptacle ready, and to stretch the fabric tightly over it, and then to pour the solutions made up as directed through the material into the basin.

In the case, however, of such stains as oil, tar, grease and asphalt, it is more efficacious to place a pad of blotting paper or other absorbent material beneath the stain and to apply the cleaning fluid recommended by means of a small tuft of cotton wool or a pad of fabric.

The successful removal of stains is largely dependent on the correct method of working. The aim should be to work from the outer edge of the stain towards the center, as this lessens the risk of leaving a ring. It is better to dab the solution on, so as to carry the stain through into the blotting paper, and not to rub, as any friction, especially on delicate fabrics, is liable to leave a slight roughening of the surface after the discoloration has been removed.

Grass stains may be treated with hot methylated spirits or with hot glycerin followed by methylated spirits. Coffee stains can be removed by a treatment with a solution of potassium permanganate, followed by a slightly acidulated solution of hydrogen peroxide. Glycerin is also a useful reagent for coffee spots, and is best applied with a small sponge or a pad of cotton wool. The glycerin is more effective when used hot, and the fabric should be left for a few minutes saturated with the liquid, and then washed out with either methylated spirits or water. This method is also useful in the case of chocolate and cocoa stains.

Perspiration stains are best removed with warm water and ammonia when the fabric is wool or silk. For cotton or rayon a treatment with a solution of one part of sodium hypochlorite to four parts of warm water is better. This latter process must be followed by a thorough rinsing, first in clear water and then in acidulated water, and must not be used for any materials made of wool or real silk.

Scorch marks may be treated by damping the place and then exposing it to the sun. Two other methods recommended are:

(1) Treat with hydrogen peroxide acidified with an addition of sulphuric acid.

(2) Bleach with potassium permanganate, and then rinse in weak sodium bisulphate.

Stains caused by soot or lampblack will be found to yield to a treatment with some organic solvent such as ether, chloroform acetone or carbon tetrachloride, followed by washing with paraffin oil and then again in soapy water. Glue stains can be dealt with by means of hot methylated spirits or by a weak solution of soap and ammonia.

Stains caused by pitch, tar and asphalt should be treated with organic solvents as listed under soot stains. Turpentine and lard are also effective. Another method which has been recommended especially for the removal of tar stains from silk is to rub butter on the marks, and allow to remain some time, and then to go over them again with benzene.

The removal of paint marks is generally attended with considerable difficulty, especially in the case of marks of long standing. Here again the organic solvents are of great value. If the stains are not dry they may be removed with turpentine and benzene; a strong solution of ammonia is sometimes effective; while an emulsion of turpentine and ammonia is often advocated for this purpose, although this should not be employed for silk. Varnish stains may be treated in the same manner, although the presence of gums and resins makes their eradication even more difficult.

Ink stains frequently prove difficult, if not impossible, to remove by ordinary laundering methods; yet they yield fairly easily to proper treatment. The most common ink stains are, of course, those caused by ordinary writing inks. A dilute warm solution of oxalic acid is generally successful with blue-black inks, but care must be taken not to leave this reagent in the material; it is also liable to remove the color from dyed fabrics. Another method is to treat with a dilute solution of potassium permanganate, and to follow this with a slightly acidified solution of peroxide of hydrogen, as was employed for coffee stains.

A third method is to lay the stained portion on a saucer, moisten with hot water, then to rub in salts of lemon until the stains disappear and finally rinse thoroughly.

Red inks can sometimes be removed by the use of methylated spirits slightly acidified with acetic acid. If this method fails, a warm solution of sodium hydrosulphite will remove the stain. This reagent, however, will decolorize most dyed goods, and should therefore be applied with caution.

Printers' ink stains are less commonly met with, which is fortunate, as they frequently prove troublesome. One course is to rub them with lard, and a second to soak in turpentine, and then to wash them out with methylated spirits. Such organic solvents as chloroform, acetone and carbon tetrachloride are also useful.

Stains due to Indian inks are often impossible to remove completely. Three solvents which are sometimes successful are:

(1) Hot methylated spirits.

(2) Glacial acetic acid.

(3) Chloroform or carbon tetrachloride.

Marking inks are of two types: One is made from an organic dye. (Goods marked with this type have to be washed before ironing, to fix the color.) These are practically impossible to remove when fixed.

The second is the silver nitrate type, where the goods have to be ironed before washing to fix. A repeated treatment with a weak solution of chloride of lime, followed by washing in ammonia water, will often remove these stains. Corrosive sublimate and nitrobenzene are two further reagents which have been recommended.

Blood stains may be removed from silk by the use of strong borax water or by soaking repeatedly with chloroform or hydrogen peroxide. Many medicine stains can be dissolved out in alcohol or methylated spirits. Iodine marks are best treated with ammoniated alcohol or ammonia or by sodium hydrosulphite. Picric acid leaves yellow stains on silk, wool and acetate silk, which can best be removed with a paste of magnesium or lithium carbonate. Ordinary laundering methods will remove these stains from cotton and rayons other than acetate.

A mixture of

Aniline Oil	1
Powdered Soap	1
Water	10

is advised for the removal of vaseline marks from fabrics. The stained portion of the goods should be allowed to remain in the solution for some time, and then be well washed with water.

Acid stains should be well washed and then neutralized with an alkali, a warm solution of soap and ammonia being particularly effective. Alkali stains, after a preliminary wash, should be immersed in dilute acid, such as acetic or formic.

Rust stains are of frequent occurrence, and can be dealt with by means of a mixture of one part of potassium bioxalate added to forty-five parts of water and one part of glycerine. A well-known remedy is to moisten the spot with lemon juice and salt and expose it to the sun. One or two applications are most successful. This method is often advocated for ink stains.

Grease stains can be dealt with by the use of organic solvents, such as petrol, benzene or carbon tetrachloride, or by covering with blotting paper and ironing with a hot iron. Black grease, however, such as is formed on axles and machinery, is more difficult to remove, as it contains minute particles of iron, removed by friction, and when the grease itself has been removed these iron particles leave a yellowish-brown stain. It is best in these cases to remove the grease as above, and then to treat as for rust stains or by the use of a warm solution of weak oxalic acid.

Color stains caused by dyestuffs generally require bleaching with one of the usual agents—hypochlorites, hydrogen peroxide or a permanganate—for their removal, although many basic dyes will yield to a treatment with ammonia and methylated alcohol.

Before applying these cleaning methods it is safer to try them on a spare scrap of material, if one is available; failing this, use some corner of the fabric which will not show. In this way one can see if the reagent is going to decolorize the fabric or have any other harmful action, as it is not possible to foretell in all cases what effect a given reagent will have on the dyestuffs which have been used.

Removal of Stains from Fabrics

Fabrics used in our daily life are made from materials such as cotton, silk, wool and artificial silk. These fabrics often get stained with substances such as butter, vegetable or animal oil, ghee, fats, tea, coffee, fruit juice, wine, beer, ink, iron, rust, iodine, paint, varnish, tar, blood, mildew, perfume, etc. These stains can be removed without damaging the fabric. But there are some stains which cannot be removed and are, therefore, better left alone. Ordinary stains such as those stated above can be removed by the application of the correct reagent, and this in its turn depends upon ascertaining the substance causing the stain and the material from which the fabric is made. If the fabric is dyed it is also necessary to know, in order to remove the stain without removing the color of the dyed fabric, whether the dye used is fast or fugitive. Certain reagents which can be safely used for removing a particular stain from a cotton fabric, if applied on silk or wool, will damage the cloth. Then again it must be borne in mind that fresh stains can be more easily removed than old stains and, therefore, efforts to remove the stain should be made as early as possible. In any case, the cloth should be washed with hot soap solution immediately on discovering the stain. Several stains can be removed easily by the mere application of warm soap solution, or by holding the cloth tight on an empty vessel and pouring boiling hot water from a kettle on the stained spot.

If the cause of the stain is not known and cannot be identified, then much difficulty will be experienced in selecting the suitable reagent for removing the stain. In such cases, it is necessary to carry out several trials with different reagents which are suitable for the nature of the fabric. If the material from which the fabric is made is not known, it is necessary to conduct a simple test. Some fibers may be taken out from the fabric and held in the hand and the gentle flame of a match applied to them. Cotton and artificial silk will burn rapidly, giving a smell of burning paper and leaving a little ash. Wool and silk will, on the other hand, burn rather slowly, emitting a disagreeable smell of burning feather, and leave a black bead at the end. If the fabric is dyed, trials will have to be carried out with different reagents or hot soap solution on a corner of the fabric to ascertain whether the color is fast or fugitive. In some cases, it will be found that the stain cannot be removed without removing the color of the cloth at the same time.

Stains are removed by different methods of treatment such as solvent, chemical and absorption.

Solvent Method

In this method the stain is dissolved so that it may pass from the fabric on to a blotting paper pad placed underneath the fabric. The stain is absorbed by the blotting paper and the solvent in the fabric evaporates. Before proceeding to remove the stain, the portion of the fabric which is stained should be washed with hot Sunlight soap solution in preferably a solution of Igepon T. Soap. The fabric should then be dried.

Now a pad, consisting of a few sheets of blotting paper, is placed on the table and the cloth containing the stain laid on the top of the blotting paper. A small quantity of the reagent is poured in a small cup. A clean cotton muslin rag is then dipped in the reagent, the excess quantity squeezed and the rag pressed gently on the stain from the outer side of the stain to the center. The solvent will carry the stain into the blotting paper. At every application of the solvent, a clean rag and a fresh blotting paper should be used. The excessive use of reagent at every application should be avoided, in order to prevent the formation of a ring around the stain. The portion of the cloth stained may finally be washed in cold water. In removing stains, the rag dipped in the reagent should be pressed gently but not rubbed on the fabric, as it will cause the dissolved stain to spread out on the fabric. In some cases more than one solvent will have to be applied in succession in order to remove the stain. As many of the solvents are inflammable, they should be kept at a safe distance from any flame.

Chemical Method

In this method, the stain is decomposed or bleached and the chemical that remains in the fabric is removed by washing. Before proceeding to remove the stain, the portion of the fabric which is stained should be washed with hot Sunlight soap solution or preferably a solution of Igepon T. Soap. The solution of the chemicals is first poured in a small cup. A white cloth is placed on a table, and the fabric containing the stain laid on the cloth. A clean cotton muslin rag is then dipped in the solution and the rag pressed gently on the stain. After the stain has been removed, the fabric is washed in cold water to remove the chemical.

Absorption Method

This method is applicable only in the case of grease and lubricating oil stains. As magnesium carbonate has the property of absorbing oil and grease, it is rubbed on the stain and left until the oil is absorbed, and then brushed off. For removing old stains, powdered magnesium carbonate is mixed with benzol (not petrol) and then the paste is rubbed on the spot and allowed to dry and brushed off. This is useful on heavy garments such as coats.

Hot Application

This method is applicable in the case of stain produced by rain-water on silk fabrics. A white cloth is laid on the table, and the stained fabric is placed over it. A wet cloth is then placed on the stained cloth, and a dry cloth is placed over it. A hot iron is pressed on the cloth in order to remove the stain.

I. Nature of stain.—Milk, butter, ghee, vegetable and animal oils, fats.

Class of fabric—Cotton, silk, wool and artificial silk.

Reagent to be used—Carbon tetrachloride.

Method to be adopted—Solvent.

II. Nature of stain.—Tea, coffee, cocoa, chocolate, fruit juice, wine and beer.

(a) Class of fabric—White cotton and artificial silk.

Reagent to be used—Sodium hypochlorite solution (1 oz. of sodium hypochlorite in 10 oz. cold water).

Method adopted—Chemical.

(b) Class of fabric—White silk, wool and colored cotton and colored artificial silk.

Reagents to be used—Ammonia solution (1 oz. ammonia in 5 oz. water) followed by hydrogen peroxide (12 volumes strength).

(c) Class of fabric—Colored silk and wool.

Reagents to be used—Ammonia solution (1 oz. ammonia in 5 oz. water).

Method adopted—Chemical.

III. Nature of stain.—Ink and iron rust.

(a) Class of fabric—White cotton and artificial silk.

Reagents to be used—Hot oxalic acid solution (½ oz. oxalic acid in 16 oz. water) followed by sodium hypochlorite solution (1 oz. in 10 oz. water).

Method to be adopted—Chemical.

(b) Class of fabric—Colored cotton and artificial silk.

Reagent to be used—Hydrochloric acid (1 oz. acid in 15 oz. water) followed by washing and then treatment with ammonia (1 oz. ammonia in 5 oz. water).

Method adopted—Chemical.

(c) Class of fabric—White silk and wool.

Reagent to be used.—Warm oxalic acid solution (½ oz. oxalic acid in 16 oz. water) followed by washing and then treatment with ammonia (1 oz. ammonia in 5 oz. water).

Method adopted—Chemical.

IV. Nature of stain.—Iodine.

Class of fabric—Cotton, silk, wool and artificial silk.

Reagent to be used—Sodium thisulphate solution (1 oz. thisulphate in 5 oz. water).

Method adopted—Chemical.

V. Nature of stain.—Grease, tar and wax.

Class of fabric—Cotton, silk, wool and artificial silk.

Reagent to be used—Cocoanut oil followed by hot soap washing and then treatment with carbon tetrachloride.

Method adopted—Solvent.

VI. Nature of stain.—Paint and varnish.

Class of fabric—Cotton, silk, wool and artificial silk.

Reagent to be used—(1) Methylated spirit followed by turpentine and soap washing finally after drying.

(2) Carbon tetrachloride.

(3) Mixture of acetone and amyl acetate (1 oz. of acetone and 1 oz. of amyl acetate).

VII. Nature of stain.—Blood.

(a) Class of fabric—White cotton and artificial silk.

Reagent to be used—Sodium hypochlorite solution (1 oz. in 10 oz. water).

Method adopted—Chemical.

(b) Class of fabric—Wool and silk and colored cotton and artificial silk.

Reagents to be used—Acetic acid (2 oz. concentrated acetic acid ⅛ oz. common salt in 19 oz. water) followed by washing and treatment with ammonia (1 oz. ammonia in 5 oz. water).

Method adopted—Chemical.

VIII. Nature of stain.—Mildew.

Class of fabric—Cotton, silk, wool and artificial silk.

Reagent to be used—Hydrochloric acid (1 oz. in 15 oz. water) followed by wash and treatment with hydrogen peroxide (12 volumes strength).

Method adopted—Chemical.

IX. Nature of stain.—Perfume.

Class of fabric—Cotton, silk, wool and artificial silk.

Reagents to be used—Ammonia (1 oz. in 5 oz. water) and oxalic acid solution (½ oz. oxalic acid in 16 oz. water) alternately.

Method adopted—Chemical.

X. Nature of stain.—Perspiration.

Class of fabric—Cotton, silk, wool and artificial silk.

Reagent to be used—Hydrogen peroxide (12 volumes strength).

Method adopted—Chemical.

Method of preparing hypochlorite solution:

Take 1½ oz. of soda ash and dissolve it in 5 oz. of cold water. Then take 2 oz. of bleaching powder (containing 35% available chlorine) and dissolve it in 15 oz. of cold water. Mix the two solutions, stir and leave it undisturbed for 20 minutes. The clear solution is sodium hypochlorite and this should be poured into another vessel without disturbing the sediment. As the sodium hypochlorite solution is not stable, it should be prepared fresh every time.

The following materials may be stocked for removal of stains: Methylated spirit, acetone, amyl acetate, carbon tetrachloride, turpentine, cocoanut oil, soap, soda ash, liquor ammonia, hydrogen peroxide (12 volumes), bleaching powder, acetic acid (strong), oxalic acid crystals, hydrochloric acid (strong), sodium thiosulphate.

Textile Rust Remover
Dutch Patent 50,110

Phosphoric Acid (65%)	25
Glycerin (28° Bé)	75

Rug Cleaner

Cocoanut Oil Soap	12.0
Ammonia (28%)	2.8
Glycerin	7.9
Water	77.3

Cloth Top Shoe Cleaner

Trisodium Phosphate	2	oz.
Borax	2	oz.
Diglycol Stearate	1	oz.
Sulfatate (Wetting Agent)	½	oz.
Water, To make	1	gal.

Warm and mix to dissolve. Apply with a sponge or medium-stiff brush.

Dry Cleaning Leather Gloves

The best, i.e., cleanest, gloves are treated with dry cleaning solvent only and the badly soiled goods in solvent plus dry soap. For lightly soiled gloves the following compound is used:

Stock Solution

Cocoanut Oil	40
Castor Oil	20
Neatsfoot Oil	20
Olive Oil	20

Three parts of Stoddard's solvent is added to one part of the above stock solution and 70 to 80 pairs are processed in 10 gallons in an ordinary brushing machine for 35 to 40 minutes. After machining the gloves are hydro-extracted and then dried at 100° F.

In the case of badly soiled gloves

1½ gallons of dry soap are added to 40 gallons of solvent, usually white spirit or Stoddard's solvent. The fingers of particularly badly soiled gloves, usually due to sweat marks, are first soaked for ½ hour in a neat wet soap made up as follows:

To 1 gallon of strong oil bath (compound) is added ½ gallon wet soap, ½ pint water and 1½ pints butyl "cellosolve."

They are then cleaned in the usual way in the compound. These soiled grades, after cleaning, are reworked in white spirit for 30 minutes, extracted and then further reworked for five minutes in a solution containing:

Castor Oil	12½ oz.
Neatsfoot Oil	3½ lb.

made up with 5 gallons white spirit. In summer an addition of 12½ oz. beeswax is added to the above.

They are then extracted, "knocked out" and dried, paired and sorted.

Dishwashing Compounds

(Hand Washing)

Formula No. 1

Trisodium Phosphate	30
Tetrasodium Pyrophosphate	60
Borax	10

No. 2

Trisodium Phosphate	88
Soap, 92% Anhydrous	2
Borax	10

No. 3

Sodium Carbonate	25
Sodium Bicarbonate	35
Tetrasodium Pyrophosphate	20
Trisodium Phosphate (Monohydrated)	20
Trisodium Phosphate	10
Borax	8
Soap, Anhydrous	2

No. 4

Sodium Hydroxide	3
Sodium Carbonate	27
Sodium Bicarbonate	20
Trisodium Phosphate	40
Soap	8
Wetting Agent (Sulfatate)	2

(Machine Washing)

Formula No. 5

Sodium Carbonate	40
Sodium Metasilicate	15
Tetrasodium Pyrophosphate	45

No. 6

Sodium Metasilicate	20
Trisodium Phosphate	15
Tetrasodium Pyrophosphate, Crystalline	65

No. 7

Sodium Carbonate	20
Sodium Metasilicate	80

No. 8

Sodium Carbonate	10
Trisodium Phosphate	20
Sodium Metasilicate	30
Tetrasodium Pyrophosphate	40

No. 9

Sodium Carbonate	10
Sodium Hydroxide	5
Sodium Metasilicate	45
Tetrasodium Pyrophosphate	40

No. 10

Sodium Carbonate	15
Sodium Metasilicate	40
Tetrasodium Pyrophosphate	30
Trisodium Phosphate	15

No. 11

Sodium Carbonate	25
Sodium Metasilicate	54
Trisodium Phosphate	20
Wetting Agent, Low Foam (Sulfatate)	1

No. 12

Sodium Carbonate	30
Trisodium Phosphate	55
Sodium Metasilicate	5
Tetrasodium Pyrophosphate	10

No. 13

Trisodium Phosphate	85
Tetrasodium Pyrophosphate	15

No. 14

Sodium Carbonate	25
Trisodium Phosphate	60
Sodium Metasilicate	3
Tetrasodium Pyrophosphate	12

No. 15

Sodium Metasilicate	40
Trisodium Phosphate	20
Tetrasodium Pyrophosphate	40

No. 16

Sodium Hexametaphosphate	40
Sodium Metasilicate (Pentahydrate)	40
Trisodium Phosphate (Monohydrate)	14
Sodium Hydroxide	5
Sulfatate (Wetting Agent)	1

Dish Washing Powder
British Patent 528,964

Trisodium Phosphate	40
Soda Ash	26
Sodium Silicate	25
Sodium Sulphite	5
Sodium Hexametaphosphate	2
Soap, Powdered	2

Glass Cleaner
Formula No. 1

Glycerin	1.25
Oxalic Acid	56.25
Potassium Chromate	2.50
Water	940.00

No. 2

Aerosol O. T. (Wetting Agent)	1
Methanol	2
Ethylene Glycol	1
Water	96

No. 3

Isopropyl Alcohol	9
Diglycol Laurate	1
Ethylene Glycol	3
Alcohol, Denatured	28
Water	59

Glass Cleaning Jelly
British Patent 532,816

Hydrofluoric Acid	9	gal.
Hydrochloric Acid	3⅓	gal.
Water	13⅓	gal.
Glue	9½	lb.
Glycerin	12½	lb.

Cleaning Organic Smears—Tars, Etc.—from Apparatus

Put the dirty beakers, test tubes, distilling flasks, etc., into a large evaporating dish containing concentrated commercial sulphuric acid heated to 200–225° C. (400–435° F.). Small quantities of nitric acid should be added whenever the heated acid becomes black in color. Large beakers can be turned around sufficiently to bring all parts into the hot acid. When the apparatus is clean, remove (use a glass rod) and allow the articles to cool before rinsing in water. This cleaning solution is not satisfactory for petroleum products. This solution is much more economical than chromic acid solutions—the same acid can be used many times.

Caution: Have beakers, flasks, tubes, etc., pointed away from the operator while they are being put into the acid. Sometimes the reaction with wet, very dirty apparatus causes a spattering of acid from the open end of the vessel. The acid should be kept in a glass-stoppered bottle when stored.

Cleaning Flasks or Other Apparatus from Baked-In Carbon Deposits

Put a small quantity of potassium chlorate in the dry flask; heat the flask gently in a Bunsen burner flame until the chlorate is barely melted; rotate the flask so that the molten chlorate comes in contact with the carbon. The quantity of chlorate to contact the carbon residue is surprisingly small. After cooling, dissolve the remaining mixture in water.

Caution: Use this only on the solid carbon stain that will not wash out or that cannot be removed by ordinary mechanical means.

Chromic Acid Cleaning Mixture

This is needed for apparatus such as burettes, etc., which cannot be immersed conveniently in a vessel of hot sulphuric acid.

Add 1 l. of conc. commercial sulphuric acid to 50 g. chromic acid dissolved in 25 cc. of warm water. There will be no trouble from crystallization of salts from this mixture. This mixture, without heating, removes organic material after a few hours of contact. More chromic acid can be added whenever the solution loses its red color.

Caution: Keep in a glass-stoppered bottle. The reason: absorption of water from the air reduces the activities.

Cleaning Laboratory Glassware

Iodoform stains and odor may be removed by washing glassware with a solution of potassium or sodium hydroxide and rinsing it with a small amount of alcohol. Of course, the final step in this cleaning operation, as well as in all others to be described, is thorough washing with soap and water.

Ferrocyanide or iron stains are easily and rapidly removed with a solution of potassium hydroxide.

Lime deposited by lime water or similar preparations can be removed from glassware with diluted solutions of acetic acid or nitric acid. Deposits from lead subacetate solution can also be removed with these cleansing agents.

Deposits of soluble metallic salts are usually readily removed by thorough rinsing with water; however, in some cases, a small amount of hydrochloric acid may facilitate the cleansing. Insoluble or practically insoluble salts may be removed from glassware by dissolving them with the appropriate solvent which differs for each salt.

Metallic soaps, such as oleates and lead plaster, may be removed by oil of turpentine.

Oils, resins, balsams, and similar resinous bodies can usually be removed with soap, but in some cases a solution of potassium hydroxide may be required.

Collodion may be removed by peeling it off the glassware. If, however, the film adheres firmly, a mixture of ether and alcohol will remove it. In the case of adherent gutta-percha film, chloroform should be used as the solvent.

Sawdust is one of the best materials for removing petrolatum, lard, or other greasy substances from mortars and ointment tiles. Paper cleansing tissues are also excellent for this purpose. After the greater part of the grease has been removed with the sawdust or paper, the glassware should be washed with soap and water.

Bottle Cleaner

U. S. Patent 2,241,984

Caustic Soda	2½– 6%
Sodium Aluminate	¼– 1%
Water	93 –97%

Window Washing

The best solution consists of one part alcohol to two parts water. In freezing weather the alcoholic content should be increased to 50%. The same solution can be used until the water is almost thick with dirt, since its efficiency is not materially reduced until it has assumed the approximate consistency of cream. If a chamois is used it should be merely dampened with the solution, not soaked in it. After the window is dry it should be wiped with a silk rag.

Stain Remover for Polished Glass

Hydrofluoric Acid	0.25
Water	99.75

To Remove Scratches on Glass

Scratches on glass may be eradicated by using a paste made of glycerin, water and rouge (iron oxide) mixed to the desired consistency. A hard felt pad is dipped in this paste and rubbed briskly back and forth over the scratched surface until the markings disappear. The paste can be washed away by simply flushing with water. This paste is particularly suitable for removing shallow scratches. Deeper gougings require more specialized treatment and coarser abrasives to start with. For this latter purpose emery powder will often serve, and glycerin makes a satisfactory medium in which to suspend the powdered abrasive.

Dairy Utensil Cleaners

General Work

Formula No. 1

Sodium Carbonate	46
Sodium Metasilicate	46
Trisodium Phosphate, Monohydrated	8

No. 2

Sodium Carbonate	58
Sodium Bicarbonate	42

No. 3

Sodium Carbonate	60
Sodium Bicarbonate	40

No. 4

Sodium Carbonate	60
Trisodium Phosphate	40

No. 5

Sodium Carbonate	60
Trisodium Phosphate	30
Sodium Hydroxide	10

No. 6

Sodium Carbonate	66.6
Trisodium Phosphate	14.0
Sodium Metasilicate	19.4

No. 7

Sodium Carbonate	5
Trisodium Phosphate	20
Sodium Metasilicate	65
Sodium Hydroxide	5
Sodium Aluminate	5

No. 8

Trisodium Phosphate	90
Sodium Tetraborate	10

No. 9

Sodium Carbonate	55
Sodium Hydroxide	10
Soap	35

No. 10

Trisodium Phosphate	75
Sodium Sesquicarbonate	25

No. 11

Sodium Carbonate	30
Sodium Metasilicate	50
Sodium Hydroxide	6
Soap	14

Dairy Bottle and Glassware Cleaners

Formula No. 1

Sodium Carbonate	55
Sodium Hydroxide	45

No. 2

Sodium Carbonate	40
Sodium Hydroxide	60

No. 3

Sodium Carbonate	30
Sodium Hydroxide	70

No. 4

Sodium Carbonate	20
Sodium Hydroxide	80

No. 5

Sodium Carbonate	25
Sodium Metasilicate	10
Sodium Hydroxide	65

No. 6

Sodium Carbonate	3
Sodium Metasilicate, Anhydrous	4
Trisodium Phosphate, Anhydrous	6
Sodium Hydroxide	87

Melt together. Especially suited for washing machine.

Formula No. 7

Sodium Metasilicate, Anhydrous	8
Caustic Soda	92

Melt together.

For Use in Soaker-Washing Machine

Formula No. 8

Sodium Carbonate	5
Trisodium Phosphate, Anhydrous	7
Sodium Metasilicate, Anhydrous	5
Sodium Hydroxide	83

No. 9

Sodium Carbonate	20
Trisodium Phosphate	15
Sodium Hydroxide	65

No. 10

Sodium Carbonate	3
Trisodium Phosphate	6
Sodium Metasilicate	8

No. 11

Sodium Metasilicate	15
Sodium Hydroxide	85

Dairy Milk Stone Removers

Formula No. 1

Trisodium Phosphate	45
Tetrasodium Pyrophosphate	55

No. 2

Sodium Carbonate	20
Sodium Metasilicate	15
Trisodium Phosphate	65

No. 3

Phosphoric Acid 50%	30
Monosodium Phosphate	45
Water	20
Sulfatate (Wetting Agent)	5

Jelly Detergent for Porcelain

British Patent 532,816

Hydrofluoric Acid	8	lb.
Hydrochloric Acid	3	lb.
Water	10	gal.
Glue	4¾	lb.
Flour	37⅛	lb.
Glycerol	6⅛	lb.

Use with rubber gloves or swab only.

Removing Carbon Grime from Porcelain Crucibles

Place crucible in a dish of potassium bisulphate and heat until latter is melted. Allow to remain for 5 minutes, cool slowly and wash with hot water.

Removing Tobacco Stains from Marble

An excellent preparation for removing tobacco stains from marble is made as follows: Dissolve 2 pounds of trisodium phosphate crystals in 1 gallon of hot water. Mix 12 ounces of chlorinated lime to a paste in an enameled pan by adding water slowly and stirring. Pour the two solutions into a stoneware jar and add water until about 2 gallons are obtained. Stir thoroughly, cover the jar and permit the lime to settle. Add some of the liquid to powdered talc until a thick paste is formed and apply a layer ¼ inch thick with a trowel. To apply with a brush instead, add about one teaspoonful of sugar to each pound of powdered talc. If working on polished marble, scrape off with a wooden paddle; if dull marble, scrape off with a trowel. This mixture is a strong bleaching agent and corrodes metals, hence care must be exercised to prevent its dropping on metal fixtures or colored fabrics.

Removing Ink Stains from Marble

Ink marks on marble may be removed with a paste made by dissolving an ounce of oxalic acid and half an ounce of butter of antimony in a pint of rain water, and adding sufficient flour to form a thin paste. Apply this to the stains with a brush; allow it to remain on 3 or 4 days and then wash it off. Make a second application, if necessary.

Stripping Compounds

(For removing paints, lacquers, enamels, etc., from metals, without corrosion.)

Formula No. 1

Sodium Hydroxide	60
Potassium Hydroxide	30
Wetting Agent (Sulfatate)	5
Rosin	5

No. 2

Caustic Soda	75
Caustic Potash	15
Trisodium Phosphate, Monohydrated	5
Wetting Agent (Sulfatate)	2½
Rosin	2½

No. 3

Caustic Soda	70
Trisodium Phosphate, Monohydrated	25
Rosin	5

No. 4

Caustic Soda	50
Sodium Carbonate	35
Trisodium Phosphate, Monohydrated	10
Rosin	5

No. 5

Caustic Soda	50
Sodium Carbonate	35
Trisodium Phosphate, Monohydrated	10
Sodium Abietate	5

No. 6

Sodium Carbonate	25
Caustic Soda	67
Trisodium Phosphate, Monohydrated	5
Wetting Agent (Sulfatate)	3

No. 7

Caustic Soda	50
Sodium Carbonate	15
Trisodium Phosphate, Monohydrated	25
Soap (20° C. Titre)	5
Wetting Agent (Sulfatate)	2
Rosin	3

No. 8

Caustic Soda	70
Sodium Carbonate	22
Trisodium Phosphate, Monohydrated	5
Wetting Agent (Sulfatate)	3

All above mentioned compositions, when used for the stripping of paint, lacquer or enamel from iron and steel (only) are employed by making a hot, nearly boiling, solution of a concentration of from 6 to

8 oz. per gallon of water. The painted articles are either suspended in the solution until the paint is loosened sufficiently so that it can be rinsed off. The "milder ones," that is the compositions with a lower caustic content, can be sprayed on with a steam gun (*not* of aluminum), or a poultice can be made from a paste of these compounds and placed on painted surfaces to remove the paint. In the latter case, care must be employed not to get the paste in contact with the bare skin.

Paint Stripper
(*From Aluminum*)
Formula No. 1

Sodium Metasilicate	47.5
Sodium Silicate, "G" Brand	47.5
Rosin	5.0

No. 2

Sodium Carbonate	8.0
Trisodium Phosphate	35.0
Sodium Metasilicate	40.0
Sodium Silicate, "G" Brand	10.0
Sodium Sulphite	1.5
Soap, Neutral (30° C. Titer)	5.0
Wetting Agent (Wetanol)	0.5

No. 3

Oleic Acid	20.0
Cresol	67.2
Water	8.0
Potassium Hydroxide	4.8

Dilute 1 part of the compound with 4 parts of water; heat to not more than 140° F. and submerge articles until paint, enamel or lacquer is completely loosened, when it will come off in coherent large films.

This material can be used successfully for the removal of lithographic inks from tinned sheets without attacking the tin.

Paint Stripper
(*From Steel*)
Formula No. 1

Caustic Soda	80
Sodium Carbonate	15
Trisodium Phosphate, Monohydrated	5

No. 2

Sodium Silicate, Anhydrous	50
Supersilicate	20
Sodium Carbonate	10
Trisodium Phosphate	20

Enamel Stripper
(*From Steel*)
Formula No. 1

Caustic Soda	86
Soda Ash	14

No. 2

Potassium Hydroxide	75
Sodium Fluoride	25

No. 3

Potassium Hydroxide	20
Sodium Hydroxide	35
Sodium Fluoride	20
Alum	25

Removing Metallic Stains from Painted Surfaces
Formula No. 1

Sodium Metasilicate	2 oz.
Water	1 gal.

Stained areas on weathered surfaces should be thoroughly wetted with water before using this solution. Gentle rubbing with a cloth wetted with the solutions will generally remove the stains. In severe cases, two applications may be required. After the stains have been removed steps should be taken to remove the cause of staining by painting the exposed metal surfaces.

No. 2

These unsightly brown stains will usually be found beneath copper screening, but they may be caused by rusted leaders and gutters, or by exposed or improperly puttied nail heads, hinges, etc. Rain water, passing over these parts, takes a small amount of metal into solution and deposits it on the painted surface below. Staining takes place as a result of these metal salts reacting with or being adsorbed by the paint film.

An inexpensive, easily prepared solution which removes metal stains, particularly copper, without damage to the paint film is obtained by dissolving 1½ to 2½ ounces of sodium metasilicate in a gallon of water. The surface to be cleaned is first wet with plain water, using a soft sponge or cloth. The solution is then applied with another sponge, rubbing lightly until the stain disappears. This is followed by a thorough washing of the cleaned surface with plain water.

Ordinary rust stains may also be cleaned with a 2% solution of oxalic and/or a 5% solution of phosphoric acid, rinsing with clean water after application.

The best practice to avoid metal staining is to apply a coat of zinc dust-zinc oxide paint to all exposed metal parts, particularly copper screens, and thus protect them from the weather.

Removing Mildew from Painted Surfaces

Mildewed exterior surfaces should be washed prior to repainting. Abrasive soaps and water or alkali cleaning soap solutions containing trisodium phosphate or sodium metaphosphate may be used to advantage. After thorough scrubbing, the siding should be flushed with clean water and allowed to dry. Laps and seams or portions which have shown badly mildewed areas may be treated with a solution of one part of bichloride of mercury in 300 parts of water. Precaution should be taken to prevent this solution from coming in contact with the skin. Rubber gloves could be employed. After drying, the finishing coats of house paint may then be applied.

Removing Varnish from Furniture

Some varnish removers contain wax. This type of remover works best on pieces of furniture that have been finished for months or years, as the wax slows up the evaporation of the remover, making it stay wet longer, hence its softening action is quicker on the film to be removed. Much care must be exercised when using this wax-type remover to avoid leaving any of the wax on the surfaces to be refinished. If wax were left on such surfaces it would cause the finish to remain soft and not dry as it should—and probably the finish would peel off or scale off later.

After removing the finish with a wax-type remover, all cleaned parts should be washed clean with denatured alcohol. This will remove all of the wax residue.

Perhaps the best method of removing the old finish is to use a vat in which to put the articles to be washed. Put the pieces to be washed in the vat and let them soak a few minutes; then use a rag or pad of steelwool, dipping this in the remover and then washing with the grain of the wood while the piece remains in the vat. After all of the old finish is off, remove the piece of furniture from the vat and wipe it clean with a soft cloth, being sure that all remover is wiped from the piece washed. Then allow the washed piece of furniture to dry for about two hours. It is now ready to refinish.

Where the piece of furniture is too large to be put into a vat, use a bench the same height as the sides of the vat. The piece of furniture can be put partly over the vat, so as to wash it and allow the unused "rundown" to drain back into the vat.

Whenever or wherever a vat is used, a cover should be kept on the vat when it is not in service, so as to avoid any evaporation of the remover material in the vat.

Another method is to brush the remover on the finish to be removed. Keep brushing it on so as to keep the surface wet until the old film gets soft; then use a putty knife or cabinet scraper to lift the film to be removed. This should be followed by a sponging of all the surface with some of the remover on a soft rag. This will remove all streaks left when scraping.

Still another method of removing the old finish from furniture is to spread several layers of old cloth or burlap on the surfaces to be cleaned and then pour the remover on this covering. This will keep the surface wet until the old film becomes soft. Then follow by scraping and sponging the surfaces.

Painted Woodwork Cleaner

Kerosene (about 15%) emulsified in a mixture of soap (6%), colloidal clay (7%), and water (72%). The kerosene dissolves grease while the clay adsorbs dirt and keeps it dispersed. The soap exerts its usual powerful detergent action.

Painted Woodwork and Wall Cleaner

Starch	69
Trisodium Phosphate	15
Soda Ash	7
Sulfatate	1
Water	8
Preservative, As needed	

The above is made into a slurry with water before use.

Paint Brush Cleaner

Kerosene	2
Diglycol Oleate	1
Ammonia	1/4
Alcohol	1/4

Place brush in above over-night and then wash well with water.

Cleaning and Conserving Paint Brushes

Strong alkalies should never be used to clean paint brushes. Hardened paint (oil-base) can be softened by soaking the brush over night in any good grade lacquer thinner or (if not too hard) in gasoline or turpentine. Softened paint can be removed with a putty knife or steel painter's comb—a step which, properly done, should clean out all of the paint pigment right down to the rubber setting.

In the washing operation *thorough* scrubbing to remove remaining paint particles is important. A good grade of white floating soap, or any other neutral soap may be used. When the brush is cleaned, rinse first in hot water, then in cold water. Repeat until all trace of soap is removed. Shake out excess water.

Combing the bristles immediately after, removes all twists and curls; wrapping in paper keeps the bristles straight. Thorough drying is especially important, for if the brush is used before the bristles are completely dry, paint will seal the moisture inside of the porous bristles, causing them to lose their inherent resiliency that is so important to quality painting.

Sweeping Compound

Sand (Clean)	44
Paraffin Oil	20
Sawdust, Kiln Dried	36

Garage Concrete Floor Cleaner

Water	132
Copper Sulphate	16
Cotton Seed Oil	1½
Muriatic Acid	100
Powdered Pumice	9
Water, To make	400

Degreasing Detergents

Formula No. 1

Sulphated Higher Alcohol or "Sulfatate"	30–35
Trichloroethylene	65–70

Pour the trichloroethylene into the sulphated higher alcohol under constant stirring. Stir again before use.

No. 2

Sulphated Higher Alcohol	30–35
Turpentine	10–15
Tetrahydronaphthalene	20
Water	30

Add methylhexalin until dissolved.

No. 3

Soda Ash	8 oz.
Trisodium Phosphate	2 oz.
Caustic Soda	2 oz.
Water	1 gal.

No. 4

Sodium Metasilicate	95
Nacconol NR (Wetting Agent)	5
Water	3300

Use at 140° F.

Organic, Emulsifiable Grease and Dirt Solvents

For use on metals, metallic articles, on automotive equipment (chassis, wheels, engines), in aviation industry, and anywhere, where large amounts of old grease (mineral, etc.) have to be removed:

Metal Degreasers

Formula No. 1

Cocoanut Fatty Acid	15 fl. oz.
Oleic Acid	28 fl. oz.
Water	34 fl. oz.
Potassium Hydroxide	9.8 oz.
Diethylene Glycol Monobutyl Ether	20 fl. oz.
Naphtha, Hydrogenated	26 fl. oz.
"Chlorasol"	2 gal.
Cresol	2 gal.
Kerosene	16 gal.

This formula can also be changed to a jelly-like paste and then used in a water solution (emulsion) by stirring into the formula a low titer, highly hydrated jelly soap, made from low titer vegetable oils.

No. 2

Sodium Carbonate	18
Trisodium Phosphate	20
Sodium Hydroxide	55
Sodium Abietate	7

No. 3

Sodium Carbonate	20
Sodium Silicate (40° Bé)	30
Caustic Soda	40
Sodium Abietate	10

Above two compounds are efficient metal cleaners, which can be used for many metals, but are especially suitable for the removal of heavy oils and greases from ferrous metals prior to plating such as zincking, tinning, etc., or prior to enameling and lacquering.

But buffing compounds and lighter oils and greases can also be removed very efficiently from unbuffed brasses, bronzes, copper articles, etc.

No. 4

Sodium Carbonate	25
Trisodium Phosphate, Monohydrated	25
Caustic Soda	37
Sodium Abietate	3
Caustic Potash	10

No. 5

Sodium Metasilicate	20
Trisodium Phosphate, Monohydrated	20
Sodium Carbonate	10
Caustic Soda	58
Cresol	2

The cresylated cleaner should not be used as an electric cleaner, because it produces stains later.

No. 6

Sodium Carbonate	25
Sodium Metasilicate	25
Sodium Silicate (G Brand)	5
Trisodium Phosphate, Monohydrated	15
Trisodium Phosphate	15
Caustic Soda	10
Sodium Abietate	5

(From 1–2% wetting agent (Wetanol) can be added to above.)

No. 7

Another milder, but very efficient metal cleaner is composed of:

Tetrasodium Pyrophosphate, Anhydrous	10
Trisodium Phosphate, Monohydrated	15
Trisodium Phosphate	10
Sodium Carbonate	38
Sodium Hydroxide	15
Potassium Hydroxide	10
Sodium Abietate	2

Iron and Steel Wire Cleaner

Formula No. 1

Sodium Carbonate	30
Trisodium Phosphate, Monohydrated	5
Trisodium Phosphate	10
Sodium Hydroxide	45
Sodium Abietate	10

1% of Wetanol (wetting agent) may be added.

No. 2

Trisodium Phosphate, Monohydrated	20
Trisodium Phosphate	15
Sodium Carbonate	35
Caustic Soda	30

No. 3

Trisodium Phosphate, Monohydrated	20
Trisodium Phosphate	25
Sodium Carbonate	40
Caustic Soda	15

No. 4

Above Cleaner	95
Dextrose	5

Soak Cleaner

Formula No. 1

Sodium Carbonate	40

Tetrasodium Pyrophosphate, Anhydrous	4
Trisodium Phosphate, Monohydrated	10
Trisodium Phosphate	5
Sodium Abietate	5
Rosin, Powdered	4
Caustic Soda	30
Wetting Agent (Wetanol)	2

No. 2

Sodium Metasilicate	50
Sodium Orthosilicate	25
Sodium Aluminate	5
Caustic Soda	20

No. 3

Sodium Carbonate	15
Sodium Metasilicate	40
Sodium Orthosilicate	30
Trisodium Phosphate, Monohydrated	10
Aluminum Phosphate	5

No. 4

Sodium Carbonate	35
Sodium Metasilicate	30
Sodium Orthosilicate	25
Sodium Abietate	10

No. 5

Sodium Carbonate	20
Sodium Metasilicate	10
Sodium Orthosilicate	5
Trisodium Phosphate, Monohydrated	8
Trisodium Phosphate	10
Sodium Abietate	5
Sodium Oleate	5
Wetting Agent (Wetanol)	2
Caustic Soda	35

No. 6

Sodium Carbonate	10
Sodium Orthosilicate	15
Sodium Metasilicate	50
Trisodium Phosphate, Monohydrated	10
Sodium Aluminate	5
Caustic Soda	10

No. 7

Sodium Orthosilicate	80
Trisodium Phosphate, Monohydrated	20

No. 8

Sodium Carbonate	55
Sodium Metasilicate	20
Caustic Soda	15
Sodium Abietate	10

Light Metal Alloy Degreaser

Trisodium Phosphate	5
Sodium Silicate	3
Caustic Soda	1–2
Water, To make	100

Treat at 50–60° C. for 5 minutes.

Boiler Tube Grease Remover

U. S. Patent 2,248,656

Tetrasodium Pyrophosphate	30
Sodium Sulphate	40
Sulphite Liquor (Solids)	30

Metal Cleaner

Sodium Silicate	42.7%
Sodium Metasilicate	8.0%
Sulfatate (Wetting Agent)	2.0%
Water, To make	100 %

Metal Cleaning Emulsion

Kerosene	70
Pine Oil	6
Wetanol (Sulphated Fatty Alcohol)	12
Water	12

The above is diluted with

Water	100

Metal Cleaner and Sterilizer

British Patent 533,265

Sodium Hypochlorite	10.0%
Sodium Hexametaphosphate	15.0%
Caustic Soda	2.4%
Water	72.6%

Metal Test Panel Cleaner

Phosphoric Acid	3
Tergitol Penetrant No. 4	1
Water, To make	1000

Metal Utensil and Appliance Cleaner

Australian Patent 111,451

Soft Soap	2
Soda Ash	1
Crude Sugar	1.5
Magnesium Carbonate	3
Diatomaceous Earth	18
Water	30 5/8

Dissolve the soap in half the water, the soda ash and the sugar in the other half. Mix the earth and magnesium carbonate and knead well with the strained solutions.

Magnesium Alloy Cleaners

Formula No. 1

Soda Ash	3	oz.
Caustic Soda	2	oz.
Diglycol Stearate	1	oz.
Water	1	gal.

No. 2

Sodium Bichromate	1½	lb.
Nitric Acid	1½	lb.
Water	1	gal.

Use at 70–90° F.

Metal Cleaners

Mild metal cleaners are such without added caustic soda or potash, or with caustic constituents not higher than 10% and, at the same time, low in sodium carbonate and orthosilicates.

Medium strong metal cleaners may contain caustic alkalies up to 25–30% depending upon the percentage content of other strong alkalies such as sodium carbonate and orthosilicates.

Strong metal cleaners contain, generally, high concentrations of caustic substances, and also sodium carbonate, orthosilicates.

The higher the percentage concentration of abietates, soaps, etc., or silicates plus these organics, mentioned in cleaning compounds, the more are such cleaners suited as soak cleaners or still cleaners. Cleaning solutions are used to surround or penetrate oils, greases, emulsify and loosen them and dirt from the metal surfaces. Then their removal by rinses and their subsequent complete cleaning to a chemically pure surface is easily and quickly attainable in an electric cleaner.

Wetting agents, soaps, abietates, solvents, etc., are generally to be avoided in electric cleaners, while their presence in larger amounts is preferable in many instances in still cleaning solutions (soak cleaners), especially, where this preliminary cleaning operation is followed up and completed in a second and preferably electric cleaner.

None of the alkaline cleaning compositions is particularly or at all suitable for the cleaning of aluminum or of tin articles. Specially buffered and composed cleaners are used for these two metals and for their alloys.

Mild Plater's Cleaner

Sodium Carbonate	40
Trisodium Phosphate	45
Caustic Soda	13½
Sodium or Potassium Abietate	1½

Sometimes, coloring gums (e.g. gum guaiac) are used instead of abietates whenever the presence of abietates is objectionable. The above is used for copper, brasses, white metals and diecast metals, and also for the cleaning of stamped steel articles (but not for tin). It is also quite efficient in removing buffing compounds, light oils, etc., from such metals where a pure surface is required prior to enameling or lacquering.

Concentrations of Cleaning Solutions

For brass, copper, die-castings, etc.: (still tank) 4–6 oz./gal. Temperature of operation: 180–200° F.

For iron and steel cleaning: (still) 6–8 oz./gal. Temperature: 190–212° F.

For electric cleaning of iron and steel: 4–6 oz./gal. Temperature: 212° F.

For electric cleaning of brasses, copper, die-castings: 2–4 oz./gal. Temperature: 180–200° F.

When used as electric cleaner, use as counter electrodes: iron or steel sheets about 1/8 to 1/4 inch thickness, or connect the tank to the positive electrode. The articles to be cleaned are connected cathodically. The cleaning time is from 1/2 minute to 3 minutes or sometimes more, depending upon the amount and type of grease and dirt to be removed. Following this cleaning operation, the work can be made the anode for from 5–10 seconds. To make this change instantly and without removing the articles from one pole and switching them with the steel plates, a double-throw switch should be installed. A voltage of from 3–5 volts is required for soft metals and of 4–6 volts can be used for steel and iron.

After cleaning, the articles are removed and immediately rinsed in a good warm and clean rinse water or spray, followed immediately by a second rinsing in running cold water. Thereafter, the articles are dipped in a hydrochloric acid dip (concentration from 5–15% volume for brasses, etc.; of 40–60% volume for iron and steels) for from 2–10 seconds, rinsed in cold water, thoroughly, and plated in the desired solution. If in copper cyanide, brass cyanide or silver or other cyanide, both, the articles can be dipped momentarily in a sodium cyanide dip (2–4 oz./gal. or more) and then introduced into the plating solution.

When making up the cleaning solution, fill the tank with about 2/3 of the total desired volume of cleaning solution, heat the water to about 160° F., dissolve the measured quantity of cleaner in it and bring the solution to the boiling point and keep it there for about 15 minutes with intermittent stirring. Now, add water to bring the volume of the cleaning solution to the proper measure and the solution is ready for use.

Above outlined procedure is, generally, the same for all inorganic metal or plater's cleaners, except for concentrations, temperatures, time of cleaning and intensity and quantity of electric current used.

Brass, Copper and Soft Metal Cleaners

Formula No. 1

Sodium Carbonate	40
Trisodium Phosphate	35
Caustic Soda	15
Sodium Abietate	10

No. 2

Sodium Carbonate	30
Sodium Silicate (60%)	25
Borax	5
Caustic Soda	35
Sodium Abietate	5

No. 3

Sodium Carbonate	20
Sodium Silicate (40%)	35
Caustic Soda	35
Sodium Abietate	10

No. 4

Sodium Carbonate	25
Sodium Metasilicate	50
Trisodium Phosphate	25

No. 5

Sodium Carbonate	24
Sodium Metasilicate	48
Trisodium Phosphate	24
Sodium Abietate	4

No. 6

Formula No. 4 or No. 5 with 5–10% caustic soda.

Soak Cleaner for Removal of Tripoli and Buffing Compounds from Brass, etc.

Cocoanut Oil	18.5
Tallow Fatty Acids	30.0
Potassium Hydroxide	8.3
Water	23.5
Sodium Hydroxide	2.7
Water	4.5
Pine Oil	1.0
Boric Acid	1.5
Water	1.0
Cane Sugar	9.0

Melt fatty matter; add all of the lyes as usual at a temperature of about 60°–70° C. and saponify. After complete saponification, add pine oil; thereafter, add the sugar and the boric acid in paste form, and finally stir thoroughly.

If desired, from 1–5% of wetting agent (Wetanol) can be successfully added to the total compound. It may also be beneficial to use other combinations of fatty matter but the titer of the total fatty matter should be kept at around 30–35° C.; for example:

Corn Oil
Cottonseed Fatty Acids
Red Oil (small amount that can be substituted for part of the tallow fatty)
Cocoanut Oil

In any such case, the total amount of hydroxides must be adjusted to the mean saponification number of the combined fatty acids and oils, to obtain a neutral soap.

From 2–6 oz./gal. of the above compound are dissolved at a temperature of from 140–160° F. The articles are now suspended, either in baskets, on hooks, or on wires, in the solution, until the polishing and buffing materials are at least completely penetrated. They may now be removed and either rinsed in hot water, or they may be dipped in a second cleaner solution of mild alkalinity, similar to previously mentioned brass cleaners, and then rinsed in hot and cold water.

Tin Cleaners

Formula No. 1

Sodium Carbonate	30
Tetrasodium Pyrophosphate, Anhydrous	30
Trisodium Phosphate, Monohydrated	25
Sodium Metasilicate	10
Bichromate of Soda	5

No. 2

Sodium Carbonate	5
Tetrasodium Pyrophosphate, Anhydrous	50

Trisodium Phosphate, Monohydrated	30
Sodium Silicate (40° Bé)	10
Bichromate of Soda	5

No. 3

Sodium Carbonate	35
Sodium Silicate (40° Bé)	25
Sodium Metasilicate	25
Disodium Phosphate	15

Similar alkaline compositions as above plus lasting reducing substances, which are without physiological or toxic or poisonous characteristics may also be used.

No. 4

British Patent 537,136

Sawdust (1/32″)	14
Wood Flour	85
Whiting	1

How to Clean Stainless Steel

The necessity for cleaning stainless steel is not usually due to the formation of any rust or corrosion products from the metal itself, but to surface contamination and deposits. The equipment may be set up adjacent to sewers where vapors may develop an iridescent film that is difficult to remove. Stainless decorative trim is frequently exposed to corrosion products from less corrosion resistant metals and washings from acid-treated insulating materials, which stain the surface or form hard deposits. Heat exchanger tubes often become coated with mineral deposits from cooling water. Such deposits, unless uniform over the entire surface, are a source of localized pitting or corrosion.

Because stainless steel is a solid metal and not a plating, frequent cleaning is more beneficial than harmful. If proper cleaning methods and materials are employed, the original luster remains unaffected indefinitely. Probably the simplest procedure for removing ordinary deposits is frequent washing with soap and water. This prevents the deposits from becoming hard and tightly adherent, and subsequent cleanings are less difficult.

The luster of the polished finish can usually be restored with any of the stainless polishing pastes or powders available. In some cases, the use of vinegar or dilute acetic acid instead of water increases the effectiveness of the dry powders. When vinegar or acetic acid is used, thorough washing with water should follow.

Iron-free emery, of 180 grit or its equivalent, with kerosene as a lubricant, should be used when any abrasive is necessary.

More caution is required in sterilizing stainless steel equipment than in cleaning it. Most sterilizing compounds consist of, or contain, sodium or calcium hypochlorites. These hypochlorites should never be permitted to remain in contact with stainless steel for more than six continuous hours, and a thorough washing with water should follow each period of exposure. The type of corrosion usually encountered with the hypochlorites is entirely confined to small localized areas.

The cleaning of condenser or heat exchanger tubing or brine coils presents quite a difficult problem. In cases where construction permits, solutions of nitric, sulphuric or muriatic acid can be used for removing water scale or other deposits that may form on the walls of the tubing. The most desirable acid solution for cleaning such stainless steel equipment is 10 to 20% nitric acid. Stainless steel is not attacked by solutions of nitric acid. Consequently, the acid solution can remain in the equipment for long periods of time without any harmful effects. Nitric acid cannot be used, however, if ordinary steel, brass, bronze, nickel or any material that is soluble in this acid is present in the system.

Steel Plate Cleaner

Phosphoric Acid (Sp. Gr. 1.557)	840 cc.
Ethylene Glycol Monomethyl Ether	150 cc.
Sodium Chromate Solution (200 g. per l.)	10 cc.
Isopropyl Naphthalene Sodium Succinate (Aerosol OS)	15 g.
Water	2000 cc.

Dissolve 15 g. of the solid wetting agent in the glycol, add the chromate solution, and pour slowly with constant agitation into the phosphoric acid. Final gravity should be 1.48.

Cannon Sponging Fluid

Castile Soap	2 lb.
Sulfatate	3 g.
Water	8 gal.

Rifle Fouling Cleaner

Ammonia	3/8 pt.
Ammonium Persulphate	1 oz.
Ammonium Carbonate	5 dr.
Water	1/4 pt.
Sulfatate (Wetting Agent)	1 dr.

Rifle Swabbing Solution

Ammonia	300
Water	200
Sulfatate (Wetting Agent)	2

Aircraft Cleaning Compound

Formula No. 1

Alcohol	1 gal.
Pine Oil	1 qt.
Naphtha	5 pt.
Soap	4 lb.
Sulphonated Castor Oil	1 lb.

No. 2

Fusel Oil	7 lb.
Naphtha	4½ lb.
Pine Oil	2 lb.
Soap	4–5 lb.
Castor Oil, Sulphonated	1 lb.
Kerosene	76 lb.

Automobile Soap

Corn Oil, Crude	54.7 lb.
Caustic Potash, Flaked	9.2 lb.

Dissolve caustic in 30 lb. water and add it to the oil slowly with stirring. Bring to a boil and then add water to bring to strength desired.

To color green add 5 g. alizarine green per 100 lb. soap.

Windshield Cleaner

U. S. Patent 2,296,097

Feldspar	12
Calcium Carbonate	8
Sodium Bicarbonate	¾
Bentonite	3
Water, To make a paste.	

Engine Carbon Remover

Formula No. 1

Neutral Coal Tar Oil	39.5 cc.
O-Toluidine	15.0 cc.
Diethanolamine	15.0 cc.
Oleic Acid	15.0 cc.

Ethylene Glycol	15.0	cc.
Phosphoric Acid	½	cc.
Ethyl Silicate	½	cc.

No. 2

Neutral Coal Tar Oil (Cresol Free)	40	cc.
O-Toluidine	15	cc.
Diglycol Oleate	28	cc.
Ammonia	2	cc.
Ethylene Glycol	15	cc.
Phosphoric Acid	½	cc.
Ethyl Silicate	½	cc.

No. 3

Neutral Coal Tar Oil	40	cc.
O-Toluidine	15	cc.
Ethylene Glycol	15	cc.
Emulsifier A 3076A	30	cc.

Rust Remover

U. S. Patent 2,209,291

Metal rust may be removed by a composition of 36 parts phosphoric acid, 3 parts zinc phosphate, 2 parts gum arabic, 1 part manganese chloride, 30 parts butyl propionate and 28 parts water.

Watch Cleaning Preparation

Formula No. 1

Ammonia (10%)	12.0
Castor Oil	1.2
Glycerin	0.2
Water, To make	100.0

No. 2

White Oleic Acid	800
Potassium Hydroxide (35° Bé)	300
Saponin	16
Water, To make	1 gal.

Mix the potassium hydroxide solution with most of the water, heat to 80° C. Add the oleic acid slowly with agitation. Dissolve the saponin in the remaining water, add to the above, making to 1 gallon, and let cool. To use, mix 1 tablespoon of the above with 1 pint water.

No. 3

Sulfatate (Wetting Agent)	90
Trisodium Phosphate	5
Tetrasodium Pyrophosphate	5

Mix intimately. To use, add 1–2 tablespoons to 1 pint of water.

Chapter Twenty

TEXTILES AND FIBERS

Rotproofing Sandbags

Formula No. 1

A copper/ammonium solution containing approximately 4% of metallic copper can be prepared by dissolving 20 pounds of copper sulphate in water and making up the liquor to 10 gallons. To the solution so prepared is now added 2.1 gallons of strong ammonia (Sp. Gr. 0.88), care being taken to protect the eyes and respiration from the ammonia fumes. The resulting cuprammonium solution contains approximately 4% of copper and can be used for purposes of immersion when correctly diluted with water to give the required copper content.

An exceedingly useful water emulsion can be prepared from this liquor in combination with creosote:

(a) Prepare a cuprammonium solution as described above. (Solution A.)

(b) Prepare a 14% solution of carpenter's glue by soaking the glue in water overnight and then warming the mixture until a clear solution is obtained ready for use. (Solution B.)

(c) Neutralize the acids in the creosote with a trace of caustic alkali, to ensure stability of the ensuing emulsion. This is done by mixing 5 gallons of creosote with 0.5 gallon of caustic soda solution of 10% concentration. (Solution C.)

To make 10 gallons copper/creosote emulsion containing 50% creosote and 1.5% metallic copper we proceed in the following manner. The creosote caustic soda mix is first taken and 5 gallons (Solution C) measured out. To 3.8 gallons of Solution A add 0.7 gallon of Solution B. Now add the latter mixture (copper/ammonia/glue) gradually to the former liquor Solution C with continuous stirring. Warm the whole mixture to about 50°, stir vigorously, or better still, put through an emulsifier machine, and the composition readily becomes an aqueous emulsion. Sandbags treated with this by immersion or spraying are not sticky and quite dry to the touch.

No. 2

Soda Ash	4½	lb.
Soda Crystals	11½	lb.
Water	6	gal.

Stir the above solution, slowly, into a solution of:

Copper Sulphate	10	lb.
Water	36	gal.

When well mixed, add water to make up to 48 gal. and

Wetting Agent (Sulfatate)	2	oz.

Immerse bags in above until saturated and wring out excess solution. Dry at low temperature.

Mothproofing Cloth
U. S. Patent 2,127,252

Cloth is immersed in an emulsion of

Magnesium Silicofluoride	4–10%
Water	96–80%
Liquid Hydrocarbon	10–20%
Sulphonated Fatty Alcohol	1%

After saturation of cloth, it is squeezed and dried.

Destroying Moths in Garments
Expose to

Ethylene Dichloride	10
Carbon Tetrachloride	10

Use 20–25 lb. of above per 1000 cu. ft. of air for 24 hr. at 65–75° F.

Detecting Beginning of Mildew
The following fluorescence test is valuable for the detection of the start of mildew formation on wool, cotton, and rayon. Usually it is very difficult to detect the start of mildew formation, but this test solves this problem. When examined under ultra-violet light, the fiber shows a brilliant blue to greenish fluorescence at the point where mildew is starting.

Mildew Proofing

Copper Naphthenate	9 lb.
Zinc Bromide	5 lb.
Amyl Acetate	1 lb.
Water	20 gal.

(1) Mix and dissolve thoroughly.
(2) Filter.
(3) Make working solution with 1 part of above and 200 parts of water.
(4) Soak for 5 minutes.
(5) Dry.

Mildewproofing Spray
U. S. Patent 2,292,423

Mineral Oil, Light	95
p-Chlorophenol	5

Spray until textile has absorbed about 0.2% by weight.

Fireproofing for Textiles and Paper
U. S. Patent 2,097,509

Zinc Chloride	2.5%
Sodium Dichromate	0.5%
Sodium Silicofluoride	0.5%
Ammonium Dihydrogen Phosphate	9.6%
Urea	1.9%

Flameproofing Wadding (Cellulose)
U. S. Patent 2,132,016

The pulpweb is impregnated on forming cylinder with 0.7–1.1 lb. per gal. of water of following mixture:

Ammonium Sulphate	80
Ammonium Phosphate	10
Borax	10

Flameproofing Kapok
British Patent 535,058

Kapok is sprayed with or immersed in

Casein	1
Borax	5
Polyvinyl Acetate in Benzol (10%)	10
Triphenyl Phosphate	5
Water	85

Then squeeze through rollers; pass through 5% aluminum sulphate solution and dry.

Water and Alkali Resistant Textile Finish

Ceresin Wax	90
Bitumen (m.p. above 100°)	45
Tricresyl Phosphate	45
Turpentine	820

Water-Repellent for Textiles

Paraffin Wax (130° F.)	140

Stearic Acid	60
Naphthenic Acids	10
Gelatin	20
Aluminum Formate (30% Solution)	200
Water	570

Melt the wax and stearic acid and add the naphthenic acids, following with 150 g. of boiling water. Stir until an emulsion is produced. Dissolve the gelatin in 180 g. of water. Add the wax mixture to the gelatin solution at 60° C. Run through a colloid mill if desired. Finally add the aluminum formate solution very slowly and with much stirring. Now add the remaining water and filter if necessary.

Water Repellent for Cotton or Rayon Textiles
British Patent 474,403

A 0.5% solution of decyl isocyanate, hexadecyl isocyanate, or octadecyl isocyanate, is made in benzol. The rayon or cotton fabric is dipped and dried. It is then heated to 100° C. for one minute, followed by two minutes at 140° C. The result is a water repellent material which can be dry cleaned.

Water-Repellent for Felts, Etc.

Vistanex (Medium)	2.0
Toluene	60.0
Paraffin Wax (130° F.)	2.0
Aluminum Distearate	6.0
Aluminum Monostearate	4.0
Amyl Alcohol	10.0
Trichlorethylene	15.9
Triethanolamine	0.1

Dissolve the Vistanex in the toluene, add all the ingredients except the triethanolamine and disperse by stirring and heating to 60° C. Cool and work in the triethanolamine.

To use, dissolve in trichlorethylene and immerse the material; wring or centrifuge dry. Use a 10% solution. This may also be sprayed on.

Waterproofing Duck or Canvas

Waterproofing is of two general kinds, one, when chemical treatment of the fabric results in the deposit of chemical substance that is water repellent, and the other is when a water-repellent is applied directly. The first named process is as follows: Prepare a solution of acetate of alumina standing at 9° Tw., and heated to 100° F. Prepare a second solution of 25 gallons of water in which is dissolved about 25 pounds of soap, heated 100° F. Pass the goods first through the acetate solution, squeeze, and then pass through the soap solution, and dry. This causes an insoluble deposit of an alumina soap on the goods, and fabric is absolutely water-repellent.

The other process consists of impregnating the duck with a solution of paraffin wax and stearic acid in benzene (petroleum benzene-naphtha). About two or three ounces of equal parts of paraffin and stearic acid for gallon of benzene makes a most satisfactory proportion. After impregnating the goods, the fabric is dried in a machine provided with a condenser, so as to recover the benzene for re-use. Fabrics so waterproofed are practically impervious to water and the "proofing" lasts even longer than the utility of the duck.

Waterproofing Umbrellas

Tannic Acid	10
Sulphonated Oil	200
Lead Acetate	5
Aluminum Sulphate	10

Olive Oil	20
Pyridine	4
Water	1000

Pass fabric rapidly through the above solution at 50–60° C. and immediately centrifuge. Repeat until required degree of waterproofness is attained.

Flameproof, Waterproof and Mildewproof Coating for Textiles

Pigment Composition

	Formula No. 1
Antimony Oxide	12.0
Zinc Oxide	1.5
Black Iron Oxide	1.2
Olive Drab Composite	15.3

Non-Volatile Vehicle

	Formula No. 1	No. 2
Chlorinated Paraffin Wax	16.2	16.2
Polychlorphenol	1.8	1.8
Nevtex 10 Resin	10.2	..
Acrawax C	1.8	..
Nevtex 90 Resin	..	11.1

Volatile Solvent

	Formula No. 1	No. 2
Neville Hi-Flash Solvent	40.0	..
Xylol	..	32.0
Isopropanol	..	8.0

Resin

Polyvinylbutyral Resin	10.0
Xylol	90.0

The Acrawax C is melted into the chlorinated paraffin wax or Nevtex 10 Resin. All the ingredients except the solvents are combined at room temperature and ground on a 3 roll differential speed paint mill. When a uniform grind is obtained, thin with the solvent. Grinding the thinned pigment dispersion is not recommended—stirring is sufficient.

A 10% by weight solution of Vinyl Butyral Resin is prepared by wetting it with xylol, stirring to a slurry, and then adding the alcohol. About 4 hours agitation is required to dissolve it. A tumbling drum or other closed means of agitation is preferred.

The pigments, chlorinated paraffin wax, Nevtex 90 Resin, and polychlorphenol are mixed together and ground on a 3 roll paint mill. This ground mixture is then thinned with the remainder of the solvent not used in preparing the Vinyl Butyral Solution. After the thinned pigmented composition is made uniform by stirring, it is added slowly to the requisite Vinyl Butyral Resin Solution or *vice versa*. The whole is stirred together until uniform. Grinding the final solution is not recommended.

Identification of Naphthol Dyes on Fibers

Reds produced on yarn or pieces with Naphthol AS may be detected in most cases by treating the dyeings with chlorine at 3° Tw., the solution weakly acidulated with hydrochloric acid. Direct, developed and sulphur dyestuffs, also Alizarine Red, are destroyed within a short time. If the color withstands this test, it is reduced by heating with hydrosulphite (conc.) and caustic soda. If the color disappears and then returns on exposure to air it may be assumed to be a vat color. If the color does not return and the material remains white it most likely is Para Red but if the material retains a slight yellowish coloration, the dyeing may be assumed to have been produced with Naphthol AS. To verify, it is advisable to make a sublimation test. Para Red sublimes, Naphthol Reds do not.

Glycerin Aid in Micro-identification of Textiles

In the microscopic analysis of textile fibers glycerin is a most useful temporary mountant. Glycerin is the best all-around mounting medium; it is easily removed from the used slides with either warm water or soap, and is in addition without odor. However, since the index of refraction of acetate fibers is almost the same as that of glycerin, bromonaphthalene alpha is advocated as the mountant when such textiles are being studied.

In the identification process, a drop of glycerin is placed upon the slide, and the fiber sample (about 3/16 inches long) is brought on the drop and teased apart with mounted (dissecting) needles, and finally covered with a cover glass. Practice will soon enable one to determine the correct size of the drop of glycerin so as to avoid smearing and waste.

If desired, it is easy for a technician to convert these temporary mounts into permanent slides for filing for future reference and comparison. All that is necessary is to seal around the cover glass with hot paraffin which is applied with a soft camel's hair brush such as is used for water colors. Properly labeled, in time such a collection will form an invaluable textile fiber slide library.

Fiber Identification Stain

Acid Fuchsin (Color Index No. 692)	6 g.
Picric Acid	10 g.
Tannic Acid	10 g.
National Soluble Blue 2B Extra (Color Index No. 707)	5 g.

The dyes may be ground together and dissolved together, or dissolved separately in any order and diluted to 1 liter, with water. While the over-all concentration may not be very critical, these ratios of components appear to give the best differentiation. The dye mixture can be dissolved readily only in hot water, but the solution may be used hot or cold. Momentary immersion in the hot solution is sufficient, but commonly something over 2 minutes is allowed for cold dyeing. A thorough rinsing in water completes the test. Some dyed textiles may be identified without previous bleaching. The fibers are treated as usual, and then rinsed. When pressed wet (after rinsing) between white absorbent papers, a dye mixture characteristic of the color which would have been shown by the undyed fibers is transferred to the papers.

The colors shown by common fibers are:

Vegetable fibers	Cotton or Linen	Light blue
Synthetic fibers	Acetate or Nylon	Pale greenish yellow
	Cuprammonium	Dark blue
	Viscose	Lavender
	Vinyon	Very pale blue
Animal fibers	Wool	Yellow
	Silk (Raw)	Black
	Silk (Degummed)	Brown

IDENTIFICATION OF RAYONS AND NATURAL SILK

		CHARDONNET	GLANZSTOFF COLLODIUM	XANTHOGENATE LUNA	
REAGENT	ACETATE	NITRO	CUPRAMMONIUM (BEMBERG)	VISCOSE	TRUE SILK
Burning	Melts hard, black beady residue	No odor, burns slower than viscose	Easiest to ignite, no odor, slight ash	No odor, burns slower than Cupra.	Odor of burning feathers, black ash
Water	No swelling	All swell	All swell	All swell	Very slight swelling
Acetone	Soluble	All insoluble	All insoluble	All insoluble	All insoluble
Acetic Acid (Conc.)	Soluble	All insoluble	All insoluble	All insoluble	Dissolves on boiling
Chloroform	Softens or dissolves	All insoluble	All insoluble	All insoluble	Insoluble
Ammoniacal Copper Oxide	Swells but not dissolved	Swells quickly, dissolves slowly	Swells slowly, dissolves slowly	Swells and slowly dissolves	Fibroin dissolves
Chromic Acid	Swells, not dissolved	Dissolves cold	Dissolves cold	Dissolves cold	Dissolves slowly
Million's Reagent	Unstained	Unstained	Unstained	Unstained	Colored red
Decolorized Magenta	Unstained	Unstained	Unstained	Unstained	Colored pink
Ammoniacal Nickel Oxide	Swells, not dissolved	Swells, not dissolved	Swells, not dissolved	Swells, not dissolved	Dissolves quickly
Alkaline Glycerol Copper	No change	No change	No change	No change	Dissolves quickly
P-Nitraniline	Uncolored	Uncolored	Uncolored	Uncolored	Red
Diphenylamine	No color	Blue	No color	No color	
Brucine Sulphate	No color	Red	No color	No color	
Fehling's solution	Blue	Green	Blue	Blue	
Alkaline Silver Nitrate			No change	Brown	
Sulphuric Acid	Dissolves slowly	Quickly dissolves yellowish solution	Dissolves slowly, yellowish brown solution	Quickly dissolves, rusty brown solution	Quickly dissolves
Ruthenium Red	Colorless	Violet	Slight pink	Pink	Rose
Iodine Zinc Chloride	Yellow	Reddish violet	Brown	Bluish green	Pale yellow
Iodine Sulphuric Acid	Yellow	Violet	Light blue	Blue	Yellow

Nylon is difficultly soluble in most reagents. Phenols and Formic Acid are active solvents. Nylon materials are practically non-inflammable.

The presence of this new synthetic among acetate and other fibers is indicated by a treatment with potassium permanganate followed by the usual stripping of the brown coloration in a bath of sodium bisulphite. Nylon retains some of the brown stain while other fibers are bleached.—Nylon shows affinity for many classes of color but best results so far are obtained with the SRA types.

The stain is also useful in the identification of films of cellulose acetate or viscose (cellophane), giving the above colors. The colors realized with any such dye mixture will depend somewhat on the history of the sample tested, and increased confidence follows a check by other dye mixtures.

Distinguishing Linen and Cotton

The unknown fiber and one each of cotton and linen are soaked in a 5% solution of ortho hydroxy quinoline sulphate for ten minutes. Wash well and place in a 5% soda ash solution. Examine the samples under ultra-violet light. Linen fluoresces a brilliant canary yellow while cotton does not fluoresce and appears dark violet.

Sizing

Starch	14.0
Wheat Flour	48.5
Mineralite	64.0
Water	350.0
Oil and Drier (Linseed)	55.0
Paraffin Wax, Amorphous	33.0

Adhesive Sizing for Rayon Warps

U. S. Patent 2,231,050

Starch	100
Lactic Acid	25
Sodium Chlorate	1

Heat to 110° C. to solubilize starch.

Nylon Yarn Size

Polyvinyl Alcohol	6%
Boric Acid	2%
Water	92%

Textile Sizing, Thermoplastic

U. S. Patent 2,230,792

Five hundred grams of (ethylcellulose 16.6 plus hydrogenated methyl abietate 3.4 plus toluene 56.0 plus butyl alcohol 16.0 plus "high-flash" naphtha 8.0 g.) is emulsified in a colloid mill with 300 g. of water containing sodium lauryl sulphate 1.5 and (if the m.p. of the size is not more than approximately 205°) sulphonated castor oil 3 g. The fabric is immersed in 1% aqueous aluminum chloride and squeezed, and the emulsion applied from paddle rolls; excess is squeezed out, and the ethylcellulose dried at approximately 138° and pressed at a temperature which incipiently fuses the size (e.g., approximately 171°).

Logwood Textile Sizing

For black shirtings, glaces and similar materials with a hard finish, the finishing size is frequently stained with logwood. For 100 gallons size, the usual quantities are:

Logwood Extract (52° Tw.)	10 gal.
Eustic Extract (52° Tw.)	1¼ gal.
Nitrate of Iron (77° Tw.)	3½ qt.
Bichromate of Potash	3¾ lb.

Sizing Awning Duck

When sizing 10s skein mineral dyed duck yarn the fact that the dye deposits a stiffening mineral matter in the fiber should be considered in selecting a suitable size that will have a tendency to make a boardy finished product. A suitable formula which should give an efficient weaving quality to the yarn is as follows:

Water	100 gal.
Corn Starch	80 lb.
Potato Starch	40 lb.
Beef Tallow	15 lb.

Stir thoroughly in cooker for 8

to 10 minutes before applying heat, then bring to a boil as quickly as possible. Then shut off steam and reduce the mixture to 175° F. and maintain same for use. If the finish is too harsh, decrease the corn and potato starch in equal proportions and increase temperature to 185° F.

This procedure carefully adhered to should give a satisfactory sizing to the yarn, and the dye in question should not be affected with regard to brilliancy in the finish.

Rayon Fabric Desizer
German Patent 703,497

Soda Ash	1–5 g.
Sodium Bicarbonate	1–5 g.
Hydrogen Peroxide (40%)	½–1 g.
Water	1 l.

Lusterizing Finish for Rugs

Diglycol Laurate	1 pt.
Dry Cleaners' Solvent	5 gal.

Apply to pile of rug or carpet with a clean, soft cloth.

Increased Flexibility and Luster for Cotton and Rayon Braid

A mixture of Glyceryl tristearate and Flexo Wax C gives high luster and flexibility to cotton and rayon braids, shoe laces, etc. The blended waxes are melted and the braid is run through in continuous lengths and then over brushes on polishing rollers. The addition of oil soluble colors to the wax mixture will enable the coating to be used for different colored braids.

Lusterizing Delustered Rayon

Ammonium Adipate	10
Gum Tragacanth Solution (8%)	51
Water	39

This is printed on the delustered acetate rayon fabric and then dried. Afterwards the fabric is steamed for about one-half hour at 100° C. and then rinsed. A clearly defined lustrous pattern is thus produced on a matt ground.

Textile Duller (Delusterant) for Hosiery

Formula No. 1

Barium Chloride	40
Anhydrous Zinc Sulphate	26
Dextrin	22
Starch	12

All ingredients should be finely powdered and intimately mixed.

No. 2

Barium Chloride	48.0
Corn Starch	8.5
Hydrous Aluminum Sulphate	43.5

Bleaching Textiles

For fabrics made of rayon taffeta and cotton yarn the following bleaching procedure is recommended: (1) washing at 95–100° for 80 minutes in a solution containing 20 g. of sodium hydroxide and 5 g. of sodium ricinoleate per liter of water, (2) washing with hot and cold water, (3) treatment with alcohol for 2–5 minutes at 40–50° in a solution containing 3 g. of active chlorine and 2 g. of sodium silicate per liter, (4) washing with cold water, (5) acidification with sulphuric acid (5 g. per liter) and washing with cold water.

Bleaching Cotton with Perborate

For yarn the process is as follows: 2000 lb. cotton yarn is soaked in boiling water for about half an hour, rinsed, centrifuged, and packed in a boiler provided with a circulating pump. The goods are then heated in the bleach lye for four hours at a pressure of about 22 lb. with constant circulation. The goods are then rinsed in the boiler, taken out, soured with ½% of sulphuric acid, rinsed again, blued if necessary, and dried. The bleach lye is made up as follows: 100 lb. caustic soda and 50 lb. soap or 25 lb. Turkey-red oil are dissolved in 2000 gal. of water. In the meantime another solution is prepared of 15 lb. sodium perborate and 12 lb. aluminum sulphate in a convenient but not excessive quantity of water. The caustic soda and soap solution is then mixed with the perborate solution.

Piece goods are freed from dressing the first place, and then well rinsed. Scouring is unnecessary except in the case of very dirty goods. The pieces are then steeped in the bleach liquor and heated in the caustic soda lye for four to five hours while the whole liquid circulates under a pressure of 22 to 30 lb. The subsequent processes are as for yarn.

Bleaching Textiles with Textone

Cloth—1200 yards cotton and acetate collar interlining.

Water	1000 gal.
Textone	8 lb.
Wetting Agent	8 lb.
Acetic Acid (56%)	12 q.
Phosphoric Acid (85%)	½ p.
Tetra Sodium Pyrophosphate	1 lb.

The temperature is maintained at 185° F. or higher for a period of two hours. It is possible that a satisfactory bleach may be obtained in a shorter period of time. When the desired bleach is obtained, the cloth is given one hot water and one cold water wash and then dried.

Textone is also being successfully used on the acid side for bleaching and preparing spun rayons, rayon taffeta crepe and other rayon fabrics.

Another use of acid Textone is for pre-treating "Nylon" or rayon hosiery. The following is a typical procedure:

Solution

Cloth Ratio	35:1
Textone	0.1%
Wetting Agent	0.1%
Acetic Acid (56%)	0.3%

The solution is made up at a temperature of 135° F. The hosiery is immersed and during a period of 20 minutes the temperature is raised to 155–160° F. The hose are then washed and dyed in accordance with customary practice. Besides giving the hose a greater dye amenability, fewer tiger stripes are encountered.

In practically all cases where acid Textone is employed the equipment must be made of acid resistant material. The process can be used safely in vessels made of wood, stainless steel or nickel. The presence of copper, bronze, brass, or Monel metal may cause greenish or yellowish stains on the goods. In many cases stainless steel clad equipment will show apparent excessive corrosion. Investigations generally reveal, however, that this is caused by the liquor penetrating a pin hole in the stainless steel lining and corroding the underlying metal. The addition of small quantities of tetra sodium pyrophosphate is effective in minimizing such corrosion.

Scouring Rayon Circular Knit Fabric

1. Run water in kettle (80° to 120° F.) using minimum amount that will enable the fabric to run freely over the reels. A properly loaded kettle of the correct type requires approximately a 20 to 1 bath.

2. Load kettle with fabric.

3. Add 2 lb. soda ash or trisodium-phosphate (depending upon water conditions).

4. Turn on steam and run goods for 10 minutes.

5. Add 3 lb. high grade neutral soap—olive or red oil base.

6. Add 2 lb. "soluble pine oil" or a similar solvent containing material. If desired this solvent material and soap can be added simultaneously in order to aid solvent dispersion.

7. Raise bath to boil. Observe condition of bath at all times. If bath does not show a good, clean, sudsy condition, add more soda soap and pine oil. It is impossible to accurately predict the amount of soda soap and solvent or the exact proportions of the same that will be required under an unknown set of conditions.

8. Run the kettle at or near the boil for one hour.

9. Drop bath and proceed with bleaching or dyeing operation.

Dyeing "Black-Out" Cloth

Pad Liquor

Aniline Hydrochloride (16° Tw.)	20	gal.
Prussiate Liquor	44	gal.
Acetic Acid (56%)	½	gal.
Aniline Oil	1	gal.
Gum Tragacanth (6 oz.)	2½	gal.
Water	32	gal.

Prussiate Liquor

Sodium Chlorate	80	lb.
Sodium Yellow Prussiate	140	lb.
Water	50	gal.

Bulk to 100 gal.

Prepare in adequately cooled apparatus a mixture of the following proportions:

Aniline Oil	275	lb.
Muriatic Acid (32° Tw.)	300	lb.
Water	355	lb.

Run the cloth through the pad liquor, then through the squeeze rolls, and directly on to the dry cans.

After padding, the goods are dried on dry cans and should come off slightly greenish in shade, and after running in the rapid aniline ager for one minute they should come out a green black. Passage through hot soda ash solution to neutralize the acidity, together with hot sodium dichromate to develop the oxidation to the ungreenable black, and a good soaping completes the process.

Common practice is to utilize the open range, but a rope soaper is more thorough in washing effectively.

Brown (Vegetable) Dyeing of Cotton

Cutch	35	lb.
Hypernic Extract	16	lb.
Logwood Extract	3½	lb.

Add to dye bath and boil until dissolved, then add 3 lb. copper sulphate, add cold water, rake well and enter yarn. Give 6 turns and put down over night. Take up, give 6 turns, introduce into a solution of 4 lb. Chrome at 160° F. and give 6 hours. Remove, wash well in cold water, put back in cutch liquor, 6 turns; into chrome, 4 turns; into cutch, 4 turns; into chrome, 4 turns. Wash off each time after chrome. Start new kettle with

Fustic Extract	7 lb.
Logwood Extract	3½ lb.

Boil well for 2 hours.

Dyeing Cotton with Chrome Logwood

Dissolve 3.3 lb. of bichromate of potash in a small quantity of water, mix the solution with 100 gal. of logwood decoction at 3° Tw., and add 7.7 lb. hydrochloric acid, 34° Tw. The cotton is introduced into the cold solution, and the temperature is very gradually raised to boiling point. The cotton acquires at first a deep indigo-blue shade, which changes to a blue-black on washing with a calcareous water.

A slight modification of this process consists in working the cotton in a solution containing at first only the bichromate of potash and hydrochloric acid, and adding the decoction of logwood to the dye-bath in small portions from time to time, gradually raising the temperature as before.

Logwood Speck Dye

Logwood Extract (51° Tw.)	48 lb.
Soda Ash	30 lb.
Copper Sulphate	12 lb.

This should be diluted to about 2°–3° Tw.

Dyeing Vat Colors on Thread and Yarn

Package Dyeing Procedure

Shade: Khaki. Pigment Method

Soap Flakes—2%

Dye—X%

Circulate the above liquor at 180° F. on the dry packages for ten minutes, or until the entrapped air has been displaced.

Add.

Caustic Soda—¼ oz. per gal.

Sodium Hydrosulphite—¼ oz. per gal.

Circulate 20 min. at 180° F.

Rinse inside out with running cold water, overflowing to sewer, until neutral to Clayton Yellow Paper.

Add:

(100 vol.) Hydrogen Peroxide—1%

Circulate 10 min. at 140° F.

Add:

Soap Flakes—1½%

Circulate 20 min. at 200° F.

Rinse twice at 200° F., 10 minutes each.

Cascade Machine Procedure

Shade: Khaki. Reduced Method

Soap Flakes—¼ oz. per gal.

Dye—X%

Caustic Soda—1 oz. per gal.

Sodium Hydrosulphite—1 oz. per gal.

The dye is reduced for five minutes at 200° F. The liquor is then circulated on the dry yarn and dyeing proceeds for six minutes. Cold water is then circulated until oxidation is complete.

The dyeing is oxidized at 140° F. with 1/16 oz. of hydrogen peroxide per gallon, and soaped ten minutes at 200° F. with ¼ oz. of soap per gallon. Rinsing completes the operation.

Hand Tub Procedure

Shade: Khaki. Reduced Method
Dyebath: 120° F.
Caustic Soda—½ oz. per gal.
Deceresol AS—1/5 oz. per gal.
Stock vat: 140° F.
Dye—X%
Caustic Soda—4 oz. per gal.
Sodium Hydrosulphite—4 oz. per gal.
Soft Water—1 gal. per 4 oz. of double strength dye
Reduce ten minutes at 140° F.

The scoured thread is worked in the dye bath and then laid up while the reduced dye is added. The thread is worked in the dye until the desired shade is attained.

The dye bath is drained, and the thread is rinsed in cold water, oxidized with 1/16 oz. of hydrogen peroxide per gallon, and soaped at a boil with ¼ oz. of soap per gallon.

Buhlman Machine Procedure

Shade: Khaki. Pigment Method
Deceresol OS—1/10 per gal.
Dye—X%
Caustic Soda—¾ oz. per gal.
Temperature—140° F.

Enter dry thread and run five minutes. Add:

Sodium Hydrosulphite — ¾ oz. per gal.

Run until shade is correct, adding salt if necessary.

Drain, oxidize at 140° F. with 1/16 oz. of hydrogen peroxide per gallon, and soap ten minutes at 200° F. with ¼ oz. of soap per gallon.

Klauder Weldon Machine Procedure

Shade: Khaki. Pigment Method
Deceresol OS—1/10 oz. per gal.
Dye—X%
Temperature—140° F.

Run for ten minutes at 140° F.
Add:
Caustic Soda—¾ oz. per gal.
Sodium Hydrosulphite — ¼ oz. per gal.
Run five minutes. Add:
Sodium Hydrosulphite — ¼ oz. per gal.
Run five minutes. Add:
Sodium Hydrosulphite — ¼ oz. per gal.
Common Salt—2 oz. per gal.

Run ten minutes. Run in cold water and rinse ten minutes. Drain. Refill machine with cold water and rinse ten minutes. Drain.

Oxidize ten minutes at 140° F. with 1/16 oz. of hydrogen peroxide per gallon. Add ¼ oz. of soap per gallon and run ten minutes at a boil. Drain and rinse.

Vat Dye Reducing Bath

Caustic Soda	5 lb.
Sodium Hydrosulphite	5 lb.
Deceresol AS	1 lb.
Water (140° F.)	80 gal.

Vat Dye Oxidizing Bath
Formula No. 1

Sodium Bichromate	2 lb.
Acetic Acid (56%)	3 qt.
Water (140° F. or 212° F.)	40 gal.

No. 2

(100 vol.) Hydrogen Peroxide	2 lb.
Acetic Acid (56%)	3 qt.
Water (140° F.)	40 gal.

No. 3

Ammonium Persulphate	2 lb.
Acetic Acid (56%)	3 qt.
Water (80° F.)	40 gal.

The cloth is then rinsed several ends at a boil after which it is after-treated in a soaping bath prepared as follows:

Soap Flakes	2 lb.
Soda Ash	2 lb.
Water (212° F.)	80 gal.

After two to four ends in the boiling soap bath, the cloth is rinsed several ends at a boil and is shelled up, dried, and finished.

Mineral Khaki Color

This form of dyeing is used on army field equipment such as heavy duck, canvas, etc., for tents and army cots. It is not permitted for clothing or anything coming into close bodily contact.

This dyeing method is simple and productive and consists, actually, of producing a mixture of iron and chrome oxides within and upon the fabric fibers from a bath of black iron liquor and acetate of chrome to which is added an agent for leveling and penetration. A standing bath is used and replenished as needed.

A red shade of khaki is produced with:

Black Iron Liquor	10 gal.
Acetate of Chrome (8%)	5 gal.
Levelene	1 gal.

which is brought to a volume of 50 gallons with cold water.

The material is simply padded cold in this solution, then dried at about 200° F. to drive off the acetic acid formed. Following this the material is soaped at an elevated temperature to develop the shade. About 1 oz. soap and 2 oz. of soda ash is used per each gallon of bath.

Greener shades of khaki or olive drab can be made by increasing the chrome acetate and decreasing the quantity of iron liquor.

Dyeing of Synthetic Wool Fibers

Aralac, a casein product, one of the new synthetic wool fibers, may be successfully dyed with many acid colors.

Some of those colors applied and found satisfactory are:

Brilliant Milling Yellow 5G
Amacid Yellow R Supra
Amacid Fast Yellow RS
Amacid Milling Yellow O
Amacid Neutral Orange G
Amacid Neutral Orange SGS
Polar Orange GS
Polar Orange R
Amacid Milling Scarlet G
Amacid Red 3B
Amacid Milling Scarlet 3R
Amacid Azo Phloxine G
Formyl Violet 3B
Amacid Blue A
Sulphon Cyanine 5R
Alizarine Green CE Ex.
Brilliant Milling Green B Ex.
Amacid Black 10BR Conc.
Sulphon Cyanine Blacks

These may be dyed at the boil with 10–15% Glauber salt and 3–5% acetic acid.

In mixtures with real wool and synthetic fibers, the latter draws the color faster.

Fugitive Wool Colors

(For tinting wool for identification purposes, to be later cleared by soaping.)

Indigotine Conc.	Orange II
Ultramarine Blue	Guinea Green B
Amacid Chinoline Yellow Ex.	Carmoisine B
Kiton Fast Violet 10B	Amacid Blue A
Amacid Fuchsine 10B Conc.	Amacid Black 10BR Conc.

The tints are applied for 10 minutes at 100° F., extracted and dried. When required the tint is removed by soaping at 140° F. with a clear, hot rinse.

Dyeing Brush Bristles

When dyeing fiber materials to be used for the manufacture of brushes, etc., and necessitating the material being dyed through well, it is best to use a combination of about 2–3% of a Direct Black and 2–4% logwood extract.

Charge the starting bath with 2% ammonia and ¼–½% soda ash, add 2–3% dye previously well dissolved in condensed water, and then about 5% cryst. Glauber's salt; boil up well, enter the material, work for 5–10 minutes, cover with a lattice frame weighted with stones, boil for 2–3 hours, and allow to feed for ½–1 hour in the cooling bath. Then lift the material, allow it to lie exposed to the air for several hours, and enter into a fresh bath heated to 30°–40° C. (85°–105 F°.) containing pyrolignite of iron of 4°–7° Tw.; leave in this bath for ½–1 hour, throw out and leave exposed to the air for several hours, rinse well and dry.

If so-called patent or luster-fiber is to be produced, the method of working is exactly as described above; only the fiber is finally taken through a bath of 40°–50° C. (105°–120° F.) charged as follows:

Liquor	10	gal.
Gelatin Glue	2	lb.
Soft Soap	2	lb.
Logwood Extract	2	lb.
Fustic Extract	½	lb.
Pyrolignite of Iron	½	lb.

Treat the goods in this bath for thirty minutes, allow to drain, and brush dry with suitable brushing machines. If the fiber is not lustered, 8 oz. of whitening per 10 gallons liquor are added to the bath of pyrolignite of iron.

The dye liquors may be used repeatedly; dyeing in the standing bath requires about ½–⅔ of the stated quantities of dye and logwood extract, equal quantities of soda and ammonia, and about 3% salt calculated on the weight of the goods.

Lubrication for Nylon, Vinyon and Other Textiles

In many cases, expensive olive oil can be replaced by a soluble oil consisting of the following:

Diglycol Laurate	12.0	lb.
White Mineral Oil	87.9	lb.
Tergitol	0.1	lb.

This solution is easily made in the cold. The color is almost water-white, the emulsions in water are stable and easily removed. This preparation is suggested as a worsted lubricant, rayon coning oil, spun rayon stock lubricant, Nylon and Vinyon lubricant, mercerizers lubricant, finishing and luster oil for cotton, rayon and other natural and synthetic materials. It is also recommended for silk-soaking and for lubrication of needles and sinkers. When used as a worsted stock lubricant, it is applied in combing, carding or spinning operations as an oil, or in the emulsified condition. The amount of oil required in the finishing yarn determines the proportions used. In rayon stock lubrication, the straight soluble oil is sprayed. A water emulsion should not be used unless other fibers in addition to rayon are present. In such a case, the emulsion must be sprayed on a layer of the non-rayon fiber, then lay down a layer of rayon, then another layer of compounding fiber, etc. In mercerizing, an emulsion of the proper proportions is used in the finishing bath. The 100% oil is applied on the coning machines. For finish and luster the emulsion is used in the last rinse or finishing bath.

For needle and sinker lubrication, the straight oil can be applied by spraying or brushing on needles and sinkers. Metal parts are not discolored or corroded by this preparation.

The unusually high dye solvency of Diglycol laurate permits incorporation of tints in the soluble oil, so that the tinting operation may be carried on simultaneously with any one of the above operations. In addition to the fact that this oil is very much cheaper than olive oil, it has the advantage of being non-oxidizing, stainless and will not carbonize or become rancid. It is anhydrous and free of acids, completely self-emulsifying and easy scouring.

Synthetic Textile Fiber Lubricant
British Patent 539,333

Mineral Oil (Colorless)	64.3
Tricresyl Phosphate	11.6
Oleic Acid	10.9
Triethanolamine	5.7
Sulphonated Olive Oil	5.0
Sperm Oil	2.5

Dye Printing Paste

Induline B Powder	1.5
Ethyl Tartaric Acid (20° Tw.)	6.0
Acetin	4.2
Water	5.3
Acid Starch Paste	60.0
Acetic Acid (9° Tw.)	8.0
Tannic-Acetic Acid Solution (50%)	15.0

Printing on Plush
(Block Printing Dye)

Acid Dye	310 g.
British Gum, Powdered	325 g.
Glycerin	75 g.
Potassium Chlorate	25 g.
Sodium Oxalate	130 g.
Ammonium Hydroxide	130 cc.
Water	5 l.

Direct colors (for cotton), may be printed similarly on wool pile, by means of this recipe.

But when direct colors are to be printed on mixed fabrics, phosphate of soda is added, and the salt which frees the acid (oxalate of ammonia), is suppressed.

For printing on mixed fibers of wool, cotton and silk, the color is prepared with, for example:

Substantive Dye	6
Phosphate of Soda	3
Brown Glycerin, Thickened According to the Quality of the Fabric	1 (to 1.5)

Chrome colors may be used also for printing plush, for example, according to the following recipe:

Chrome Dye in Paste Form (20%)	454	g.
Acetate of Chromium	283	g.
Oxalate of Ammonia	273	g.

Designs of large area, of brown or taupe, are enriched by printing their center with campeche black. The color (to be printed), is applied with an ordinary stenciling brush, and the borders of the central design are rubbed, in order to blend them more or less imperceptibly with the rest of the original design. Campeche is used for super-printing fabrics of great value, which have been printed, steamed and washed. The campeche black is prepared in the following manner:

Campeche Black

A

Pyrolignite of Iron (20° Bé)	4.5	l.
Tapioca	1.350	kg.
Wheat Starch	680	g.
Water, To make	20	l.

B

Hematine (Paste)	9	l.
Acetic Acid (6° Bé)	1.5	l.
Iron Sulphate	2.040	g.
Nitrite of Soda	425	g.
Bisulphite of Soda (35° Bé)	125	cc.
Water, To make	16	l.

Equal parts of A are mixed with equal parts of B.

After printing, the fabric must be lightly dried and steamed as in the printing of ordinary wools.

To prepare printing colors for mixed plushes, composed of silk and mohair, or cotton and mohair, it is often necessary, in the desire to obtain a solid color on all the fibers, to use, with substantive dye, an acid color which is fixed with medium neutral fixing, excluding, however, the dyes which reduce very easily, such as marine blue and sulphocyanine blacks. In the presence of oxalate of ammonia, acid dyes of combination resist washing better which follows the steaming. To the contrary, few of the substantive dyes color vegetable fibers with dark shades, when the dye contains acid. It is necessary, therefore, to try the colors before combining them.

Brown pigment of manganese is used to produce certain effects, as much on mohair plush as on alpaca plush. The printing composition is prepared as follows:

Permanganate of Potash	267	g.
Magnesium Sulphate	200	g.
Gray China Clay	133	g.
Boiling Water	400	cc.

The mixtures may be used after recooling.

To fix and develop this color, it is not necessary to steam it; it is sufficient to rinse the printed fabrics in cold water. To raise the points of white nap, the nap points are brushed with a cutting solution made of:

Hydrogen Peroxide (12 Vol.)	9	l.
Acetic Acid (30%)	850	cc.
Oxalic Acid (20%)	850	cc.

If black effects are to be produced on the brown pigment of manganese, it is sufficient to treat the fabric for some minutes in a bath containing aniline salt and sharpened with sulphuric acid. The development of the aniline black which is thus formed, is completed by finally soaping the pieces in a hot bath containing ammonia.

For other kinds of lifting, acid colors are used in place of brown pigment of manganese. The necessary acid color is printed, and, after being steamed and washed, the fabric is extended on a table and then the points of the nap are brushed with a discharge, such as:

Rongol NC	285
Oxide of Zinc	140
British Gum Thickening	575

After having been lightly dried, the fabric is steamed for three minutes, without pressure, then washed and allowed to dry.

Ordinarily, mohair plushes are made brilliant or glazed before printing. For this effect, the down is coated with a heavy sizing of white flour or British gum boiled in water. The sizing is allowed to dry, and the pieces are then steamed for an hour in 3–4.5 kg. per cc. of goods, then washed with lukewarm water. The brilliancy of finish is added to by repeating this treatment several times.

To obtain "uniformity" on mohair plush, the pieces are treated in a bath of carbonate at 6° Bé, for nearly one hour at 75–80° C., then rinsed freely with cold water. Good imitations of young animals and others of the same kind are obtained.

Further, these unions are used to produce waves, by setting the nap by means of rotating brushes, kept constantly humid by a spray of water across their width. The water is sometimes replaced by a fluid paste of flour or British gum, serving to lubricate and adhere on the nap. In this case, the waves obtained by the brushing are fixed by steaming. The action of the water, in the presence of diastases, adds further to the luster of the waved nap.

On imitation fur plushes of short hair, special effects may be produced by printing with metallic powders. To do this, the plush pieces are first steamed on a rotating cylinder, in order to get the best possible uniformity of the nap.

Here is a recipe for a metallic printing composition, capable of being applied by block as well as by stencil:

Phenol	2
Acetone	4
Cellulose Acetate	3
Metallic Powder	1

Printing effects (mechanical), or embossing, may be obtained on certain qualities of plush. The best effects of this kind are obtained by passing the pieces between engraved and heated rollers.

The hair may also be curled and shrunk in places, by printing a strong solution of sulphocyanide of calcium, thickened with tragacanth, on the plush; afterwards, the pieces are steamed, washed and dried. The hairs are also curled and relooped, to contrast with the rest of the cover of the plush.

To imitate the pelt of a leopard, tiger and other savage animals, printing is carried out by means of a block, of which the engraving is very hard and also deeper than that of the other blocks of the print. Notice that dull rayon possesses ideal properties for producing this article, thanks to its appearance, to the subduing of its luster, to its soft, impressive feel, resembling that of fur, and also because this fiber permits the production of the particular colors of the fur. Further, dull rayon is especially well adapted to the operations of manufacture.

The skin of the ocelot is imitated, for example, with a plush of two wefts, with a mercerized raw cotton back (102/2 for the warp, and 40/1 for the weft). Dull Celta (woven 120/1), unbleached, is used to form the hair. As a height for the nap, a height of 4 to 6 mm. is the one which is most adaptable, with a count of 50 to 60 hairs per square cm., the tuft being formed with W. (tungsten). The pieces are 120 cm. wide and to 35 meters long; in sorting and inspecting the weave, they are uniformly dampened with lukewarm water on a squeezing spray (moistening machine), then they are brushed on a machine of a single cylinder, with the cylinder cold, to smooth and straighten the curls. The pieces enter the machine through their leader. Care must be taken to preserve the same direction of the nap during the process of printing; for if the direction of the nap is reversed, the outlines of the printed design will become waved to the detriment of the colored effect.

The pieces are then dried in a hot chamber, then pressed on the cylinder. The object of the last treatment is to compress the nap strongly on the foundation of the fabric, a condition indispensable to the success of the printing. Thus prepared, the pieces are ready for printing. If a good oily sulphonate is added to the moistening water, the printings will be more uniform.

In printing on plush, in particular, in the printing of several colors, the screen method has given the very best results, even in those which concern the cost of operation. Small lots of pieces, more frequently, as brought out in the beginning, are printed preferably by hand. There exists, for large production and for very large designs which are in demand, apparatus for screen printing.

Take, for example, a print of three colors. The piece is put on the printing table, stretched in the direction of the warp by means of gripping rollers, governed by a hand wheel; in the direction of the length, the pieces are extended by nailing first a narrow strip on the table at intervals of 30 to 50 cm., then stretching the piece strongly in the direction of the weft and fixing solidly the other strip in the same manner (with nails), on the table. It is essential that the pieces remain firmly stretched, especially if they are to be printed with several colors, in order that each application of the screen will not disarrange the surface to be reprinted, and not interfere with the superplacement of the following color. It is necessary that the table should be sufficiently long in order to permit the stretching of each piece to its full length.

When the table is too small, screen printing becomes difficult and does not have the desirable continuity, because of the dislocation of the design at the part of the piece that is resting on the extremity of the table. The printing is carried out by means of frames enclosing the design, on which the color is poured, and which two men on each side of the table spread equally with the aid of a wooden blade. The imitation of ocelot is composed of two colors, and therefore, it is necessary for two pairs of printers to execute the application, the first pair impressing the clear color and the second the deep color when the first pair has raised the frame (screen), to carry it back to its new position. Mordant colors are used for this kind of printing.

The preparation of the colors is as follows:

Beige

Anthracene Brown RD Paste	20	g.
Alizarine Yellow R Paste	2.5	g.
Water	300	cc.
Starch-Tragacanth Thickening	600	g.
Acetic Acid (6° Bé)	30	g.
Chrome Green Acetate (20° Bé)	40	g.

Deep Brown

Anthracene Brown RD Paste	150	g.
Alizarine Yellow R Paste	70	g.
Alizarine Black S Paste	100	g.
Water	70	g.
Starch-Tragacanth Thickening	400	g.
Acetic Acid (6° Bé)	60	g.
Chrome Green Acetate (20° Bé)	150	g.

In preparing these colors, it is necessary to watch that they always have the same consistency, in order that the tones may vary in intensity, much as when one carefully weighs the quantity of color. The proportion of starch must be more or less large, according to whether the prints are large or small in size. For the ocelot article, it is preferable not to print the hair as far as the base, in order to better imitate the fur. It is necessary, therefore, to regulate the quantity of thickening. The number of displacements to be given to the scraper to spread out the dye on the plush depends on the quantity of color to be printed. Four displacements are calculated for the clear color, and six for the deep color. It is necessary that the pressure which the two operators exert on the blade is uniform so that the impression will not become deeper or lighter on one part of the fabric than on the other. The designs to be effected must be of such kind that the dyes will not encroach one on the other, or not mutually color themselves. In order that the colored designs may retain all their clarity, it is necessary that the outlines of the different motives be separated by at least a millimeter.

The printed pieces, as is customary, are either transported in an ordinary drier, or extended on poles fixed to a movable frame and dried in an ordinary warm room at a temperature of between 40–50° C.

The pelts of the ocelot and leopard present a gradation of color, going from beige, nearly white, from the middle of the back to the sides. Naturally, if this diminution of shade is to be reproduced by screen printing, one has, of course, recourse to a spraying pistol. The third color to be applied for diminishing of the shade, is sprayed on the table. It is composed of:

Anthracene Brown RD Paste	5	g.
Alizarine Yellow R Paste	0.5	g.
Water	950	g.
Acetic Acid (6° Bé)	25	g.
Chrome Green Acetate (6° Bé)	20	g.

This diminishing color must also be well pulverized before printing. But, in this case, care must be taken that, at the moment of stretching and fixation of the pieces on the printing table, the parts where the colored effect is deepest must be placed to correspond with the corresponding parts of the design.

After spraying the color, the piece is dried again on the frame, and if possible, left during the night.

The following operation, or steaming, to fix the dyes on the rayon fiber, deserves the most careful attention. Chrome colors must be steamed with humid steam, for one hour at 0.1 kilo. It is carried out in a circular steamer of which the bottom is filled with water. The steam is introduced by means of a system of perforated tubes which boil and vaporize the water. Constantly humid steam is thus produced, which is a very important factor in obtaining uniform printings. The pieces are deposited on a frame, rolled up under tension in order that the different layers of the fabric will not come in contact with one another and so that the steam will pass freely between them.

If some parts of the nap of one fold touch on the back of the neighboring fold, the steam will not produce its effect here, and the color, insufficiently fixed or not fixed, will more or less totally fade when washed. The steaming must be well clear of air before the introduction of the frame loaded with the pieces. During the steaming, the steam must circulate perfectly; it is necessary that it enters in a continuous manner when it is ejected through the outlet valve. This is evidently not a means of steaming cheaply; the procedure must seem crude, but, at least, it assures the success of the printing. Once steamed, the pieces are left suspended in the stretching frame for about an hour, until completely cooled, then they are removed and suspended again on movable frames and left there for 6 to 8 hours before rinsing.

It remains to eliminate the thickening and the excess of color which has not been fixed. This is accomplished best at width in a washer provided with pressing rolls operated on top of a trough situated in the middle of the tank, 60 cm. from the bottom, in order to directly throw off the water that is squeezed out. At the farthest part, the vessel contains an obstruction, on top of which is found a winding roll, the action of which is completed by another roll placed at a lower level and in the squeezing trough. In this latter, two faucets sprinkle water copiously on the two sides of the piece. Naturally, the fabric circulates in the direction that is not against the direction of the nap.

For washing, simply cold water is used which is renewed constantly in order that the parts not printed will not become tinted. The washing machine is particularly suited for cleansing printed plushes drawn at width. The system of sprays of water under pressure, in combination with the pressing rollers, free the fabric of the thickener and unfixed dye. The soiled water is collected in the central trough where it is thrown off through a pipe. In this manner, the rinsing water of the large rinsing vessel is not worthlessly polluted and does not color the base of the fabric. Washing at width for half an hour to an hour, follows the deepening of the color; the speed of the passage of the fabric in the spraying and rinsing trough is about 10 meters a minute. Then the pieces are squeezed at the end and dried.

It is then that the foundation of the surface of the fabric shows clearly the gradations of color characteristic of the natural fur, from yellowish white of the stomach to beige of the back, as well as the light and dark spots.

It remains to bring about the completion of the fabric, and the first precaution to take is to be assured that the pieces are completely dried, in which case the colors are stained or become soiled in the brushing machine.

The pieces are placed again in a moistening machine, and pressed out with cold water sharpened with acetic acid, then passed through a brushing machine of one cylinder, consisting essentially of a rotating cylinder carrying the piece off on its back, on top of which cylinder is situated a brushing roller, also rotating, which sets and smooths the fur. The guiding rolls and an obstruction complete the depositing brush. The pieces pass once in the direction of the fur, once against the direction of the fur, and again in the first direction. They must be subjected to a large number of passages, according to the density of the fabric, but the number of these passages is always odd, since the direction finally to be given to the fur must be the same as the first passage.

Notice that the damp fabric, when removed from the machine after its first passage, is suspended on the frame to be dried in a warm room. Once dried, it is repassed three or four times in the same brushing machine, and finally it is clipped on an ordinary clipping cylinder.

After stretching and mangling on the back, the pieces are examined, measured and folded before being sent out.

Cotton Piece Goods Printing

Preliminary preparation of cotton piece material for printing consists of:

(1) Singeing, to remove the nap and produce a smooth surface. This operation is necessary to insure a uniform application of color in the intended design. Gas singeing is still in use for plain cotton goods which is passed quickly, full width, over a series of open gas flames.

(2) De-sizing. This is accomplished by saturating the cloth in a dilute but hot caustic soda solution, or liquor from a previous kier-boil and allowed to lie in this state for several hours. Malt extracts are more efficient and faster. A thorough washing follows.

(3) Kier Boiling comes next which is usually given under a mild pressure for 5–7 hours in a 5° Tw. caustic soda liquor. Natural waxes and seed motes are thus removed. A thorough rinse and scouring in 0.3° Tw. sulphuric acid follows and then a further rinse.

(4) Bleaching is carried out by passing the goods through a cold solution of sodium hypochlorite of 2 grams available chlorine per liter and then piled for 1–3 hours. A

thorough rinse and, if necessary, an anti-chlore plus a rinse completes the bleach.

Hydrogen peroxide is sometimes used in place of chlorine and when conditions require it both bleaches are applied, separately.

Cotton and materials of allied fibers are printed with selected groups from among the acid, chrome, and basic colors, as well as the directs, vats and those colors regularly used on cotton.

Basic Colors—Direct Printing

The printing paste is prepared with the requisite basic color, thickening gums, tannic acid for fixing the color and an agent for retarding lake formations. Acetic acid serves well as a retardant.

Victoria Blue B	15 g.
Acetic Acid (30%)	75 g.
Kromfax Solvent	40 g.
Hot Water	170 g.
Gum Tragacanth—Starch Thickener	600 g.
Tannic Acid Fixer (50:50 Tannic and Acetic Acid)	100 g.

After printing and drying, the material is steamed for one hour without pressure and then treated with tartar emetic—10 g. per l. at about 100–120° F.

Direct Printing with Chrome or Mordant Colors

Alizarine Yellow 2G Conc.	40 g.
Kromfax Solvent	40 g.
Hot Water	140 g.
Starch-Tragacanth Thickener	500 g.
Acetic Acid	50 g.
Olive Oil	30 g.
Acetate of Chrome (32° Tw.)	200 g.

The acetate of chrome should be added only after cooling the paste.

Print, dry and steam one hour, no pressure. Then rinse warm and soap hot.

A few typical items applicable by this method are:

Alizarine Yellow 2G Conc.
Alizarine Yellow 4G Conc.
Mordant Yellow R
Alizarine Red S Conc.
Alizarine Brown HD
Chromogene Violet B
Chromaven Green G
Alizarine Blue Black B

Acid Colors (Direct Printing)

Orange II	40 g.
Hot Water	200 g.
Gum Tragacanth Thickener	430 g.
British Gum	100 g.
Acetic Acid	50 g.
Aluminum Acetate (15° Tw.)	180 g.

Print, steam for 30 minutes, no pressure and treat warm to remove gum. No further washing.

Colors suitable are:

Chinoline Yellow N
Naphthol Yellow S
Amacid Yellow T Ex.
Crystal Orange
Orange II—RO
Amacid Fast Orange LW
Brilliant Croceines
Scarlet 2R
Brilliant Indocyanines
Soluble Blues
Amacid Fast Green 3G
Brilliant Milling Green B

Printing of Direct Colors

Practically all of the substantive colors may be printed but only a selected few have sufficient resistance to steaming, wash fastness and solubility to make them suitable for satisfactory direct printing.

Direct Color	20 g.
Hot Water	310 g.
Glycerin	50 g.
Sodium Phosphate	20 g.
Starch—Tragacanth Thickener	500 g.
Egg Albumen	100 g.

Print, dry, steam one hour without pressure then rinse in water containing salt. Aftertreating with ¼ oz. Fastogene per one gallon of water at 100° F. for 15 minutes increases water fastness.

A few suitable colors are:
Chrysophenine G
Direct Orange WS
Benzo Orange R
Amanil Toluylene Orange R
Pluto Brown GG
Amanil Brown MR
Amanil Brown D3G
Amanil Fast Brown RLH
Benzo Fast Scarlet 8BA
Amanil Fast Scarlet 4BA
Amanil Fast Scarlet 4BS
Benzo Purpurine
Amanil Fast Violet RRL
Amanil Fast Violet 4B
Benzo Azurine 3R
Benzo Chrome Black Blue B
Amanil Sky Blue FF Conc.
Benzo Green FFG
Amanil Green B Conc.
Direct Deep Black E

Vat Color—Direct Printing

A great many of the vat colors when satisfactorily prepared in smooth paste form or very finely ground (micropulverized) powders are suitable for direct printing. The dyestuffs are thoroughly mixed into a thickener of "Vat Gum" prepared with the necessary reducing and fixing agents, i.e. Hydrosulphite AWC and a carbonate, usually potassium carbonate also accelerating materials such as glycerin and Kromfax Solvent.

So called Suprafix colors may be printed immediately upon mixing with the thickener. Others require a few hours in a warm temperature to effect a partial-reduction. In any case, a pre-reduced color fixes more readily.

Vat Color	40 g.
Glycerin	40 g.
Kromfax Solvent	40 g.
Starch—Tragacanth Thickener	480 g.
Potassium Carbonate	80 g.
Hydrosulphite AWC	60 g.
Water	200 g.

After printing and drying, the material is steamed without pressure for 7–10 minutes in moist steam. Then, soaped at boil and finished.

Textile Printing Lacquer

Nitrocellulose (Wet with 30% Butanol)	200 g.
Castor Oil	200 g.
Glyceryl Sebacate	170 g.
Secondary Hexyl Acetate	210 g.
Octyl Acetate	220 g.
Xylol	100 g.
Butanol	100 g.

and into this is then stirred a blue suspension consisting of 20 grams of freshly precipitated copper phthalocyanine and 80 grams of water. In this manner there is obtained a pigmented water-in-lacquer emulsion. By the further addition of 770 grams of water containing 6 grams of a wetting agent (for example, sodium lauryl sulphate) and 12 grams of Turkey Red oil, and passing the mixture through a colloid mill to disperse the ingredients more completely there results a stable blue lacquer-in-water emulsion (the reversion of the two phases should be noted) at once suitable for printing on fabric.

Textile Printing Resist

Formula No. 1

Precipitated Chalk	200
Potassium Sulphite (90° Tw.)	50
Acetate of Soda	50
Water	365
Dark British Gum	325

Beat the whole into a smooth paste, heat until the gum is dissolved, and cool.

No. 2

Zinc Oxide	200
Water	170
Glycerin	25

Beat into a paste, and add:

Dark British Gum	200
Gum Senegal (50% Solution)	150
Turpentine	30

Starch—Tragacanth Thickener

(*Textile Printing Thickener*)

Wheat Starch	15 lb.
Water (Cold)	15 lb.
Gum Tragacanth (60% Paste)	25 lb.
Olive Oil or Glaurin	3 lb.
Water	42 lb.

Starch is pasted with 15 lb. water then gum tragacanth paste added plus olive oil followed by balance of water to make 100 lb. Then the whole is gradually heated while continually stirring; boil for 20 minutes and stir until cold.

Textile Printing Paste Thickener

Formula No. 1

British Gum	5	oz.
Wheat Starch	14	oz.
Potassium Carbonate	1¾	lb.
Glycerin	¼	pt.
Olive Oil	¼	pt.
Formosul (Sodium Formaldehyde Sulphoxylate)	1¼	lb.

Water, Sufficient to make 1 gallon of thickening.

The above is for thickening a strong printing paste; one suitable for a weak paste has the following composition:

No. 2

British Gum	5	oz.
Wheat Starch	12	oz.
Potassium Carbonate	8	oz.
Glycerin	3/16	pt.
Olive Oil	3/16	pt.
Formosul	6	oz.

Water, Sufficient to make 1 gallon of thickening.

In making the printing paste itself, a mixture is made (for a strong color) of 3 parts of the thickening and 1 part of the vat dye paste, or (for a weak color) the strong color paste is suitably diluted with the thickening.

Artificial Velvet

Of the several methods of production the simplest is the use of a concentrated rubber latex dispersion. Practically any cloth is suitable for the fabric base, but, of course, certain weaves are better than others. It is not unusual to give the fabric a preliminary light dressing of latex on one side. This promotes a smooth surface in the case of cloths which lack the required smoothness, and also prevents excessive shrinkage. Where a light weight fabric forms the base, or a waterproof product is required, a thorough initial coating is required. A representative formula for a suitable compound is as follows:

Concentrated Latex	100
Sulphur	3
Zinc Oxide	5
Casein Solution (10%)	5
Antioxidant	½
Accelerator	½

Modifications of above may be necessary to suit particular cases. The initial coating is effected by a spreading machine, during which penetration of light weight fabrics is avoided by stretching and not exerting counter pressure. The spreading knife should always be fairly sharp, and set in the opposite direction to that in which the cloth is running.

The mix is then diluted with water, and poured into the trough through which the fabric is running on a rubber roller. The level of the mix is adjusted so that it just reaches the surface of the fabric. The artificial velvet base then goes forward to the machine which applies the dust on the still moist surface. Rotating sieves, preferably hexagonal in shape and of 1 millimeter gauge, carry the dusting media. They should cover the same area. As the amount of dust they release varies with the amount they contain, both sieves should be in operation. When they are full or nearly empty, and only one operating when they are half empty, and, therefore releasing most dust. About 5/6 times as much dust is shaken on the base as is finally required. Meanwhile the fabric is beaten quickly and regularly from underneath with flat instruments to insure that each individual hair assumes as perpendicular a position on the cloth as possible. The material is then left for about 10 minutes on a hot plate, or pressed over drums heated by approximately 15 lb. of steam to dry it.

Brushing with soft cylindrical brushes takes place when the artificial velvet is quite cold. The superfluous dust removed by this process may be recovered and used again. Vulcanization, the final stage, is best carried out in a hanging position and heating for 10–30 minutes at a temperature of 140° C. to 150° C.

Different effects may be obtained by varying the dusting media, silk, artificial silk, wool, and cotton dust are all used. The depth of pile may also be varied, and it is possible to emboss these artificial velvets. A similar process is used to produce cloths hardly distinguishable from moquettes and suede.

Crush Proof Finish for Silk Velvet

The material after dyeing is treated cold for 15 seconds with 0.13% formaldehyde and 0.3% ammonium chloride and then, without rinsing, is baked at 310° F. for 5 minutes or until dry. Next wash five minutes at 150° F. with a little ammonia in the bath, follow with a hot, then a cold rinse in clear water.

Of a number of colors checked, the following were found to withstand the treatment:

Brilliant Milling Yellow 5G
Milling Yellow O
Amacid Neutral Yellow GNS
Amacid Fast Yellow RS
Polar Orange GS
Amacid Neutral Orange SGS
Amacid Milling Scarlet G
Amacid Milling Scarlet 3R Conc.
Supramine Red 2G
Polar Red RS
Amacid Fast Red A Conc.
Brilliant Croceine 3BA Conc.
Amacid Green G
Alizarine Green CE Ex.
Brilliant Milling Green B
Amacid Silk Brown R
Amacid Fast Blue BL
Amacid Fast Blue GL
Amacid Silk Black RW

Scrooping Compounds
Formula No. 1

Casein	4½
Formic Acid	1
Boric Acid	1
Ocenol KD	17
Odorant (Coverene), To suit	

Use up to 2% in water.

No. 2

Ocenol	9
Gelatin	6
Aluminum Formate (Basic)	24
Odorant (Coverene), To suit	
Water	2000

No. 3
Viscose

Lactic Acid	5–6%
Malt Extract	4%
Oil Emulsion (1%), To make	100%

No. 4
Mercerized Poplin

Lactic Acid	1
Paraffin Wax Emulsion (1%)	5
Gelatin	5
Water	1000

Weighting for Textiles

Glycerin	6.0
Urea	2.0
Potato Starch	2.0
Sodium Benzoate	0.2
Water	27.8

Cook the starch in the water until swollen. Add the remaining ingredients and cool with stirring.

Textile Back-Filling Stiffener

Starch	1	lb.
Talc	1¼	lb.
Tallow Softener	1	lb.
Water	1	gal.

Textile Fiber Preservative
U. S. Patent 2,119,525

Vegetable fibers, cordage, thread, or nets are soaked in an aqueous solution of a tanning agent (approximately 6% solution of catechu extract) at about 100° for about 12 hours, boiled for about 15 minutes in an aqueous solution of an oxidizing agent (0.5% solution of potassium dichromate), rinsed in water, dried and immersed for about 3 minutes in coal tar at about 95°.

Preventing Sticking of Resin Coated Textiles
Formula No. 1

The addition of ¾% Duponol ME or solids in synthetic resin finished on textiles prevents sticking together of cloth.

No. 2

The incorporation of ½% of Acrawax C in polyvinyl butyral resin compounds, used in rainproofing textiles, prevents sticking, even under tropical conditions.

Textile Fibers from Maize Straw
U. S. Patent 2,271,218

A method of producing spinnable textile fibers from corn straw, comprises the steps of boiling the straw with 1–5% soda solution, washing and squeezing the material to remove the leafy parts, then boiling for a short time with a 1–3% urea solution, leaving the material in said urea solution for about 3 to 4 hours, adding to the urea solution a solution of 1–3% alkali, boiling once more for 1 to 3 hours, thereby evolving free ammonia, and washing, drying and disintegrating so as to form fibers.

Casein Fibers
U. S. Patent 2,225,198

Casein	24
Water	198
Ethyl Glycolate	5
Triethanolamine	6
Calcium Chloride (10% Solution)	7

The above is extruded, dried and finished.

Glazed Thread

Bring 50 gallons of soft water to 80° F. Then add 15 pounds potato starch which has previously been dissolved in cold water. After thoroughly mixing, add 1 pound paraffin wax and 2 pounds cocoanut oil. After all is dissolved, empty into clean pails and use when cold. Keep the tension off the yarn on the machine as much as possible. This will help to get a good, soft, brilliant luster. With too much tension the thread will be wiry.

Cotton Cord Glaze Finish

Water	50	gal.
Potato Starch	15	lb.
Castile Soap	½	lb.
Paraffin Wax	4	lb.
Beeswax	1	lb.
Lard Oil	2½	pt.

Mix the starch in cold water and add to the boiler when the water is at the boiling point. Add the other ingredients and boil 15 minutes. Use when cold.

Giving Natural Finish to Smooth Calico

Potato Starch	5
Wheat Flour	7½
Boiled with Water	250
Then add	
China Clay Paste	10
French Mineral White	10
Boil and add	
Cocoanut Oil	¾
White Soap	½
Carbonate of Soda	¼
Water	3
Add to a vat containing	
Potato Starch	15
Water	75

Stir thoroughly and then slowly add

Potato Starch	5
Water, With a Trace of Ultramarine	5

Refinishing Fiber or Grass Rugs

Spread a thick mat of newspapers on the floor and lay the rug out flat. For the stain simply use flatting oil to which the necessary colors-in-oil have been added.

The amount of color required will depend on how deep a tone is wanted —ordinarily a pint of the color to each gallon of flatting oil is sufficiently strong to freshen up the faded colors of the rug. The stain should be applied freely, with a wide brush, so that it penetrates the fibers and imparts a uniform appearance to the rug. If something more decorative than an over-all single-tone is wanted, stencils or masks may be used for certain patterns in various colors. Give the rug time to dry through before putting in service.

Immunized Wool

Wool	1000	lb.
Tannic Acid	200	lb.
Color (Direct) to Tint	2½	oz.
Acetic Acid (28%)	150	lb.
Tin Crystals	40	lb.
Formic Acid	60	lb.

Prepare bath with color, enough water to work wool in easily. Add 5% acetic plus 20% of tannic acid; dye at 120° F. Bring slowly to boil and boil ¾ hr. Add 5% acetic, boil ½ hr. longer and add 5% acetic. Boil ½ hr. longer.

Without washing off treat in a fresh liquor for 1 hr. at 195° F. made up of 4% tin crystals and 6% formic acid. Wash off, do not pole too much as wool will be felted.

Chlorine Treatment of Wool

Scoured goods are worked ½ hour in a cold bath containing 10 lb. hydrochloric acid.

After allowing to drain, the goods are laid upon the reel and carried to a second cask and worked cold for ½ hour in a bath of chlorinated lime ¾° Tw. The lime (bleaching powder) should contain about 35% active chlorine. Meanwhile the first cask is drained and a bath of 8 lb. sulphuric acid (66° Bé) made up. The goods are worked for ½ hour cold, rinsed well and dried for printing.

Chlorinating leaves a yellowish tint which can be removed by stoving. The machinery used for production is a 16 line dye kettle with studded reel.

Non-Shrinking Wool
Formula No. 1
British Patent 541,965

Woolen fabric is immersed for 2½ hr. at 70° C. in

Stannous Chloride	10 g.
Hydrochloric Acid (2N)	100 cc.

It is then squeezed, washed and dried.

No. 2
British Patent 540,052

An 18:82 wool-spun viscose fabric is dried, treated for 1 hr. at 20° in beta propyl alcohol (100 cc.) containing potassium hydroxide (5 g.), centrifuged, rinsed in dilute alcoholic potash, washed, and dried.

Non-Felting Wool
British Patent 540,613

Woolen yarn (15) is immersed for 1 hr. at 30° in n-sterocarbamide (1 part) and sulphuryl chloride (6 vol.) in carbon tetrachloride (200 vol.), rinsed in cold water, immersed in 0.2% aqueous ammonia, washed, and dried. It is then non-felting and softer to handle than the untreated yarn.

Glass Wool Coating Emulsion

Hydrocarbon Oil	24.5
Stearic Acid	18.5
Ammonia	4.0
Water	43.2

Preparing Metal-Coated Fabrics

Impregnate the cloth surface with a prepared graphite-in-oil, which provides uniform distribution of the fine carbon particles. Then run the cloth through rollers, allow to dry and place in the plating bath. The cloth must be properly stretched to obtain uniform deposition. For nickel plating the cloth use 37 ounces nickel sulphate, 8 ounces nickel chloride, and 5 ounces boric acid per gallon of electrolyte. In addition use brightening agents. Bath temperature should be 43–48 degrees; current density, 40 amperes per square foot; then wash and dry the cloth.

Metal Coated Fabric
U. S. Patent 2,125,341

The fabric is coated with a film of cellulose nitrate 26–23%, a plasticizer (castor oil or ortho dibutyl phthalate) 46–52%, and a pigment 28–25%, then with an adhesive coating of cellulose nitrate not greater than 10%, ethyl acetate 40–20%, ethyl alcohol 60–40%, and dibutyl tartrate not more than 30%, and finally, when the adhesive is tacky, with a dusted-on coating of metal (aluminum) powder. The metal particles are brushed on so as to orient the flakes and produce a mirror-like luster, which is preserved by applying a thin coating of a clear lacquer.

Paraffin Coating for Balloons and Airships

From one to two pounds of paraffin wax, depending upon the weight of coating, and the condition of the surface to which it is to be applied, is melted over a steam bath or upon a large electric plate to prevent overheating.

The melted paraffin is allowed to come to a temperature 5 to 10° C. above its melting point, when it is mixed with one gallon of solvent which has been previously heated to the same temperature. Care should be taken that the paraffin and solvent are the same temperature, as the addition of hot paraffin will cause the solvent to boil violently.

The solvent should never be heated over an open flame due to the fire risk.

The air shall be between 50% and 75% relative humidity at approximately 70° F. as there is danger of fire due to the building up of static charges from the evaporation of the solvent which will be dissipated if the humidity is in excess of 50%, while if the humidity is over 75% there is a tendency to precipitate moisture. The application of paraffin should not be accomplished on days when there is an approaching thunderstorm or the surrounding temperature is under high electrical tension.

The envelope shall not be inflated and coated from the inside.

The envelope shall be opened at the mid-section above ⅓ of the circumference when about 20% inflated, and reversed by rolling the envelope on the air in the interior of the envelope. After removing the original tape, a tape approximately four inches wide shall be applied at the opened seam on both sides. This will prevent the application of paraffin at this point, which would cause a poor union when the envelope is reassembled. The tape should not be removed after paraffining until the envelope is turned and ready to be reassembled.

The paraffin shall be applied with a standard sweeping broom in a uniform coating so that it flows smoothly over the fabric, leaving no broom marks on the surface. Insofar as practicable, the surface to which the solution is to be applied should be flat in order that a uniform coating may be applied.

The increase in weight due to the application of the solution shall be approximately 0.5 ounce per square yard.

The envelope shall be required to dry for at least 48 hours to permit complete evaporation of the solvent from the paraffined surface before being rolled up.

It is very necessary to thoroughly ventilate the room in which the operation is being performed. Care should be exercised to prevent the formation of pockets of gas from the solvent, especially around the ballonet. When a section of the envelope has been doped it should be allowed to dry thoroughly before being turned over.

After the paraffin solution has been applied to the inside surface, the envelope should be inflated about 20% and returned to its original position. The tape shall then be removed from the seams and a careful inspection made to determine if paraffin is on the cemented surface. If there is any paraffin on the cemented surface, it shall be removed by scraping, buffing and washing with the solvent before any cement is applied. The seams shall then be cemented, sewed and re-taped in accordance with standard practice.

Retting of Istle and Sisal Fibers

In order to remove the gum from sisal and istle fibers, it is necessary, at first, to consider the chemical structure of this gum. Upon analysis it will be readily found that the gum consists of bastose, a form of sugar, and tannin matter.

It is necessary that, in order to decompose the gum, the bastose be attacked.

The bastose can be successfully attacked by dissolving, in boiling water, 4 oz. by weight of the material of tri-sodium phosphate for every 5 lb. of raw fiber. The fiber should be kept in the boiling solution for about 10 minutes, after which it should be thoroughly rinsed with cold water, to remove any phosphate from the fiber.

Softening Sisal Fibers
British Patent 541,383

The fibers are boiled for 2 hours in 5% soft soap, washed in hot water, immersed for 5 minutes in aqueous calcium hypochlorite (pH 4.5 and containing not less than 1% available chlorine), washed with dilute aqueous alcoholic potash, then with water, lubricated, and dried.

Sisal, Aloe and Other Plant Fibers, Separating
British Patent 542,017

The leaves are crushed and the juice is removed. Then boil in 1% soda ash solution for 30 min. and scrape off fleshy covering of fibers. If desired the fibers may be boiled with 2% sodium sulphite solution for 1 hr. and then dipped in 1% cold hydrochloric acid. Finally they are washed with water and dried.

Kier Boiling of Cotton Goods

1. Type of Goods: Single and ply yarn and cables.

Water: 90 ppm. hardness.

Conditions: Boil 3 hours at 214° F.

	%
Caustic Soda	1.6
Sodium Silicate	3.84
T.S.P.P.*	0.8
Wetting Agent	1.6

2. Type of Goods: Woven piece goods—flannels, sheetings, duck, etc.

Water: 80 ppm. hardness.

Conditions: Boil 14 hours at 25 lb. pressure.

	%
Caustic Soda	4.0
Sodium Silicate	1.5
T.S.P.P.*	1.0
Wetting Agent	0.075

3. Type of Goods: Turkish toweling.

Water: 18 ppm. hardness.

	%
Caustic Soda	3.0
Sodium Silicate	1.0
T.S.P.P.*	1.0
Wetting Agent	small amt.

4. Type of Goods: Light weight woven goods.

Water: 100 ppm. hardness.

	%
Caustic Soda	3.0
Sodium Silicate	0.25
T.S.P.P.*	0.5

Gets cleaner kier and eliminates resist spots in dyeing, which are due to insoluble calcium soaps formed in the kier.

Doubled Fabrics

Fabrics doubled with latex have many additional properties. It is not surprising, therefore, that such cloths are finding an increasing use in the production of shoe linings, boot tops, wind jackets, printers' blankets, motor car hoods, miners' clothing and balloon fabrics. In this case a standard doubling mix consists of:

Concentrated Latex	100
Whiting	50
Zinc Oxide	3
Sulphur	2
Du Pont Accelerator	0.2
Casein Solution (10%)	5
Water	15

French chalk or barytes may be substituted for whiting, which serves to promote permeability to air as well as to cheapen the spreading mix. If more weight or a better handle is required the quantity can be increased, in which case it may be advisable to increase the quantities of water and casein solution. Curing is effected by the addition of zinc oxide, sulphur, and accelerators under heat. If Du Pont accelerator is taken, it is sufficient to run the fabric over a heating plate or cylinder heated to about 100–120° C. Vulcanization of the mix sets in and continues at ordinary room temperatures.

The quality of the material and the degree of waterproofness determine the amount of mix to be spread. The doubling mix will adhere to untreated material better than to a smooth dressed cloth. Too pronounced penetration of the mass into the fabric may be prevented by using a sharp spreading knife, set at an acute angle to the direction in which the fabric is traveling. The fabric should also pass under the knife as quickly as possible and without friction.

With light weight fabrics it frequently happens that a light colored mix is visible through the new ma-

*T.S.P.P. = tetrasodium pyrophosphate.

terial. This has the effect of causing an apparent change in the shade. In such cases suitable coloring agents should be added to the mix. When dealing with pure white cloths, it is possible to avoid discoloration by displacing 10 parts of whiting in the above mentioned mix, and substituting 20 parts of titanium dioxide along with a further 5 parts of 10% casein solution. Here the addition of a water soluble blue is also helpful to tint the mix slightly.

Where good adhesion only is required, as in shoe fabrics, the cloths may be doubled in one operation by the aid of a combined spreading and doubling machine. It is also possible to use a calender having a spreading knife in front and drying apparatus behind. One fabric, after passing under a pulley, passes a spreading knife without counter pressure, and joins the second length at the doubling roller. The doubled fabric is then dried and rolled up. Speaking generally, when about 1 ounce of the mix is used for 1 square yard the material is sufficiently permeable to air. Penetration during calendering must be avoided, and the mix must not have time to dry out before doubling. For this reason it is recommended that the spreading knife and doubling rollers be set at 18 inch to 1 yard apart. When uniting the cloths, the highest possible pressure should be exerted to insure equally strong adherence to both cloths.

Where waterproof doublings or heavier cloths are treated, it is preferable to carry out process in more than one operation. Here the whiting may be reduced or even omitted altogether. Each length of cloth is coated separately on the spreading machine, either one or several times. To prevent the subsequent applications penetrating into the base of the fabric a sharp spreading knife is preferable. A blunt knife may be used for subsequent coats, however. Three to four ounces of the mix per square yard should give good waterproof qualities. The spreads of each length of cloth must be sufficiently dry before they are doubled on the calender to avoid penetration.

Fluorescent American Flag

By fabricating a flag from fluorescent textiles or ordinary stains which often contain fluorescent dyes an American flag of unusual beauty may be obtained, for display purposes in ultraviolet light.

White fluorescing fabric can be prepared by dipping satin or other cloth with a solution made up of buffered fluorescein and quinine or cinchonine.

Red fluorescing fabric can be prepared by dyeing cloth with eosine or rhodamine.

Blue fluorescing fabric is prepared by dyeing cloth with a solution of umbelliferone or Fluorescent Blue G.

Hat Conditioning Material
(used after cleaning hats)

This material is used to restore some of "life" and body to a felt hat after cleaning or much wearing. It adds some water-repellency properties and makes it more serviceable in other ways.

Water: 98.

Carnauba or Beeswax: Between 0.2 and 0.5.

Disodium Phosphate: Sufficient to give a buffered solution of pH between 7.50 and 7.80 in final mixture.

Aluminum Stearate: 0.2.

Titanium Dioxide (Finely Ground): 0.5.

A trace of an emulsifying agent to help ingredients stay in solution and suspension. This may be triethanolamine, hexalin or other common ones depending upon availability. Maintenance of proper pH is more important than the particular emulsifying agent. This solution will probably be purplish. However, various dyes may be added to either mask or make provision for hats on which it would not be suitable. Under certain conditions, this purple color is never evident. The material is applied with a soft cloth pad preferably while hat is on a block. Rub in direction of felt layer, let dry slowly (overnight) and finish in usual manner of pressing, etc.

Chapter Twenty-One

MISCELLANEOUS

Fireproofing Straw

Straw can be satisfactorily fireproofed by immersion in a solution of 45 g. boric acid and 65 g. borax per liter. Another fireproofing treatment consists in immersing the straw in a solution of 50 g. ammonium phosphate, 25 g. ammonium sulphate, and 25 g. ammonium chloride per liter. Another consists in the use of a solution containing (per liter) 75 g. sodium acetate, 75 g. trisodium phosphate, and 20 cc. pale neutral 28° Bé glycerin. This last mentioned method of treatment has been employed for fireproofing the straw packing for carboys of nitric acid.

Anti-leak

For Hot-Water Heating Systems

Soapmakers' silicate of soda (1 qt. per 1000 gal. water) is used.

Anti-corrosion

For Hot-Water Heating Systems

Formula No. 1

(where considerable gas and air must be blown from radiator)

Caustic Soda	⅓
Chestnut-Oak Extract	⅓
Water	⅓

One pound of this solution is used to 1000 gal. water.

No. 2

(where waters are very soft—and where salt is present)

Sodium Bichromate	30
Sodium Carbonate—Soda Ash	10
Caustic Soda	1

In natural water use 2 lb. to 1000 gal. water.

In brine use 1 lb. to 5 gal. brine.

To Remove Carbonate Scale

(scale that foams when acid is added)

Hydrochloric Acid	2
Chestnut-Oak Extract	1

One part of this mixture to 25 parts of water makes a little over 1% acid strength solution. If stronger solutions are used, some supervision by a competent chemist should be arranged for. For boiling out boilers 2 to 2½ lb. per rated horsepower should be added; all the vents are left open and foaming anticipated.

Treatment for Vacuum or Vapor Heating Systems

Same as for hot-water systems, save that 2 lb. per year is enough—1 lb. added at outset and 1 lb. about three months later.

Leak Sealing, Rust Preventing Composition

For Hot-Water Systems

U. S. Patent 2,129,459

Flaxseed Meal	5
Potassium Dichromate	7

Stop Leaks for Radiators

To stop leaks in automobile or

tractor radiators the following preparation is helpful:

Ground Flaxseed	½ oz.
Glycerin	1 oz.
Water	5 oz.

Anti-leak Anti-freeze
U. S. Patent 2,264,387

Less than 0.1% of sodium alginate is added to alcohol.

Non-corrosive Anti-freeze
Canadian Patent 403,486

To the alcohol is added:

Sodium Nitrate	0.05–1.0%
Sodium Molybdate	0.05–1.0%

Aircraft Engine Anti-freeze

Ethylene Glycol	94½ g.
Water	3 g.
Trihydroxyethylamine Phosphate	2½ g.

Windshield and Window Anti-freeze
U. S. Patent 2,258,184

Ethylene Glycol	20–60
Glycerol	8–10
Formaldehyde	2– 5

The above prevents ice formation on any exposed surface.

Dielectric, Transformer
Formula No. 1
British Patent 534,143

o-Tolyl Diphenyl Phosphate	34–73%
m-Tolyl Diphenyl Phosphate	66–27%

This is liquid above 15° C.

No. 2
U. S. Patent 2,214,877

Chlorinated Paraffin	45–50
Trichlorbenzene	25–40
1, 2, 3, 4 Tetrachlorbenzene	15–25

Electric Resistor with Negative Temperature Coefficient
U. S. Patent 2,274,592

Nickel Oxide	23
Manganic Oxide	47
Cobaltic Oxide	30

Composite Resistance
British Patent 527,687

Steatite Dust	2
Copper Dust	1

Mix together and press into shape with or without firing at a temperature below melting point of copper.

Heat Resistant Stencil Film
U. S. Patent 2,242,313

Cellophane or other tough, nonabsorbent transparent foil is coated with

Mineral Black	5
Aluminum Powder	34
Cottonseed Oil	12
Vegetable Wax	49

Adhesion Preventing Compounds
U. S. Patent 2,262,689
Formula No. 1

Talc	64
Glycerin	½
Water	35½

No. 2

Soapstone, Powdered	99.8
Calcium Chloride, Powdered	0.2

Insulating Materials
British Patent 526,510–11
Formula No. 1

41.4 parts by weight of a paste containing slaked lime (7.4), powdered dehydrated silica gel (10.0), and transformer oil (14.0 parts) is mixed with 100 parts by weight of a mixture containing rosin (15.0), oleic acid (18.0), petroleum residual pitch (45.0), and transformer oil (22.0 parts).

No. 2

A mixture containing rosin (20.0), oleic acid (20.0), castor oil (40.0), and coumarone resin (20 parts) is mixed with a paste containing litharge (20.7), slaked lime (2.6), ground quicklime (7.0), and castor oil (10.0 parts).

No. 3

Alternatively, a mixture composed of oleic acid (25.0), ester gum (50.0), and castor oil (25.0 parts) is mixed with a paste containing slaked lime (3.3) and castor oil (3.3).

Electrical Insulation

Formula No. 1

U. S. Patent 2,259,134

Portland Cement	12 to 18
Asbestos, Long Fiber	7 to 11
Montan Wax	¼ to 2
Colloidal Clay	Up to 7
Graphite	Up to 1
Water sufficient to form a slurry.	

Preferably colloidal clay is used within the limits of 3 to 7 parts and graphite, within the limits of 0.5 to 1 part.

In molding the mixed ingredients a screen of approximately 20 mesh size is placed in the bottom of the mold, the material is flowed thereon, and another screen of approximately the same size is placed on top of the wet mass. Upon the application of pressure the water is freed from the mass, flowing out from a suitable opening in the bottom of the mold, and also from the top of the mold between the punch and the jacket. By using screens in the manner described, the larger part of the excess water is readily removed from the mass. If water be allowed to remain in the panel, the finished board has poor impact and flexural strength characteristics. On the other hand, it is very desirable to have an excess of water in the mix as it is introduced into the mold. By so doing, the asbestos fibers are not knotted or broken during the molding operation. Instead, they form an interwoven mass which has been found to be essential in obtaining optimum mechanical strength.

The molded parts are air-dried for a suitable period, for instance for about 4 to 5 hours. Thereafter they are placed in an oven in which steam is constantly introduced. The panels are cured in this oven for a suitable period, for example for about 24 hours. After the steam cure they are immersed in warm water (100° to 140° F.) for another suitable interval, for instance for about 24 hours. Following this treatment the panels are baked in a suitable oven, such for example as an electrical oven, at a gradually rising temperature up to approximately 500° F. over a period of, for instance, about 12 hours. All the uncombined water is removed from the panels by this baking treatment, and the components are bonded together to form a hard, rigid mass. The baked panels are sanded to a smooth finish, after which they usually are sprayed or otherwise coated with a suitable lacquer or varnish.

No. 2

British Patent 533,520

Isobutylene Polymer (Molecular Weight Above 100,000)	1–2
Bitumen (High Melting)	4–1

A little wax may be added to improve working characteristics.

No. 3

British Patent 523,383

Rubber	75
Polystyrene	25
Zinc Oxide	¾
Zinc Laurate	¾
Tetramethylthiuram Disulphide	2

This gives an extremely low power factor and specific inductive capacity.

Condenser Electrolyte

British Patent 531,706

Formula No. 1

Furfuryl Alcohol	64
Ammonium Borate	36

No. 2

Ammonium Borate	10–40
Acetamide	16–20
Tetrahydrofurfuryl Alcohol	40–80

Dry Cell Electrolyte

Ammonium Chloride	15.0%
Zinc Chloride	30.0%
Calcium Chloride	35.0%
Mercuric Chloride	0.2%
Water, To make	100.0%

This gives minimum corrosion of zinc, maximum depolarizing properties and does not dry out on exposure.

Temporary Gas Shut Off Composition

Gelatin	6
Molasses	2
Phenol (15% Solution)	¼
Water	91¾

Warm and mix until homogeneous. Apply while just fluid.

Revivifying Spent Gas Oxide

Add to:

Spent Iron Oxide	44
Soda Ash	52
Peat	11

Non-dusting Coal Treatment

Formula No. 1

British Patent 514,671

Calcium Chloride	100 lb.
Cornstarch (6% Gluten)	8 lb.
Water	50 gal.

No. 2

U. S. Patent 2,242,398

Rosin	1
Petroleum	1–9

Coloring Coal

U. S. Patent 2,129,901

Coal free from dust (washed or freshly broken) is treated separately with each of the three following solutions:

1. Calcium Ferricyanide	0.413%
2. Ferric Chloride	0.500%
3. Sodium Bisulphite	10.000%

Flare or Lamp Illuminant

U. S. Patent 2,258,910

Kerosene	95
Naphthalene	1
Rosin	1
Methanol	1
Caustic Soda (10% Solution)	2

Lamp Kerosene Rectifier

This powder is intended for house lamps, etc., to prevent the oil from smoking and having a disagreeable smell, and to increase and brighten the light.

Powdered Naphthalene	52
Fine Dry Salt	10
Powdered Camphor	2

Mix thoroughly and put up in tins or packets. A little is to be added to the oil in the lamp or stove, and renewed when this is consumed.

Frosting for Inside of Lighting Bulbs

U. S. Patent 2,278,257

Hydrofluoric Acid	23
Ammonium Bifluoride	20
Sodium Bicarbonate	4
Alcohol	10
Water	10

Motor Carbon Remover

Formula No. 1

British Patent 542,589

Ethylene Dichloride	80
Acetone	15
Cellulose Acetate	4
Stearin	1

No. 2

Neutral Coal Tar Oil (Free from Cresols)	40 cc.
o-Toluidine	15 cc.
Diethanolamine	15 cc.
Oleic Acid	15 cc.
Ethylene Glycol	15 cc.
Phosphoric Acid	½ cc.
Ethylsilicate	½ cc.

Hydraulic Fluid

U. S. Patent 2,232,581

	Formula No. 1	No. 2	No. 3	No. 4	
Propylene Glycol Ricinoleate	35.5	..	..	..	fl. oz.
Propylene Glycol	4.5	..	35.0	..	fl. oz.
Isobutanol	60.0	60.0	..	..	fl. oz.
Sodium Nitrite	9.0	..	..	..	g.
Potassium Soap	71.0	103.0	..	83.0	g.
Ethylene Glycol Ricinoleate	..	32.7	..	27.3	fl. oz.
Ethylene Glycol	..	7.3	..	6.0	fl. oz.
Calcium Nitrite	..	19.0	..	19.0	g.
Glyceryl Ricinoleate	..	..	10.0	..	fl. oz.
Butanol	..	..	55.0	66.7	fl. oz.
Potassium Nitrite	..	..	12.0	..	g.

No. 5

Castor Oil, Blown	50
Ethylene Glycol	16
Alcohol	26
Hydroxyethylbutylether	11
Tricresyl Phosphate	½
Diphenylamine	2½

No. 6

U. S. Patent 2,249,800

Isobutanol	75
Glyceryl Monoricinoleate	25

No. 7

Isobutanol	56.25
Glycerin	13.50
Propylene Glycol	20.50
Glyceryl Monoricinoleate	10.00

No. 8

U. S. Patent 2,255,208

Polypropylene Glycol	3
Polypropyleneglycolmono-ricinoleate	3
Propylene Glycol	3
Butanol	11

No. 9

U. S. Patent 2,102,825

Butanol	55%
Glycerin	13–15%

Protection Against War Chemicals

General Information Concerning the Characteristics, Effects, and Counteraction of the Agents an Enemy Might Use in War

By Walter P. Burn, Lt. Col., C.W.S.

Group	Names and Symbols	Form	Odor	Physiological Effect	Tactical Class	Protection	First Aid [*After removal from gassed area*]	Persistence	Field Neutralization
Vesicants	MUSTARD $S(CH_2CH_2)_2Cl_2$ Dichlorethyl sulphide	Liquid and vapor	Garlic, mustard, horse-radish	Delayed effect. Burns skin or membrane. Inflammation of respiratory tract leading to pneumonia. Eye irritation, conjunctivitis	Hospital case	Gas mask and protective clothing	Undress; remove liquid mustard with protective ointment, bleach paste, or kerosene; bathe; wash eyes and nose with soda solution	One day to one week; longer if dry or cold	Cover with unslaked lime and earth; 3% solution of Na_2SO_3
Vesicants	LEWISITE $CHClCH\text{-}AsCl_2$ Chlorvinyl-dichlorarsine	Liquid and vapor	Geraniums	Burning or irritation of eyes, nasal passages, respiratory tract, skin. Arsenical poison	Hospital case	Gas mask and protective clothing	Undress; remove liquid Lewisite with hydrogen peroxide, lye in glycerine, or kerosene; bathe; wash eyes and nose with soda. Rest and doctor	One day to one week; longer if dry or cold	Wash down with water. Cover with earth. Alcohol. NaOH spray
Vesicants / Lung irritants	ETHYLDICHLORARSINE $C_2H_5\text{-}AsCl_2$	Liquid and vapor or gas	Stinging, like pepper in nose	Causes blisters, sores, paralysis of hands, vomiting. Severe on long exposure	Hospital case	Gas mask and protective clothing	Undress; remove liquid with hydrogen peroxide, lye in glycerine, or kerosene; bathe; wash eyes and nose with soda. Rest and doctor	One hour	Cover with earth, caustic
Lung irritants	CHLORINE Cl_2	Gas	Highly pungent	Lung irritant	Hospital case	Gas mask	Remove from gassed area. Keep quiet and warm. Coffee as stimulant	10 minutes	Alkaline solution
Lung irritants	CHLORPICRIN CCl_3NO_2 Nitrochloroform	Gas	Fly-paper, anise	Causes severe coughing, crying and vomiting	First-aid; and hospital case	Gas mask	Wash eyes, keep quiet and warm. Do not use bandages	Open 6 hrs.; woods 12 hrs.	$NaSO_3$ — sodium sulphite in alcohol solution
Lung irritants	DIPHOSGENE $ClCOOC\text{-}Cl_3$ Trichlormethyl chloroformate	Gas	Ensilage, acrid	Causes coughing. Breathing hurts, eyes water, toxic	Hospital case	Gas mask	Keep quiet and warm. Give coffee as a stimulant	30 minutes	Alkali
Lung irritants	PHOSGENE $COCl_2$ Carbonyl chloride	Gas	Musty hay, green corn	Irritation of lungs, occasional vomiting, tears in eyes, doped feeling. Occasionally symptoms delayed; later, collapse, heart failure	Hospital case	Gas mask	Keep quiet and warm; rest in bed. Coffee as a stimulant. Loosen clothing. No alcohol or cigarettes	10 to 30 minutes	Alkali

	Names and Symbols	Form	Odor	Physiological Effect	Tactical Class	Protection	First Aid [*After removal from gassed area*]	Persistence	Field Neutralization
LACRIMATORS	CLORACETOPHENONE $C_6H_5CO\text{-}CH_2Cl$	Gas	Apple blossoms	Makes eyes smart. Shut tightly. Tears flow. Temporary	First-aid treatment	Gas mask	Wash eyes with cold water or boric-acid solution. Do not bandage. Face wind. For skin, sodium sulphite solution	10 minutes	Strong, hot solution of sodium carbonate
	BROMBENZYLCYANIDE $C_6H_5CH\text{-}BrCN$	Gas	Sour fruit	Eyes smart, shut, tears flow. Effect lasts some time. Headache	First-aid treatment	Gas mask	Wash eyes with boric acid. Do not bandage	Several days; weeks in winter	Alcoholic sodium hydroxide spray
STERNUTATORS	ADAMSITE $(C_2H_4)_2\text{-}NHAsCl$ Diphenylaminechlorarsine	Gas	Coal smoke	Causes sneezing, sick depressed feeling, headache	First-aid treatment	Gas mask	Keep quiet and warm. Loosen clothing. Reassure. Spray nose with neo-synephrin or sniff bleaching powder. Aspirin for headache	10 minutes	Bleaching powder solution
	DIPHENYLCHLORARSINE $(C_6H_5)_2\text{-}AsCl$	Smoke	Shoe polish	Causes sick feeling and headache	First-aid treatment	Gas mask	Remove to pure air, keep quiet. Sniff chlorine from bleaching-powder bottle	Summer 10 minutes	Bleaching powder solution
SMOKES	HC MIXTURE $ZN+C_2Cl_6$	Smoke	Sharp-acrid	Harmless	Smoke	None needed	Produces no effect requiring treatment	While burning	None needed
	SULPHUR TRIOXIDE SO_3+SO_3HCl In chlorsulphonic acid	Smoke	Burning matches	Cause prickling of skin, flow of tears	Smoke	Gas mask	Wash with soda solution	5 to 10 minutes	Alkaline solution
	TITANIUMTETRACHLORIDE $TiCl_4$	Smoke	Acrid	Harmless	Smoke	None needed	Produces no effect requiring treatment	10 minutes	None needed
INCENDIARIES	WHITE PHOSPHORUS P	Smoke	Burning matches	Burning pieces adhere to skin and clothing	First-aid treatment	None needed	Pack in cloths wet with copper sulphate (blue vitriol) or water or immerse in water. Pick or squeeze out particles. Treat for burn	10 minutes	Burns out
	THERMIT $8Al+3FeO_4$	Incendiary	None	5,000° heat ignites materials	Incendiary	None needed	Treat for severe burn	5 minutes	Quickly cover with dry earth or sand

From *Army Ordnance*, March-April, 1942.

General Instructions—Protective masks suitable for fire fighting and for mine rescue work are not suitable anti-gas devices. The only masks that are effective against all common war gases are those manufactured according to the Chemical Warfare Service specifications and procurable through the Office of Civilian Defense. General manufacture and public sale of gas masks is prohibited by War Production Board's General Limitation Order L-57, March 3, 1942.

The importance of proper first aid for gas victims cannot be overemphasized. The following are general rules which apply in all cases.

A. Act promptly and quietly; be calm.

B. Put a gas mask on the patient if gas is still present or, if he has a mask on, check to see that his is properly adjusted. If a mask is not available, wet a handkerchief or other cloth and have him breathe through it.

C. Keep the patient at absolute rest; loosen clothing to facilitate breathing.

D. Remove the patient to a gas-free place as soon as possible.

E. Summon medical aid promptly; if possible, send the victim to a hospital.

F. Do not permit the patient to smoke, as this causes coughing and, hence, exertion.

Diethylene Glycol	20–25%
Methyl Ricinoleate	10%

No. 10

Castor Oil No. 3	40 gal.
Alcohol	60 gal.
Sodium Chromate	1 oz.

No. 11

Castor Oil	40 gal.
Amyl Alcohol	60 gal.
Sodium Chromate	1 oz.

No. 12

Sulphonated Castor Oil (80%)	25 gal.
Alcohol	60 gal.
Water	15 gal.
Sodium Chromate	1 oz.

In place of the above alcohols, the following may be used: methanol, isopropanol, "Cellosolve" or isopropylether.

No. 13
German Patent 692,303
Formula No. 1

Tritolyl Phosphate	61
Diphenyl Oxide	22
Biphenyl	5

No. 2

Tritolyl Phosphate	55
α Chlornaphthalene	45

No. 14
(Non-corrosive)
U. S. Patent 2,232,581

Ethylene Glycol Ricinoleate	32.7 gal.
Ethylene Glycol	7.3 gal.
Butyl Alcohol	60.0 gal.

To each gallon of the above add:

Calcium Nitrate	19 g.
Potassium Soap	103 g.

Anti-mist (Fogging) Compounds

British Patent 524,987
Formula No. 1

Sodium Oleyl Sulphate	8
Titanium Dioxide	24
Water	68

No. 2

Sodium Oleyl Sulphate	12
Titanium Dioxide	36
Water	12
Cyclohexanol	40

Auto Sound Deadening Compound

U. S. Patent 2,265,770

Reclaimed Rubber	12
Asphalt	50
Wood Flour	30
Rosin	8

Air Raid Window Protection

Strips of old bedsheets or similar fabric will do a fine job of window-saving if pasted on with a home-made paste like this:

Wheat Flour	6 oz.
Alum, Powdered	½ oz.
Corn Syrup	¼ pt.
Water	2 pt.

The syrup and the water are mixed and used to make a smooth batter of the flour and alum; this is heated quickly to the active bubbling point with constant stirring to prevent scorching; or it may be cooked on a double boiler. The paste may be preserved with a level teaspoonful of benzoate of soda.

Don't use silicate of soda as a window stickum; it will eventually etch the glass.

Incendiary Bomb Extinguisher

Formula No. 1
British Patent 543,703

Sand	40–60%
Borax	60–40%

No. 2

Ordinary powdered talc is recommended as a cheap efficient extinguisher. Its covering and adhesive power is superior to sand or other suggested materials. It will also extinguish oil and phosphorus incendiaries by preventing access of air.

No. 3

Sprinkled on a bomb, just as sand has been employed, feldspar quickly melts and forms a protective coating which cuts off the supply of air and stops the bomb from burning.

It is superior to mixtures containing salt, pitch, ashes or fine powders, as it does not burn, give off smoke, blow out or scatter appreciably from the intense heat of the incendiary material.

No. 4

Hardcoal-tar pitch, flaked or granulated, has been recommended by the Bureau of Mines, Department of the Interior, Washington, for extinguishing incendiary bombs. The method is as follows: Let the thermite burn out; about one minute. If the bomb is on metal or concrete or other non-inflammable material spread a layer of the pitch on the bomb with a long handled shovel. Don't throw the pitch. If a flame persists, put on another layer of pitch. Allow the bomb to cool and take it out of the way in a bucket or metal container. If the bomb has fallen on something inflammable like wood, cover the bomb with a layer of pitch, again with a long handled shovel. Then spread a layer near the bomb and roll the bomb on the spread pitch; spread a layer entirely over the bomb. Fires already started may (if the bomb is covered with pitch and out of the way) be put out with a hose or chemical extinguisher. Some bombs contain explosive charges; so protective clothing, goggles, and long handled implements are recommended.

No. 5

For Explosive Incendiary Bombs

Spray on water, staying behind an iron shield or wall.

Gun Recoil Fluid

Glycerin C.P.	3 gal.
Water, Distilled	3 gal.
Caustic Soda, C.P.	2 oz.

Transfer Liquid for Newspaper Pictures

Tincture Green Soap	1/8 fl. oz.
Turpentine	1/8 fl. oz.
Water, To yield	16 fl. oz.

Moistener for Mimeograph Rolls

Glycerin	2 fl. oz.
Water, To yield	1 gal.

Preventing Printing Press Static

Use finely powdered mica to dust off tympan. This works better than talcum and eliminates offset as well as static electricity. This is especially useful on high speed presses.

Preventing Printing Offset

Canadian Patent 373,457

The following solution is sprayed through currents of air to facilitate concentration and rapid drying on paper:

Gum Arabic	18 oz.
Glucose	12 oz.
Water	35 oz.
Alcohol	35 oz.

Making Small Animals Transparent

For best results use such vertebrates as mice, fish, salamanders, frogs, and snakes. Anything larger than a rat should not be used. It is also important that throughout this technique chemically pure materials are used. Impure chemicals result in discoloration.

1—The animal is first fixed and posed in 40% formalin or alcohol for two days. In this technique the animals are cleared with their skin, hair, and insides unmolested, but in the case of fish and snakes the scales must be removed because they take on the stain. By first fixing the animal in formalin or alcohol the tissues are thus hardened, a process similar to that used by the ancient Egyptians in preparing mummies.

2—Wash the specimen in water and place into a 2% solution of potassium hydroxide (KOH) to which a few drops of 3% hydrogen peroxide has been added. The 2% solution of potassium hydroxide is prepared by adding 2 grams of C.P. KOH pellets to 98 cc. of water. Whenever the solution becomes cloudy or discolored it should be changed. Continue this step until the flesh appears jelly-like and the bones become visible.

3—The animal is now stained with a selective dye, one that stains the bones rather than the tissues of the specimen. The stain for this purpose is alizarine monosulphonate. The stain is prepared by adding very small amounts of the powdered dye to a solution of 2% KOH until a light wine color is obtained. The animal is kept in this stain for two days or until all the bones are stained red and are clearly visible. Do not leave specimen in the stain too long, otherwise overstaining will take place and the tissues as well as the bones will be colored. If this does occur the animal may be destained by placing into a 2% solution of KOH plus a few drops of 3% hydrogen peroxide.

4—At this stage clearing of the specimen is begun by means of glycerin. This may be accomplished by placing the animal into each of the following solutions for at least three days:

(a) 25 parts glycerin: 75 parts 2% KOH
(b) 50 parts glycerin: 50 parts 2% KOH
(c) 75 parts glycerin: 25 parts 2% KOH
(d) 100% glycerin

5—The animal is now preserved in a fresh bath of pure glycerin to which a crystal of thymol is added to prevent the growth of molds.

6—The final step is the sealing of the container. A cork stopper should never be used. A Bakelite screw cap, glass stopper, or paraffin will do very well. Mount on a white glass background.

Nutrient Broth for Yeast

Sucrose	100 g.
Peptonum Siccum	20 g.
Dipotassium Phosphate	5 g.
Magnesium Sulphate	1 g.
Ammonium Sulphate	5 g.

The growth of brewers' yeast increases at least six times in 48 hours in this medium.

Fermentation Mash Sterilizer

0.7 liter of formaldehyde is used per 100 liters of mash.

Medium for Drosophilia Culture

Water	1000
Tomato Paste	100
Corn Syrup, White	100
Agar, Granulated	20
Moldex	1

Killing Spray Pond Algae

Use 30 p.p.m. sodium pentachlorophenate.

Preventing Mold Growth on Barrels

Hungarian Patent 124,005

Wooden wine or other barrels are coated with

Shellac	12 g.
Alcohol	1 l.

Heating Mixture for Food Containers

U. S. Patent 2,289,007

Sodium Monoxide	30
Aluminum Powder	20
Pumice	20
Water	50

The purpose of this mixture is to create a reaction in the inner can of a container holding one quart of food in its outer compartment, which will raise the temperature of the food from 0° F. to 110° F., thus thoroughly warming the food. The sodium monoxide, aluminum powder and pumice are placed at the bottom of the inner can while the water is placed in the upper section of the can, the two areas being separated by a plate segment. The inner can here should not occupy more than one third of the total volume of the container. Perforation of the plate segment allows the water to drip on to the mixture thus initiating the action. The heat evolution during the reaction is sufficient to raise the temperature of the contents of the can to 110° F.

Self Heating Pad Composition

Iron Filings (150 Mesh)	1 lb.
Cupric Chloride	5 g.

Before use add

Water	15 cc.

and shake well.

Boiler Compounds

Formula No. 1

Soda Ash	76
Trisodium Phosphate	10
Starch	1

and sufficient cutch or dry extract of hemlock, oak or chestnut bark to yield 2% tannin.

No. 2

Soda Ash	44
Disodium Phosphate, Anhydrous	47
Corn Starch	9

No. 3

British Patent 513,386

Graphite, Powdered	85
Aluminum Bronze	9
Zinc Dust	3
Talc	3

No. 4

U. S. Patent 2,291,146

Sodium Carbonate	7%
Trisodium Phosphate	5%
Borax	3%
Water, To make	100%

Clarification of Water

U. S. Patent 2,251,748

The following composition, in powdered form, is added to the water. It forms a floc with suspended and dissolved solids which settles to the bottom. The clear water is drawn off from the top of the settling tank.

Coal	70
Sodium Aluminum Tannate	20
Ammonium Ferrous Sulphate	5
Magnesium Tannate	5

Water Softener

U. S. Patent 2,102,219

Trisodium Phosphate	20 lb.
Borax	3 lb.
Alum	3 lb.
Water, Hot	6 qt.

Dissolve the above and mix with following solution:

Sodium Aluminate	½ lb.
Water	1 pt.

Boil mixture until it sets to a gel on cooling.

Conditioned Well Drilling Mud

U. S. Patent 2,271,696

Formula No. 1

Add Flour	1 %

No. 2

Add Flour	1 %
Glycol Oleate	0.1%

No. 3

Add Flour	1 %
Glyceryl Monoricinoleate	0.1%

Sculptor's Modeling Composition

U. S. Patent 2,214,126

Sand, Fine	37.5
Aluminum, Powdered	12.5
Talc	12.5
Lacquer	37.5

Artificial Cat Fish Bait

Obtain several dozen earthworms and pack them in a tin can in the following order. First, place a layer of cut up bath sponge in the bottom of the can, then a layer of earthworms, then another layer of sponge, then earthworms, and so on until the can is over half full. Cut up the sponge in pieces about ½ inch by ½ inch.

Now set the can away in a warm place and let the whole can of worms and sponge pieces rot. After several weeks, there will be an almost unbearable stench. Use these sponge pieces to bait your hooks when you go fishing for cat fish, and you will be surprised how the fish will take your bait.

Sensitized Paper to Indicate Exposure of Sterile Solutions, etc., to High Temperatures

Silver Nitrate	15
Potassium Bitartrate	15
Ammonia (28%)	60
Sugar	4
Powdered Acacia	4
Carmine, (No. 40)	1

This yields a red preparation that is applied by brush, to white filter papers. After drying in air, the papers are cut into strips, about 2 or 3 inches long, and ¼ inch wide. These sensitized strips are to be stored in glass containers kept in a dark, cool location.

The red color is changed to black when strip has been exposed to the moist heat of an autoclave or steam sterilizer.

A strip of the sensitized paper is placed over the cotton plug in the mouth of a flask of solution subject to sterilization. A strip of coarse gauze is then fastened over cotton and strip. The presence of a black strip, subsequently indicates the contents of flask to have been exposed to high temperature necessary to insure sterility.

Colors Indicating Temperatures

The following inorganic combinations with methenamine (hexamethylene tetramine) and water are agents for indicating temperatures based on color changes:

	Color Change at ° C.	Color Change
Cobalt Chloride : Methenamine : water	35	Rose Blue
Cobalt Bromide : Methenamine : water	40	Rose Blue
Cobalt Iodide : Methenamine : water	50	Rose Green
Cobalt Thiocyanate : Methenamine : water	60	Rose Blue
Cobalt Nitrate : Methenamine : water	75	Rose Purple
Nickel Chloride : Methenamine : water	60	Bright Green Violet
Nickel Chloride : Methenamine : water	100	Yellow Violet
Nickel Bromide : Methenamine : water	60	Bright Green Blue
Cobalt Sulphate : Methenamine : water	60	Rose Violet

Universal pH Indicator

Five mg. of thymol blue, 25 mg. of methyl red, 60 mg. of bromothymol blue and 60 mg. of phenolphthalein are dissolved in 75% alcohol to make 100 cc. of solution; this is then neutralized with hundredth-normal sodium hydroxide to produce a green color, but for high and low pH values it can be used directly. This combination shows at whole pH values between 4 and 10 the following distinct color changes: red, orange, yellow-green, blue, indigo and violet. Intermediate color changes are recognizable with an accuracy of 0.5 pH.

To Dissolve Methyl Cellulose Quickly

Methyl Cellulose	1
Propylene Glycol	1
Water	1
Water, To desired concentration	

Add the propylene glycol diluted with an equal part of water to the methyl cellulose. Mix intimately for about 10 mins. Add half of the water heated to boiling and continue to stir for another 5 minutes. Now add the remaining water and cool as much as possible.

Activator for Zinc Printing Plates

U. S. Patent 2,215,551

Asphalt	5.42
Carbon Black	5.42
Pitch	2.46
Petrolatum	29.90
Turpentine	50.80
Phenol	6.00

Sodium Hexametaphosphate

Russian Patent 54,794

Sodium Dihydrogen Phosphate	70
Sodium Hydrogen Phosphate	30
Sodium Carbonate	5–7

Fuse at 650–800° C.

Substitute for Acetic and Formic Acids

Ammonium Sulphate	15 lb.
Sulphuric Acid (66°)	25 lb.
Water	40 gal.

In textile practice the above replaces 56% acetic or formic acid pound for pound.

Water Resistant Alkali Silicate
U. S. Patent 2,234,646

Sodium Silicate (40–42° Bé)	½	gal.
Water	½	gal.
Formaldehyde	1–6	fl. oz.
Sodium Aluminate (30% Solution)	⅓–1½	oz.
Sodium Abietate	1–3	oz.

Mix well before use.

Clear Starch Solution

Water, Distilled	100	cc.
Thymol	0.1	g.

Bring to a boil and add

Starch	1	g.

mixed to a cream with cold water.
Boil for 2–3 minutes. Add

Mercuric Iodide	0.1	g.

if solution is to be kept for a long time.

Syrupy Gasoline

1. Aluminum Cetyl Acetate	⅓
2. Oleic Acid, White	⅔
3. Gasoline	9

Rub 1 and 2 into a paste and then stir in 3 warming gently.

Wax Beakers for Use with Hydrofluoric Acid

Glass beakers or bottles, having outside diameters equal to the inside diameters, desired for the wax beakers and filled with cold water, are dipped into the molten wax (paraffin wax, old hydrofluoric acid bottles, damaged wax beakers, etc.). Then the beakers are held in the air until the wax solidifies from one dipping before the next dipping. Repeated dippings will build up the wax to any desired thickness. The depth to which the beakers are immersed should be greater than the desired height of the beaker. Reinforcement, by a paper band, can be pasted around the wax before the dippings are completed.

The formed vessel is removed from the mold by putting boiling water in the glass vessel, whereby an inside film of the wax is melted. The wax vessel will slip off easily if two holes are made in the bottom of the wax vessel, one at the center and one at the edge; these holes admit air and prevent a vacuum suction. The wax vessel is ready for use after closing the two holes with a warm rod and trimming the top to give the desired height. If the top part of the newly formed wax vessel is too thin, the vessel can be held by the bottom and dipped some more in the molten paraffin. Two vessels can be made simultaneously—one in each hand. Manufacture is faster: (1) out of doors in cold weather; (2) if the temperature of the molten wax is just above its melting point; and (3) if the water in the beakers, when it gets warm, is changed for cold water.

Artificial Perspiration

	Acid perspiration	Alkaline perspiration
Sodium Chloride	3.5	3.5
Urea	0.5	0.5
Glucose	0.15	0.15
Di-Sodium Hydrogen Phosphate	0.2	0.2
Ammonium Chloride	Nil	0.8
Ammonia (.880)	Nil	0.4
Acetic Acid (1% Solution)	20 cc.	50 cc.
Distilled Water	to 1000 cc.	to 1000 cc.
pH	5.0	7.5

The amount of 1% acetic acid to be added is that just required to give the pH values indicated.

DYEING COROZO (IVORY NUT) BUTTONS

One-Color or Uni-Dyeing

First the buttons are soaked. Wooden containers are preferable to metal tanks, since the latter may give off iron causing the buttons to appear spotty. The buttons are soaked in cold water for 12 hours. By this time the albumin contained in the ivory nut is dissolved. Then the buttons are boiled, by way of a steam jet. The albumin will form a white frothy mass which should be removed in order to avoid the forming of veins on the buttons during the dyeing process. In order to check the developing of veins about 3% Monopol-Brilliant-Oil may be added.

For dyeing ivory nut buttons it is advisable to dissolve the dyes in advance.

Here are the proportions:

Colorant	Water
1 kg. Logwood Extract	8 l.
1 kg. Yellow Wood (Fustic Wood)	8 l.
1 kg. Catechu	6 l.
1 kg. Sumac Leaves	8 l.
1 kg. Logwood Shavings	10 l.
2 kg. Yellow Wood Shavings	10 l.
1 kg. Red or Brazil Wood or Barwood Shavings	10 l.
1 kg. Yellow Oak (Quercitron Bark) Extract	8 l.
1 kg. Potash Chromate	15 l.
1 kg. Ferric Sulphate	15 l.

The following formulae are based upon 10 l. of water:

Light Blue-Gray

3/8 liter of logwood shavings, 1/8 liter of yellow wood shavings, 1/4 liter of sumac leaves, 10 liter of water. Boil buttons in this mixture for 20 minutes, take them out, put them into solution of 1/2 liter of ferric sulphate to 10 liter of cold water for 4 minutes, take them out; do not rinse. Rub buttons in saw dust.

Medium Blue-Gray

3/4 liter of logwood shavings, 1/4 liter of yellow wood shavings, 1/2 liter of sumac leaves, 10 liter of water. Boil buttons in this mixture for 20 min., take them out, rub them in solution of 1 liter of ferric sulphate to 10 liter of water for 4 min.

Dark Blue-Gray

1.5 liter of logwood shavings, ½ liter of yellow wood shavings, 1 liter of sumac leaves, 10 liter of water. Boil buttons for 20 min., take them out, put them into solution of 1.5 liter of ferric sulphate for 6 min., take them out, rub them.

Light Yellowish Gray

⅛ liter of logwood shavings, ⅜ liter of yellow wood shavings, ⅛ liter of sumac leaves, 10 liter of water. Boil buttons for 20 min., take them out, put them into ferric sulphate solution (10 liter of water to ½ liter of ferric sulphate) for 5 min., rub them.

Medium Yellowish Gray

10 liter of water, ¼ liter of logwood shavings, ¾ liter of yellow wood shavings, ¼ liter of sumac leaves. Boil buttons for 20 min., take them out, put them into solution of 10 liter of water to 1 liter of ferric sulphate for 5 min. Rub them.

Dark Yellowish Gray

10 liter of water, ½ liter of yellow wood shavings, 1 liter of sumac leaves. Boil for 20 min. Take them out, put them into solution of 10 liter of water to 1.5 liter of ferric sulphate for 6 min. Rub them.

Reddish Gray

10 liter of water, 1 liter of sumac leaves. Boil buttons for 20 min. Take them out, put them into solution of 10 liter of water to 1 liter of ferric sulphate for 10 min. Rub them.

For darker or lighter coloring add more or less of dyes.

Fashion Colors

For these shades a combination of sumac leaves and yellow oak or yellow oak extract is used. For instance:

10 liter of water, 1.5 liter of sumac leaves, ⅛ liter of yellow wood extract. Boil buttons for 20 min., take them out, put them into solution of 10 liter of water to ⅛ liter of ferric sulphate for 2 min., take them out, put them into hot water, rub them. More or less yellow wood extract makes the buttons more or less yellowish. The longer the buttons are allowed to remain in the solution the grayer they become.

Olive Shades

For these colors yellow oak and yellow wood extract in combination with sumac leaves and logwood extract are widely used. For instance:

10 liter of water, ½ liter of yellow oak or yellow wood extract. In using the above recipes buttons must be thoroughly boiled. Also stirring must not be overlooked, otherwise buttons get spotted. All the buttons dyed in the same bath should be of the same material if the same color is desired. It is, e.g., impossible to dye uniformly small waist coat buttons made of peeled guayaquil and coat buttons made of large hollanders in one batch. Different sorts of nuts give different dyeing effects.

Multi-Colored Buttons

We distinguish between

1. Speckled or jaspe.
2. Imitation Buffalo Horn.
3. Fabric imitation, produced by spraying stripes and checkers.

1. Speckled or Jaspe Buttons

There are two kinds. Those which are sprayed directly with dyes and those which are first sprayed with insulating varnish and later on dyed.

Sprayers used for both methods consist of two thin tubes arranged at right angles like atomizers for flower sprinkling. Buttons are soaked in water for 12 hours, then boiled and finally thoroughly cooled by rinsing in cold water. Then, wet as they are, they are placed on boards, tin sheets or the like and again sprayed with water. Then excess water is removed with a sponge so that all the buttons are equally moist. Now the dye is sprayed on; several different dyes may be sprayed on at the same time and very nice shades may thus be obtained.

For the varnishing method the buttons are boiled, dried in saw dust, placed upon boards and covered with fine dots of varnish. The varnish which is also extensively used in other button spraying methods is a solution of colophony or rosin in alcohol. It is of great importance that it does not peel off.

When the varnish is dry, buttons are dyed in different shades like unicolored ones excepting that all baths must not be warmer than 50° C. After dyeing, varnish is removed in a hot solution of sodium carbonate or ammonia.

2. Spraying for Imitation Buffalo Horn

This is done with a spraygun.

There is a great variety of material for spraying stencils. Some are etched or pressed copper sheets, some are of braided wire or turpentine soaked string. For imitation buffalo horn fret sawed zinc stencils are most frequently used.

The working method is as follows:

Buttons are soaked for 12 hours, then thoroughly boiled and cooled in cold water. Part of them is now first rubbed in saw dust then well cleaned by shaking in a bag of porous material. It is advisable not to prepare more buttons than can be sprayed within an hour. If buttons are exposed to dry air for more than an hour the pores of the ivory nut contract again preventing proper infiltration of the dye.

Boards to place the buttons on should be 1.5 cm. thick. The boards have holes on their upper side into which the buttons fit. The stencils have as many patterns as there are holes in the board. Care must be taken that stencils and button boards match precisely. After spraying the buttons remain on their boards for at least 12 hours allowing the dye to penetrate deeply into the body of the buttons.

Then the buttons, according to the color desired, are put into a ferric sulphate or a potash chromate bath in order to fasten the dye and to prevent its getting off in the polishing process.

If multi-colored buttons are to be made they are sprayed a second time, this time with varnish. For this process stencils must fit particularly well, to the holes, allowing the varnish to penetrate in very delicate lines so that the finished buttons look exactly like natural horn buttons, showing only a few white spots. When the varnish is dry, buttons are dyed in a little lighter shade than that of the spraying dye. Compositions of shades such as brown spray on gray ground or olive spray on greenish ground are very popular and very attractive.

Numerous variations and combinations are possible. We can spray simply with varnish and then dye or we may spray with a lighter shade first and then with a darker one. There are so many methods that there is ample room for the inventiveness of the expert.

In some factories they produce excellent imitations of genuine buffalo horn buttons by scratching off the varnish with a wire brush. The very fine lines thus produced look deceptively like end grains of genuine horn.

3. Producing Stripes and Checkers

Box spraying method:

For this method an airtight wooden box, 3 meters high, 2 meters deep and 2 meters wide is needed. Its bottom is removable. Into this box the boards, covered with buttons (after soaking and boiling), are inserted. The shape of the boards depends on the shape of the stencils available. Then the box is closed airtight. There is a hole in front of the box which can be closed with a flap. Through this hole dye is sprayed in a very fine rain or even as mist. It spreads inside the box and slowly settles upon the buttons. After 20 min. the bottom is pulled out and the stencils are removed.

This method results in a very attractive and uniform coloring. By repeated spraying most beautiful shades and patterns may be obtained that look deceptively like woven fabrics.

Dyeing of Nacre or Mother-of-Pearl

Quickly and easily a great variety of shades can be laid on nacre. Besides natural shades most beautiful artificial shades and lusters may be produced.

First the objects are treated with a weak pickling solution in order to remove grease. For this purpose they are put in a solution of 1 part of pure white potash to 10 water; kept therein at a steady temperature for about ¼ of an hour, drained and repeatedly rinsed and dried. Treated this way, nacre will eagerly absorb dyes, particularly those dissolved in alcohol; in a short time it will be colored throughout. On account of the density of the material, coloration turns out rather dark. Weak solutions are therefore advisable. The longer the material is allowed to stay in the bath the deeper the dye will penetrate. It is therefore up to the dyer to dye the whole body equally and uniformly. After dyeing the objects are rinsed in pure water and dried in saw dust.

Often light or irregularly colored gray or gray-black mother-of-pearl is to be blackened. Many shells are colored only around the edge or a few cm. toward the center, the center itself being white.

Nacre is treated in such a way that silver chloride is formed and then exposed to sunlight. A saturated solution of silver chloride in ammonia is used. Put the nacre in and leave the well sealed container for several days in a dark room, shaking the same from time to time. Then put the pieces on blotting paper, expose them to sunshine and after 3 days dark colored nacre is obtained.

Now the finished and polished articles of nacre are put into a glass container and the above solution is poured over them. From time to time the container is shaken in order to change the position of the pieces so that not always the same spots are covered. After 24–60 hours the pieces are taken out, put on blotting paper and exposed to direct sunlight.

The blackening effect goes rather deep. Disks of about 10 mm. thickness which have been in the solution for 48–60 hours, when broken apart, appear, inside, a uniform dark gray. They may be ground or polished and no white spots will appear. The outer appearance of nacre, dyed this way, is exactly like natural black nacre. The longer the solution and the sunlight are allowed to act on the material, the darker it will turn. Iridescence will be the higher, the more brilliant the original piece of nacre.

It has recently been asserted that admission of air to the solution will produce an even more attractive coloring than exclusion of air.

Bronzelike Iridescent Colors

Iridescent shades may be produced by way of coal tar dyes which, besides iridescence, will give the material a very attractive appearance.

Aniline blue will produce a brownish; fuchsine a greenish bronze tone.

Dissolve 1 part of aniline blue in 40 parts of alcohol or 1 part of fuchsine in 45 parts of alcohol and put articles in such solution for ½ hour.

The finished objects should actually look blue or red as the case may be and the bronzelike iridescence should be seen only when they are struck by oblique rays of light.

Coloring Mother of Pearl Buttons

Buttons or other items made of Mother of Pearl are advisedly prepared for dyeing by first cleansing of grease and other repellent substances. This is best effected by immersing the material in a 10% solution of potassium hydroxide for 10 minutes at 120° F. and followed by a thorough hot rinsing. Thereafter dyeing is carried out by boiling mildly for 1 to 2 hours in water solution of the selected dyestuffs:

Out of hundreds checked, the following colors give shades of good depth and satisfactory brightness:

Acid Colors

- Amacid Yellow M Conc.
- Amacid Azo Yellow G Conc.
- Amacid Yellow GNS
- Naphthol Yellow S
- Brilliant Milling Yellow 5G
- Orange Y Conc.
- Milling Orange PGS Conc.
- Amacid Neutral Orange G
- Orange RO
- Amacid Neutral Orange SGS
- Brilliant Croceine 3BA Conc.
- Amacid Milling Scarlet G
- Fast Red A Conc.
- Amacid Neutral Red BW Conc
- Brilliant Milling Red P2B
- Acid Rhodamine B
- Eosine Y
- Amacid Resorcine Brown G
- Amacid Neutral Brown RN Conc.
- Formyl Violet 3B Ex. Conc.
- Brilliant Indocyanine 6B
- Alizarine Leveling Blue CA
- Amacid Fast Blue BL
- Brilliant Indocyanine G
- Amacid Blue A
- Amacid Fast Green BBF
- Amacid Black 10BR

The following colors of the Basic class give excellent results as to depth of shade and brightness:

- Auramine O
- Thioflavine T
- Phosphine 2G
- Acridine Oranges

Rhodamine 6GDN Ex.
Safranine 6B
Safranine Y
Methyl Violets
Rhoduline Blue 6G
Brilliant Rhoduline Blue R
Methylene Blue ZF
Astrophloxin FF Ex.

Basic colors are best applied at 160° F. and never in conjunction with the acid colors.

Manufacture of Artificial Gems

The fact that pieces of suitably colored glass can be made to show a superficial resemblance to precious stones, has led to the manufacture of imitation jewels of all descriptions. The glass used for the purpose possesses greater hardness and density than the ordinary product, and colored to simulate the precious stones. Finally the external shapes of gems are, of course, readily imitated by cutting and grinding the glass.

All the properties mentioned above are imparted to the flux, partly by special treatment, partly by admixtures, but principally by the purity of the substances used. Besides the essential components, lead oxide, minium, etc., are added to the fluxes. These impart greater density to the glass, more luster and specific gravity. But too much lead oxide must be avoided, as it disintegrates the surface and spoils the luster. A great degree of hardness can be obtained by using large proportions of silica, but the flux becomes refractory, to prevent which borax is added.

The raw materials which are necessary for mixing a good flux are as follows:

1. Pure silica. It is best to use for this finely powdered rock crystals.
2. Pure potash or sodium carbonate.
3. Borax.
4. Lead oxide, carbonate or minium.
5. A little potassium nitrate, partly to promote the fusion, but especially to destroy by oxidation any carbonaceous impurities which might injure the color.
6. A metallic oxide to give color to the flux.

In melting a specific composition an ordinary glass-melting furnace can be employed. The best type of furnace is furnished with a gas-conducting pipe from which four pipes branch off. The upper end of these pipes is bent inward. The gas flame plays under the fire-brick furnace, the thick walls of which form the hearth. The bottom is provided with an opening through which the gases enter into the crucible placed exactly over it, so as to circulate around the actual crucible containing the flux, and resting upon a support of fire-clay and a movable rod. The gases, after playing around the crucible pass out through a hole in the cover and then around and down the walls towards the escape pipe.

The ingredients are separately ground to impalpable powder, sifted through a fine sieve and then weighed. These are then intimately mixed in a mortar and put in a new Hessian crucible and covered with a clay plate. The crucible is next introduced into the furnace and heated by gas flame from bottom. When the mass has been fused the crucible is withdrawn from the hearth and the clay plate is removed to pour out the molten mass. After allowing to cool to normal temperature the glass is cut into desired shape and polished.

Fluxes

Rock Crystal	20.23
Sodium Carbonate	14.61
Calcined Borax	10.96
Minium	7.20
Potassium Nitrate	3.65

Powder the ingredients separately and mix. Then fuse the mixture in the manner stated under general process.

A harder flux is obtained by mixing the substances in the following proportions:

Rock Crystal	43.84
Sodium Carbonate	14.61
Calcined Borax	10.96
Minium	7.20
Potassium Nitrate	1.21

A flux so hard that it will spark when struck with a piece of steel can be prepared from the following substances:

Rock Crystal	10.96
Powdered Glass	29.23
Minium	10.96
Calcined Borax	2.20
Potassium Nitrate	2.43
Arsenic	0.60

Pure flint finely powdered may be used instead of rock crystal, or white powdered glass, but in the latter case some white arsenic must be added to obtain the frit entirely colorless.

These fluxes furnish the "Strass" which is the basis for the manufacture of artificial gems.

Ruby

	I	II
Rock Crystal	29.23	29.23
Dry Sodium Carbonate	14.61	14.61
Calcined Borax	10.96	4.84
Saltpeter	5.47	2.43
Copper Oxide	3.65	0.91
Antimony Trisulphide	0.50	..
Manganese Dioxide	0.50	..
Minium	10.92	..
Sal-Ammoniac	..	3.65

Sapphire

	I	II
Rock Crystal	43.84	29.23
Sodium Carbonate	21.92	14.61
Borax	7.20	10.96
Minium	7.20	5.47
Potassium Nitrate	3.65	1.82
Cobalt Carbonate	0.06	..
Copper Carbonate	..	1.82

Emerald

Rock Crystal	43.84
Dry Sodium Carbonate	14.61
Calcined Borax	7.20
Minium	7.20
Saltpeter	2.43
Cobalt Carbonate	0.09
Chrome Green	0.30

The above mixture produces a beautiful green colored emerald.

As a general rule uranic oxide, which gives yellow colors shading only slightly into green, furnishes an emerald green when used in the following proportions:

Rock Crystal	36.43
Sodium Carbonate	10.96
Minium	7.20
Saltpeter	3.65
Uranic Oxide	2.43
Green Copper Carbonate	0.18
Stannic Oxide	0.18
Bone Calcined	0.18

Chrysoprase

The following mixture is decidedly the best for imitating the transparent, apple-green color of this stone.

Rock Crystal	43.84
Sodium Carbonate	14.61
Calcined Borax	10.96
Minium	7.20
Saltpeter	1.21
Calcined Bones	7.20
Copper Carbonate	0.12
Ferric Oxide	0.24
Chrome Green	0.36

The above mixture gives dark colored stone.

Opal

Rock Crystal	32.29
Sodium Carbonate	10.96
Calcined Borax	7.20
Minium	5.47
Saltpeter	0.91
Copper Oxide	0.06
Calcined Bones	0.09
Silver Chloride	0.12

Garnet

Rock Crystal	32.29
Sodium Carbonate	10.96
Calcined Borax	7.93
Minium	5.47
Saltpeter	2.43
Manganese Dioxide	0.36
Ferric Oxide	0.18

If a brighter color is desired add 0.06 part of copper oxide to the mixture.

Amethyst

Amethyst may be prepared by using manganese dioxide, but not more than 0.06 part of it must be taken for a frit producing about 30 parts of flux. Powdered glass in the proportion of 30 parts, 3.65 parts of potassium nitrate, and some borax and minium give also a good imitation of the amethyst.

Lapis Lazuli

Rock Crystal	21.92
Sodium Carbonate	7.20
Calcined Borax	5.47
Minium	3.65
Potassium Nitrate	1.00
Calcined Bones	3.65
Cobalt Oxide	0.12

Agate

Agate can be imitated by allowing fragments of different fluxes to run together, stirring the mass in the meanwhile.

Several varieties of agate are obtained by mixing about 1.82 parts of ferric oxide with 43.84 parts of flux.

The following is another method of producing artificial gems: 2 parts of pure rock crystal, 1 of calcined soda, ¾ of anhydrous borax, 1–16 of lead oxide are rubbed together as intimately as possible, and heated in a crucible for one hour without allowing the mass to become liquid. It is then brought into fusion and kept so for one hour, when it is allowed to congeal. It is then moderately heated for 24 hours, and the resulting flux taken from the crucible, cut and grained.

This forms the base for the flux of the artificial gems.

The following minerals are added as coloring substances:

Blue—Cobalt oxide.

Yellow—Antimony pentoxide.

Green—Cupric oxide.

Red—Purple of cassius.

Violet—Black oxide of manganese.

Fluorescent Gems

Artificial gems which fluoresce a bright greenish or yellow color under ultraviolet light may be cut from ordinary green glass which contains uranium as a pigment.

By adding small amounts of manganese, chromium, or uranium to fused glass and allowing the metal to dissolve, in the glass, a highly fluorescent substance is produced. This can be cut into cabochon and facet gems suitable for theatrical and advertising effects.

Artificial Fluorescent Minerals
Formula No. 1

Ordinary and slightly porous minerals can be dipped into solutions of fluorescent chemicals, such as sodium salicylate, and allowed to dry. These artificially prepared fluorescent minerals can be used for display purposes and will fool many experts on the subject.

No. 2

The following mixture is fused in a crucible:

Sodium Nitrate	100
Uranyl Nitrate	0.5

This solid fluoresces a bright yellow to green under ultraviolet light.

No. 3

Agates can be soaked in kerosene in order to produce a blue-fluorescing stone. After several weeks the kerosene will evaporate and the treatment will have to be repeated.

No. 4

The following solution can be used for making ordinary stones fluorescent:

Quinine	10
Water	999
Sulphuric Acid	5–8

This solution should not be used on calcareous rocks such as limestone.

Fluorescent Sand and Stones

Sand used in advertising, display, and for models may be dyed with fluorescent dyes and illuminated with ultraviolet light. This is especially effective when used in store windows and for small scale demonstrations.

Pebbles of different fluorescent minerals, such as wernerite, dakeite, willemite, calcite, and others, can be mixed and used for a large number and wide variety of show window effects.

Fluorescent Signs

Glass signs made from tubing filled with a solution of umbelliferone, uranine, or other dye, and under ultraviolet light, may be used in show windows and for advertising as a "cold" source of colored light, as well as to substitute for neon signs.

Small particles of fluorescent minerals can be embedded in a wax backing and made into signs. These are very colorful under ultraviolet light.

Fluorescent Prospecting Screen

A piece of cardboard is coated with powdered willemite or other brightly fluorescing mineral. When ore containing mercury is heated and placed between the screen and a source of short wave-length ultraviolet light dense dark fumes of mercury vapor will be seen rising from the ore. Other ores and metals do not respond in this way.

Fluorescent Smoke

When a finely divided phosphor, such as zinc orthosilicate, is blown into a beam of ultraviolet light in an otherwise darkened room, the particles will light up with a bright green light.

Billows simulating fire can be made by blowing zinc beryllium silicate or other red-fluorescing phosphor or powdered calcite into a beam of ultraviolet light. Dyes should not be used since they stain garments and skin.

"Invisible Thief-Catching Powder"

A substance which is white and unnoticeable in ordinary light, but brightly fluorescent in ultraviolet light may be used to catch thieves. Anthracene, zinc sulphide phosphor, fluorescein, or other such substance may be employed to advantage.

Ordinary powdered tumeric can be dusted on money, food, or other objects. When handled by a guilty person traces of powder will adhere to the hands and clothes and this may be seen in ultraviolet light by the bright yellow fluorescence which is enhanced by treatment with water or chemical reagents.

Old X-ray fluoroscopic screens may be scraped off and the brightly fluorescent powder used for tagging objects. This powder usually glows brightly under short wavelength ultraviolet light, but it should not be used on articles intended for human consumption.

Multi-Purpose Fingerprint Powder

When anthracene is dusted over a finger, particles adhere and dissolve slightly in the oil secretion. This fingerprint is self-luminous under ultraviolet light and a photograph may be taken on a multi-colored or irregular object.

The fingerprint can be "lifted" as usual with colorless scotch tape and photographed under ultraviolet light. The tape itself may fluoresce slightly.

"Hot Rocks"

In a darkened room lumps of red-fluorescing calcite may be used to simulate hot rocks and glowing coals when ultraviolet light (short wavelength) is directed toward the calcite. A person can walk across the red-fluorescing calcite in his bare feet and create a great deal of interest since, in all appearance, he is walking on glowing coals.

Fluorescent Animals

A small chick or other living animal may be dipped in a solution of neutral acriflavine and allowed to dry. Under long wavelength ultraviolet light, the animal will emit its own light with a striking effect on the audience.

A dog's tail, for example, can be dyed with an organic and fluorescent dye and, under ultraviolet light in darkness, all that may be seen is a luminous wagging tail.

Fluorescent Moustache

A moustache may be dyed with acriflavine or other fluorescent dye and made to glow with a weird light when illuminated with ultraviolet rays. Long wavelength and filtered ultraviolet light should be employed.

Naturalizing False Teeth and Eyes

The life-like luster of natural teeth may be simulated in false teeth by incorporating a small amount of metallic activator and according the mixture heat treatment so that a phosphor-like system is set-up. Porcelain may be used as the base.

Likewise, the life-like luster of the sclera of natural living eyes may be imitated in the same manner.

Dental Investment Composition
U. S. Patent 2,247,395

Cristobalite	50–80%
Calcium Sulphate ($\frac{1}{2}$ H_2O)	20–50%
Calcium Chloride	2%
Ammonium Chloride	$\frac{1}{4}$–3%

Imitation Meerschaum
U. S. Patent 2,245,489

Hydrous silicate of magnesium is finely powdered in a ball mill or other suitable grinding device. The grinding should be carried on until an exceedingly fine impalpable powder is produced.

This powder is then thoroughly mixed with sufficient raw egg whites to form a thick, damp paste which can be easily pressed or molded into any desired shape. The preferred portions are 100 grams of egg whites to 144 grams of magnesium silicate. The dampened powder or paste is then placed into molds and pressed into its final form such as a smoking pipe; a smoking pipe lining; or a cigar or cigarette holder.

The damp pressed articles are then placed in a suitable drier and maintained at a temperature of from 100 to 130° F. for a period of 24 hours. When removed from the drier, it is a hard, light, form-retaining product which can be polished, ground, machined, or carved as desired to produce the finished article.

Stylus
German Patent 682,661

Red beechwood is impregnated at 50° C. with

Sodium Acetate	25 g.
Sodium Nitrate	20 g.
Potassium Aluminum Sulphate	35 g.
Sodium Nitrite	25 g.
Water	3 l.

This stylus will not break or scratch in ordinary use.

Preservative for Feathers

Salicylic Acid	10 lb.
Salt	40 lb.
Boric Acid	45 lb.
Potassium Nitrate	65 lb.

Chapter Twenty-Two

SUBSTITUTES *

Selecting the proper substitute is no easy task. Since no material has all the same properties as the original, it cannot be expected that the replacement will give the finished product all of the same characteristics that it originally possessed. A replacement, therefore, that will produce a finished product which will perform almost the same function as the original, without too great a difference, is ordinarily considered satisfactory. For example glycerin, in an anti-freeze, has been satisfactorily replaced by ethylene glycol even though the two products differ in chemical and certain physical properties.

A substitute material, excluding price and availability, must be considered from many angles before it can qualify as a good substitute. Since it cannot have *all* the same physical and chemical properties as the original material, a compromise must be made. Thus, corn syrup may be suitable, as a glycerin replacement, in a suspending medium, where the former's viscosity is primarily desired; as in certain tooth-pastes. It is not of importance that corn syrup does not lower the freezing point of water or that it is not as hygroscopic as glycerin. Where, however, the last two factors, or others, are important, the use of corn syrup in place of glycerin, is not advisable.

Even when a suitable substitute is found, it may be necessary to modify the original formula by using a smaller or larger amount of the substitute and often, to add one or more other ingredients to balance it. Thus, because corn syrup is more viscous and less hygroscopic than glycerin, it may be necessary to reduce its viscosity by the addition of water and increase its hygroscopicity by means of a compatible hygroscopic salt. Introduction of these two additional ingredients may require considerable testing and aging to avoid subsequent undesirable effects.

Because of the uncertainty of the continued availability of any substitute material, it is advisable to experiment with a number of materials on each problem, so as to have a substitute ready for the substitute used. It means additional work, but is a worth while insurance for continuance in business.

Sometimes it may be desirable to change the composition of a formulation radically or entirely because a suitable substitution cannot be made. Thus flavoring extracts depend on the use of pure alcohol as the solvent for the flavoring ingredients. Since there is no good substitute for alcohol (in food products) available, a formulation without alcohol is indicated, e.g.,

* This section is a condensation of the book, *Substitutes,* H. Bennett. Chemical Publishing Co., Inc., Brooklyn, N. Y. (1942).

an emulsion of the flavoring ingredient (e.g. lemon oil) made with an edible gum (gum tragacanth) and water. Of course, the finished product does not look like the original lemon extract, but it can replace it in most of its uses.

Price should not be too great a deterrent in selecting a substitute. Sometimes a substitute will so alter a product as to make it more useful, desirable and salable. An example of this is the use of monoglycollin in place of glycerin. Although the former is more than twice as expensive as the latter, its much greater solvency for certain dyes makes it far more economical to use than the cheaper material which it replaces. In electrolytic condenser manufacture mannitol, at about three times the cost of glycerin, is replacing the latter because it gives a much more desirable product.

In getting outside assistance in finding a substitute, it is important to disclose a problem in its entirety. Reputable manufacturers and consultants hold all communications in strict confidence. Therefore give them the complete formulation, method of manufacture, packaging and a sample of the finished product. Also inform them how and where the finished product is to be used. Only with such complete information can an intelligent recommendation be made.

Product	Substitute or Alternative
Accroides, Gum	Rosin
	Seed-Lac
	"Vinsol"
Acetaldehyde	Aldol
	Formaldehyde
	Furfuraldehyde
	Glyoxal
Acetamide	Ammonium acetate
	Ethanolamine Acetate
	Formamide
	Urea
Acetanilide	Alum
	Pyramidone
	Zinc sulphocarbolate
Acetic Acid	Ammonium sulphate with dilute sulphuric acid
	Boric acid
	Citric acid
	Formic acid
	Glycollic acid
	Lactic acid
	Levulinic acid
	Phosphoric acid
	Propionic acid
	Pyroligneous acid

Product	Substitute or Alternative
Acetic Acid	Saccharic acid
	Salt
	Sodium diacetate
Albumen	Agar
	Alum, potash
	Casein
	Emulsifiers
	Protein, fish
	Protein, soybean
	Resin, natural
	Resins, synthetic
	Thickeners
Alcohol	See Ethyl alcohol
Alkalies	Amines, primary, secondary, tertiary, quaternary
	Aminoalcohols
	Ammonium hydroxide
	Barium hydroxide
	Borax
	Calcium hydroxide
	Calcium oxide
	Lithium hydroxide
	Magnesium hydroxide
	Magnesium oxide
	Nephelin
	Potassium carbonate
	Potassium hydroxide
	Potassium silicate
	Sodium carbonate
	Sodium hydroxide
	Sodium metasilicate
	Sodium orthosilicate
	Sodium pyrophosphate
	Sodium silicate
	Trisodium phosphate
Alkyd Resins	"Flexoresins"
	"Piccolyte" resins
	Resins, synthetic
Almond Oil	Apricot kernel oil
	Benzaldehyde
	Cherry kernel oil
	Mineral oil, refined
	Peach kernel oil
	"Persic" oil
	Vegetable oils

Product	Substitute or Alternative
"Alperox"	Hydrogen peroxide
Alpha Protein	Casein
	"G"-protein
Alum	Acetanilide
	Alum, potash
	Aluminum chloride
	Aluminum hydrate
Aluminum Hydrate	Alum
	Ammonia alum
	Copperas
	Ferric sulphate
	Lime
	Potassium alum
	Sodium aluminate
Aluminum Oleate	Calcium oleate
	Lead oleate
	Magnesium oleate
Aluminum Phosphate	Calcium phosphate
Aluminum Powder	Graphite
	Mica
	Pearl essence
	Sericite
	Slate powder
Aluminum Resinate	Calcium resinate
	Lead resinate
	Magnesium resinate
Aluminum Silicate	Alum, potash
	Talc
Ammonium Compounds	Alkalies
	Amides
	Amines
	Ammonium thiocyanate
	Cyanamide
	Dicyandiamide
	Urea
Ammonium Chloride	Manganese chloride
	Zinc chloride
Ammonium Hydroxide	See Alkalies
Ammonium Lactate	Ammonium glycollate
	Glycerin
Ammonium Phosphate	Calcium cyanamid
	Guano
	Sodium nitrate

Product	Substitute or Alternative
Ammonium Sulphamate	"Abopon"
	Borax with boric acid
	Sodium chlorate
Ammonium Sulphate	Ammonium phosphate
Ammonium Sulphite	Potassium bisulphite with ammonia
	Sodium bisulphite with ammonia
Ammonium Thiocyanate	Ammonium compounds
	Sodium chlorate
Amyl Acetate	Butylacetate
	Fusel oil
Amyl Alcohol	Capryl alcohol
	Fusel oil
	Hexyl alcohol
	Octyl alcohol
	Solvents
	Tetrahydrofurfuryl alcohol
Aniline	σ-Aminodiphenyl
	Furfural
	Pyridin
Anise Oil	"Annol"
Annatto	Dyes, aniline
Antimony	Cadmium
	Calcium
	Selenium
	Tellurium
Antimony Lactate	Tartar emetic
Antimony Oxide	Tin oxide
	Titanium oxide
Apricot Kernel Oil	Almond oil
Beeswax	"B-Z Wax"
	Ceresin with soap
	Coffee wax
	Flax wax
	"Flexo Wax"
	"Isco 662, 663"
	"Norco Wax 36"
	Sugar cane wax
	Wax, synthetic
Belladonna	Stramonium
Bentonite	Alum, potash
	Clay, colloidal
	Emulsifiers
	Fillers
	Gums, water dispersible
	Thickeners

Product	Substitute or Alternative
Benzaldehyde	Bitter almond oil
	Nitrobenzol
Benzene Sulphonic Acid	Phenolsulphonic acid
Benzine	Petroleum ether
Calcium Sulphide	Barium sulphide
Camel's Hair	"Nylon" fleece
Camphor	Benzyl benzoate
	Camphene
	Dibutyl tartrate
	"Dehydranone"
	Diethyl phthalate
	Esparto wax
	Hexachloroethane
	Menthol
	Naphthalene
	Phenol
	Plasticizers
	Resorcinol
	"Tetralin"
	Triphenyl phosphate
Camphor Oil	"Japp-O"
	"Terpesol"
	Turpentine
Candelilla Wax	"Norcowax 72"
Cane Sugar	See Sugar
Capric Acid	"Alox" acids
	Cocoanut oil fatty acids
Capryl Alcohol	See Octyl alcohol, normal
Carbon Dioxide	Ammonium bicarbonate
	Carbon tetrachloride
	Methyl chloride
	Nitrogen
	Sodium bicarbonate
Carbon Tetrachloride	Carbon dioxide
	Chloroform
	Ether, petroleum
	Ethylene dichloride with sulphur dioxide
	Methyl bromide
	Methyl chloride
	Solvents
	Trichlorethylene
"Carbowax"	Glycerin
	Polymerized glycol stearate
Cardamom Oil	"Card-O-Mar"

Product	Substitute or Alternative
Carnauba Wax	"Acrawax"
	Candelilla wax
	Cotton wax, green
	Esparto wax
	Hydrogenated castor oil
	"Norcowax 350"
	Ouricouri wax
	"Rezowax"
	"Santowax M"
	Stearamides, substituted
	"Stroba Wax"
	Sugar cane wax
Carob Gum	Sodium alginate
Carragheen	Sodium alginate
Casein	Albumen
	Alum, potash
	Alkyd resins
	"Alpha" protein
	Cellulose esters
	Emulsifiers
	Gluten
	Gums, water dispersible
	"Proflex"
	"Prosein"
	Resins, natural
	Resins, synthetic
	Shellac
	Thickeners
	Zein
Castor Oil	"Dipolymer"
	"Flexoresin L 1"
	Glycol hexaricinoleate
	Grapeseed oil
	Vegetable oils
Cetyl Alcohol	Lanolin alcohols
	Monostearin
	Oleyl alcohol
	Stearyl alcohol
Cherry Kernel Oil	Almond oil
China Clay	Barytes
	Talc
	Whiting
China Wood Oil	See Tung oil
Chinese Blue	See Iron blue

Product	Substitute or Alternative
Chinese Wax	See Insect wax
Chloramin	Hydrogen peroxide
	Sodium chlorite
Chlorine	Bleaching powder
	Bromine
	Catalysts
	Hydrogen peroxide
	Iodine
	Nitric acid
	Sulphur dioxide
Chloroform	Carbon tetrachloride
Chlorophyll	Dyes, aniline
Chloropicrin	"Dry-Ice" with 10% ethylene oxide
	"Ethide"
	Furoylchloride
	Insecticides
	Methyl bromide
Chlororubbers	"Halowax"
	Rubbers, synthetic
Cholesterol	Lanolin alcohols
	Phytosterols
Chondrus	Agar
Chrome Alum	Alum, potash
Chrome Orange	Ochre
	Orange mineral
	Sienna
Chromic Acid	Nitric acid
Chromium Acetate	Aluminum sulphate
Chromium Plating	Cadmium plating
	Nickel and silver plating
	Pearl lacquer
Chromium Sulphate	Ferric sulphate
Cinnamon	Cinnamaldehyde and eugenol with powdered nut shells
Citral	Lemongrass oil, Florida
Citric Acid	Acetic acid
	Gluconic acid
	Glycollic acid
	Lactic acid
	Levulinic acid
	Malic acid
	Phosphoric acid
	Propionic acid
	Saccharic acid

Product	Substitute or Alternative
Citric Acid	Sodium acid sulphate
	Sodium bisulphite
	Sodium diacetate
	Tartaric acid
	Sulphuric acid, dilute
	"Tartex"
	Vinegar
Citronella Oil	"Andro"
	"Javonella"
Clay	Alum, potash
	Whiting
Clay, Colloidal	Bentonite
	Kaolin
Clove Oil	"Clovel"
	Eugenol
Cobalt	Lead
	Manganese
	Tantalum
Cobalt Chloride	Barium chloride
Cochineal	Dyes, aniline
Cocoanut Oil	Babassu oil
	Castor oil
	Castor with cottonseed oils
	Cocoanut oil fatty acids
	Cohune oil fatty acids
	Confectioners' oil, "Crystal"
	Carozo oil
	Coyol oil
	Glyceryl myristate with castor oil
	Hydrogenated vegetable oils, partially
	Lard oil
	Macanilla oil
	Mineral oil with lard oil
	Murumuru oil
	Myristic with ricinoleic acid
	Neatsfoot oil
	"Neo-Fat 13"
	Oleic with ricinoleic acid
	Olive oil with mineral oil
	Oxidized paraffin wax with red oil
	Palm kernel oil
	Peanut oil, blown
	Polyhydric alcohol, fatty acid esters, e.g., Diglycol ricinoleate

Product	Substitute or Alternative
Cocoanut Oil	Rosin with linseed oil
	Tucum oil
	Vegetable oils
	Vegetable oils, blown
Cod Liver Oil	Rice bran oil, purified
	Sardine oil
	Shark liver oil
	Sterols, irradiated animal
	Tuna liver oil
Cod Oil	Degras
	Herring oil, blown
	Menhaden oil, blown
	Pilchard oil
	Sardine oil, blown
	Whale oil
Copper Naphthenate	Copper carbonate, basic
	Copper "mahogany" sulphonate
	Copper oleate
	Creosote
Copper Oxide	Manganese dioxide
	Mercuric chloride
Copper Sulphate	Aluminum sulphate
Cork	Asbestos fiber with asphalt or resin binder
	Bark fiber with asphalt or resin binder
	Bran fiber with asphalt or resin binder
	"Cushiontone"
	Felt, hair or wool, impregnated
	"Fiberglas"
	"Foamglass"
	"Joinrite"
	Linseed meal
	Millboard soft
	Mineral wool
	"Naturazone"
	Oatmeal
	Palmetto wood
	Paper pulp
	Peat moss
Cream of Tartar	Adipic acid
	Ammonium sulphate
	Mucic acid
	Saccharolactic acid

Product	Substitute or Alternative
Creosote	Coal tar
	Copper chromate
	Copper naphthenate
	Copper oleate
	Copper phosphate
	Copper sulphate
	Cresylic acid
	Pentachlorphenol
	Tar oils
	Zinc chloride
Cresol	Coal tar acids
	Creosote
	Furfural
	Phenol
Cresylic Acid	Creosote
	See Cresol
Degras	Cod oil
	Petrolatum
	"Sublan"
Derris	Devil's shoe-string root
	"Thanite"
Dextrin	"Abopon"
	Adhesives
	Glycerin
	Malt extract
	Sodium silicate
	Sodium sulphate
	Sugar
	Urea
Dextrose	See Sugar
Diacetone	Acetone
	Solvents
Diamond, Industrial	Abrasives
	Boron
	Boron carbide
	"Corundum"
	Silicon carbide
Diatomaceous Earth	Carbon, activated
	"Dicalite"
Dibutyl Phthalate	Butyl oleate
	Castor oil
	Castor oil blown
	"Dipolymer"
	"Glaurin"
	Glycol hexaricinoleate

Product	Substitute or Alternative
Dibutyl Phthalate	2, 5 Hexanediol
	Monoglycollin
	Plasticizers
	"Theop"
Dichloramine	Chlorine
	Hydrogen peroxide
Dichlorethylene	Methyl chloride
Dicresyl Carbonate	Glycerin
Dyes, Aniline	Amaranth
	Annatto
	Caramel coloring
	Chlorophyll
	Cochineal
	Coffee grounds
	Cudbear
	Cutch
	Dragon's blood
	Fustic
	Hypernic
	Indigo
	Logwood
	Madder
	Orchil extract
	Osage orange extract
	Pigments, mineral, e.g. sienna
	Precipitates, chemical e.g. antimony sulphide
	Quercitron bark
	Saffron
	Tannin
	Turmeric
Dyes, Vat	"A.A.P. Naphthols"
	Dyes, aniline
	Precipitate, chemical e.g. lead chromate
Ether, Petroleum	Benzol
	Carbon tetrachloride
	Ether, ethyl
	Ethyl chloride
	Isopropyl ether
	Pentane
Ethyl Acetate	Acetone
	Isopropyl acetate
	Methyl acetate

Product	Substitute or Alternative
Ethyl Alcohol	Methyl alcohol
	Pentane
	Propylene glycol
	Rum
	Solubilizers or emulsifiers "Glyco S533"
	"Stago CS"
	Sulphonated oils
	Solvents
	Tetrahydrofurfuryl alcohol
	Wine
Fluorine	Iodine
Fluorspar	Ammonium bifluoride
	Cryolite
	Sodium silicofluoride
Formaldehyde	Acetaldehyde
	Aluminum chloride
	Aluminum sulphate
	Furfural
	Glyoxal
	Potassium bichromate
	Sodium bichromate
	Tannin
Gluconic Acid	Citric acid
	Lactic acid
	Mucic acid
Glucose	See Sugar
Glue	Casein
	Emulsifiers
	Gelatin
	Gums, water dispersible
	Latex
	Resins, synthetic
	Rosin soap
Gluten	Casein
Glycerin	Aminoalcohols
	Ammonium lactate
	Apple syrup
	"Aquaresin"
	Butylene glycol
	Calcium chloride
	"Carbowax"
	Corn syrup
	Dextrin
	Dicresyl carbonate

Product	Substitute or Alternative
Glycerin	Diglycol oleate
	Ethylammonium phosphate
	Glycols
	"Glycopon"
	"Glucarine B"
	Glucose
	Invert sugar
	Kerosene
	Lactic acid
	Magnesium chloride
	Methyl cellulose
	Methyl sodium potassium phosphate
	Mineral oil
	Nonaethylene glycol ricinoleate
	Polymerized glycol oleate
	Sorbitol syrup
	Sugar
	Sulphonated castor oil
	"Yumidol"
Glyceryl Chlorhydrin	Ethylene chlorhydrin
Glyceryl Phthalate	Glyceryl maleate
Glycol Diacetate	Monoglycollin
Glycol Monoacetate	Monoglycollin
Glycollic Acid	Acetic acid
	Citric acid
	Lactic acid
Glycols	Ethyl potassium phosphate
	Glycerin
	Sorbitol
Glyoxal	Formaldehyde
Grapeseed Oil	Castor oil
Graphite	"Acrawax"
	Bone black with talc
	Iron oxide
	Metals, powdered
	Mica, powdered
	Paraffin wax
	Red lead
	Silica black
	Talc
Gum Arabic	See Gum Acacia
Gum Benzoin	See Benzoin
Gum Karaya	Gums, water dispersible
Gum, Locust Bean	Gums, water dispersible

Product	Substitute or Alternative
Gum Tragacanth	Gums, water dispersible
	Thickeners
Gums, Water Dispersible	"Abopon"
	Agar
	Algin
	"Algaloid"
	Ammonium alginate
	Carragheen
	Casein
	Cherry gum
	Dextrin
	"Diglycol" stearate
	Emulsifiers
	"G" protein
	Gum acacia
	Gum karaya
	Gum tragacanth
	Locust bean gum
	Methyl cellulose
	Pectin
	Quince seed
	Sodium alginate
	Sodium borophosphate
	Soap
Gut	Fiber
	Metal wire
	"Nylon"
	Protein
	Resins, synthetic
Gutta Percha	Balata
	Resins, synthetic
	Rubbers, synthetic
"Halowax"	"Arochlor"
	Chlorinated mineral oil
	Chlorinated paraffin wax
	Chlororubbers
	Resins, synthetic
Hydrofluoric Acid	Aluminum chloride, anhydrous
	Ammonium bifluoride
	Phosphoric with chromic acid
Hydrofuramide	Hexamethylenetetramine
Hydrogen	Acetylene
	Helium
Hydrogen Peroxide	"Alperox"
	Benzoyl peroxide

Product	Substitute or Alternative
Hydrogen Peroxide	Bleaching powder
	Calcium peroxide
	Chloramine
	Chlorine
	Dichloramine
	Magnesium peroxide
	Oxalic acid
	Oxygen
	Ozone
	Potassium bichromate
	Potassium chlorate
	Potassium chromate
	Potassium perchlorate
	Potassium permanganate
	Selenium dioxide
	Sodium chlorate
	Sodium chlorite
	Sodium hypochlorite
	Sodium hydrosulphite
	Sodium perborate
	Sodium perchlorate
	Sodium peroxide
	Sulphur dioxide
	Zinc hydrosulphite
	Zinc peroxide
Hydroquinone	Maleic acid
	Naphthol, beta
	Pyrogallol
	Resorcinol
	Selenium
Hydroxycitronellal	Cyclamen aldehyde
Hypernic	Dyes, aniline
Iceland Moss	Gums, water dispersible
	Thickeners
Indigo	Dyes, aniline
Indol	"Indolene"
Infusorial Earth	See Diatomaceous Earth
Insecticides	Amines, higher fatty
	Bordeaux mixture
	Calcium arsenate
	Castor leaf extract
	Chloropicrin
	Cryolite
	"Derex"

Product	Substitute or Alternative
Insecticides	"Ethide"
	Hydrocyanic acid
	Ketones, higher fatty
	Lead arsenate
	"Lethane"
	Methyl bromide
	Naphthalene
	Nicotine
	Paradichlorobenzene
	Paris green
	Phenothiazin
	Phthalonitrile
	Pyrethrum
	Rotenone
	Sodium fluoride
	Sodium silicofluoride
	Sulphur
	Tetrahydrofurfuryl lactate
	"Thanite"
	Tobacco dust
Invert Sugar	Glycerin
Iodine	Bromine
	Chlorine
	Fluorine
Irish Moss	Gum, water dispersible
	Thickeners
Lard	Hydrogenated vegetable or fish oils
	Tallow, refined
Lard Oil	Fish oil
	Mineral oil
	Mustard seed oil
	Polyhydric alcohol fatty acid esters, e.g. Diglycol oleate
	Rosin oil
	Vegetable oil
Latex	Blood albumen
	"Dispersite"
	"Emulsion 58-8"
	Gelatin
	Glue
	Methyl cellulose
	Resin emulsions
	"Seatex"
	Vinyl copolymer emulsions
Monoacetin	Monoglycollin

Product	Substitute or Alternative
Monoglycollin	"Carbitol"
	Glycol diacetate
Montan Wax	Lignite wax
	"Monten" wax
	"Norcowax 12A"
	Peat wax
	"Rezo Wax"
	"Santowax"
Morpholine	Ammonia
	Ethylenediamine
	Methylamine
Mucic Acid	Adipic acid
	Gluconic acid
	Saccharic acid
Mucin	Agar
"Nipagen"	"Moldex"
	"Parasept"
	Preservatives
Nitre Cake	Hydrochloric acid
Nitric Acid	Chlorine
	Chromic acid
	Hydrochloric acid
	Hydrogen peroxide
Nitrocellulose	Cellulose esters
	Plastics
	Resins, synthetic
	Vinyl copolymers
Nitrogen	Ammonia, anhydrous
	Carbon dioxide
Nitromannite	Mercury fulminate
Nutgalls	Gall apples
	Oak galls (oak apples)
	Tannin
Nux Vomica	Strychnine hydrochloride
Ochre, French	"Witco Yellow"
Octyl Alcohol, Normal	Hexyl alcohol
	Tributyl phosphate
Oleic Acid	Fatty acids
	"Indusoil"
	Talloil
Oleyl Alcohol	Cetyl alcohol
Olive Oil	Apricot kernel oil
	"Lenolene"
	Corn oil with crushed green olives

Product	Substitute or Alternative
Olive Oil	Diglycol laurate
	"Glaurin"
	Grapeseed oil
	Lard oil with mineral oil
	Mineral oil with cocoanut oil
	"Nopco C.P."
	"Olev-ol"
	Peach kernel oil
	Peanut oil, destearinated
	Rice oil
	Vegetable oils
Platinum	Iron containing 42–50% nickel
"Pliofilm"	"Cellophane"
	"Ethocel"
	Plastic films
	Parchment paper
Polystyrene	Cellulose acetopropionate
	Glass, tempered
Polyvinyl Alcohol	"Abopon"
	Gums, water dispersible
	"Hevealac"
	Methyl cellulose
	Synthetic resin emulsion
	Urea-formaldehyde resins
Pyrogallic Acid	Hydroquinone
Pyrogallol	Hydroquinone
	Naphthol, beta
	Pyrogallic acid
	Resorcinol
	"S A 326"
Quartz	Garnet
	Silica, fused
	"Vycor"
Quercitron Bark	Dyes, aniline
Quince Seed	Gums, water dispersible
	Psyllium seed
Quinine	"Atabrin"
	Pamaquine naphthoate
	"Promin"
	Quinarine hydrochloride
	Salicin
	Sulphadiazin
Rubber	"Foamglas"
	"Jointite"

Product	Substitute or Alternative
Rubber	Lead
	Lead oleate with carbon black
	Polyvinyl butyral resin with 1% Acrawax C
	Resins, synthetic
	"Resistoflex PVA"
	Rubbers, synthetic
	"Saflex"
	Shellac
	"Tygon"
	Varnished cambric
	Vinyl acetate and chloride copolymers
	"Vinylite"
Rubber Cement	Blood albumen
	"Cement E C-226"
	Polyvinyl acetate and copolymer emulsions
	Resins, synthetic, solutions of
	Rubber, synthetic solutions
	"Vinylite" solutions
Rubber, Chlorinated	Cumarone resin
Rubber, Hard	Ceramics, pressed
	"Densite"
	"Electrite"
	Vulcanized fiber
Saponin	"Foamapin"
	Soap bark
	"Virifoam"
Sapphire	Glass, fused hard
Sardine Oil	Cod oil
	Rice bran oil
	Vegetable oils
Sassafras Oil	"Cam-O-Sass"
	"S-O-Frass"
Sebacic Acid	Maleic acid
Seed-Lac	Accroides, gum
	Ester gum, alcohol soluble
	Resins, synthetic
Selenium	Antimony
	Hydroquinone
	Manganese dioxide
	Sulphur
	Tellurium
Selenium Dioxide	Hydrogen peroxide

Product	Substitute or Alternative
Sesame Oil	Diglycol dilaurate
	Peanut oil
	Sunflower seed oil
Shellac	Alkyd resins
	"Bullzite"
	Batavia gum
	Casein
	Copal, alcohol soluble
	"Elastolac"
	Ester gum with plasticizer
	Gelatin
	Glass with "Vinylite" coating
	Glyceryl phthalate
	Gum accroides
	Gum kauri
	Polyvinyl chloride
	"Protoflax"
	Resins, synthetic
Silica Gel	Agar
	Aluminum hydroxide
	Calcium chloride
	Carbon, activated
	Sodium alumino-silicate
Sodium Alginate	Agar
	Emulsifiers
	"G" protein
	Gelatin
	Gums, water dispersible
	Thickeners
Sodium Alkyl Sulphate	"Wetanol"
Sodium Aluminate	Alum, potash
	Aluminum hydrate
	Copperas with slaked lime
Sodium Antimony Fluoride	Tartar emetic
Sodium Benzoate	Preservatives
Sodium Bichromate	Aluminum sulphate
	Formaldehyde
	Hydrogen peroxide
	Tannin
Sodium Bisulphite	Acetic acid
	Catalysts
	Potassium metabisulphite
	Sodium hyposulphite
Suet	Tallow, edible

Product	Substitute or Alternative
Sugar	Apple juice, concentrated
	Calcium chloride
	Dextrin
	"Diglycol" stearate with water and saccharin
	Glycerin
	Glycols
	Glucose
	Gum, water dispersible
	Honey
	Invert sugar
	Lactose
	Magnesium chloride
	Malted barley
	Malt syrup
	Molasses
	"Nulomoline"
	Saccharin
	Sorghum
	"Sweetose"
	Urea
Sugar Coloring, Burnt	See Caramel coloring
Sulphated Fatty Alcohol	Emulsifiers
	Wetting agents ("Wetanol")
Sulphonated Castor Oil	Emulsifiers
	Glycerin
	Naphthenic soaps
	Polyglycol fatty acid esters with or without wetting agents, e.g. nonaethylene glycol oleate
	Sulphonaphthenic soaps
	Sulphonated olive oil
	Sulphonated tall oil
	Sulphonated vegetable oil
Sulphonated Cocoanut Oil	Sulphonated castor oil
Sulphonated Olive Oil	Diglycol monoricinoleate
	Diglycol oleate
	Emulsifiers
	Glyceryl mono-oleate
	Sulphonated castor oil
Sulphonated Red Oil	Sulphonated castor oil
Sulphonated Tallow	Sulphonated castor oil
Sulphur Dioxide	Chlorine
	Hydrogen peroxide
	Methyl chloride

Product	Substitute or Alternative
Talc	Soapstone
	"Stroba" wax
	Wax, synthetic
	Zinc, stearate
Tallow	Fatty acids
	Garbage grease
	Glyceryl oleo-stearate
	Hydrogenated vegetable oil
	Lard
	Lubricating grease
	Petrolatum
	Soap
	Stearin
	Vegetable oils
	Whale oil with stearin
Tallow Oil	Menhaden oil
Tannic Acid	Alum
	Ammonium bichromate
	Dyewoods
	Formaldehyde
	Potassium bichromate
	Sodium bichromate
	"Syntans"
Tannin	Dyes, aniline
	Formaldehyde
	Lignin sulphonates
	"Maratan"
Tartaric Acid	Acetic acid
	Citric acid
	Saccharolactic acid
	Sodium acid sulphate
	Sulphuric acid
Tellurium	Antimony
	Sulphur
Tetrachlorethylene	Trichlorethylene
Tetrahydrofurfuryl Alcohol	Ethyl alcohol
	Trichlorethylene
"Tetralin"	Turpentine
Thickeners	Agar
	Albumen
	Ammonium caseinate
	Ammonium stearate
	Arrowroot
	Bentonite
	Blood, dried

Product	Substitute or Alternative
Thickeners	Casein
	Clay
	"G" protein
	Glue
Titanium Dioxide	"Celite No. 340"
	Tin oxide
Titanium Tetrachloride	Silicon tetrachloride
Toluol	"Enn Jay" solvents
	Hydrogenated petroleum fractions
	"Nevsol"
	"Notol 1"
	Solvent
	"Solvesso 1"
	Tollac solvent
Tonka Beans	Coumarin
	"Tonka-Mel"
Triacetin	Butyl "Carbitol"
	"Carbitol"
	Plasticizers
	Triglycollin
Trichloracetic Acid	Salicylic acid
Trichlorethylene	"Dresinate"
	Mineral spirits
	Naphtha, petroleum (340–410° F.)
	Soap with solvent
	Solvents
	Tetrahydrofurfuryl alcohol
Tricresyl Phosphate	Diglycol oleate
	"Glaurin"
	Sperm oil
Triethanolamine	Alkalies
	Alkyl amines e.g. amylamine
	Amino alcohols e.g. aminomethyl propanol
	Emulsifiers
	Glycerin
	"Glyco S489"
	"Trigamine"
Tripoli	Diatomaceous earth
Trisodium Phosphate	Alkalies
Tritolyl Phosphate	See Tricresyl phosphate
Tung Oil	Castor oil, dehydrated
	"Kellsey"
	"Kellin"
	Linseed oil, polymerized

Product	Substitute or Alternative
Tung Oil	Synthetic resin solutions
	Vegetable oil
Zein	Casein
	"G" protein
Zinc Oleate	Calcium oleate
	Lead oleate
	Magnesium oleate
Zinc Oxide	Barium sulphate
	"Bolted King White"
	Titanium dioxide with talc or kaolin
	Whiting
Zinc Perborate	Sodium perborate
Zinc Peroxide	Hydrogen peroxide
	Sodium peroxide
	Zinc perborate
Zinc Stearate	Aluminum stearate
	Barium sulphate, purified
	Graphite
	Magnesium stearate
	Stearic acid
	Talc
Zinc Sulphate	Alum, potash

TABLES

Weights and Measures

Troy Weight

24 grains = 1 pwt.
20 pwts. = 1 ounce
12 ounces = 1 pound

Apothecaries' Weight

20 grains = 1 scruple
3 scruples = 1 dram
8 drams = 1 ounce
12 ounces = 1 pound

The ounce and pound are the same as in Troy Weight.

Avoirdupois Weight

27 11/32 grains = 1 dram
16 drams = 1 ounce
16 ounces = 1 pound
2000 lbs. = 1 short ton
2240 lbs. = 1 long ton

Dry Measure

2 pints = 1 quart
8 quarts = 1 peck
4 pecks = 1 bushel
36 bushels = 1 chaldron

Liquid Measure

4 gills = 1 pint
2 pints = 1 quart
4 quarts = 1 gallon
31½ gals. = 1 barrel
2 barrels = 1 hogshead
1 teaspoonful = ⅙ oz.
1 tablespoonful = ½ oz.
16 fluid oz. = 1 pint

Circular Measure

60 seconds = 1 minute
60 minutes = 1 degree
360 degrees = 1 circle

Long Measure

12 inches = 1 foot
3 feet = 1 yard
5½ yards = 1 rod
5280 feet = 1 stat. mile
320 rods = 1 stat. mile

Square Measure

144 sq. in. = 1 sq. ft.
9 sq. ft. = 1 sq. yard
30¼ sq. yds. = 1 sq. rod
43,560 sq. ft. = 1 acre
40 sq. rods = 1 rood
4 roods = 1 acre
640 acres = 1 sq. mile

Metric Equivalents

Length

1 inch = 2.54 centimeters
1 foot = 0.305 meter
1 yard = 0.914 meter
1 mile = 1.609 kilometers
1 centimeter = 0.394 in.
1 meter = 3.281 ft.
1 meter = 1.094 yd.
1 kilometer = 0.621 mile

Capacity

1 U. S. fluid oz. = 29.573 milliliters
1 U. S. liquid qt. = 0.946 liter
1 U. S. dry qt. = 1.101 liters
1 U. S. gallon = 3.785 liters
1 U. S. bushel = 0.3524 hectoliter
1 cu. in. = 16.4 cu. centimeters
1 milliliter = 0.034 U. S. fluid ounce
1 liter = 1.057 U. S. liquid qt.
1 liter = 0.908 U. S. dry qt.
1 liter = 0.264 U. S. gallon
1 hectoliter = 2.838 U. S. bu.
1 cu. centimeter = .061 cu. in.
1 liter = 1000 milliliters or 100 cu. c.

Weight

1 grain = 0.065 gram
1 apoth. scruple = 1.296 grams
1 av. oz. = 28.350 grams
1 troy oz. = 31.103 grams
1 av. lb. = 0.454 kilogram
1 troy lb. = 0.373 kilogram
1 gram = 15.432 grains
1 gram = 0.772 apoth. scruple
1 gram = 0.035 av. oz.
1 gram = 0.032 troy oz.
1 kilogram = 2.205 av. lbs.
1 kilogram = 2.679 troy lbs.

Approximate pH Values

The following tables give approximate pH values for a number of substances such as acids, bases, foods, biological fluids, etc. All values are rounded off to the nearest tenth and are based on measurements made at 25° C.

pH Values of Acids

Hydrochloric, N	0.1
Hydrochloric, 0.1N	1.1
Hydrochloric, 0.01N	2.0
Sulphuric, N	0.3
Sulphuric, 0.1N	1.2
Sulphuric, 0.01N	2.1
Orthophosphoric, 0.1N	1.5
Sulphurous, 0.1N	1.5
Oxalic, 0.1N	1.6
Tartaric, 0.1N	2.2
Malic, 0.1N	2.2
Citric, 0.1N	2.2
Formic, 0.1N	2.3
Lactic, 0.1N	2.4
Acetic, N	2.4
Acetic, 0.1N	2.9
Acetic, 0.01N	3.4
Benzoic, 0.1N	3.1
Alum, 0.1N	3.2
Carbonic (saturated)	3.8
Hydrogen Sulphide, 0.1N	4.1
Arsenious (saturated)	5.0
Hydrocyanic, 0.1N	5.1
Boric, 0.1N	5.2

pH Values of Bases

Sodium Hydroxide, N	14.0
Sodium Hydroxide, 0.1N	13.0
Sodium Hydroxide, 0.01N	12.0
Potassium Hydroxide, N	14.0
Potassium Hydroxide, 0.1N	13.0
Potassium Hydroxide, 0.01N	12.0
Lime (saturated)	12.4
Sodium Metasilicate, 0.1N	12.6
Trisodium Phosphate, 0.1N	12.0
Sodium Carbonate, 0.1N	11.6
Ammonia, N	11.6
Ammonia, 0.1N	11.1
Ammonia, 0.01N	10.6
Potassium Cyanide, 0.1N	11.0
Magnesia (saturated)	10.5
Sodium Sesquicarbonate, 0.1N	10.1
Ferrous Hydroxide (saturated)	9.5
Calcium Carbonate (saturated)	9.4
Borax, 0.1N	9.2
Sodium Bicarbonate, 0.1N	8.4

pH Values of Foods

Apples	2.9–3.3
Apricots	3.6–4.0
Asparagus	5.4–5.8
Bananas	4.5–4.7
Beans	5.0–6.0
Beers	4.0–5.0
Beets	4.9–5.5
Blackberries	3.2–3.6
Bread, white	5.0–6.0
Butter	6.1–6.4
Cabbage	5.2–5.4
Carrots	4.9–5.3
Cheese	4.8–6.4
Cherries	3.2–4.0
Cider	2.9–3.3
Corn	6.0–6.5
Crackers	6.5–8.5
Dates	6.2–6.4
Eggs, fresh white	7.6–8.0
Flour, wheat	5.5–6.5
Gooseberries	2.8–3.0
Grapefruit	3.0–3.3
Grapes	3.5–4.5
Hominy (rye)	6.8–8.0
Jams, fruit	3.5–4.0
Jellies, fruit	2.8–3.4
Lemons	2.2–2.4
Limes	1.8–2.0
Maple Syrup	6.5–7.0
Milk, cows	6.3–6.6
Olives	3.6–3.8
Oranges	3.0–4.0
Oysters	6.1–6.6
Peaches	3.4–3.6
Pears	3.6–4.0
Peas	5.8–6.4
Pickles, dill	3.2–3.6
Pickles, sour	3.0–3.4
Pimento	4.6–5.2
Plums	2.8–3.0
Potatoes	5.6–6.0
Pumpkin	4.8–5.2
Raspberries	3.2–3.6
Rhubarb	3.1–3.2
Salmon	6.1–6.3
Sauerkraut	3.4–3.6
Shrimp	6.8–7.0
Soft Drinks	2.0–4.0
Spinach	5.1–5.7
Squash	5.0–5.4
Strawberries	3.0–3.5
Sweet Potatoes	5.3–5.6
Tomatoes	4.0–4.4
Tuna	5.9–6.1
Turnips	5.2–5.6
Vinegar	2.4–3.4
Water, drinking	6.5–8.0
Wines	2.8–3.8

pH Values of Biologic Materials

Blood, plasma, human	7.3–7.5
Spinal Fluid, human	7.3–7.5
Blood, whole, dog	6.9–7.2
Saliva, human	6.5–7.5
Gastric Contents, human	1.0–3.0
Duodenal Contents, human	4.8–8.2
Feces, human	4.6–8.4
Urine, human	4.8–8.4
Milk, human	6.6–7.6
Bile, human	6.8–7.0

Interconversion Tables and Chart
for Units of Volume and Weight, and Energy

MULTIPLY BY

TO CONVERT FROM	To Cu. In.	To Cu. Ft.	To Cu. Yd.	To Fl. Oz.	To Pint	To Quart	To Gallon	To Grain	To Oz. Troy	To Oz. Av.	To Lb. Troy	To Lb. Av.	To CC. or G.	To Ltr. or Kg.	To Cu. M.
Cu. in.	1.00000	$.0_{3}5787$	$.0_{4}2143$	.554112	.034632	.017316	.004329	252.891	.526857	.578037	.043905	.036127	16.3871	.016387	$.0_{4}1639$
Cu. Ft.	1728.00	1.00000	.037037	957.505	59.8442	29.9221	7.48052	436996	910.408	998.848	75.8674	62.4280	28316.9	28.3169	.028317
Cu. Yd.	46656.0	27.0000	1.00000	25852.6	1615.79	807.896	201.974	117990_{3}	24581.0	26968.9	2048.42	1685.56	764556	764.556	.764556
Fl. Oz.	1.80469	.001044	$.0_{4}3868$	1.00000	.062500	.031250	.007813	456.390	.950813	1.04318	.079234	.065199	29.5736	.029573	$.0_{4}2957$
Pint	28.8750	.016710	$.0_{3}6189$	16.0000	1.00000	.500000	.125000	7302.23	15.2130	16.6908	1.26775	1.04318	473.177	.473177	$.0_{4}4732$
Quart	57.7500	.033420	.001238	32.0000	2.00000	1.00000	.250000	1460.45	30.4260	33.3816	2.53550	2.08635	946.354	.946354	$.0_{3}9463$
Gallon	231.000	.133681	.004951	128.000	8.00000	4.00000	1.00000	58417.9	121.704	133.527	10.1420	8.34541	3785.42	3.78542	.003785
Grain	.003954	$.0_{6}2288$	$.0_{7}8475$	.002191	$.0_{3}1369$	$.0_{4}6850$	$.0_{4}1712$	1.00000	.002083	.002286	$.0_{3}1736$	$.0_{3}1428$	.064799	$.0_{4}6479$	$.0_{7}6479$
Oz. Troy	1.89805	.001098	$.0_{4}4068$	1.05173	.065733	.032867	.008217	480.000	1.00000	1.09714	.083333	.068571	31.1035	.031104	$.0_{4}3110$
Oz. Av.	1.72999	.001001	$.0_{4}3708$	.958608	.059913	.029957	.007489	437.500	.911457	1.00000	.075955	.062500	28.3495	.028350	$.0_{4}2835$
Lb. Troy	22.7766	.013181	$.0_{3}4882$	12.6208	.788800	.394400	.098600	5760.00	12.0000	13.1657	1.00000	.822857	373.242	.373242	$.0_{3}3732$
Lb. Av.	27.6799	.016018	$.0_{3}5933$	15.3378	.958611	.479306	.119826	7000.00	14.5833	16.0000	1.21528	1.00000	453.593	.453593	$.0_{3}4536$
CC or Gram	.061024	$.0_{4}3531$	$.0_{5}1308$	.033814	.002113	.001057	$.0_{3}2642$	15.4323	.032151	.035274	.002679	.002205	1.00000	.001000	.000001
Liter or Kcg.	61.0237	.035315	.001308	33.8140	2.11337	1.05669	.264172	15432.3	32.1507	35.2739	2.67923	2.20462	1000.00	1.00000	.001000
Cu. M.	61023.7	35.3146	1.30795	33814.0	2113.37	1056.69	264.172	154320_{3}	32150.7	35273.9	2679.23	2204.62	1000000	1000.00	1.00000

Note. The small subnumeral following a zero indicates that the zero is to be taken that number of times; thus, $.0_{3}1428$ is equivalent to .0001428.

Values used in constructing table:
1 inch = 2.540001 cm.
1 cu. in. = 16,387083 cc. = 16.387083 g H_2O at 4°C. = 39°F.

1 lb. av. = 453.5926 g.
∴ 1 gal. = 8.34541 lb.
∴ 1 lb. av. = 27.679886 cu. in. H_2O at 4°C.

1 lb. av. = 7000 grains.
∴ 1 gallon = 58417.87 grains.
231 cu. in. = 1 gallon = 3785.4162 g.

TO CONVERT FROM	MULTIPLY BY B. T. U.	P. C. U.	Cal.	Ft. Lbs.	Ft. Tons.	Kg. M.	HP Hrs.	KW Hrs.	Joules	Lbs. C	Lbs. H_2O
B. T. U.	1.00000	.555556	.251996	778.000	.389001	107.563	$.0_{3}3929$	$.0_{3}2931$	1055.20	$.0_{4}6876$	.001031
P. C. U.	1.80000	1.00000	45.3593	1400.40	.700202	193.613	$.0_{3}7072$	$.0_{3}5276$	1899.36	$.0_{3}1238$	.001855
Calories	3.96832	2.20462	1:00000	3091.36	1.54368	426.844	.001559	.001163	4187.37	$.0_{3}2729$	.004089
Ft. Lbs.	.001285	$.0_{3}7141$	$.0_{3}3239$	1.00000	.000500	.138255	$.0_{6}5050$	$.0_{6}3767$	1.35625	$0_{7}8840$	$.0_{6}1325$
Ft. Tons	2.57069	1.42816	.647804	2000.00	1.00000	276.511	.001010	$.0_{3}7535$	2712.59	$.0_{3}1768$	.002649
Kg. M.	.009297	.005165	.002343	7.23301	.003617	1.00000	$.0_{5}3653$	$.0_{5}2725$	9.81009	$.0_{6}6394$	$.0_{5}9580$
HP Hrs	2544.99	.141388	641.327	1980000	990.004	273747	1.00000	.746000	2685473	.175044	2.62261
KW Hrs.	3411.57	1895.32	859.702	2654200	1327.10	366959	1.34041	1,00000	3599889	.234648	3.51562
Joules	$.0_{3}9477$	$.0_{3}5265$	$.0_{3}2388$	.737311	$.0_{3}3687$	.101937	$.0_{6}3724$	$.0_{6}2778$	1.00000	$.0_{7}6518$	$.0_{6}9766$
Lbs. C.	14544.0	8080.00	3665.03	113150_{2}	5657.63	1564396	5.71434	4.26285	153470_{2}	1.00000	14.9876
Lbs. H_2O	970.400	539.111	244.537	754971	377.487	104379	.381270	.284424	1023966	.066744	1.00000

"P. C. U." refers to the "pound-centigrade unit." The ton used is 2000 pounds. "Lbs. C" refers to pounds of carbon oxidized, 100% efficiency equivalent to the corresponding number of heat units. "Lbs H_2O" refers to pounds of water evaporated at 100°C. = 212°F. at 100% efficiency

C1

By the use of the foregoing table[1] about 330 interconversions among twenty-six of the standard engineering units of measure can be directly estimated from the alignment chart to three significant figures or calculated by simple multiplication to six figures. The multiplier factor given in the table is located on the center scale "A" giving the point which when aligned with any number point on "C1" determines the product on "C." Imperfections in the scale due to lack of precision in printing should be checked at intervals along "A" scale by actual division of "C" by "C1," the lines being left out so that the reader can do this. A line scratched on a transparent celluloid triangle gives the best medium for making alignments.

When volume and weight interconversions are given, water is the medium the calculations are based upon. By the introduction of specific gravity factors the medium can be changed, giving the weight of any volume

---

[1] Courtesy of [illegible]

CONVERSION OF THERMOMETER READINGS

F°	C°	F°	C°	F°	C°	F°	C°	F°	C°	F°	C°
—40	—40.00	30	—1.11	80	26.67	250	121.11	500	260.00	900	482.22
—38	—38.89	31	—0.56	81	27.22	255	123.89	505	262.78	910	487.78
—36	—37.78	32	0.00	82	27.78	260	126.67	510	265.56	920	493.33
—34	—36.67	33	0.56	83	28.33	265	129.44	515	268.33	930	498.89
—32	—35.56	34	1.11	84	28.89	270	132.22	520	271.11	940	504.44
—30	—34.44	35	1.67	85	29.44	275	135.00	525	273.89	950	510.00
—28	—33.33	36	2.22	86	30.00	280	137.78	530	276.67	960	515.56
—26	—32.22	37	2.78	87	30.56	285	140.55	535	279.44	970	521.11
—24	—31.11	38	3.33	88	31.11	290	143.33	540	282.22	980	526.67
—22	—30.00	39	3.89	89	31.67	295	146.11	545	285.00	990	532.22
—20	—28.89	40	4.44	90	32.22	300	148.89	550	287.78	1000	537.78
—18	—27.78	41	5.00	91	32.78	305	151.67	555	290.55	1050	565.56
—16	—26.67	42	5.56	92	33.33	310	154.44	560	293.33	1100	593.33
—14	—25.56	43	6.11	93	33.89	315	157.22	565	296.11	1150	621.11
—12	—24.44	44	6.67	94	39.44	320	160.00	570	298.89	1200	648.89
—10	—23.33	45	7.22	95	35.00	325	162.78	575	301.67	1250	676.67
— 8	—22.22	46	7.78	96	35.56	330	165.56	580	304.44	1300	704.44
— 6	—21.11	47	8.33	97	36.11	335	168.33	585	307.22	1350	732.22
— 4	—20.00	48	8.89	98	36.67	340	171.11	590	310.00	1400	760.00
— 2	—18.89	49	9.44	99	37.22	345	173.89	595	312.78	1450	787.78
0	—17.78	50	10.00	100	37.78	350	176.67	600	315.56	1500	815.56
1	—17.22	51	10.56	105	40.55	355	179.44	610	321.11	1550	843.33
2	—16.67	52	11.11	110	43.33	360	182.22	620	326.67	1600	871.11
3	—16.11	53	11.67	115	46.11	365	185.00	630	332.22	1650	898.89
4	—15.56	54	12.22	120	48.89	370	187.78	640	337.78	1700	926.67
5	—15.00	55	12.78	125	51.67	375	190.55	650	343.33	1750	954.44
6	—14.44	56	13.33	130	54.44	380	193.33	660	348.89	1800	982.22
7	—13.89	57	13.89	135	57.22	385	196.11	670	354.44	1850	1010.00
8	—13.33	58	14.44	140	60.00	390	198.89	680	360.00	1900	1037.78
9	—12.78	59	15.00	145	62.78	395	201.67	690	365.56	1950	1065.56
10	—12.22	60	15.56	150	65.56	400	204.44	700	371.11	2000	1093.33
11	—11.67	61	16.11	155	68.33	405	207.22	710	376.67	2050	1121.11
12	—11.11	62	16.67	160	71.11	410	210.00	720	382.22	2100	1148.89
13	—10.56	63	17.22	165	73.89	415	212.78	730	387.78	2150	1176.67
14	—10.00	64	17.78	170	76.67	420	215.56	740	393.33	2200	1204.44
15	— 9.44	65	18.33	175	79.44	425	218.33	750	398.89	2250	1232.22
16	— 8.89	66	18.89	180	82.22	430	221.11	760	404.44	2300	1260.00
17	— 8.33	67	19.44	185	85.00	435	223.89	770	410.00	2350	1287.78
18	— 7.78	68	20.00	190	87.78	440	226.67	780	415.56	2400	1315.56
19	— 7.22	69	20.56	195	90.55	445	229.44	790	421.11	2450	1343.33
20	— 6.67	70	21.11	200	93.33	450	232.22	800	426.67	2500	1371.11
21	— 6.11	71	21.67	205	96.11	455	235.00	810	432.22	2550	1398.89
22	— 5.56	72	22.22	210	98.89	460	237.78	820	437.78	2600	1426.67
23	— 5.00	73	22.78	215	101.67	465	240.55	830	443.33	2650	1454.44
24	— 4.44	74	23.33	220	104.44	470	243.33	840	448.89	2700	1482.22
25	— 3.89	75	23.89	225	107.22	475	246.11	850	454.44	2750	1510.00
26	— 3.33	76	24.44	230	110.00	480	248.89	860	460.00	2800	1537.78
27	— 2.78	77	25.00	235	112.78	485	251.67	870	465.56	2850	1565.56
28	— 2.22	78	25.56	240	115.56	490	254.44	880	471.11	2900	1593.33
29	— 1.67	79	26.11	245	118.33	495	257.22	890	476.67	2950	1621.11

ALCOHOL PROOF AND PERCENTAGE TABLE

U. S. Proof at 60° F.	Per cent Alcohol by Volume at 60° F.	Per cent Alcohol by Weight
0	0.0	0.00
1	0.5	—
2	1.0	0.80
3	1.5	—
4	2.0	1.59
5	2.5	—
6	3.0	2.39
7	3.5	—
8	4.0	3.19
9	4.5	—
10	5.0	4.00
11	5.5	—
12	6.0	4.80
13	6.5	—
14	7.0	5.61
15	7.5	—
16	8.0	6.42
17	8.5	—
18	9.0	7.23
19	9.5	—
20	10.0	8.05
21	10.5	—
22	11.0	8.86
23	11.5	—
24	12.0	9.68
25	12.5	—
26	13.0	10.50
27	13.5	—
28	14.0	11.32
29	14.5	—
30	15.0	12.14
31	15.5	—
32	16.0	12.96
33	16.5	—
34	17.0	13.79
35	17.5	—
36	18.0	14.61
37	18.5	—
38	19.0	15.44
39	19.5	—
40	20.0	16.27
41	20.5	—
42	21.0	17.10
43	21.5	—
44	22.0	17.93
45	22.5	—
46	23.0	18.77
47	23.5	—
48	24.0	19.60
49	24.5	—
50	25.0	20.44
51	25.5	—
52	26.0	21.28
53	26.5	—
54	27.0	22.13
55	27.5	—
56	28.0	22.97
57	28.5	—
58	29.0	23.82
59	29.5	—
60	30.0	24.67
61	30.5	—
62	31.0	25.52
63	31.5	—
64	32.0	26.38
65	32.5	—
66	33.0	27.24
67	33.5	—
68	34.0	28.10
69	34.5	—
70	35.0	28.97
71	35.5	—
72	36.0	29.84
73	36.5	—
74	37.0	30.72
75	37.5	—
76	38.0	31.60
77	38.5	—
78	39.0	32.48
79	39.5	—
80	40.0	33.36
81	40.5	—
82	41.0	34.25
83	41.5	—
84	42.0	35.15
85	42.5	—
86	43.0	36.05
87	43.5	—
88	44.0	36.96
89	44.5	—
90	45.0	37.86
91	45.5	—
92	46.0	38.78
93	46.5	—
94	47.0	39.70
95	47.5	—
96	48.0	40.62
97	48.5	—
98	49.0	41.55
99	49.5	—
100	50.0	42.49
101	50.5	—
102	51.0	43.43
103	51.5	—
104	52.0	44.37
105	52.5	—
106	53.0	45.33
107	53.5	—
108	54.0	46.28
109	54.5	—
110	55.0	47.24
111	55.5	—
112	56.0	48.21
113	56.5	—

U. S. Proof at 60° F.	Per cent Alcohol by Volume at 60° F.	Per cent Alcohol by Weight
114	57.0	49.19
115	57.5	——
116	58.0	50.17
117	58.5	——
118	59.0	51.15
119	59.5	——
120	60.0	52.15
121	60.5	——
122	61.0	53.15
123	61.5	——
124	62.0	54.15
125	62.5	——
126	63.0	55.16
127	63.5	——
128	64.0	56.18
129	64.5	——
130	65.0	57.21
131	65.5	——
132	66.0	58.24
133	66.5	——
134	67.0	59.28
135	67.5	——
136	68.0	60.32
137	68.5	——
138	69.0	61.38
139	69.5	——
140	70.0	62.44
141	70.5	——
142	71.0	63.51
143	71.5	——
144	72.0	64.59
145	72.5	——
146	73.0	65.67
147	73.5	——
148	74.0	66.77
149	74.5	——
150	75.0	67.87
151	75.5	——
152	76.0	68.92
153	76.5	——
154	77.0	70.10
155	77.5	——
156	78.0	71.23
157	78.5	——
158	79.0	72.38
159	79.5	——
160	80.0	73.53
161	80.5	——
162	81.0	74.69
163	81.5	——
164	82.0	75.86
165	82.5	——
166	83.0	77.04
167	83.5	——
168	84.0	78.23
169	84.5	——
170	85.0	79.44
171	85.5	——
172	86.0	80.62
173	86.5	——
174	87.0	81.90
175	87.5	——
176	88.0	83.14
177	88.5	——
178	89.0	84.41
179	89.5	——
180	90.0	85.69
181	90.5	——
182	91.0	86.99
183	91.5	——
184	92.0	88.31
185	92.5	——
186	93.0	89.65
187	93.5	——
188	94.0	91.02
189	94.5	——
190	95.0	92.42
191	95.5	——
192	96.0	93.85
193	96.5	——
194	97.0	95.32
195	97.5	——
196	98.0	96.82
197	98.5	——
198	99.0	98.38
199	99.5	——
200	100.0	100.00

Buffer Systems

The following table gives some common buffer systems and the approximate pH of maximum buffer capacity. The zone of effective buffer action will vary with concentration but the general average will be ± 1.0 pH from the value given, for concentrations approximately 0.1 molar.

Buffer System	pH
Glycocoll - Sodium Chloride - Hydrochloric Acid	2.0
Potassium Acid Phthalate-Hydrochloric Acid	2.8
Primary Potassium Citrate	3.7
Acetic Acid-Sodium Acetate	4.6
Potassium Acid Phthalate-Sodium Hydroxide	5.0
Secondary Sodium Citrate	5.0
Carbonic Acid-Bicarbonate	6.5
Primary Phosphate-Secondary Phosphate	6.8
Primary Phosphate-Sodium Hydroxide	6.8
Boric Acid-Borax	8.5
Borax	9.2
Boric Acid-Sodium Hydroxide	9.2
Bicarbonate-Carbonate	10.2
Secondary Phosphate-Sodium Hydroxide	11.5

Courtesy of W. A. Taylor & Company

REFERENCES AND ACKNOWLEDGMENTS

Abrasive & Cleaning Methods
Agr. Gaz. N. S. Wales
Allg. Oes. v. Gettzeitung
Aluminum Co. of Amer.
Amer. Cyanamid & Chem. Corp.
Amer. Druggst
Amer. Dyestuff Reporter
Amer. Electrop. Society
Amer. Gum Importers' Ass'n
Amer. Paint Jol.
Amer. Perfumer
Amer. Photography
Amer. Wool & Cotton Reporter
Analyst
Anal. Fis. Quim.
Ault & Wiborg Varnish Wks. Handbook

Baker's Helper
Bakers Review
Baker's Weekly
Behr Manning Corp.
Better Enameling
Boonton Molding Co.
Bottler & Packer
Boyce Thompson Inst.
Brewers' Tech. Review
Brick & Clay Record
Br. Jol. Dent. Science
Brit. Jol. of Photography
Brit. Medical Jol.
Bull. Imp. Hyg. Lab.
Bulletin of Imperial Institute
Bull. Soc. Franc. Phot.

Camera
Camera (Luzern)
Canadian Jol. of Med. Technology
Canadian Textile Jol.
Canner
Cement & Cement Mfr.
Ceramic Age
Chemical Abstracts
Chemical Analyst
Chemical Industries
Chemical Products
Chemical Weekblad
Chem. Zent.
Chemist & Druggist
Chr. Hansen's Lab.
Cleaning & Dyeing World
Combustion
Confectioner's Jol.
Consumers' Guide
Cramer's Manual

Dairy World
Damsk. Tids. Farm
Dental Items
Dental Lab'y Review
Devt. Part. Zeitung
Drug & Cosmetic Industry
Druggists Circular
Drugs, Oils, & Paints
DuPont Rubber Bulletins

Eastman Kodak Co.
Electric Journal

Farbe u. Lacke
Farben Zeitung
Farming S. Africa
Fein Mechanic v. Prazision
Fettchem, Umschan
Fils & Tissus
Flavours
Focus
Food Manufacture
Fruit Products Jol.

Gelatin, Leim, Klebstoffe
General Abrasive Co.
Glass Industry
Graphic Arts Monthly

Hawaiian Planters' Record
Hercules Powder Co. Bulletins
Hide & Leather

Ice Cream Review
India Rubber World
Indian Lac Research Inst.
Indian Soap Jol.
Indiana Acad. of Sciences
Industrial Chemist
Industrial Finishing
Instruments
Intern'l Salt Co.
Int'l Tin Res. & Dev. Council
Iowa State College Bull.

J. Amer. Dental Assn.
J. Amer. Medical Assn.
J. Chem. Eng.
J. Chinese Chem. Soc.
J. Federation Curriers
J. Federation Light Leather Tanners
J. Res. Nat. Bur. Standards
J. Rubber Industry
J. Russ. Rubber Ind.

Jol. Soc. Leather Trades
Jol. Soc. Rubber Ind. Japan
Jol. Tech. Physics
Jol. of Technical Methods (I.A.M.M.)

Keram Steklo
Khimstroi
Kozhevenna-Obuvnaya Prom.
Kunstdunger, Und Leim

Lakokras, Ind.
Leather Trades Review
Leather Worker
Les Mat. Grasses
Lithographic Tech. Foundation

Malayan Agric. Jol.
Manufacturing Chemist
Meat
Meat Merchandising
Melliand
Metal Industry
Metall und Erz
Metallurg
Metallurgist
Metals & Alloys
Mich. Agric. Exp. Sta.
Milk Dealer
Mineralogist
Monatschr. Textil-Ind.
Monsanto Chem. Co.
Munic. Eng. San. Record

Nat'l Butter & Cheese Jol.
Nat'l Provisioner
Nickelsworth
Nitrocellulose

Ober Flachen Tech.
Oil & Color Trades Jol.
Oil & Soap

Pacific Rural Press
Paint Technology
Paper Trade Jol.
Parfum Mod.
Peinture, Pigments, Vernis
Phar. Acta Helva
Pharmaceutical Jol.
Phot. Abstracts
Photo Art Monthly
Phot. Ind.
Phot. Korr.
Photog. Kronik
Phot. Rev.
Photo Rundschau
Physics
Phytopathology
Pix
Plater's Guide Book
Portland Cement Assn.
Power
Practical Druggist
Practical Everyday Chemistry
Printing Industry
Prob. Edelmetalle
Process Engr. Mo.
Proc. World Petroleum Congress

Rayon & Mell. Tex. Monthly
Refiner & Nat. Gas Mfr.
Rev. Aluminum
Rev. Amer. Electro Society
Rock Products

Science
Sharpless Solvents Corp.
Silver Technologist
Shoe and Leather Journal
Soap
Soap Gazette & Perfumer
Solvent News
Sovet-Sakhar
Spirits
Steel
Synthetic & Applied Finishes

Textile Colorist
Textile Mfr.
Textile Recorder

Univ. Nebr. Agric. Coll. Bull.
U. S. Department of Agriculture
U. S. Bureau of Mines
U. S. Bureau of Standards

Veneers and Plywood

Z. Elektrochem.
Zeit. Unters. Lebensm.

ADDENDA

American Colloid Co.
Army Ordnance
Cowles Laundry Tips
Electrochemical Society
Indiana Farmer's Guide
Industrial & Eng. Chemistry
Manufacturing Confectioner
New York Physician
Phila. Quartz Co.

TRADE-NAME CHEMICALS

During the past few years, the practice of marketing raw materials, under names which in themselves are not descriptive chemically of the products they represent, has become very prevalent. No modern book of formulae could justify its claims either to completeness or modernity without numerous formulae containing these so-called "Trade Names."

Without wishing to enter into any discussion regarding the justification of "Trade Names," the Editors recognize the tremendous service rendered to commercial chemistry by manufacturers of "Trade Name" products, both in the physical data supplied and the formulation suggested.

Deprived of the protection afforded their products by this system of nomenclature, these manufacturers would have been forced to stand helplessly by while the fruits of their labor were being filched from them by competitors who, unhampered by expenses of research, experimentation and promotion, would be able to produce something "just as good" at prices far below those of the original producers.

That these competitive products were "just as good" solely in the minds of the imitators would only be evidenced in costly experimental work on the part of the purchaser and, in the meantime, irreparable damage would have been done to the truly ethical product. It is obvious, of course, that under these circumstances, there would be no incentive for manufacturers to develop new materials.

Because of this, and also because the "Chemical Formulary" is primarily concerned with the physical results of compounding rather than with the chemistry involved, the Editors felt that the inclusion of formulae containing various trade name products would be of definite value to the producer of finished chemical materials. If they had been left out many ideas and processes would have been automatically eliminated.

As a further service the better known "trade name" products are included with the list of chemicals and supplies.

CHEMICALS AND SUPPLIES: WHERE TO BUY THEM*

Numbers on right refer to list of suppliers on pages directly following this list. Thus to find out who supplies borax look in left hand column, alongside borax, on page 591. The number there is 34. Now turn to page 603 and find number 34. Alongside is the supplier, American Potash & Chemical Corp., New York, N. Y.

Product	*No.*
A	
A-Syrup	423
Aacagum	251
Abalyn	276
Abietic Acid	276
Abopon	251
Accelerator 808	195
Accelerator 833	195
Accelerators, Vulcanization	195
Acceloid	245
Accroides	463
Acelose	18
Acetaloid	2
Acetamide	20
Acetic Acid	140
Acetic Anhydride	16
Acetoin	335
Acetol	445
Acetone	150
Acetphenetidin	360
Acetyl Cellulose	see Cellulose Acetate
Acetyl Salicylic Acid	368
Acidolene	349
Acids, Fatty	540
Acimul	251
Acrawax	251
Acriflavine	1
Acrolite	157
Acrxyeol	442
Acrylic Resins	450
Acryloid	450
Acrysol	450
Acto	507
Adeps Lanae	see Lanolin
Adheso Wax	251
A.D.M. No. 100 Oil	44
Aerogel	368
Aerosol	23
Agar	14
Agene	399
Agerite Powder	559
Akcocene	23
Aktivin	8
Albacer	251
Alba-Floc	550
Albalith	387
Albasol	381
Albatex	137
Alberit	471
Albertol	471
Albinol	464a
Albolit	53
Albolith	387
Albone C	195
Albron	11
Albumen	515
Albusol	344
Alcohol, Denatured	449
Alcohol, Pure	551
Aldehol	312
Aldol	392
Aldydal	294
Alfalate	322
Alframine	365
Alginic Acid	313
Alizarin	595
Alkalies	148
Alkaloids	360
Alkanet	285
Alkanol	195
Alkyd Resins	409
Alloxan	88
Almond Oil	342
Aloes	413
Aloin	398
Aloxite	109
Alperox	335
Alpha Naphthol	286
Alphasol	23
Altax	559
Alugel	376
Alumina	11
Aluminum	11
Aluminum Acetate	392
Aluminum Bronze Powder	549
Aluminum Chloride	99
Aluminum Hydrate	121
Aluminum Oleate	479
Aluminum Silicate	580
Aluminum Stearate	229
Aluminum Sulphate	513
Alums	256

* Please see addenda p. 602.

B

Product	*No.*
Beckolin	440
Beckophen	440
Beckosol	440
Beeswax	192
Beetle Resin	23
Bellaphan	311
Bensapol	588
Bentonite	22
Benzaldehyde	279
Benzidine	236
Benzine	32
Benzocaine	1
Benzoic Acid	111
Benzol	68
Benzoyl Peroxide	335
Benzyl Cellulose	6
Bergamot Oil	407
Beryllium	72
Beryllium Salts	223
Beta Naphthol	99
Beutene	384
Bicarbonate of Soda	136
Bismuth	122
Bismuth Subnitrate	388
Bitumen	see Asphaltum
Black Leaf	539
Blanc Fixe	291
Blandol	494
Blankit	240
Bleaching Powder	208
Blendene	251
Blood Albumen	370
Bludtan	123
Boea	297
Bole	580
Bonderite	411
Bondogen	559
Bone Ash	176
Bone Black	484
Bone Glue	170
Bone Meal	60
Bone Oil	532
Borax	34
Bordeaux Mixture	358
Bordow	188
Boric Acid	83
Borol	193
Botanical Products	414
Bresin	276
Brisgo	276
Bromine	181
Bromo "Acid"	436
Bromo-Fluorescein	251
Bronze Powder	189
Brosco	466
Burgundy Pitch	331
Butacite	195
Butalyde	149
Butex	402
Butoben	360
Butyl Acetate	149
Butyl Acetyl Ricinoleate	150
Butyl Alcohol (Normal)	434
Butyl Aldehyde	150
Butyl Amine	150
Butyl Carbitol	106
Butyl Cellosolve	106
Butyl Lactate	150
Butyl Propionate	96
Butyl Stearate	316
Butyric Ether	396

C

Product	*No.*
Cadalyte	256
Cadmium	555
Cadmolith	128
Cajuput Oil	292
Calagum	251
Calcene	148
Calcium Arsenate	85
Calcium Carbonate	330
Calcium Carbonate (Precipitated)	360
Calcium Chloride	366
Calcium Chloride (Anhydrous)	213
Calcium Cyanamid	23
Calcium Fluoride	238
Calcium Hydroxide	see Lime, Hydrate
Calcium Oxide	see Lime
Calcium Phosphate	433
Calcium Polysulphide	98
Calcium Propionate	251
Calcium Stearate	524
Calcium Sulphate	see Plaster of Paris
Calcium Sulphide (Luminous)	29
Calcocid	99
Calcolac	99
Calcoloid	99
Calcozine	99
Calgon	94
Calgonite	100
Calomel	590
Calorite	505
Camphor	69
Camphor Oil	342
Candelilla Wax	298
Capillary Syrup	see Glucose Syrup
Captax	563
Caramel Color	231
Caraway Oil	336
Carbasol	107
Carbitol	106
Carbofrax	109
Carbolac	97
Carbolic Acid	see Phenol
Carbolic Oil	441
Carbolineum	108
Carbon, Activated	305
Carbon Bisulphide	62
Carbon Black	547
Carbon, Decolorizing	169
Carbon Tetrachloride	395
Carbonex	68

Product	*No.*
Cresophan	258
Cresylic Acid	68
Cromodine	19
Cryolite	568
Cryptone	387
Crysalba	368
Cumar	68
Cupric Chloride ..see Copper Chloride	
Curbay Binder	552
Curgon	471
Cuttle Fish Bone	228
Cyclamal	239
Cycline	368
Cyclohexylamine	195
Cyclohexanol	195
Cyclonol	195
Cymanol	296

D

Product	*No.*
Dammar Gum	331
Dapol	247
Darco	169
Darvan	559
Daxad	178
Deceresol	23
Degras	27
Deo-base	494
Deramin	251
Derris Extract	472
Derris Root	73
Devolite	471
Dextrins	370
Dextrose	162
Diacetin	316
Diafoam	442
Diakonn	295
Diamond K Linseed Oil	499
Diamyl Phthalate	552
Diastafor	504
Diastase	526
Diatol	552
Diatomaceous Earth	586
Dibutyl Cellosolve Phthalate	368
Dibutyl Tartrate	312
Dibutylphthalate	316
Dicalite	180
Dichlorbenzol	284
Dichlorethylene	195
Dichlorethylether	106
Diethanolamine Lactate	251
Diethylcarbonate	552
Diethylene Glycol	106
Diethyl Phthalate	560
Digestase	345
Diglycol Laurate	251
Diglycol Oleate	251
Diglycol Stearate	251
Dilecto	157
Dinitrophenol	379
Diolin	195
Dionin	360
Dioxan	106
Dipentene	276
Diphenyl	521
Diphenyl, Chlorinated	368
Diphenyl Oxide	248
Diphenyl Phthalate	368
Discolite	454
Disodium Phosphate	565
Disperso	586
Distoline	587
Dowicide	188
Dow-metal	188
Dow plasticizers	188
D. P. G.	368
Drierite	265
Driers (Paint and Oil)	401
Driers, Varnish	401
Drop Black	582
Dry Ice	351
Dulux	195
Duolith	320
Dupanol	195
Duphax	501
Duphonol	195
DuPont Rubber Red	195
Duprene	195
Duraplex	442
Durez	245
Durite	516
Durophene	471
Dutox	195
Dyestuffs	379
Dynax	195

E

Product	*No.*
East-India Gum	23
Eastman Products	202
Egg, Dried	430
Egg Yolk	515
Elaine	210
Elemi	552
Emulphor	240
Emulsifier L83A	251
Emulsone	251
Emulsone B	251
Eosin	440
Ephedrine	360
Epsom Salt	238
Erinoid	33
Erio Chrome Dyes	235
Escolite	163
Essential Oils	152
Esso	507
Ester Gum	329
Esterol	409
Estersol	552
Ethavan	368
Ether	106
Ethox	561
Ethyl Acetate	361
Ethyl Cellulose	276
Ethyl Lactate	23
Ethyl Parasept	279

Q

Product	*No.*
Quince Seed	285
Quinine Bisulphate	258
Quinine Hydrochloride	360
Quinoline	68

R

Product	*No.*
Raisin Seed Oil	444
Rancidex	251
Rapeseed Oil	63
Rapidase	572
Rauzene	438
Rayox	559
Red Oil	119
Red Squill	285
Redmanol	56
Reogen	559
Resin DA 1	251
Resin R-H-35	195
Resinox	442a
Resins, Natural	23
Resins, Synthetic	440
Resipon	45
Resoglaz	6
Resorcin	416
Revertex	443
Rezidel	251
Rezinel	251
Rezyl	23
Rhodium	58
Rhonite	450
Rhoplex	450
Rhotex	450
Rochelle Salts	422
Rodo	559
Rose Water	336
Roseol	342
Rosin	244
Rosin Oil	383
Rosoap A	251
Rotenone	535
Rubber	201
Rubber Hydrochloride	347
Rubber Latex	334
Rubber Resin	251
Rubber, Synthetic	195, 254, 255
Rubidium Salts	340

S

Product	*No.*
"S" Syrup	423
Saccharine	279
Sal Soda	136
Salicylic Acid	188
Salt	371
Salt Cake	23
Saltpetre	164
Santicizers	368
Santobane	368
Santomask	368
Santox	368
Sapamine	137
Saponin	309
Savolin	251
Schultz Silica	124
Selenium	31
Sellatan A	235
Sepia	see Cuttle Fish Bone
Serinol	310
Serrasol	130
Shellac	595
Shellac Wax	291
Sherpetco	482
Sicapon	251
Siennas	217
Silex	580
Silica	67
Silica Black	304
Silicon	72
Silvatol	137
Silver	266
Silver Cyanide	160
Silver Nitrate	202
Slaked Lime	see Lime
Soap	511
Soda Ash	179
Soda, Caustic	351
Soda, Sal	154
Sodium Acetate	542
Sodium Alginate	313
Sodium Aluminate	376
Sodium Arsenite	268
Sodium Benzoate	284
Sodium Bicarbonate	136
Sodium Bichromate	431
Sodium Bisulphite	256
Sodium Borate	see Borax
Sodium Borophosphate	251
Sodium Carbonate	493
Sodium Chlorate	404
Sodium Chlorite	351
Sodium Choleate	183
Sodium Cyanide	195
Sodium Fluoride	23
Sodium Hydrosulphite	454
Sodium Hydroxide	360
Sodium Hypochlorite	174
Sodium Hypochlorite Liquid	447
Sodium Hyposulphite	256
Sodium Lauryl Sulphate	251
Sodium Metaphosphate	94
Sodium Metasilicate	423
Sodium Nitrate	70
Sodium Nitrite	493
Sodium Oleate	251
Sodium Oxalate	565
Sodium Perborate	195
Sodium Phosphate	521
Sodium Propionate	251
Sodium Pyrophosphate	565
Sodium Resinate	408
Sodium Silicate	423
Sodium Silico Fluoride	256
Sodium Stannate	269
Sodium Stannate	354
Sodium Sulphate	238

T

ADDENDA

SELLERS OF CHEMICALS AND SUPPLIES

No.	*Name*	*Address*
1.	Abbott Laboratories	North Chicago, Ill.
2.	Acetate Products Corp.	London, England
3.	Acheson Colloids Corp.	Port Huron, Mich.
4.	Acheson Graphite Corp.	Niagara Falls, N. Y.
5.	Acme Oil Corp.	Chicago, Ill.
6.	Advance Solvents & Chem. Corp.	New York, N. Y.
7.	Ajax Metal Co.	Philadelphia, Pa.
8.	Aktivin Corp.	New York, N. Y.
9.	Allied Asphalt & Mineral Corp.	New York, N. Y.
10.	Alpha Lux Co., Inc.	New York, N. Y.
11.	Aluminum Co. of America	Pittsburgh, Pa.
12.	Amecco Chemicals, Inc.	Rochester, N. Y.
13.	American Active Carbon Co.	Columbus, O.
14.	American Agar Co., Inc.	San Diego, Calif.
15.	American Aniline Products, Inc.	New York, N. Y.
16.	American-Brit. Chem. Supplies, Inc.	New York, N. Y.
17.	American Catalin Corp.	New York, N. Y.
18.	American Cellulose Co.	Indianapolis, Ind.
19.	American Chemical Paint Co.	Ambler, Pa.
20.	American Chemical Products Co.	Rochester, N. Y.
21.	American Chlorophyll, Inc.	New York, N. Y.
22.	American Colloid Co.	Chicago, Ill.
23.	American Cyanamid & Chem. Co.	New York, N. Y.
24.	American Dyewood Co.	New York, N. Y.
25.	American Fluoride Corp.	New York, N. Y.
26.	American Insulator Corp.	New Freedom, Pa.
27.	American Lanolin Corp.	Lawrence, Mass.
28.	American Lecithin Corp.	New York, N. Y.
29.	American Luminous Products Co.	Huntington Park, Calif.
30.	American Maize Products Co.	New York, N. Y.
31.	American Metal Co.	New York, N. Y.
32.	American Mineral Spirit Co.	New York, N. Y.
33.	American Plastics Corp.	New York, N. Y.
34.	American Potash & Chem. Corp.	New York, N. Y.
35.	American Smelting & Refining Co.	New York, N. Y.
36.	American Zinc Co.	New York City
37.	Amido Products Co.	New York, N. Y.
38.	Anchor Chemical Co.	Manchester, England
39.	Anderson Prichard Oil Corp.	Oklahoma City, Okla.
40.	Ansbacher-Siegle Corp.	Rosebank, New York
41.	Ansul Chem. Co.	Marinette, Wis.
42.	Antwerp Naval Stores Co., Inc.	Boston, Mass.
43.	Apex Chem. Co.	New York, N. Y.
44.	Archer-Daniels-Midland Co.	Minneapolis, Minn.
45.	Arkansas Co.	New York, N. Y.
46.	Armour & Co.	Chicago, Ill.
47.	Asbury Graphite Mills	Asbury Park, N. J.
48.	Atlantic Gelatine Co.	Woburn, Mass.
49.	Atlantic Refining Co.	Philadelphia, Pa.
50.	Atlantic Research Associates	Newtonville, Mass.
51.	Atlas Import Co.	Chicago, Ill.
52.	Atlas Powder Co.	Wilmington, Del.
53.	Augsburger, Kunst Fabrik	Augsburg, Germany

No.	Name	Address
54.	Autoxygen, Inc.	New York, N. Y.
55.	Badcock, Robert & Co.	New York, N. Y.
56.	Bakelite Corp.	New York, N. Y.
57.	Baker Castor Oil Co.	Jersey City, N. J.
58.	Baker & Co., Inc.	Newark, N. J.
59.	Baker, Franklin Co.	Hoboken, N. J.
60.	Baker, H. J. & Bro.	New York, N. Y.
61.	Baker, J. E., Co.	York, Pa.
62.	Baker, J. T. Chem. Co.	Philipsburg, N. J.
63.	Balfour, Guthrie & Co., Ltd.	New York, N. Y.
64.	Barada & Page, Inc.	Kansas City, Mo.
65.	Barber Asphalt Co.	Philadelphia, Pa.
66.	Barium Reduction Corp.	Charleston, W. Va.
67.	Barnsdall Tripoli Corp.	Seneca, Mo.
68.	Barrett Co.	New York, N. Y.
69.	Barry, E. J., Inc.	New York, N. Y.
70.	Battelle & Renwick	New York, N. Y.
71.	Battleboro Oil Co.	Battleboro, N. C.
71a.	Beck, Koller & Co.	Detroit, Mich.
72.	Belmont Smelting & Refining Wks.	Brooklyn, N. Y.
73.	Benkert, W. & Co., Inc.	New York, N. Y.
74.	Benzol Products Co.	Newark, N. J.
75.	F. C. Bersworth Labs.	Framingham, Mass.
76.	Beryllium Corp. of America	New York, N. Y.
77.	Bick & Co., Inc.	Reading, Pa.
78.	Bilhuber-Knoll Corp.	New York, N. Y.
79.	Binney & Smith	New York, N. Y.
80.	Bisbee Linseed Co.	Philadelphia, Pa.
81.	Bohme, A. G., H. Th.	Chemnitz, Germany
82.	Bopf-Whittam Corp.	Linden, N. J.
83.	Borax Union, Inc.	San Francisco, Calif.
84.	The W. H. Bowdlear Co.	Syracuse, N. Y.
85.	Bowker Chem. Corp.	New York, N. Y.
86.	Bradley & Baker	New York, N. Y.
87.	Brazil Oiticica, Inc.	New York, N. Y.
88.	British Drug Houses, Ltd.	London, England
89.	British Xylonite Co.	London, England
90.	Brooke, Fred L., Co.	Chicago, Ill.
91.	Bud Aromatic Chemical Co., Inc.	New York, N. Y.
92.	Buffalo Electro Chem. Co., Inc.	Buffalo, N. Y.
93.	Burkard-Schier Chem. Co.	Chattanooga, Tenn.
94.	Buromin Corp.	Pittsburgh, Pa.
95.	Bush, W. J. & Co., Inc.	New York, N. Y.
96.	C. P. Chemical Solvents, Inc.	New York, N. Y.
97.	Cabot, Godfrey L., Inc.	Boston, Mass.
98.	Calcium Sulphide Corp.	Damascus, Va.
99.	Calco Chemical Co.	Bound Brook, N. J.
100.	Calgon, Inc.	Pittsburgh, Pa.
101.	Calif. Fruit Growers' Exchange	Ontario, Calif.
102.	Campbell, C. W. Co., Inc.	New York, N. Y.
103.	Campbell, John & Co.	New York, N. Y.
104.	Campbell Rex & Co.	London, England
105.	Carbic Color & Chemical Co.	New York, N. Y.
106.	Carbide & Carbon Chem. Corp.	New York, N. Y.
107.	Carbogen Chemical Co.	Garwood, N. J.
108.	Carbolincum Wood Preserving Co.	Milwaukee, Wis.
109.	Carborundum Co.	Niagara Falls, N. Y.
110.	Carey, Philip Co.	Lockland, Ohio
111.	Carus Chem. Co., Inc.	La Salle, Ill.
112.	Casein Mfg. Co.	New York, N. Y.
113.	The Casein Mfg. Co. of Amer. Inc.	New York, N. Y.
114.	Celanese Corp. of America	New York, N. Y.

No.	*Name*	*Address*
115.	Cellonwerke	Charlottenburg, Germany
116.	Celluloid Corp.	Newark, N. J.
117.	Celluloid Corp.	New York, N. Y.
118.	Central Scientific Co.	Chicago, Ill.
119.	Century Stearic Acid Wks.	New York, N. Y.
120.	Century Stearic Acid & Candle Wks.	New York, N. Y.
121.	Ceramic Color & Chem. Mfg. Co.	New Brighton, Pa.
122.	Cerro de Pasco Copper Corp.	New York, N. Y.
123.	Champion Paper & Fibre Co.	Canton, N. C.
124.	Chaplin-Bibbo	New York, N. Y.
125.	Chazy Marble Lime Co., Inc.	Chazy, N. Y.
126.	Chesebrough Mfg. Co.	New York, N. Y.
127.	Chemical & Pigment Co.	Baltimore, Md.
128.	Chemical & Pigment Co., Inc.	Scranton, Pa.
129.	Chemical Publ. Co., Inc.	Brooklyn, N. Y.
130.	Chemical Solvents, Inc.	New York, N. Y.
131.	Cheney Chem. Co.	Cleveland, Ohio
132.	Chicago Apparatus Co.	Chicago, Ill.
133.	Chicago Copper & Chem. Co.	Blue Island, Ill.
134.	Chipman Chem. Co., Inc.	Bound Brook, N. J.
135.	Chrystal, Charles B. Co., Inc.	New York, N. Y.
136.	Church & Dwight Co., Inc.	New York, N. Y.
137.	Ciba Co., Inc.	New York, N. Y.
138.	Cinelin Co.	Indianapolis, Ind.
139.	Clarke, John & Co.	New York, N. Y.
140.	The Cleveland-Cliffs Iron Co.	Cleveland, Ohio
141.	Climax Molybdenum Co.	New York, N. Y.
142.	Clinton Co.	Clinton, Ia.
143.	Coleman & Bell Co.	Norwood, Ohio
144.	Colgate-Palmolive-Peet Co.	Chicago, Ill.
145.	Colgate-Palmolive-Peet Co.	Jersey City, N. J.
146.	Colledge, E. W., Inc.	Cleveland, Ohio
147.	Colonial Beacon Oil Co.	Everett, Mass.
148.	Columbia Alkali Corp.	New York, N. Y.
149.	Commercial Solvents Corp.	New York, N. Y.
150.	Commercial Solvents Corp.	Terre Haute, Ind.
151.	Commonwealth Color & Chem. Co.	Brooklyn, N. Y.
152.	Compagnie Duval	New York, N. Y.
153.	Conewango Refining Co.	Warren, Pa.
154.	Consolidated Chem. Sales Corp.	Newark, N. J.
155.	Consolidated Feldspar Corp.	Trenton, N. J.
156.	Conti Products Corp.	New York, N. Y.
157.	Continental Diamond Fibre Co.	Bridgeport, Pa.
158.	Continental Oil Co.	Ponca City, Okla.
159.	Cook Swan Co., Inc.	New York, N. Y.
160.	Cooper, Charles & Co.	New York, N. Y.
161.	Coopers Creek Chem. Co.	W. Conshohocken, Pa.
162.	Corn Products Refining Co.	New York, N. Y.
163.	Cowles Detergent Co.	Cleveland, Ohio
164.	Croton Chem. Corp.	Brooklyn, N. Y.
165.	Crowley Tar Products Co.	New York, N. Y.
166.	Crystal, Charles B. Co., Inc.	New York, N. Y.
167.	Cudahy Packing Co.	Chicago, Ill.
168.	Danco, Gerard J.	New York, N. Y.
169.	Darco Sales Corp.	New York, N. Y.
170.	Darling & Co.	Chicago, Ill.
171.	Davison Chem. Corp.	Baltimore, Md.
172.	Deep Rock Oil Corp.	Chicago, Ill.
173.	C. P. De Lore Co.	St. Louis, Mo.
174.	Delta Chem. Mfg. Co.	Baltimore, Md.
175.	Delta Chem. & Iron Co.	Wells, Mich.
176.	Denver Fire Clay Co.	Denver, Colo.

No.	*Name*	*Address*
177.	Devoe & Raynolds Co.	New York, N. Y.
178.	Dewey & Almy Chem. Co.	Boston, Mass.
179.	Diamond Alkali Co.	Pittsburgh, Pa.
180.	Dicalite Co.	New York, N. Y.
181.	Dickinson, E. E. Co.	Essex, Conn.
182.	Dickinson, J. Q. & Co.	Malden, W. Va.
183.	Difco Laboratories, Inc.	Detroit, Mich.
184.	Digestive Ferments Co.	Detroit, Mich.
185.	Marshall Dill	San Francisco, Calif.
186.	Distributing & Trading Co.	New York, N. Y.
187.	Dodge & Olcott Co.	New York, N. Y.
188.	Dow Chemical Co.	Midland, Mich.
189.	Drakenfeld, B. F. & Co.	New York, N. Y.
190.	Dreyer, P. R. Co.	New York, N. Y.
191.	Dreyfus Co., L. A.	Rosebank, N. Y.
192.	Drury, A. C. & Co., Inc.	Chicago, Ill.
193.	Ducas, B. P. Co.	New York, N. Y.
194.	Duche, T. M. & Sons	New York, N. Y.
195.	DuPont, E. I., de Nemours & Co.	Wilmington, Del.
196.	E. I. DuPont de Nemours & Co., Inc.	Parlin, N. J.
197.	Durite Plastics	Philadelphia, Pa.
198.	Dynamit, A. G.	Troisdorf, Germany
199.	The Eagle-Picher Lead Co.	Cincinnati, Ohio
200.	Eakins, J. S. & W. R., Inc.	Brooklyn, N. Y.
201.	Earle Bros.	New York, N. Y.
202.	Eastman Kodak Co.	Rochester, N. Y.
203.	Economic Materials Co.	Chicago, Ill.
204.	Eff Laboratories, Inc.	Cleveland, Ohio
205.	Egyptian Lacquer Co.	Kearney, N. J.
206.	Eimer & Amend	New York, N. Y.
207.	Elbert & Co.	New York, N. Y.
208.	Electro Bleaching Gas Co.	New York, N. Y.
209.	Electro-Metallurgical Co.	New York, N. Y.
210.	Emery Industries, Inc.	Cincinnati, Ohio
211.	Empire Distilling Corp.	New York, N. Y.
212.	Enterprise Animal Oil Co.	Philadelphia, Pa.
213.	Fales Chem. Co., Inc.	Cornwall Landing, N. Y.
214.	Falk & Co.	Pittsburgh, Pa.
215.	Fansteel Metallurgical Corp.	No. Chicago, Ill.
216.	Felton Chemical Co.	Brooklyn, N. Y.
217.	Fezandie & Sperrle, Inc.	New York, N. Y.
218.	Fiberloid Corp.	Indian Orchard, Mass.
219.	Filtrol Co.	Los Angeles, Calif.
220.	Fishbeck, Chas. Co.	New York, N. Y.
221.	Fisher Scientific Co.	Pittsburgh, Pa.
222.	Florasynth Laboratories	New York, N. Y.
223.	Foote Mineral Co.	Philadelphia, Pa.
224.	Formica Insulation Co.	Cincinnati, Ohio
225.	Fougera, E. & Co.	New York, N. Y.
226.	France, Campbell & Darling	Kenilworth, N. J.
227.	Franco-American Chemical Wks.	Carlstadt, N. J.
228.	Frank-Vliet Co.	New York, N. Y.
229.	Franks Chem. Products Co., Inc.	Brooklyn, N. Y.
230.	French Potash Co.	New York, N. Y.
231.	Alex Fries & Bro.	Cincinnati, Ohio
232.	Fries Bros.	New York, N. Y.
233.	Fritzchie Bros.	New York, N. Y.
234.	Garrigues, Stewart & Davies, Inc.	New York, N. Y.
235.	Geigy Co., Inc.	New York, N. Y.
236.	General Aniline Works, Inc.	New York, N. Y.
237.	General Atlas Carbon Co.	New York, N. Y.
238.	General Chemical Co.	New York, N. Y.

No.	*Name*	*Address*
239.	General Drug Co.	New York, N. Y.
240.	General Dyestuffs Corp.	New York, N. Y.
241.	General Electric Co.	Pittsfield, Mass.
242.	General Electric Co.	Schenectady, N. Y.
243.	General Magnesite & Magnesia Co.	Philadelphia, Pa.
244.	General Naval Stores Co.	New York, N. Y.
245.	General Plastics Corp.	London, England
246.	General Plastics, Inc.	No. Tonawanda, N. Y.
247.	Girdler Corp.	Louisville, Ky.
248.	Givaudan-Delawanna, Inc.	New York, N. Y.
249.	Glidden Co.	Cleveland, Ohio
250.	Globe Chem. Co.	Cincinnati, Ohio
251.	Glyco Products Co., Inc.	Brooklyn, N. Y.
252.	Goldschmidt, A. G., Th.	Essen, Germany
253.	Goldschmidt Corp.	New York, N. Y.
254.	Goodrich, B. F., Co.	Akron, Ohio
255.	Goodyear Tire & Rubber Co.	Akron, Ohio
256.	Grasselli Chemical Co.	Cleveland, Ohio
257.	W. S. Gray Co.	New York, N. Y.
258.	Greeff, R. W. & Co.	New York, N. Y.
259.	Griffith Laboratories	Chicago, Ill.
260.	Gross, A. & Co.	New York, N. Y.
261.	Hall, C. P. & Co.	Akron, Ohio
262.	Halowax Corp.	New York, N. Y.
263.	Hammil & Gillespie, Inc.	New York, N. Y.
264.	Hamilton, A. K.	New York, N. Y.
265.	Hammond Drierite Co.	Yellow Springs, Ohio
266.	Handy & Harman	New York, N. Y.
267.	Hardy, Charles, Inc.	New York, N. Y.
268.	Harrison Mfg. Co.	Rahway, N. J.
269.	Harshaw Chemical Co.	Cleveland, Ohio
270.	Hart Products Corp.	New York, N. Y.
271.	Haskelite Mfg. Corp.	Chicago, Ill.
272.	Haveg Corp.	Newark, Del.
273.	Hegeler Zinc Co.	Danville, Ill.
274.	Heine & Co.	New York, N. Y.
275.	Hercules Powder Co.	New York, N. Y.
276.	Hercules Powder Co.	Wilmington, Del.
277.	Heveatex Corp.	Melrose, Mass.
278.	C. B. Hewitt & Bro.	New York, N. Y.
279.	Heyden Chemical Works	New York, N. Y.
280.	Hill Bros. Chem. Co.	Los Angeles, Calif.
281.	Hillside Fluor Spar Mines	Chicago, Ill.
282.	Holland Aniline Dye Co.	Holland, Mich.
283.	O. Hommel Co.	Pittsburgh, Pa.
284.	Hooker Electro-Chemical Co.	New York, N. Y.
285.	Hopkins, J. L. & Co.	New York, N. Y.
286.	Hord Color Products	Sandusky, Ohio
287.	Horn Jefferys & Co.	Burbank, Calif.
288.	Horner, James B., Inc.	New York, N. Y.
289.	Huisking, Chas L. & Co., Inc.	New York, N. Y.
290.	Hummel Chemical Co., Inc.	New York, N. Y.
291.	Hurst, Adolph & Co., Inc.	New York, N. Y.
292.	D. W. Hutchinson & Co., Inc.	New York, N. Y.
293.	Hymes, Lewis Associates	New York, N. Y.
294.	I. G. Farbenindustrie	Frankfurt, Germany
295.	Imperial Chem. Industries	London, England
296.	Industrial Chem. Sales Co.	New York, N. Y.
297.	Innes, O. G., Corp.	New York, N. Y.
298.	Innis Speiden Co.	New York, N. Y.
299.	International Pulp Corp.	New York, N. Y.
300.	International Selling Corp.	New York, N. Y.

No.	Name	Address
301.	Interstate Color Co., Inc.	New York, N. Y.
302.	Iowa Soda Products Co.	Council Bluffs, Ia.
303.	Jackson, L. N. & Co.	New York, N. Y.
304.	Jacobson, C. A.	W. Va. University, Morgantown, W. Va.
305.	The Jennison-Wright Co.	Toledo, Ohio
306.	Johns-Manville Corp.	New York, N. Y.
307.	Jones & Laughlin Steel Corp.	Pittsburgh, Pa.
308.	Jones, S. L. & Co.	San Francisco, Calif.
309.	Jungmann & Co.	New York, N. Y.
310.	Kali Mfg. Co.	Philadelphia, Pa.
311.	Kalle & Co.	Wiesbaden Bierich, Germany
312.	Kay Fries Chem., Inc.	New York, N. Y.
313.	Kelco Co.	San Diego, Calif.
314.	Kentucky Clay Mining Co.	Mayfield, Ky.
315.	Kentucky Color & Chem. Co.	Louisville, Ky.
316.	Kessler Chem. Corp.	Philadelphia, Pa.
317.	Kinetic Chem., Inc.	Wilmington, Del.
318.	H. Kohnstamm & Co.	New York, N. Y.
319.	Koppers Products Co.	Pittsburgh, Pa.
320.	Krebs Pigment & Color Corp.	Newark, N. J.
321.	Kuhlman, Etabls.	Paris, France
322.	Kurt, Albert, G. M. B. H.	Amoneburg, Germany
323.	Lattimer-Goodwin Chem. Co.	Grand Junction, Ohio
324.	Laxseed Co.	New York, N. Y.
325.	Leghorn Trading Co., Inc.	New York, N. Y.
326.	Lehn & Fink Corp.	New York, N. Y.
327.	Theo. Leonhard Wax Co., Inc.	Haledon, Paterson, N. J.
328.	Lewis, C. H. & Co.	New York, N. Y.
329..	Lewis, John D., Inc.	Providence, R. I.
330.	Limestone Products Corp. of Amer.	Newton, N. J.
331.	Lincks, Geo. H.	New York, N. Y.
332.	Liquid Carbonic Corp.	Chicago, Ill.
333.	Litter, D. H., Co.	New York, N. Y.
334.	Littlejohn & Co., Inc.	New York, N. Y.
335.	Lucidol Corp.	Buffalo, N. Y.
336.	Geo. Lueders & Co.	New York, N. Y.
337.	Lundt & Co.	New York, N. Y.
338.	Maas & Waldstein	Newark, N. J.
339.	MacAndrews & Forbes Co.	New York, N. Y.
340.	Mackay, A. D.	New York, N. Y.
341.	Magnetic Pigment Co.	New York, N. Y.
342.	Magnus, Mabee & Reynard, Inc.	New York, N. Y.
343.	Makalot Corp.	Boston, Mass.
344.	Mallinckrodt Chemical Works	St. Louis, Mo.
345.	Malt Diastase Co.	Brooklyn, N. Y.
346.	Manchester Oxide Co.	Manchester, England
347.	Marbon Corp.	Gary, Ind.
348.	Marine Magnesium Prod. Corp.	S. San Francisco, Calif.
349.	Martin, Dennis Co.	Newark, N. J.
350.	Martin, L. Co.	New York, N. Y.
351.	Mathieson Alkali Co.	New York, N. Y.
352.	Maywood Chem. Works	Maywood, N. J.
353.	McCormick & Co.	Baltimore, Md.
354.	The McGean Chem. Co.	Cleveland, Ohio
355.	McKesson & Robbins, Inc.	New York, N. Y.
356.	McLaughlin, Gormley, King & Co.	Minneapolis, Minn.
357.	Mearl Corp.	New York, N. Y.
358.	Mechling Bros. Chem. Co.	Camden, N. J.
359.	E. Meer & Co., Inc.	New York, N. Y.
360.	Merck & Co.	Rahway, N. J.
361.	Merrimac Chemical Co.	Boston, Mass.
362.	Metro-Nite Co.	Milwaukee, Wis.

No.	*Name*	*Address*
363.	Meyer & Sons, J.	Philadelphia, Pa.
364.	Mica Insulator Co.	New York, N. Y.
365.	Michel Export Co.	New York, N. Y.
366.	Michigan Alkali Co.	New York, N. Y.
367.	Miller, Carl F., Co.	Seattle, Wash.
368.	Monsanto Chem. Works	St. Louis, Mo.
369.	Moore-Munger	New York, N. Y.
370.	Morningstar, Nicol, Inc.	New York, N. Y.
371.	Morton Salt Co.	Chicago, Ill.
372.	Murphy Varnish Co.	Newark, N. J.
373.	Mutual Chem. Co. of Amer.	New York, N. Y.
374.	Mutual Chem. Co. of America	New York, N. Y.
375.	Mutual Citrus Products Co.	Anaheim, Calif.
376.	National Aulminate Corp.	Chicago, Ill.
377.	Nat'l Ammonia Co., Inc.	Philadelphia, Pa.
379.	Nat'l Aniline & Chem. Wks.	New York, N. Y.
380.	National Lead Co.	New York, N. Y.
381.	National Oil Products Co.	Harrison, N. J.
382.	Nat'l Pigments & Chem. Co.	St. Louis, Mo.
383.	National Rosin Oil & Size Co.	New York, N. Y.
384.	Naugatuck Chem. Co.	Naugatuck, Conn.
385.	Neville Co.	Pittsburgh, Pa.
386.	N. J. Laboratory Supply Co.	Newark, N. J.
387.	N. J. Zinc Co.	New York, N. Y.
388.	The N. Y. Quinine & Chem. Wks., Inc.	Brooklyn, N. Y.
390.	Newmann-Buslee & Wolfe, Inc.	Chicago, Ill.
391.	Newport Industries, Inc.	New York City
392.	Niacet Chem. Co.	Niagara Falls, N. Y.
393.	Niagara Alkali Co.	New York, N. Y.
394.	Niagara Chemicals Corp.	Niagara Falls, N. Y.
395.	Niagara Smelting Corp.	Niagara Falls, N. Y.
396.	The Northwestern Chem. Co.	Wauwatosa, Wis.
397.	Norton Co.	Worcester, Mass.
398.	Norwich Pharmacal Co.	Norwich, N. Y.
399.	Novadel-Agene Corp.	Newark, N. J.
400.	Nulomoline Co.	New York, N. Y.
401.	Nuodex Products, Inc.	Elizabeth, N. J.
402.	Ohio-Apex, Inc.	Nitro, W. Va.
403.	Oil States Petroleum Co.	New York, N. Y.
404.	Oldbury Electro-Chem. Co.	New York, N. Y.
405.	Olive Branch Minerals Co.	Cairo, Ill.
406.	Onyx Oil & Chem. Co.	Passaic, N. J.
407.	Orbis Products Corp.	New York, N. Y.
408.	Papermakers' Chem. Corp.	Wilmington, Del.
409.	Paramet Chem. Corp.	Long Island City, N. Y.
410.	Parke, Davis & Co.	Detroit, Mich.
411.	Parker Rust Proof Co.	Detroit, Mich.
412.	Patent Chemicals, Inc.	New York, N. Y.
413.	Peek & Velsor, Inc.	New York, N. Y.
414.	Penick, S. B. & Co.	New York, N. Y.
415.	Penn. Alcohol Corp.	Philadelphia, Pa.
416.	Penn. Coal Products Co.	Petrolia, Pa.
417.	Penn.-Dixie Cement Corp.	New York City
418.	Penn. Industrial Chem. Corp.	Clairton, Pa.
419.	Penn. Refining Co.	Butler, Pa.
420.	Penn. Salt Mfg. Co.	Philadelphia, Pa.
421.	Pfaltz-Bauer, Inc.	New York, N. Y.
422.	Pfizer, Chas. & Co., Inc.	New York, N. Y.
423.	Phila. Quartz Co.	Philadelphia, Pa.
424.	Philipp Bros.	New York, N. Y.
425.	Pittsburgh Plate Glass Co.	Pittsburgh, Pa.
426.	Plaskon Corp.	Toledo, Ohio

No.	*Name*	*Address*
427.	Plymouth Organic Labs.	New York, N. Y.
428.	Pollopas, Ltd.	London, England
429.	Powhatan Mining Corp.	Woodlawn, Baltimore, Md.
430.	Pray, W. P.	New York, N. Y.
431.	Prior Chem. Corp.	New York, N. Y.
432.	Procter & Gamble Co.	Cincinnati, Ohio
433.	Provident Chem. Wks.	St. Louis, Mo.
434.	Publicker, Inc.	Philadelphia, Pa.
435.	Pure Calcium Products Co.	Painesville, Ohio
436.	Pylam Products Co.	New York, N. Y.
437.	Ransom, L. E., Co.	New York, N. Y.
438.	Robert Rauh, Inc.	Newark, N. J.
439.	Read, Chas. L. & Co., Inc.	New York, N. Y.
440.	Reichhold Chemicals, Inc.	Detroit, Mich.
441.	Reilly Tar & Chem. Corp.	Indianapolis, Ind.
442.	Resinous Prod. & Chem. Co.	Philadelphia, Pa.
442a.	Resinox Corp.	New York City
443.	Revertex Corp.	Brooklyn, N. Y.
444.	Revson, R. F. Co.	New York, N. Y.
445.	Rhone-Poulene, Inc.	Paris, France
446.	Richards Chem. Works	Jersey City, N. J.
447.	Riverside Chem. Co.	No. Tonawanda, N. Y.
448.	Robeson Process Co.	New York, N. Y.
449.	Rogers & McClellan	Boston, Mass.
450.	Rohm & Haas	Philadelphia, Pa.
451.	Rosenthal, H. H., Co.	New York, N. Y.
452.	Ross, Frank B., Co., Inc.	New York, N. Y.
453.	Ross-Rowe, Inc.	New York, N. Y.
454.	Royce Chem. Co.	Carlton Hill, N. J.
455.	Rubber Service Labs. Co.	Akron, Ohio
456.	Russell, W. R. & Co.	New York, N. Y.
457.	Russia Cement Co.	Gloucester, Mass.
458.	Ryland, H. C., Inc.	New York, N. Y.
459.	Saginaw Salt Products Co.	Saginaw, Mich.
460.	Salomon, L. A. & Bro.	New York, N. Y.
461.	Samuelson & Co., P.	London, England
462.	Sandoz Chem. Works	New York, N. Y.
463.	Scheel, Wm. H.	New York, N. Y.
464.	Schimmel & Co.	New York, N. Y.
464a.	Schliemann Co., Inc.	New York City
465.	Schofield-Daniel Co.	New York City
466.	Scholler Bros., Inc.	Philadelphia, Pa.
467.	F. E. Schundler & Co.	Joliet, Ill.
468.	Schuylkill Chem. Co.	Philadelphia, Pa.
469.	Schwabacher, S. & Co., Inc.	New York, N. Y.
470.	Scientific Glass Apparatus Co.	Bloomfield, N. J.
471.	Scott, Bader & Co.	London, England
472.	Seacoast Laboratories	New York, N. Y.
473.	Edwin Seebach Co.	New York, N. Y.
474.	Seeley & Co., Inc.	New York, N. Y.
475.	Seldner & Enequist, Inc.	Brooklyn, N. Y.
476.	Serinsky, Moses Co.	Indianapolis, Ind.
477.	Sharples Solvents Corp.	Philadelphia, Pa.
478.	Shawinigan, Ltd.	New York, N. Y.
479.	Shepherd Chem. Co.	Norwood, Cincinnati, Ohio
480.	Sherka Chem. Co., Inc.	Bloomfield, N. J.
482.	Sherwood Petroleum Co.	Englewood, N. J.
483.	Thomas J. Shields Co.	New York, N. Y.
484.	Siemon Colors, Inc.	Newark, N. J.
485.	Siemon & Co.	Bridgeport, Conn.
486.	Silica Products Co.	Kansas City, Mo.
487.	Silver, Geo., Import Co.	New York, N. Y.

No.	*Name*	*Address*
488.	Sinclair Refining Co.	Olmstead, Ill.
489.	Skelly Oil Co.	Chicago, Ill.
490.	Smith Chem. & Color Co.	Brooklyn, N. Y.
491.	Smith & Nichols, Inc.	New York, N. Y.
492.	Smith, Werner G., Co.	Cleveland, Ohio
493.	Solvay Sales Corp.	New York, N. Y.
494.	Sonneborn, L., Sons	New York, N. Y.
495.	Southern Mica Co.	Franklin, N. C.
496.	Southern Pine Chem. Co.	Jacksonville, Fla.
497.	Southwark Mfg. Co.	Camden, N. J.
498.	Sparhawk Co.	Sparkhill, N. Y.
499.	Spencer Kellogg & Sons Sales Corp.	Buffalo, N. Y.
500.	A. E. Staley Mfg. Co.	Decatur, Ill.
501.	Stamford Rubber Supply Co.	Stamford, Conn.
502.	Stanco Distributors	New York, N. Y.
503.	Standard Alcohol Co.	New York, N. Y.
504.	Standard Brands, Inc.	New York, N. Y.
505.	Standard Oil Co. of Calif.	San Francisco, Calif.
506.	Standard Oil Co. of Indiana	Chicago, Ill.
507.	Standard Oil Co. of N. J.	New York, N. Y.
508.	Standard Oil Co. of N. Y.	New York, N. Y.
509.	Standard Silicate Co.	Pittsburgh, Pa.
510.	Standard Ultramarine Co.	Huntington, W. Va.
511.	Stanley Co., John T.	New York City
512.	Starch Products Co.	New York, N. Y.
513.	Stauffer Chem. Co.	New York, N. Y.
514.	Stauffer Chem. Co. of Texas	Freeport, Texas
515.	Stein, Hall & Co.	New York, N. Y.
516.	Stokes & Smith Co.	Philadelphia, Pa.
517.	Strahl & Pitsch	New York, N. Y.
518.	Strohmeyer & Arpe Co.	New York, N. Y.
519.	Stroock & Wittenberg Corp.	New York, N. Y.
520.	Sun Oil Co.	Philadelphia, Pa.
521.	Swann Chemical Co.	New York, N. Y.
522.	Synfleur Scientific Labs.	Monticello, N. Y.
523.	Synthane Corp.	Oaks, Pa.
524.	The Synthetic Products Co.	Cleveland, Ohio
525.	Taintor Trading Co.	New York, N. Y.
526.	Takamine Laboratory, Inc.	Clifton, N. J.
527.	Tamms Silica Co.	Chicago, Ill.
528.	Tanners Supply Co.	Grand Rapids, Mich.
529.	Tannin Corp.	New York, N. Y.
530.	C. Tennant & Sons Co. of N. Y.	New York, N. Y.
531.	Tenn. Eastman Corp.	Kingsport, Tenn.
532.	Texas Chem. Co.	Houston, Texas
533.	Texas Mining & Smelting Co.	Laredo, Texas
534.	Thomas, Arthur H., Co.	Philadelphia, Pa.
535.	Thorocide, Inc.	St. Louis, Mo.
536.	Thurston & Braidich	New York, N. Y.
537.	Titanium Alloy Mfg. Co.	Niagara Falls, N. Y.
538.	Titanium Pigments Co.	New York, N. Y.
539.	Tobacco By-Products & Chem. Corp.	Louisville, Ky.
540.	Trask, Arthur C., Co.	Chicago, Ill.
541.	Trojan Powder Co.	Allentown, Pa.
542.	Turner, Joseph & Co.	Ridgefield, N. J.
543.	Uhe, George Co.	New York, N. Y.
544.	Uhlich, Paul Co.	New York, N. Y.
545.	Union Oil Co.	Los Angeles, Calif.
546.	Union Smelting & Refining Co., Inc.	Newark, N. J.
547.	United Carbon Co.	Charleston, W. Va.
547a.	United Clay Mines Corp.	Trenton, N. J.
548.	United Color & Pigment Co.	Newark, N. J.

No.	*Name*	*Address*
549.	U. S. Bronze Powder Works, Inc.	New York, N. Y.
550.	U. S. Gypsum Co.	Chicago, Ill.
551.	U. S. Industrial Alcohol Co.	New York, N. Y.
552.	U. S. Industrial Chem. Co.	New York, N. Y.
553.	U. S. Phosphoric Prod. Corp.	New York, N. Y.
554.	U. S. Rubber Products, Inc.	New York, N. Y.
555.	U. S. Smelting, Refining & Mining Co.	New York, N. Y.
556.	Utah Gilsonite Co.	St. Louis, Mo.
557.	Van Allen, L. R. & Co.	Chicago, Ill.
558.	Van-Ameringen Haebler, Inc.	New York, N. Y.
559.	Vanderbilt, R. T., Co.	New York, N. Y.
560.	Van Dyk & Co., Inc.	Jersey City, N. J.
561.	Van Schaack Bros. Chem. Co.	Chicago, Ill.
562.	Varcum Chem. Corp.	Niagara Falls, N. Y.
563.	Verley, Albert & Co.	Chicago, Ill.
564.	Verona Chem. Co.	Newark, N. J.
565.	Victor Chem. Works	Chicago, Ill.
566.	Virginia-Carolina Chem. Corp.	Richmond, Va.
567.	Virginia Smelting Works	W. Norfolk, Va.
568.	Vitro Mfg. Co.	Pittsburgh, Pa.
570.	Vultex Chem. Co.	Cambridge, Mass.
571.	Waldo, E. M. & F., Inc.	Muirkirk, Md.
572.	Wallerstein Co., Inc.	New York, N. Y.
573.	The Warner Chem. Co.	New York, N. Y.
574.	Warwick Chem. Co.	West Warwick, R. I.
575.	Welch, Holme & Clark Co.	New York, N. Y.
576.	Welsbach & Co.	Gloucester, N. J.
577.	Werk, M., Co.	Cincinnati, Ohio
578.	Western Charcoal Co.	Chicago, Ill.
579.	Westinghouse Elec. & Mfg. Co.	E. Pittsburgh, Pa.
580.	Whittaker, Clark & Daniels	New York, N. Y.
581.	Wiffen & Co., Sons, Ltd.	London, England
582.	Wilckes-Martin-Wilckes Co.	New York, N. Y.
583.	Will & Baumer Candle Co.	New York, N. Y.
584.	C. K. Williams & Co.	Easton, Pa.
585.	The Wilson Laboratories	Chicago, Ill.
586.	Wishnick-Tumpeer, Inc.	New York, N. Y.
587.	Woburn Degreasing Co.	Harrison, N. J.
588.	Wolf, Jacques & Co.	Passaic, N. J.
589.	Wood Flour, Inc.	Manchester, N. H.
590.	Wood Ridge Mfg. Co.	Wood Ridge, N. J.
591.	Wyodak Chem. Co.	Cleveland, Ohio
593.	Young, J. S. & Co.	Hanover, Pa.
595.	Zinsser, Wm. & Co.	New York, N. Y.
596.	Zophar Mills, Inc.	Brooklyn, N. Y.

ADDENDA

597.	Borne-Scrymser Co.	New York, N. Y.
598.	Hycar Corp.	Akron, O.
599.	Sherwin-Williams Co.	Cleveland, O.

WHERE TO BUY CHEMICALS OUTSIDE THE UNITED STATES

Argentina, Buenos Aires—M. Goetz, Rincón 332

Australia, Adelaide—Robert Bryce & Co., Pty., Ltd., 73-75 Wakefield Street

———, Melbourne, C1—Robert Bryce & Co., Pty., Ltd., 526 Little Bourke Street

———, Sydney—Robert Bryce & Co., Pty., Ltd., 188-190 Kent Street

Bolivia, La Paz—M. Romulo Vildoso, Calle Potosi 137

Brazil, Sao Paulo—Empresa Comercial Mercur Ltda., Caixa Postale 4232

Canada, Montreal—Chemicals, Ltd., 384 St. Paul St., W.

———, Toronto 2—Canada Colors & Chemicals, Ltd., 1090 King St., West

———, Vancouver, B. C.—Shanahan's Ltd., Foot of Campbell Avenue

Chile, Santiago—Bernardo Dornblatt, Clasificador 195B

Cuba, Havana—Raul Guillent, 215 Bank of Nova Scotia Bldg., P. O. Box 1133

England, London—Rex Campbell & Co., Ltd., 7, Idol Lane, Eastcheap, E. C. 3

India, Calcutta—Kaisers Trading Company, 159 Lower Chitpore Rd.

Mexico, Mexico D. F.—R. Koestinger, J. M. Velasco, 119, Insurgentes, Mixcoac

New Zealand, Wellington—Robert Bryce & Co., Pty., Ltd., 19 Lower Tory Street

South Africa, Johannesburg—Philip Elzas & Co., 132 London House, Loveday St.

Sweden, Huddinge—E. Landerholm

Switzerland, Zurich—Oswald E. Boll, Dufourstrasse 157

Uruguay, Montevideo—Compania Industrial Alfa, Lda., Porongos 2228

Venezuela, Caracas—A. G. Bulgaris, Apartado 1752

INDEX

G

H

I

M

N

O

Q

R

T

X

Y

Z

USEFUL REFERENCE BOOKS *

Adhesives	F. Braude	$ 3.00
Varnish Making	Symposium	6.00
Plastic Moulding	D. A. Dearle	4.00
Plastics in Industry	Plastes	5.00
Making and Moulding of Plastics	L. M. T. Bell	5.00
Rubber and Its Use	H. L. Fisher	2.25
Rubber Latex	H. P. and W. H. Stevens	2.00
Food Manufacturing	S. Blumenthal	7.50
Principles and Processes of Light Leather Manufacture	P. I. Smith	6.50
Leather Finishes	J. S. Mudd	4.25
White Shoe Dressings and Cleaners	W. D. John	4.25
Cosmetic Formulary	H. Bennett	3.75
Drug and Specialty Formulas	E. J. Belanger	6.00
Hair Dyes and Hair Dyeing	H. S. Redgrove and J. Bari-Wools	5.00
Basic Laboratory Practice	N. C. Sprague	3.50
Vegetable Dyes	E. M. Mariet	2.00
Practical Emulsions	H. Bennett	5.00
Embalming Fluids	S. Mendelsohn	4.00
Casein and Its Uses	H. Hadert	3.00
Technology of Solvents	O. Jordan	10.00
Industrial Research	F. R. Bichowsky	2.50
Ultra Violet Light and Its Applications	J. DeMent and H. C. Dake	3.25
Uranium and Atomic Power	J. DeMent and H. C. Dake	5.00
Plant Growth Substances	H. Nicol	2.00
Chemical Gardening (Hydroponics and Soilless Growing)	D. R. Matlin	2.25
Chemical Warfare	C. Wachtel	4.00
Chemical Calculations	C. M. Jones	1.25
Organic Chemistry Simplified	R. Macy	3.75
Handbook for Chemical Patents	E. Thomas	4.00
Substitutes	H. Bennett	4.00
Glue and Gelatin	P. I. Smith	3.75
Dictionary and Manual of Fireworks	G. W. Weingart	5.00
Soya Bean Technology	A. A. Horvath	*in preparation*

* These and other technical books may be obtained from The Chemical Publishing Co., Inc., Brooklyn, N. Y., U.S.A.